International Conference on Sustainable Management of Earth Resources and Biodiversity (SERBEMA-2023)

E3S Web of Conferences Volume 421 (2023)

Tashkent, Uzbekistan
26-27 September 2023

Editor:

D. Kodirova

ISBN: 978-1-7138-8383-8

Printed from e-media with permission by:

Curran Associates, Inc.
57 Morehouse Lane
Red Hook, NY 12571

Some format issues inherent in the e-media version may also appear in this print version.

For additional information, please contact EDP Sciences – Web of Conferences
at the address below.

EDP Sciences – Web of Conferences
17, Avenue du Hoggar
Parc d'Activité de Courtabœuf
BP 112
F-91944 Les Ulis Cedex A
France

Phone: +33 (0) 1 69 18 75 75
Fax: +33 (0) 1 69 28 84 91

contact-edps@webofconferences.org

Additional copies of this publication are available from:

Curran Associates, Inc.
57 Morehouse Lane
Red Hook, NY 12571 USA
Phone: 845-758-0400
Fax: 845-758-2633
Email: curran@proceedings.com
Web: www.proceedings.com

TABLE OF CONTENTS

LAND USE AND LIVESTOCK HUSBANDRY

ECOLOGY AND ENVIRONMENT

PREFACE

International Conference on "Sustainable Management of Earth Resources and Biodiversity" (SERBEMA-2023) was successfully held on 26–27 September 2023 by Tashkent State Agrarian University in collaboration with numerous academic and non-academic organizations.

SERBEMA-2023 conference was held in the hybrid format: on-site at Tashkent State Agrarian University and in the format of videoconference using ZOOM platform. Despite the cancellation of pandemic restrictions in the world, the organizers still considered proactive safety and all the necessary requirements for holding an event such as separate entrances and exits for the exhibition, and conference rooms were equipped with signs. Sessions were shortened to allow for airing and sanitizing of potential contact surfaces. There were some coffee breaks in a separate canteen and all conference participants and guests were provided with medical masks. For virtual involvement, several social network possibilities were available, as well as video conferencing with all interested participants. Contributors and listeners delivered reports and discussed subjects using digital technology.

The conference agenda comprised itself a **plenary session as well as other subject sessions**:

1. Rational Use of Earth Resources

2. Biodiversity Conservation and Agriculture

3. Land Use and Livestock Husbandry

4. Ecology and Environment

5. Sustainable Water Resources Management

The members of the **Scientific Editorial Board** were:

 https://doi.org/10.1051/e3sconf/202342100001

We thank all of the researchers who contributed to this volume, as well as the conference committee, speakers, participants, organizing committee, and sponsors, for making International Conference on "Sustainable Management of Earth Resources and Biodiversity" (SERBEMA-2023) successful.

All submitted papers have passed a rigorous scientific and technical selection. Only papers that present the results of original research were accepted. The Organizing Committee, as well as the participants of the conference, express their gratitude to journal E3S Web of Conferences for the publication articles of SERBEMA-2023.

Statement of Peer review

In submitting conference proceedings to *Web of Conferences*, the editors of the proceedings certify to the Publisher that

1. They adhere to its **Policy on Publishing Integrity** in order to safeguard good scientific practice in publishing.
2. All articles have been subjected to peer review administered by the proceedings editors.
3. Reviews have been conducted by expert referees, who have been requested to provide unbiased and constructive comments aimed, whenever possible, at improving the work.
4. Proceedings editors have taken all reasonable steps to ensure the quality of the materials they publish and their decision to accept or reject a paper for publication has been based only on the merits of the work and the relevance to the journal.

Title, date and place of the conference

International Conference on Sustainable Management of Earth Resources and Biodiversity
(Tashkent, Uzbekistan, September 26 - 27, 2023)
SERBEMA 2023

Proceedings editor(s):

Dilrabo Kodirova, Tashkent State Agrarian University, Uzbekistan

Date and editor's signature

Dilrabo Kodirova **August 28, 2023**

Investigating stimulated ripening indicators of longbeak rattlebox (*Crotalaria*) planted in the coasts of the Aral Sea

Botir Amanturdiev[1,*], *Bakhrom* Allashov[2], *Fozilbek* Toreev[3], *Nurali* Khudoyberdiev[1] and *Ulugbek* Beknaev[1]

[1]Research Institute of Plant Genetic Resources, Tashkent, 100030, Uzbekistan
[2]Research Institute of Cattle and Poultry Farming, Tashkent, 100200, Uzbekistan
[3]Tashkent State Agrarian University, Tashkent, 100140, Uzbekistan

Abstract. In recent years, the coast of the Aral Sea region has been undergoing comprehensive reforms aimed at fundamentally improving the environmental situation, accelerating the socio-economic development of the region, and increasing the lifestyle and income of the population. In Uzbekistan, great attention is paid to the effective use of selection, seed production and modern agro-technologies to increase the yield of fodder crops and improve the quality of grain. In particular, a lot of attention is paid to the productive use of saline lands in the strengthening of the feed base in animal husbandry, and various crops are planted in saline areas. In this article, samples of the longbeak rattlebox crop were planted in the coast of the Aral Sea, which has a unique natural climate. In addition, phenological measurements were carried out and the results obtained for each sample of valuable economic traits are presented. In the coast of the Aral Sea, seeds of 110 samples from the National Genbank were sown in 4 repetitions, an experimental nursery was established, and the results of valuable economic traits were studied and presented based on phenological observations and statistical analysis.
Keywords. Aral Sea, crotalaria, longbeak rattlebox, coastal region, *Crotalaria juncea*.

1 Introduction

Currently, there are about 600 species of plants of the genus *Crotalaria*, and 6-7 species are cultivated and grown in India, Australia, Africa and other tropical and subtropical countries as fiber, green manure, fodder, food, and medicine [1]. The main fiber producing countries from *Crotalaria juncea* are India, Sri Lanka, South and South East Asian countries. In these countries, in the last decade, fiber yield was 0.12-0.6 tons/ha in India and 0.45 tons/ha in Sri Lanka, and seed yield was 10-22 tons/ha depending on soil conditions [2].

The homeland of the longbeak rattlebox is South Asia. To this day, various species of longbeak rattlebox are found in India, Sri Lanka, South and Southeast Asia, and Transcaucasia [3]. Recently, there has been a great interest in Africa. Cratalaria is a

* Corresponding author: b.amanturdiev@mail.com

perennial or annual herb with a height of 30 cm to 10 m, semi-shrubs or shrubs, depending on the species.

Longbeak rattlebox stems are herbaceous or woody. The leaves are simple, entire or trifoliate, sometimes difficult to structure, the leaves consist of 2-7, and then the shoots are low in number (in this case, the stems are winged) [4]. The lateral leaves are not very large and do not grow together with the leaf band or not at all. Longbeak rattlebox (in latin *Crotalaria*) is a large genus of plants belonging to the legume family (*Fabaceae*). *Crotalaria juncea* plant is adapted to different soil and climate conditions by its biological properties. The seed as a food product; hay as a high-calorie fodder in animal husbandry; in increasing soil fertility in agriculture and improving land reclamation; in the treatment of various diseases in medicine; and a source of nectar in beekeeping; can be used as a source of fiber for light industry [5].

When Crotalaria (*Crotalaria juncea*) was planted in the optimal period (April 20-24) and at the rate (14 kg/ha), a high grain yield (18 quintals/ha) was obtained, and the additional yield was 5.9 quintals/ha. The correlation coefficient between the number of pods and the grain yield is $r = 0.566$ ($R^2 = 0.320$), the regression level is $b_{xy} = 0.496$, the difference between the correlation and its error is equal to tr=1.8, and it is noted that there is an average level of correlative dependence [6].

When crotalaria (*Crotalaria juncea*) is planted at 14 kg/ha on April 20-24 in conditions of meadow alluvial soils of Khorezm region in Uzbekistan, the highest net income is 9,057,764 UZS/ha and the rate of profitability is 105.2%, additional net income is up to 1,031,100 – 5,283,600 UZS/ha profit received. Srotalariani during the period of April 20-25, when 18 kg of fertile seed per hectare was planted, the highest hay yield was 173.9 quintals/ha, conditional net profit was 6,829,563 UZS/ha and the profitability rate was 59.7% [7, 8].

The fecundity of 110 samples of longbeak rattlebox was studied and the fecundity of the samples ranged from 7.5 to 9 days. The average of 110 samples was found to be 8.5 days. Among the 110 samples in the longbeak rattlebox catalog, 3 samples had a short day, namely, the 39th "*Crotoloria ineona*", the 60th "*Crotoloria spactabilis*" and the 65th "Crotoloria incona" samples were 7.5 days [9]. It has been found that the thicker the Crotalaria alata plant, the faster the growth of the plant, but the flowering and ripening periods are delayed by 2-3 days, even 4-5 days. Also, it has been determined in experiments that it is possible to plant srotalaria alata as a siderate crop and grow cotton or rice instead, and get an abundant harvest [10].

In order to develop an efficient agrotechnology of cultivation of alfalfa in saline areas, the seeds of the Kibray variety of alfalfa and the Tashkent-1 variety of ordinary blue alfalfa were planted in different options and at different sowing rates in 4 options for comparative study. In the following years, the yield of winter wheat in Alabama increased, and the cultivated area expanded to 97,000 hectares. In 2008, $85 million worth of wheat was sold in Alabama. Here, it was found that the reason for the higher yield than wheat was that the legume srotalaria juncea was planted in these fields before wheat [10-13].

As it can be seen from the above information, it is possible to grow crotalaria plant in saline areas and get a high yield. Srotalaria can be used in agriculture to improve soil fertility, as animal fodder, as a seed food, and in other ways, and this plant requires further study.

2 Materials and methods

It is known that the fertility of some agricultural crops and other valuable economic traits decrease in saline soil climate. In order to study planting in the natural soil and climate conditions of the coast of the Aral Sea, an experimental nursery was established in

Chimboy district based on 110 samples of the longbeak rattlebox crop available in the National Gene Bank. These samples were taken from different countries at different times.

These 110 samples of crotalaria were planted and studied in the experimental lands of the Agricultural Research Institute in Chimboy district of the Republic of Karakalpakstan in Uzbekistan.

3 Results and discussion

From the data in the table below, we can see that compared to other samples, the average germination time for *Crotoloria ineona* (Sn. 39), *Crotoloria spactabilis* (Sn. 60) and *Crotoloria incona* (Sn. 65) was 7.5 days, while for all other samples it was 8.0 to 9.0 days. was found to have germinated in the interval (Table 1).

When we determined the flowering index of the studied collection samples of crotalaria plant by day, we found that *Crotoloria ineona* (Sn. 24), *Crotoloria juncea* (Sn. 28), *Crotoloria ineona* (Sn. 40), *Crotoloria juncea* (Sn. 56), *Crotoloria juncea* (Sn. 68), *Crotoloria juncea* (Sn. 74), *Crotoloria Montana* (Sn. 78), *Crotoloria juncea* (Sn. 80) and *Crotoloria spectabilis* (Sn. 94) were 32 days. It was 36 days in the sample of *Crotoloria juncea* (Sn. 27), and in all other samples it was between 32.5 and 35.5 days.

Table 1. Data of phenological observations on crotalaria samples planted in the coast of the Aral Sea (2022).

Sample number	Varietal or specimen name	Germination, day	Blooming, day	Ripening, day
1	*Crotoloria juncea*	8.5	33.5	136.3
2	-//-	8.5	33	134.3
3	-//-	9	34	135.8
4	-//-	8.5	33.5	136.0
5	-//-	8.5	35	131.0
6	-//-	8	32.5	136.3
7	*Crotoloria ineona*	9	33.5	138.5
8	-//-	9	33	135.5
9	*Crotoloria ineona*	8.5	33.5	135.5
10	*Crotoloria juncea*	8.5	33.5	133.0
11	-//-	8.5	33.5	138.3
12	-//-	9	33	134.8
13	-//-	8	33.5	139.5
14	-//-	8.5	32.5	137.0
15	-//-	8	35	136.8
16	-//-	9	33	141.3
17	-//-	9	33	135.0
18	-//-	8	34.5	140.3
19	-//-	8	33.5	135.3
20	*Crotoloria juncea*	9	36.5	132.0
21	-//-	9	33	137.8
22	-//-	8.5	32.5	135.0
23	*Crotoloria ineona*	8.5	34	135.8
24	-//-	8	32	136.0
25	*Crotoloria juncea*	9	35.5	136.3
26	-//-	8.5	33	137.8
27	*Crotoloria juncea*	8.5	36	136.3
28	-//-	8.5	32	134.5
29	*Crotoloria ineona*	8.5	34.5	135.8

https://doi.org/10.1051/e3sconf/202342101001

30	-//-	8.5	32	131.0
31	*Crotoloria juncea*	8.5	35	134.8
32	-//-	8.5	32.5	135.0
33	-//-	8.5	33.5	137.5
34	-//-	8.5	32.5	138.3
35	-//-	8.5	34.5	134.3
36	-//-	9	33	138.8
37	-//-	8.5	34.5	130.5
38	-//-	8.5	33.5	134.3
39	*Crotoloria ineona*	7.5	34	132.8
40	*Crotoloria striota*	8.5	32.5	133.3
41	*Crotoloria ineona*	8	34.5	132.8
42	*Crotoloria juncea*	9	33	136.3
43	-//-	9	33.5	130.3
44	-//-	8.5	35	130.0
45	-//-	8.5	34	132.3
46	-//-	8.5	33	134.0
47	-//-	9	33.5	135.0
48	*Crotoloria ineona*	8.5	34.5	132.5
49	*Crotoloria juncea*	8.5	33.5	134.3
50	-//-	9	32.5	134.5
51	-//-	8	34	133.0
52	-//-	9	33	130.5
53	*Crotoloria juncea*	9	34.5	129.5
54	-//-	8.5	33.5	133.8
55	-//-	8.5	33.5	131.5
56	-//-	8.5	32	135.0
57	*Crotoloria juncea*	8.5	34	140.0
58	*Crotoloria striota*	8.5	32.5	134.5
59	*Crotoloria spactabilis*	9	33.5	135.5
60	-//-	7.5	34	137.0
61	-//-	9	33.5	134.0
62	*Crotoloria incona*	8.5	32.5	135.3
63	-//-	9	34.5	137.3
64	-//-	9	34	137.5
65	-//-	7.5	33	135.8
66	-//-	8.5	32.5	133.0
67	*Crotoloria juncea*	9	34.5	135.5
68	-//-	8.5	32	138.8
69	-//-	8.5	34.5	139.0
70	-//-	8.5	32.5	129.0
71	-//-	8.5	34	131.3
72	-//-	8	34	132.5
73	-//-	9	34.5	131.0
74	-//-	8.5	32	133.5
75	*Crotoloria incona*	8.5	35	135.3
76	-//-	9	32.5	135.8
77	*Crotoloria juncea*	8.5	33.5	133.5
78	*Crotoloria Montana*	8.5	32	133.5
79	*Crotoloria juncea*	9	34.5	136.0
80	-//-	8	32	132.3
81	*Crotoloria juncea*	8	35.5	134.0
82	*Crotoloria incona*	8.5	32.5	133.8
83	-//-	9	33.5	134.0

 https://doi.org/10.1051/e3sconf/202342101001

84	*Crotoloria retusa*	9	32.5	135.3
85	*Crotoloria spactabilis*	8.5	33.5	133.0
86	*Crotoloria Montana*	8.5	33	131.5
87	*Crotoloria juncea*	9	33.5	135.0
88	*Fu-e-e-way-ha*	8.5	33	137.3
89	*Sha-shi-luy-fai*	8	33.5	133.0
90	*Za – sheo-e-Bay-ha*	8	33.5	129.8
91	*-//-*	9	34.5	132.5
92	*Crotoloria Yaon-e-e-way-he*	8.5	33	135.3
93	*Crotoloria juncea*	8	33.5	136.3
94	*Crotoloria spectabilis*	9	32	132.5
95	*Crotoloria retusa*	8	34.5	132.8
96	*Crotoloria intermedia*	9	33	134.8
97	*-//-*	8.5	33	136.3
98	*-//-*	8.5	33	134.0
99	*-//-*	9	33.5	130.0
100	*-//-*	8.5	33.5	135.3
101	*-//-*	9	34	135.3
102	*-//-*	9	33.5	134.3
103	*-//-*	9	33.5	134.0
104	*-//-*	8.5	32.5	136.5
105	*-//-*	9	34.5	134.3
106	*-//-*	9	33	134.0
107	*Crotoloria juncea*	9	33	133.3
108	*-//-*	8	34	136.3
109	*-//-*	8	35	137.8
110	*-//-*	8.5	32.5	137.5

The fastest ripening samples were 129 days in the sample of *Crotoloria juncea* (Sn. 70), 129.5 days in the sample of *Crotoloria juncea* (Sn. 53), and 129.8 days in the sample of Za-sheo-e-Bay-kha (Sn. 90). It was 141.3 days in the sample of *Crotoloria juncea* (Sn. 16), 140.3 days in the sample of *Crotoloria juncea* (Sn. 18), and all other samples were in an intermediate state, ranging from 130.0 days to 140.0 days. When the longbeak rattlebox samples were planted in the island region and studied according to the maturity index, all the samples that matured the earliest by 132 days were selected, and these samples were 15. When the stem height and blue mass yield of these samples were studied, the following results were obtained (Table 2).

Table 2. Data of the selected samples according to their precociousnessamong the crotalaria samples planted in the coast of the Aral Sea (2022).

Sample number	Varietal or specimen name	Main stem height, cm	Biomass productivity, quintals/ha	Ripening, day
5	*Crotoloria juncea*	197	236	131.0
20	*Crotoloria juncea*	228	256	132.0
30	*Crotoloria ineona*	218	242	131.0
37	*Crotoloria juncea*	204	248	130.5
43	*Crotoloria juncea*	196	241	130.3
44	*Crotoloria juncea*	204	225	130.0
52	*Crotoloria juncea*	208	234	130.5

53	*Crotoloria juncea*	198	246	129.5
55	*Crotoloria juncea*	234	278	131.5
70	*Crotoloria juncea*	192	228	129.0
71	*Crotoloria juncea*	216	234	131.3
73	*Crotoloria juncea*	212	235	131.0
86	*Crotoloria Montana*	202	227	131.5
90	*Za – sheo-e-Bay-ha*	185	219	129.8
96	*Crotoloria intermedia*	198	237	130.0

When studying the selected samples according to the speed of yield of biomass, the highest indicator was observed in the 55th sample and made 278 quintals per hectare. The lowest yield was observed in the 90th sample and was 219 quintals per hectare. Intermediate yields were observed in the remaining samples.

4 Conclusions

As can be seen from the obtained data, in terms of quick ripening index, *Crotoloria juncea* (Sn. 53 and 70) and Za-sheo-e-Bay-kha samples (Sn. 90) were distinguished by their quick ripening compared to other samples. The highest index of biomass productivity was observed in the 55th sample and was 278 quintals per hectare. At the end of August (August 24), the height of the plant was 185-218 cm, and the highest reached 234 cm.

References

1. Kodiralieva, F. A., & Rakhmanberdyeva, R. K. (2011). Polysaccharides from Crotalaria alata. Chemistry of natural compounds, 47, 7-9.
2. Kodiralieva, F. K., Shashkov, A. S., & Rakhmanberdyeva, R. K. (2015). Structure of Galactomannan from Seeds of Crotalaria alata. Chemistry of Natural Compounds, 51, 405-408.
3. Qodiralieva, F. A., Rahmanberdyeva, R. K., Mezhlumyan, L. G., & Malikova, M. H. (2013). The content and accumulation dynamics of carbohydrates and amino acid composition of proteins in fruits and seeds of Crotalaria alata (Fabaceae). Rastitel'nye Resursy, 49(4), 558-564.
4. Aberkulov M, Kiderbaeva A, Tursunov Q. Possibilities of using Crotalaria plant as a siderate. //The state of breeding and seed production of cotton and the prospects for its development: materials of the international scientific and practical conference. Tashkent, 2007. p. 270-272.
5. Aitbaeva G.K. National economic importance of plants Crotalaria Alata and Guizotia Abyssinica // Teoriya i praktika sovrem. Sciences. - 2017. - No. 6(24). - p. 33-36. (The article is devoted to the economic importance of crotalaria and nougat plants).
6. Amanturdiev B., Akhmedov D., Allashov B., Toreev F., Khudoyberdiev N. Fertility index of longbeak rattlebox crop in the coast of the Aral Sea. Agroilm Scientific supplement of the journal of agriculture and water management of Uzbekistan, 2022, Special issue -2 (86). Pages 45-47
7. Negmatova S.T. Crotalaria (Srotalaria juncea L.) is an unconventional crop that cleans the land in agriculture. Materials of the republican scientific-practical conference on "Implementation of ecological startups in life" dedicated to the World Environment Day. Tashkent-2022, pp. 34-35.
8. Nurullaeva M.Sh., Planting periods and norms of crotalaria juncea l. impact on plant growth, development, productivity (under the conditions of Khorezm region, abstract

of the Doctor of Philosophy (PhD) dissertation in agricultural sciences, Tashkent, 2022. pp. 19-20.
9. Annaeva M.I., Toreev F.N., Yakubov M.M., Allashov B.D., Mavlonova N., Tursoatov S. Agrotechnology of Melilotus albus cultivation in saline area. IOP Conference Series: Earth and Environmental Science, Volume 614, 1st International Conference on Energetics, Civil and Agricultural Engineering 2020 14-16 October 2020, Tashkent, Uzbekistan.
10. Yakudzhanova N.A. Analysis of protein content of bean collection variety samples. IQRO Press. Tashkent. 2023. 414 p.
11. Jumaev, R., & Rustamov, A. (2022, July). Representatives of Lepidoptera groups in the biotecenosis of Uzbekistan and their effective parasite-entomophage types. In IOP Conference Series: Earth and Environmental Science (Vol. 1068, No. 1, p. 012026). IOP Publishing. DOI:10.1088/1755-1315/1068/1/012026
12. Jumaev, R. (2023). Invitro rearing of parasitoids. In E3S Web of Conferences (Vol. 371, p. 01032). EDP Sciences. DOI:10.1051/e3sconf/202337101032
13. Lebedeva, N., Akhmedova, Z., Kholmatov, B., & Jumaev, R. (2021). Revision of stoneflies (insecta: plecoptera) fauna in Uzbekistan. In E3S Web of Conferences (Vol. 258, p. 08030). EDP Sciences. DOI:10.1051/e3sconf/202125808030

Evaluation of fertility of irrigated soils in Uzbekistan

Sukhrob Sanakulov[1], *Jamolbek* Turdaliev[1], *Zafar* Bakhodirov[1] and *Gulchekhra* Sodikova[2,*]

[1]Institute of Soil Science and Agrochemical Research, Tashkent, 100095, Uzbekistan
[2]Tashkent State Agrarian University, Tashkent, 100140, Uzbekistan

Abstract. The article examines the current productivity status of irrigated soils, encompassing aspects such as mechanical composition, agrochemical properties, salinity levels, and soil quality evaluation. The study highlights that the prevailing condition of the researched irrigated soils primarily comprises heavy sand texture. These soils exhibit a deficiency in humus content, rendering them low in organic matter. Additionally, they are characterized by a notable scarcity of mobile phosphorus and a correspondingly low content of exchangeable potassium. The research further categorizes the soils under investigation into three distinct cadastral groups based on their quality attributes. Through this classification, a comprehensive assessment of the soils' varying characteristics is achieved. Moreover, the study quantifies the average credit score of these soils, indicating their overall quality level, which was found to be 55 points on average. The findings underscore the significance of understanding the intricate interplay between soil characteristics and agricultural productivity. This analysis provides valuable insights for optimizing soil management practices, ultimately contributing to enhanced agricultural output within the studied area.

Keywords. Irrigated soils, productivity, quality assessment, credit score, mechanical composition, humus, salinity.

1 Introduction

In the world, scientific research is being carried out in the priority areas of identifying negative processes occurring in soil layers, including salinization, glaciation, pollution with heavy metals, dehumification and other processes, their prevention and mitigating their consequences [1]. In particular, the prevention of negative processes caused by the increase of anthropogenic influence and global climate changes, the improvement of the ecological and amelioration condition of the soil during the effective use of land resources [2], the significant reduction of soil pollution, the effective use of them by identifying the changes in the properties of the soil under the influence of evolutionary processes, and special attention is paid to scientific research aimed at increasing the productivity of agricultural crops.

* Corresponding author: gulchekhra-25@mail.ru

In Uzbekistan, a number of scientific researches aimed at determining the evolutionary changes taking place in the irrigated soils, improving the ecological and land reclamation conditions, improving, restoring and increasing soil fertility are being carried out and certain results are being achieved [3, 4].

A number of decrees and decisions on soil properties, degradation, salinity, etc. are being adopted in Uzbekistan. In all of them, measures aimed at maintaining soil fertility and improving land reclamation in modern ways are being determined [5]. Also, in the decree and decision, measures are defined to increase the effectiveness of the use of land resources of Uzbekistan and the productivity of agricultural land, to establish strict control over the prevention of the use of agricultural land, especially irrigated land for other purposes, as well as to systematize the accounting of the land fund of Uzbekistan [6].

The experts of the Scientific Research Institute of Soil Science and Agrochemistry conducted soil evaluation works on the existing irrigated agricultural land areas in Boyovut district of Sirdarya region [7], and updated the information on the current state of fertility and credit scores of the irrigated soils of this area. In order to determine the productivity and production capacity of irrigated lands in this agriculture, in-depth analysis of quality indicators, i.e. score quality (soil evaluation works), classification by cadastral categories (groups) is an important stage of state land cadastre management [8-10].

The research object is alluvial soils of irrigated irrigated meadow, sedge meadow, sedge meadow and swamp meadow of Boyovut district of Sirdarya region by experts of the Scientific Research Institute of Soil Science and Agrochemistry.

2 Materials and methods

The methodology employed in the soil inspection research draws from a multifaceted approach, integrating various sources of data and techniques to comprehensively assess the soil conditions. The foundation of this methodology lies in the analysis of soil map data related to the specific study areas. This is further supplemented by a synthesis of results stemming from comparative geographical, soil-cartographic, laboratory analyses, camera-analytical studies, and soil-evaluation investigations.

Field and cartographic activities were executed following the guidelines outlined in the "Instructions for conducting soil surveys and drawing up soil maps for the maintenance of the state land cadastre" [2]. Concurrently, the soil evaluation work adhered to the methodologies detailed in the "Methodological instructions for auditing irrigated soils of Uzbekistan" [5].

In the laboratory-analytical and camera-oriented phases of the research, methodologies were employed that were both developed and widely accepted by the Research Institute of Soil Science and Agrochemistry (TAITI). These established techniques serve as a reliable foundation for conducting accurate and consistent analyses of soil samples.

By synthesizing these various approaches and adhering to well-established guidelines and methodologies, the soil inspection research methodologically ensures a robust and comprehensive assessment of the studied areas' soil conditions, paving the way for informed decision-making in agricultural practices and land management.

3 Results and discussion

Zonal-climatologically, Syrdarya region is located in the desert, semi-desert and gray soil regions of Central Asia. Climatic indicators indicate that the territory of the Syrdarya region has hot and dry summers, moderate winters, and large temperature fluctuations between daily and annual temperatures. The average annual air temperature of the region is around +12.9-

 https://doi.org/10.1051/e3sconf/202342101002

14.9 ^{0}. The soil temperature (in the driving layer) is -2.0-0.2 ^{0} in January, on average, the soil surface freezes, which causes difficulties in plowing and washing the soil. The duration of frost-free days is 200-236 days. The first frost occurs in November, and the last frost occurs in February.

Sirdarya region is located in the zone where strong winds cross, and the area is strongly affected by the north and east (Bekabad wind). The main part of the wind blows more from the east and is most common in May-June. The wind speed reaches 3.2 m per second.

The average height of Mirzachol is 250-300 m above sea level, and the highest part is in the southeast, that is, near the beginning of the irrigation canals, and its height is 350 m. The lowest land of the bogs and shorkhoks in the northwest, that is, in Mirzachol, is 230 m above sea level. The Mirzachol plain decreases to the north and northwest.

The relative humidity is not high, the air has the lowest humidity in June-August, and the annual average humidity is around 31-48%. The increase in air temperature during the summer months leads to more evaporation of moisture, which in turn is much more than the annual average of atmospheric precipitation. This phenomenon of nature leads to soil salinization and increased water demand of crops. Despite such negative aspects of the weather, the agrolandscape of the Syrdarya region is favorable for the cultivation of almost all agricultural crops.

The interaction of geomorphological, lithological, hydrogeological and climatic conditions in the territory of Boyovut district caused different directions of soil formation processes in the area. The irrigated lands of Boyovut district are mainly covered with meadow, meadow saz, saz meadow and swamp alluvial soils.

Of the arable land area used for intensive farming in the district, i.e. 32787.0 thousand (as of 2021) hectares, 51.8% is gray meadow, 20.5% is meadow, 19.5% is meadow-gray, 8.1% is swamp. -meadow and 0.1% light-colored gray soils were noted.

Irrigated gray- meadow and meadow soils are the most common in the district, and mainly consist of heavy, light, medium sandy and loamy mechanical composition. Studies devoted to the study of the mechanical composition of soils of different regions of Uzbekistan are covered in the works of L.T.Tursunov [6], I.Turapov and others [5] S.Abdullaev and others [1], R.Kurvontaev [2] and others. Physical-mechanical, hydro-physical and agro-physical properties of desert and irrigated soils of different regions of Uzbekistan were studied in these studies.

According to the mechanical composition of the above-mentioned soils, it was noted that 46.4 ha of clay, 2981.6 ha of heavy loam, 18790.6 ha of medium loam, 9631.9 ha of light loam and 1156.5 ha of sandy loam mechanical composition were noted (Figure 1).

It was observed that 32,596.2 hectares of soils distributed in all massifs in Boyovut district are very low (up to 1%) and low (up to 1.1-2.0%) in 190.8 hectares. With mobile phosphorus, 6210.1 hectares are very low (0-15 mg/kg), 24356.7 hectares are low (16-30 mg/kg), 2190.0 hectares are average (31-45 mg/kg), 30.2 hectares provided at a high (46-60 mg/kg) level. With exchangeable potassium, 285.1 hectares are very low (0-100 mg/kg), 26095.1 hectares are low (101-200 mg/kg), 6102.5 hectares are medium (201-300 mg/kg), 295.8 hectares are high (301-400 mg/kg) and 8.5 hectares were reported to be supplied at very high (>400 mg/kg) levels.

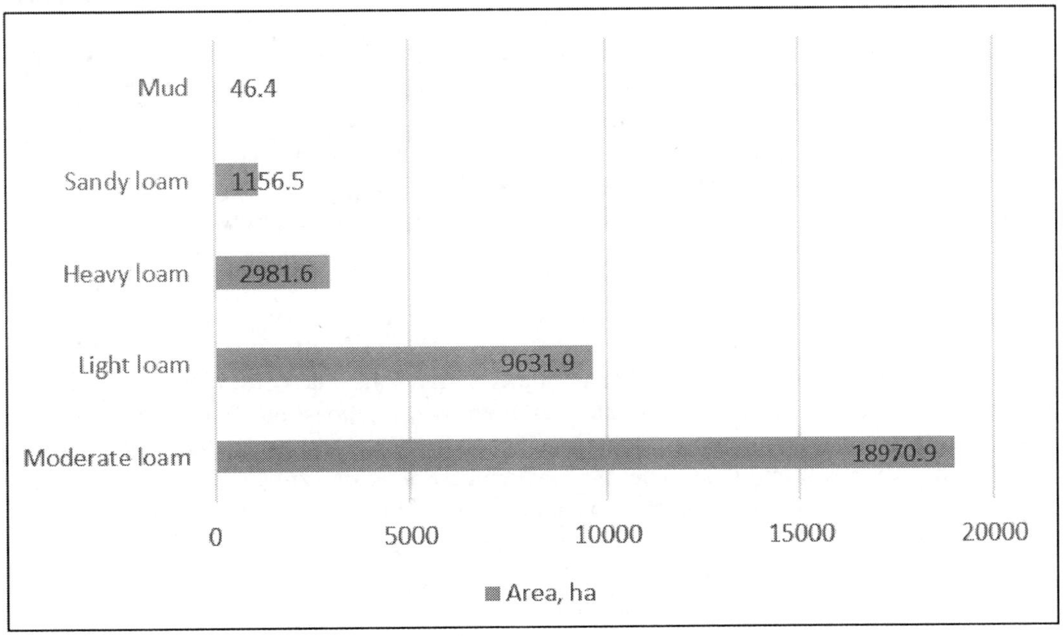

Figure 1. Mechanical composition of irrigated soils distributed in Boyovut district of Syrdarya region, per hectare.

In general, it was noted that 18.9% of the soils distributed in the district belong to the group with very little exchangeable potassium, 74.3% with low, 6.7% with medium and 0.1% with high level (Figure 2).

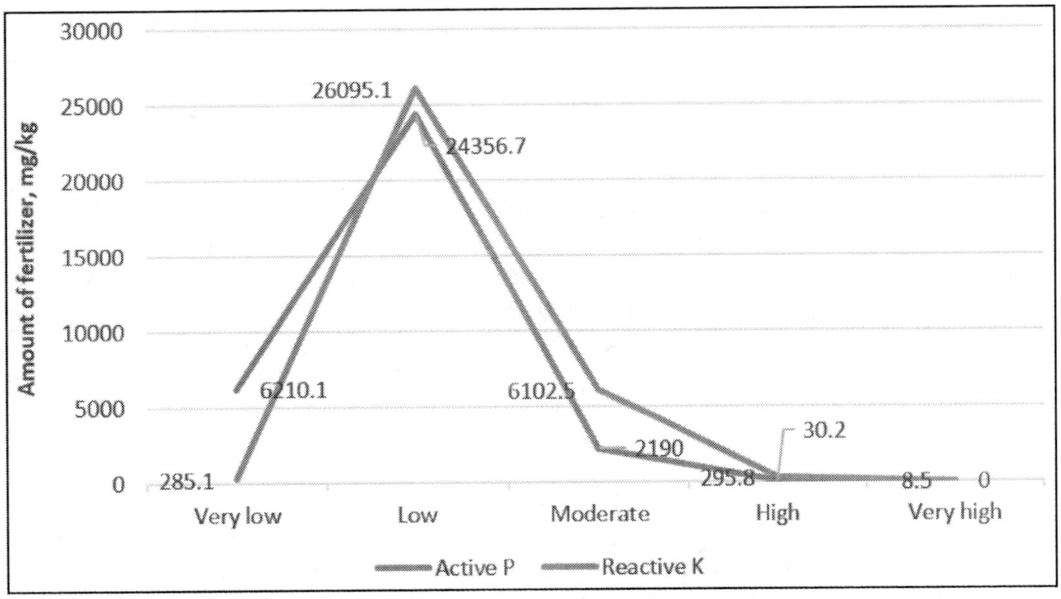

Figure 2. Amount of mobile phosphorus and exchangeable potassium in irrigated soils distributed in Boyovut district of Syrdarya region, per hectare.

In recent years, the deterioration of the ecological condition of natural components, including soils, has been observed on the territory of Uzbekistan. Increasing irrigation and saline leaching rates are leading to an increase in seepage water levels, which in turn is becoming one of the main causes of soil salinization. Soil salinity refers to the high salt content in the soil. Soil degradation due to salinity is a serious problem affecting world agriculture [4, 9, 10]. There is a total of 4.3 million hectares of irrigated land in Uzbekistan, and according to the latest data, 44.7% of irrigated land has varying degrees of salinity.

Considering the fertility and productivity of the irrigated lands, as well as the ecological-reclamation condition, salinization depends on the relief, geomorphological-lithological structure, soil-climate and human-economic conditions of the place (massives). In particular, the damage caused by the salinity of underground ground water to the national economy is extremely large, the cotton yield is reduced by 20-30% in weakly saline soils, by 40-60% in moderately saline soils, and up to 80% in highly saline soils, and cotton in highly saline and highly saline soils. It has been proven in numerous researches and field experiments that sprouts die completely at the first watering. The main reason for this is the "toxic" effect of toxic salts in the soil on plants. Therefore, when assessing the soil-melioration condition of irrigated lands, it is necessary to pay special attention to the salinity level and types of soils, the amount (%) and reserves (t/ha) of salts in the soil layer (0-30 cm) and root layer (0-1 m) will be [1].

According to the results of field-research carried out in the irrigated soils of Boyovut district of Syrdarya region and the data of soil samples taken for chemical analysis, non-saline areas are 11917.2 hectares, low salinity areas are 18031.6 hectares, moderately saline areas are 1948.3 hectares, strongly saline areas are 878, 4 hectares and 11.5 hectares of highly saline areas.

Soil assessment is a comparative assessment of soil quality and natural fertility, taking into account soil properties and characteristics that are largely related to crop productivity, and the result is expressed in points. Soil testing is carried out based on the requirements of agricultural crops, in the conditions of irrigated agriculture, it is widely used mainly in cotton and grain farming. For example, taking into account the requirements of cotton, the credit points determined by the lands are also the basis for the evaluation of the soils where other crops are grown in the cotton complex.

According to the analyzes of the main sections obtained in the data of G.S. Mirkhaydarova, soil fertility and main properties, pollution, ecological condition affect the points poured into the soil. According to the soil properties of Pop district, the soil properties are in the range of 60-46 points, with an average of 53 points. The scores for ecological condition are slightly lower, ranging from 39 to 35, with an average of 37. Total SQI scores ranged from 46 to 38, indicating that these soils are of below average quality. The overall average score for these soils is 42 points and is considered to be of below average quality [3].

In the qualitative evaluation of irrigated soils, its mechanical composition, degree and types of salinity, stony, plastering, erosion processes, leaching, humus, provision of nutrients and a number of other properties are taken into account, and the soils are evaluated (performed) on a closed scale of 100 points .

The division of soils into agroproduction (cadastre) groups - bad, below average, average, good and very good land (classes) allows, first of all, to conduct agricultural production in a scientifically based manner, to make the right choice of agrotechnical and amelioration measures. In the assessment of soils, irrigated soils with the following, i.e., the best, favorable properties, high productivity (productivity) are evaluated with 100 points, in case of deviations from the optimal indicators, reducing coefficients are used in the calculation of credit scores. The average credit rating of irrigated agricultural land types distributed in the massifs of Boyovut district is 55 points.

According to the quality of the irrigated lands suitable for agricultural use, they are grouped into ten classes, five cadastral groups, taking into account the soil fertility potential. The investigated irrigated agricultural land area is 32,787.0 hectares and was grouped into three cadastral groups by quality.

There are lands belonging to class III and IV of the first cadastral group in Boyovut district of Sirdarya region. This group is considered below-average land in terms of soil quality, and includes points assigned to soil differences with a credit score of 21 to 40 points. The characteristics of the lands belonging to this group are relatively stable, and all of them are suitable for irrigation. At the same time, the types of crops grown on these lands are limited. The soils of this cadastral group are saline, wind-eroded, very low and poorly supplied with nutrients. If the quality of soils is compared, their quality level is lower than average and includes soils with 21-40 points.

The total area of irrigated land belonging to the below-average cadastral group in the district is 1087.4 is a hectare. The average cadastral group includes irrigated lands of class V and VI. These lands make up the main part of the irrigated lands and include the types of land actively cultivated and cultivated in agriculture. The total area of land included in the average cadastral group is 21405.5 is a hectare.

The good cadastral group includes irrigated lands of class VII and VIII. These lands are lands of special value in the agriculture of irrigated lands, and the total area of lands belonging to the good cadastral group in the district is 10294.0 is a hectare.

Along with cultural-ameliorative and current works in this category of land in modern agriculture, it is possible to achieve new qualities of the soil by carrying out desalination, melioration and agrotechnical measures against erosion. If the lands of this class are used incorrectly (even temporarily), in addition to stopping the cultivation process, soil degradation will begin, the amount of humus and nutrients will decrease, erosion processes will begin on sloping lands, and soil fertility will decrease.

To determine the yield of an agricultural crop, taking into account the score credit of the field contour, that is, the value of one point according to the crop is multiplied by the score credit of the place. To determine the value of one point corresponding to the yield, the average maximum agricultural crop yield (40 t/ha) obtained for Uzbekistan is divided by the indicator of 100 points (the most fertile soil).

4 Conclusions

The outcomes of the soil analysis conducted in the Boyovut district of the Sirdayo region provide valuable insights into the soil composition of the area. The prevailing soil type in the region is primarily identified as heavy sand. This soil type is characterized by certain deficiencies in key nutrients and organic matter.

Specifically, the soil analysis reveals that the mobile phosphorus content is recorded as very low, indicating a scarcity of this essential nutrient for plant growth. Additionally, the exchangeable potassium levels are classified as low, further contributing to potential nutrient limitations in the soil. Furthermore, the supply of humus, which is crucial for soil fertility and structure, is identified as very low and low, underscoring the need for amending organic matter.

The irrigated agricultural land area that was the subject of the study encompasses a total of 32,787.0 hectares. Through a quality assessment, the area was divided into three cadastral groups based on its soil attributes. The average credit score derived from these assessments stands at 55 points on average.

The findings collectively offer a comprehensive understanding of the soil conditions in the Boyovut district. This insight is crucial for effective land management and agricultural practices in the region, aiming to address nutrient deficiencies and enhance soil fertility to optimize agricultural productivity.

References

1. Abdullaev S.A., Kurvantaev R., Khakimbaeva S. Vliyanie aroshenie na agrofizicheskie svoystva ta kyrno-lugovykh pochv Karshinskoy steppe // Pochvovedenie v Rossii: Vyzovy sovremennosti, osnovnye napravleniya razvitiya. Materialy Vserossiykoy nauchno -prakticheskoy konferentsii s mejdunarodnym uchastem k 85-letiyu Pochvennogo instituta im. VV Dokuchaeva. Moscow, 2012. - S. 276-280
2. Kurvantaev R., Kuziboev O. Izmenenie fiziko-mechanicheskikh i tekhnologicheskikh svoystv takyrno-lugovykh pochv Karshinskoy steppe pod vliyaniem orosheniya i// P ochvovedenie i agrokhimiya . Almaty , 2009. #1 . - P.40-49
3. Mirkhaydarova G.S. Characterization of soils based on quality indices. Scientific-practical conference of Uzbekistan dedicated to the 120th anniversary of the soil scientist professor Majidkhan Bahadirov on the topic " Urgent issues of the use of earth resources in agriculture ". Tashkent 2023. -B. 57.
4. Sobirova M., Murodova S. The technology of obtaining an elicitor, effectively influencing the biological properties of *Cynara Scolymus* L-M.: Scientific review. biological science, 2022. #1. c. 68-72
5. Turapov I., Kurvantaev R.K., Kutbiddinova O., Pirakhunov A. Fizicheskie i vodnye svoystva oroshaemyx pochv pravoberezhnoy chasti delty Amudari // Ecological aspects of rational use of soil: scientific-practical conference. lecture. theses of June 18-20, 1997. - Tashkent, 2017. - B.62-72
6. Tursunov L.T. Soil agrophysics and its role in soil quality // Ecological aspects of rational use of soil: scientific-practical conference. lecture. theses. Tashkent, 2017. - B.97-98 .
7. Kadirova D.A. Agribusiness development project. Methodical guide on increasing the potential of correct use of fertilizers in the territory of Uzbekistan of Uzbekistan. T-2023. 149-150 p.
8. Yuldashev G., Azimov Z. Changes in cycle chemical elements in saline landscape. Texas journal of multidisciplinary studies. 17-02-2023 V .17 38-42.
9. Khujanazarova M.K., Murodova S.S. Technology polucheniya biopreparata na osnorizobakteriy, immobilizirovannyx flokulyantom Gipan // Nauchnoe obozrenie. Biological science. – 2022. – No. 3. - S. 34-38;
10. Elena Velasquez, Patrick Lavelle, Mercedes Andrade // GISQ, a multifunctional indicator of soil quality // Soil Biology & Biochemistry 39 (2007) 3066–3080 (https://www.sciencedirect.com/science/article/abs/pii/ S0038071707002714)

Determining the algoflora of Dengizkol Lake and breeding promising species in laboratory conditions

Feruzjon Shodmonov[1], *Dilbar* Ruzibaeva[2], *Gauhar* Allamuratova[3], *Tulqin* Abdurayimov[2,*] and *Shukhrat* Abdurasulov[2]

[1]Bukhara State University, Bukhara, 200100, Uzbekistan
[2]Tashkent State Agrarian University, Tashkent, 100140, Uzbekistan
[3]Nukus State Pedagogical Institute, Nukus, 230105, Uzbekistan

Abstract. This article found that 123 species and species of microscopic algae, belonging to 4 divisions, 9 classes, 19 orders, 25 families, 32 genera, are distributed in the algal flora of the Dengizkol reservoir. When analyzing microscopic algae of the Dengizkol natural reservoir in contour sections, it was found that the number of species is high in areas where water flows into the lake, and the number of species is low in areas remote from the inflow of water. In particular, 114 species of phytoplankton were found in the 1st circuit of the Dengizkol collector, 100 species in the 2nd contour of the reservoir of the Republic of Turkmenistan, 89 of them in the 3rd circuit of the ABMK-2 outlet canal. Chlorella vulgaris Beijer on 04" nutrient medium cells up to 45.1 million/ml, Scenedesmus obliquus (Turp.) Kütz. on the basis of experiments it was proved that the number of cells increased to 46.4 million/ml.
Keywords. Dengizkol, algoflora, collector, contour, phytoplankton, *Chlorella vulgaris* Beijer., *Scenedesmus obliquus* (Turp.) Kütz., *Hypophthalmichthys molitrix* Val..

1 Introduction

In the face of climate change, depletion of natural resources, and environmental change in the world, providing the human community with quality food, especially fish products, remains one of the most urgent tasks today [1]. Changes in the hydrochemical and hydrological conditions of the inland water bodies located in the plains of Central Asia were determined, and ways of developing fisheries were developed using some hydrobionts, including algae *Chlorella vulgaris* Beijer and *Scenedesmus obliquus* (Turp.) Kütz [2]. From this point of view, increasing fish productivity as a result of using biomass obtained by breeding nutritious species of aquatic plants as additional feed for herbivorous fish is of significant scientific and practical importance [3].

* Corresponding author: abdurayimov2023@bk.ru

Numerous scientists have developed methods for breeding species rich in physiologically active substances from phytoplankton plants and using them as feed for fish in water bodies [4-7].

In the scientific sources of scientists, a number of research works have been carried out on aquatic plants and their beneficial properties, reproduction of species rich in physiologically active substances, obtaining biomass, and using the obtained biomass in various branches of agriculture [8-10].

Although several scientists have conducted scientific research on the hydrobiological condition of the world's water bodies and the distribution of aquatic plants found in them, factors such as the current global warming process, changes in the volume of water in the basins, and the increase in the amount of minerals in the water affect the hydrobiological condition of the water bodies research on the impact is ongoing.

The use of biomass of microscopic algae *Chlorella vulgaris* and *Dunaliella tertiolecta* as promising raw materials for biofuel and bioenergy production due to their high photosynthetic efficiency and rapid growth [8]. When the algoflora of reservoirs in Samarkand region was studied, 102 species and subspecies taxa belonging to 4 sections (Cyanoprocariota, Bacillariophyta, Xantophyta, Chlorophyta) were identified. When studied in terms of seasons, it was found that the richest 51 species of algoflora accounted for 42.25% in the autumn season. It was found that the diversity of algoflora depends on water temperature, environment and the amount of mineralization [11].

In the research carried out in Kara-kyr lake of Bukhara region, it was found that there are 139 species of algae belonging to 4 sections, 9 classes, 19 orders, 26 families, and 35 genera [12]. In the studies conducted on the qualitative and quantitative composition of phytoplankton in surface water bodies of Bukhara region, Bacillariophyta-34 species, Chlorophyta-20 species, Cyanophya-15 species, Dinophyta-5 species, Englenophyta - 1 species encounter has been identified [11]. 389 species and species were identified in the algoflora of Chakmoq A, Sakovich and Mavliyan ditches located in Bukhara region, 163 diatoms, 132 green algae, 72 blue-green algae, 19 euglenas and 3 dianophytes were studied. Depending on the living conditions of the identified algae, 301 species of plankton and benthos have been shown. Among the algae, 345 belong to heat-loving and 44 cold-loving algae. At the end of spring, summer and beginning of autumn, when the water temperature is 18-38°C, cold-loving species are found from the end of autumn to the beginning of spring, when the air temperature is 8-10°C [6].

2 Materials and methods

The Dengizkol reservoir is located 75-80 km southwest of the center of Bukhara, 30 km south of the center of the Olot region (coordinates: 39°06′00″ N; 64°12′00″ E), with a total with an area of 45 This is 50 thousand hectares. This natural body of water borders the Republic of Turkmenistan (Figure 1). The Dengizkol reservoir was originally considered the last runoff of the Zarafshan River, with a water saturation of 0.08 - 1.5 billion/m^3, the largest part of 30 m, an average depth of 10 m, a total water content of 5700 million/m^3. At present, the total volume of water in this reservoir is 2800 million/m^3, and the main source of water in the lake is the spillway canal ABMK-1, ABMK-2 and the collector Dengizkol [1].

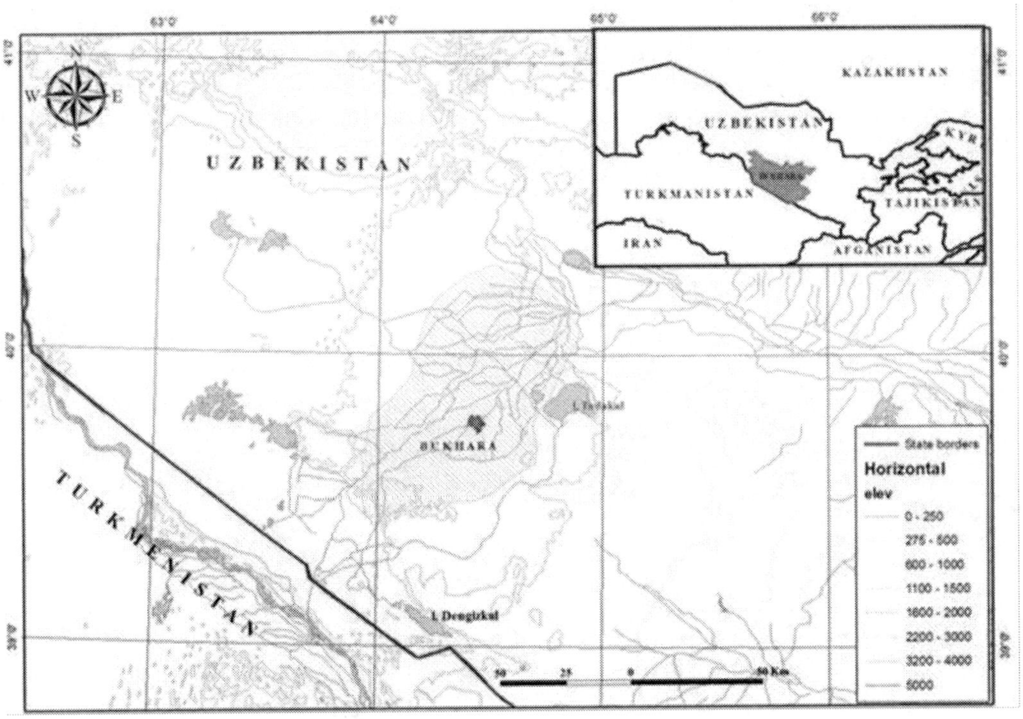

Figure 1. Map of Dengizkol.

When collecting hydrobiological samples from the Dengizkol reservoir, G.K. Plotnikov, I.G. Radchenko and N.A. Lemeza the methods of such scientists as [4-8]. In particular, two different methods were used to collect samples of phytoplankton organisms: 1. Quality; 2. Quantity. Both methods used the Epstein grid. Apshtein grid brand kapron No. 76, water inlet diameter No. 20. A specially prepared glass container was used mainly to determine the amount of microscopic aquatic plants scattered in the pond [11].

Samples of phytoplankton organisms were taken from all contours of the Dengizkol reservoir at a depth of 0.5 to 1.5 m. The collected samples were placed in a 4% formalin solution for laboratory testing and stored in a dark place, protected from sunlight. Phytoplankton in the laboratory is a common method, i.e. V.M. Katanskaya and A.M. Muzafarov's It was divided into types based on methods. Algae species isolated from the water of the lake were observed using light microscopes of the brands KHDS-3, V-380 [1-6].

To reproduce phytoplankton organisms, water samples were brought from different points of the Dengizkol reservoir and placed in nutrient medium 04 under laboratory conditions. To accelerate the reproduction of phytoplankton in the laboratory, additional special lighting lamps were installed to provide constant light (Figure 2). As a result, sufficient illumination and a comfortable temperature (not lower than 20-25^{0}C) were achieved for the growth of chlorella plants.

The temperature in special vessels where phytoplankton is grown was measured 4-5 times a day with a thermometer. After the temperature became moderate, the suspension in special containers with growing phytoplankton was mixed using a Super PUMP SP-780 compressor. The purpose of mixing the suspension is to prevent plants from sinking to the bottom of the water and speed up the process of photosynthesis.

3 Results and discussion

Studies on the analysis of phytoplankton organisms of the Dengizkol reservoir were carried out seasonally for all contours during 2020-2022. As a result, 149 samples were taken from the Dengizkol reservoir and research work was carried out. The results of the taxonomic analysis of the algae of the Dengizkol reservoir are shown in Figure 2.

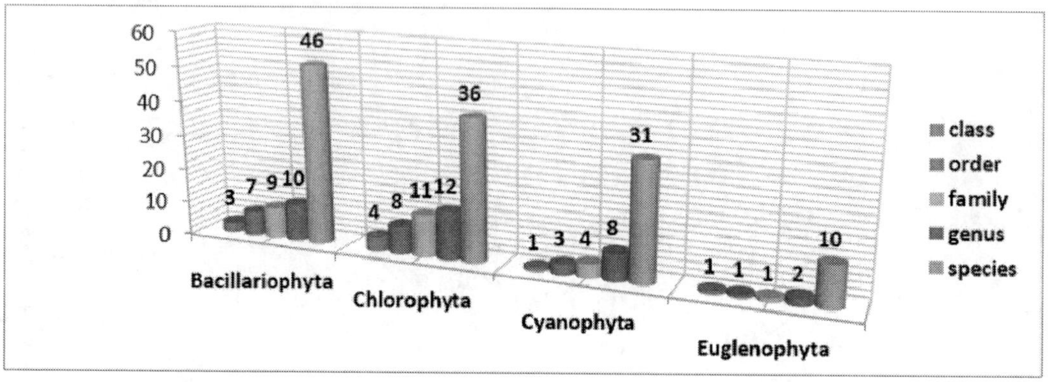

Figure 2. Histogram of taxonomic analysis of algae of the Dengizkol reservoir.

In the Dengizkol reservoir, microscopic algae include 123 species and species belonging to 4 divisions, 9 classes, 19 orders, 25 families and 32 genera.
Also, during 2020-2022, research work was carried out to determine the types of phytoplankton organisms in all contours of the Dengizkol reservoir. The results of the study are shown in Figure 3.

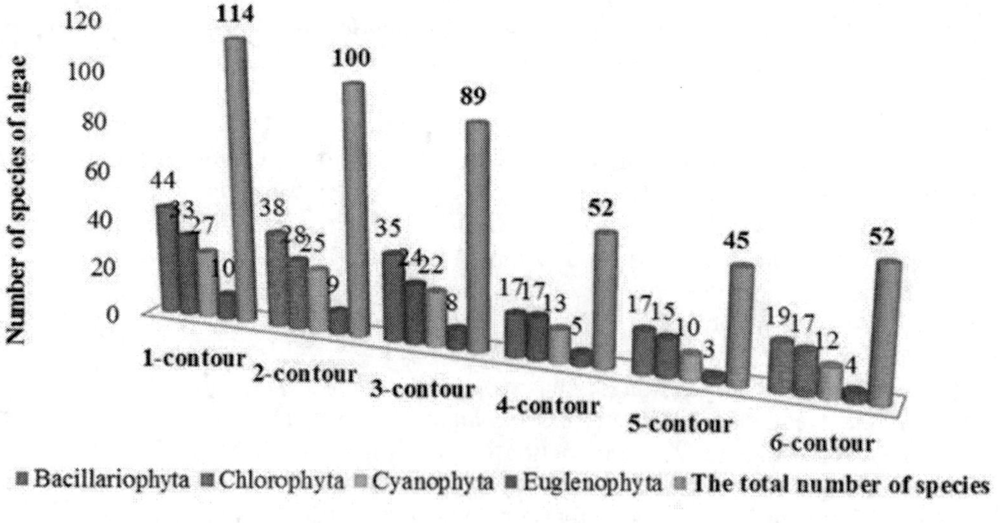

Figure 3. Distribution of microscopic algae of the Dengizkol reservoir along the contours.

When analyzing the microscopic algae of the Dengizkol natural reservoir in the contour section, it was found that the number of species is high in the areas where water flows into the lake, and the number of species is low in areas remote from the shore. In particular, 114 species of phytoplankton were found in the 1st circuit of the Dengizkol collector, 100 species in the 2nd circuit of the reservoir of the Republic of Turkmenistan, of which 89 species were found in the 3rd circuit of the ABMK-2 discharge channel.

The following common phytoplankton species of the Dengizkol reservoir have been identified:

Bacillariophyta - blue-green - Anabaena bergii, A. variabilis, A. Sphaerica, Merismopedia glauca, M. muscicola, M. Tenuissima, Oscillatoria angusta, O. geminota, O. limosa, O. acutissima, O. brevis, O. . tenuis, O.planctonica, O.cortiata. Chlorophyta - from green algae - Scenedesmys obliguus, S.quadricauda, S.acuminatus, Chlorella vulgaris, Ch.pyrenoidoza, Ankistrodesmus acicularis, A.angustus, A.arcuatus, Ulothzix zonata, U.variabilis, Cosmarium angulosum, Pediastrum baryanum. Cyanophyta - from diatoms - Diatoma ebongatum, D. anpers, D. vilgare, Cyclotella comta, C. bodanica, Melosira ambigna, Synedra acus, Navicula cari, N. basta, Cymbella affinus.

Euglenophyta - Eugleno algae - Euglina aculcate, E.deses, E.acus, E.variabilis. Chlorella vulgaris Beijer from algologically pure cells in water brought from Dengizkol reservoirs under laboratory and semi-production conditions and Scenedesmus obliquus (Turp.) Kütz. experiments were carried out to determine growth and reproduction. In laboratory conditions, experiments were carried out in bottles with a volume of 1.5-2.0 liters.

During the experiment, the temperature was 25-30 °C, the light was around 10-15 thousand lux. For active growth of cells, the suspensions were mixed using microcompressors. Phytoplankton growth rate was taken into account for 5-6 days. The data obtained on the daily reproduction of Chlorella vulgaris Beijer and Scenedesmus obliquus (Turp.) Kütz in the waters of Dengizkol 1, 3 and 5 contours are presented in Table 1.

Table 1. Daily growth rates of Chlorella vulgaris Beijer and Scenedesmus obliquus (Turp.) Kütz in Dengizkol 1st contour (collector) water, 3rd and 5th contour waters of the lake under laboratory conditions.

Algae in the experiment	**Number of cells, million/ml**					
	1st day	**2nd day**	**3rd day**	**4th day**	**5th day**	**6th day**
Control variant "04" nutrient medium						
Chlorella vulgaris Beijer.	2.0 ±0.33	5.1 ±0.03	12.3 ±0.02	23.1±0.06	37.0 ±0.03	45.1 ±0.03
Scenedesmus obliquus (Turp.) Kütz.	2.0 ±0.33	5.5 ±0.24	15.4 ±0.06	26.4±0.12	38.3 ±0.02	46.4 ±0.02
Dengizkol 1st contour (collector) water						
Chlorella vulgaris Beijer.	2.0 ±0.06	3.8 ±0.12	9.5±0.07	15.3 ±0.03	20.6 ±0.04	25.1 ±0.02
Scenedesmus obliquus (Turp.) Kütz.	2.0 ±0.03	4.0 ±0.16	10.0±0.13	14.8 ±0.08	22.4 ±0.04	28.5 ±0.04
Dengizkol 3rd contour water						
Chlorella vulgaris Beijer.	2.0±0.03	3.6 ±0.24	7.2 ±0.12	14.4 ±0.08	20.1 ±0.06	24.2 ±0.05
Scenedesmus obliquus (Turp.) Kütz.	2.0±0.03	3.8 ±0.22	8.3 ±0.08	14.9 ±0.08	21.3 ±0.06	26.6 ±0.04
Dengizkol 5th contour water						
Chlorella vulgaris Beijer.	2.0±0.03	3.4 ±0.35	5.6 ±0.22	10.5 ±0.12	17.6 ±0.06	22.3 ±0.04
Scenedesmus obliquus (Turp.) Kütz.	2.0±0.03	3.5 ±0.38	5.8 ±0.16	9.8 ±0.11	18.8 ±0.04	23.4 ±0.02

According to the results of the experiments carried out in the waters of Dengizkol under laboratory conditions, active development and reproduction of *Chlorella vulgaris* Beijer and *Scenedesmus obliquus* (Turp.) Kütz cells were determined for 5-6 days. According to the results of the experiment, at the end of the experiment, the number of *Chlorella vulgaris* Beijer cells in 1 ml of suspension in Dengizkol 1- contour (collector) water was 25.1 million, and the number of *Scenedesmus obliquus* (Turp.) Kütz. cells in 1 ml were up to 28.5 million increase was noted.

It was found that the number of *Chlorella vulgaris* Beijer cells increased to 24.2 million/ml, *Scenedesmus obliquus* (Turp.) Kütz cells increased to 26.6 million/ml in the water taken from the 3rd contour of Dengizkol.

The number of Chlorella vulgaris Beijer cells grown for 6 days in the water sample taken from the 5th contour of Dengizkol Lake increased to 22.3 million/ml, and Scenedesmus obliquus (Turp.) Kütz cells increased to 23.4 million/ml was studied.

Based on the experiments, it was proved that in the control option, i.e. nutrient medium "04", the number of Chlorella vulgaris Beijer cells increased to 45.1 million/ml, and the number of Scenedesmus obliquus (Turp.) Kütz cells increased to 46.4 million/ml.

4 Conclusions

Zooplankton and fish in a lake can use the microscopic algae found in the lake as food. *Chlorella vulgaris* Beijer from green phytoplankton, rich in physiologically active substances and *Scenedesmus obliquus* (Turp.) Kütz. isolated algologically pure cells of the species, propagated in laboratory and semi-industrial conditions and *Hypophthalmichthys molitrix* Val. used as supplementary food for fish. In the natural reservoir of Dengizkol, 123 species and varieties of microscopic algae were found, belonging to 4 divisions, 9 classes, 19 orders, 25 families and 32 genera. In the analysis of microscopic algae of the lake. Dengizkol in contour sections, it was found that the abundance of species is high in areas where water flows into the lake, and the abundance of species is low in areas remote from the watercourse. In particular, 114 species of phytoplankton were found in the 1st circuit of the Dengizkol collector, 100 species in the 2nd circuit of the reservoir of the Republic of Turkmenistan, of which 89 species were found in the 3rd circuit of the ABMK-2 discharge channel. *Chlorella vulgaris* Beijer on 04" nutrient medium cells up to 45.1 million/ml, *Scenedesmus obliquus* (Turp.) Kütz. it was found that the number of cells increased to 46.4 million/ml.

References

1. Abdullaev M.A., Urchinov D.U. Promыslovыe гыbы vodoemov nizovev p. Zarafshan. – Toshkent: Fan, 2009. 72 p.
2. Anisimova O.V., Gololobova M.A. Kratkiy opredelitel rodov vodorosley. – M.: Universitet, 2006. - 159 p.
3. Esanov H.K., Shodmonov F.Q., Kobilov A.M. High Plant Species Distributed in and aro und Dengizkul, Bukhara Region// American Journal of Plant Sciences, 2021, 12, 266-273 https://www.scirp.org/journal/ajps https://doi.org/10.4236/ajps.2021.122016
4. Hulatt C. J., Lakaniemi A.-M., Thomas D. N., Tuovinen O. H., Puhakka J. A. Biogenichydrogen and methane production from Chlorella vulgaris and Dunaliella tertiolecta biomass // Biotechnol. Biofuels. - 2011. - V. 4. 34-46 pp.
5. Lemeza N.A. Algologiya i mikologiya. Praktikum: Mn.: Vysheyshaya shkola, 2007. 6-50 p.

6. Leng R A, Stambolie J H and Bell R. Duckweed - a potential high-protein feed resource for domestic animals and fish // Livestock Research for Rural Development. - 2019. - V.7, 45-48 pp.
7. Mathew G. Srirangam. Effect of partial replacement of fish meal with duckweed (*Lemna minor* L.), and soybean meal on the growth performance of *Ctenopharyngodon Idella* (grass carp) // International Journal of Fisheries and Aquatic Studies. - 2016. - V. 4(6). 133-137 pp.
8. Watanabe M.M. Freshwater culture media/Algal culturing techniques. Elsevier Academic Press, 2005. 13-20 pp.
9. Shodmonov F.Q., Bo'riyev S.B., Okilova G.A. Fish fauna of lake Dengizkol // International Conference on Innovations in Applied Sciences, Education and Humanities, Hosted from Barcelona, Spain August 31st 2022 pp. 23-28. https://conferencea.org/index.php/conferences/article/view/1291
10. Bo'riyev S.B., Yuldoshov L.T. Role of aquatic plants in increase of fish productivity in Dengizkol lake // 6th - International Conference on Research in Humanities, Applied Sciences and Education Hosted from Berlin, Germany. Sep. 30th 2022 pp. 11-14. https://conferencea.org/index.php/conferences/article/view/1329
11. Toshov H.M., Yuldoshov L.T. The Current Condition of Ichthyofauna of DengizkulNatural Water Basin // Jundishapur Journal of Microbiology Vol. 15, No.1 (2022)pp 5933-5944. https://www.jjmicrobiol.com/index.php/jjm/issue/view/1
12. Trifonova S.N. Praktikum po sistematike rasteniy: uchebno- metodicheskoe posobie: - Arzamas: Arzamasskiy filial NNGU, 2014. - 113 p.

Promising sources for breeding golden currant (*Ribes aureum* Pursh)

Makhfurat Amanova[1,*] and *Hilola* Abdullaeva[2]

[1]Tashkent State Agrarian University, Tashkent, 100140, Uzbekistan
[2]Scientific Research Institute of Horticulture, Viticulture and Winemaking, Tashkent, 110020, Uzbekistan

Abstract. The fundamental objective of this study is to comprehensively investigate the morpho-biological attributes and valuable economic traits of hybrid samples of golden black currant. These hybrid samples are the result of crossbreeding Russian and American varieties. Beyond this, the research endeavors to establish foundational insights that can be employed as primary references in the development of novel cultivars. These cultivars are envisioned to be both fecund and endowed with larger-sized fruits, an abundance of essential vitamins and minerals, as well as adaptability to local environmental conditions. Furthermore, their suitability for extended transportation distances is an essential consideration. In essence, the study aspires to unravel the potential of these hybrid golden black currant varieties, exploring their morphological and biological characteristics, economic viability, and their potential contribution to the nutritional landscape. By assessing their adaptability, nutritional richness, and commercial prospects, we lay the groundwork for the advancement of new cultivars that hold promise not only for local agricultural systems but also for broader markets where they can meet the demands of health-conscious consumers and environmentally sustainable production practices.
Keywords. Golden currant, phenological phases, selection, productivity, bud, flower, fruit size, productivity, source material.

1 Introduction

In order to meet the demand of the population for fruit and berry products in Uzbekistan, it is envisaged to increase the volume of the product, as well as to increase its quality and expand its assortment [1, 2]. Therefore, one of the main tasks before the specialists of the field is to expand the range of fruit and berry crops, to study the biology of new crop types, to create new varieties suitable for local conditions and to introduce them into production [3, 4].

The volume of berry cultivation in the world is 7.2-8.4 million. tons, of which about 662,000 tons were blackcurrants, and the average yield was 5.5 tons per hectare. According to the statistics of FAO in 2018, the Russian Federation is the leader in the cultivation of currants (Ribes L.) in the world, and about 400,000 tons are harvested every year [5, 6].

* Corresponding author: peanut66@mail.ru

Poland (170,000 tons) and Ukraine (24,000 tons) took the next places, while the share of other countries engaged in blackcurrant cultivation did not exceed 70,000 tons [7].

In 2018, a total of 1,025 tons of blackcurrant products were obtained in Uzbekistan, including the assortment of blackcurrant varieties grown in Uzbekistan, which are high-yielding, resistant to heat, cold, drought and fungal diseases, intensive, short growing season, highly marketable and universal (for processing, freezing and keeping fresh) the reason is that there are not enough varieties [8, 9]. Therefore, enriching the assortment of berry fruit crops with promising varieties, improving the technology of creating and growing mother gardens is one of the urgent issues in the development of the horticultural sector in Uzbekistan.

Large-scale studies on systematics, selection, various resource-saving technologies of cultivation, diseases and harmful insects of the currant plant were carried out in Russia, Poland, Ukraine, Germany, England, New Zealand, Denmark, Azerbaijan, Holland, USA and other countries by K. Linney, A. Berger, A. Richard, J. Berlandier, C. Ihori, G. Burgsdorf, R. E. Regel, O. V. Panfilova, T. V. Zhidekhina, E. A. Ivanova, E. A. Gnusenkova, L. S. Sankin, I.V. Michurin, N.M. Pavlova, M.G. Abdeeva, E.I. Glebova, L.A. Tokhtar, V.S. Salykova, V.N. Sorokopudov, Yu.V. It was carried out by scientists such as Burmenko, A.E. Soloveva [10-14].

In Uzbekistan, the initial research with blackcurrant was carried out by S.I. Yagudina in 1947 (study of morpho-biological characteristics, valuable economic traits of blackcurrant varieties introduced from abroad, selection and selection), then in different years M.M. Mirzaev and R.M. These studies were continued by the Abdullaevs [15-17].

However, the physiological and economic-biological characteristics of the growth of local and newly introduced varieties of golden currant in Uzbekistan, as well as the technological issues of high-quality seedling cultivation, have not been sufficiently studied.

The purpose of the study is to study the morpho-biological and valuable economic characteristics of the hybrid samples of the golden black currant obtained by crossing Russian and American varieties, in addition to extracting primary sources for the creation of new varieties that are fertile, large-fruited, rich in vitamins and minerals, adapted to local conditions, suitable for long-distance transportation.

2 Materials and methods

During the extraction of primary data, we relied on well-established methodologies to ensure accuracy and reliability. The biochemical analysis was conducted using the method developed by A.I. Ermakov. The determination of drought resistance of varieties was carried out following the approach outlined by E.A. Goncharova. To assess heat resistance, we employed the method established by F.F. Matskov, while cold resistance of buds was determined using the method devised by M.A. Solovev [10-14].

Subsequent to data collection, we undertook meticulous mathematical and statistical processing. This critical step was executed utilizing the Microsoft Excel program, employing the guidelines recommended by B.A. Dospekhov [5]. By adhering to these recognized methodologies and protocols, we aimed to ensure the accuracy, consistency, and comprehensiveness of our research outcomes. This approach underpins the reliability and significance of the conclusions drawn from the data, providing a solid foundation for subsequent analysis and the formulation of informed recommendations.

3 Results and discussion

The golden currant *Ribes aureum* belongs to the *Ribes family of the Grossulariaceae* family.

Carl Linnaeus (1753) divided *Ribes* into two species, *Ribes inermia* and *Grossularia accileata, and 120 years later, the Russian scientist and great botanist K.I. Maksimovich (1873)* classified the plants belonging to *Ribes* divided into 6 species according to morphological characteristics, biological characteristics, distribution area. To date, scientists have identified about 190 species belonging to this genus.

Black and red currants differ in the location of the buds on the branches, the color of the bark, and whether the fruit is fragrant or odorless. The bark of the one-year-old branches of the black currant is yellow with a long gray stripe, and the bark turns brown as it grows. Freckle-like spots are visible on all branches. Buds, leaves, bark and fruit have a unique smell. Red currants have no smell and can be easily distinguished from black currants by the slow development of all branches. Blackcurrants have compound inflorescences, and red currants have simple inflorescences.

Golden currant differs dramatically in its vigorous growth compared to red and black currants. The bushes are tall. It is similar to the leaves of krijovnik, fragrant **and** the flowers are orange, with a sharp smell. The fruit has a unique aroma. Golden currant is a perennial shrub with a root system up to 2 meters deep. The leaves are alternate, three- to five-lobed, light and dark green in color, some are slightly, and some are branched. The appearance resembles the leaves of the gooseberry plant. In autumn, the leaves of this variety take on a dark red or yellow-red color.

In Uzbekistan, golden currant blooms from the end of March to the first ten days of April, and its fruit ripens in early spring, at the end of May, in the first ten days of June, i.e., on the same level as the fairy varieties of black currant. However, some varieties take a little longer. The fruit is black, red, orange, yellow, purple.

Golden currant fruit contains 6.29 to 13% sugar, 0.94 to 2.04% acids, 43.2 to 187 mg% vitamin C, 8.05 mg% provitamin A and 1.5% pectin. (on the basis of dry mass) has substance. Blackcurrant fruit is eaten fresh, as jam, juice and jelly (condensed fruit juice). 46 varieties of currants are in the gene pool of the research institute of horticulture, viticulture and winemaking named after Academician Makhmud Mirzaev, including 13 varieties of black currant (*Ribes nigrum L),* 6 varieties of red currant (*Ribes rubrum L), golden currant (Ribes aureum* Pursh) 15 varieties and 12 hybrids are kept alive.

Selection of fruit crops is fundamentally different from the selection of other agricultural crops, in some cases, several generations of scientists participate in the process of creating one variety. Scientific research completed by us was also started in 1947 by S.I. Yagudina and his followers. Between 1947 and 1979, S.I. Yagudina and a number of scientists of the institute crossed American Krondal and Russian Michurin varieties and obtained about 274 hybrids. Later, a promising hybrid isolated from these hybrids was included in the state register under the name "Plotnomyasaya" [6].

During the period of keeping the collection alive, a group of hybrids was given the names "Uzbekistanskaya krupnoplodnaya" and "Uzbekskaya sladkaya". In 2007-2009, 140 plants were cloned from Uzbekskaya sladkaya na muna . These plants gave their first crops in 2011, and in 2013 they reached full harvest and from these plants no. 48 numbered plants were selected according to valuable economic traits.

In 2008-2009, as a result of studies on valuable economic characteristics of 700 seedlings grown from the seeds of the "Uzbekistanskaya krupnoplodnaya" sample, samples No. 13-27 and No. 14-7 were selected by the method of individual selection. Complex hybrids of "Uzbekistanskaya krupnoplodnaya" obtained by free crossing of golden black currant suitable for the climatic conditions of Uzbekistan were studied according to morpho-biological and economic characteristics.

During the years 2011-2015, in the "Plotnomyasaya" variety of the golden currant, the swelling of the buds coincided with February 24, and the opening of the buds coincided with March 3. Among the studied varieties and hybrids, only the "Uzbekskaya sladkaya" variety had the swelling of the buds at the same time as the model, and in all other varieties and hybrids, it was on the 1st Late opening was observed for 3 days. Budding was observed at the latest (on March 1) in "Alyona" variety and samples #13-2 and #13-27, while in all other varieties and hybrids this physiological process was recorded on February 25-27, and opening on March 2-6.

Among varieties and hybrids of golden currant, there are differences in the phase of flowering, including: "Plotnomyasaya" variety begins on March 24, and "Kyghos" blooms on March 29, and ends on April 21, while "Altinoy", "Dostlik", "Elixir" varieties and in the hybrids, the appearance of flowers was recorded on March 21, the beginning of flowering in the varieties "Oltinoy" and "Elixir" on April 1, and in the variety "Dostlik" on March 29. The duration of flowering in the collection nursery was 28-33 days, and the longest flowering process lasted for the "Uzbekskaya sladkaya" variety (33 days).

Among the varieties and hybrids of golden black currant, 3 varieties, Oltinoy, No. 5-11, No. 13-7 and Rukhshona, were distinguished according to their early ripening characteristics, and it was observed that they ripened 4-6 days earlier than the model Plotnomyasaya variety.

Compared to the model variety, the ripening duration of the fruits lasted 5-8 days, and the varieties No. 3, Oltinoy, No. 14-7, No. 13-7 and No. 13-27 and hybrids were isolated. The following results were obtained when the cultivars and hybrids maintained in the nursery of golden currant collection samples were studied for the productivity of one plant during the years 2011-2015 (Table 1).

Table 1. Five-year average indicators of productivity of golden currant varieties and hybrids per bush (kg).

Varieties	Productivity from 1 bush, kg					
	2012	2013	2014	2015	total	average
Plotnomyasaya (control)	0.540	0.250	1.1	1.01	2.495	0.624
Uzb planted tanskaya krupnoplodnaya	0.450	0.301	0.65	0.89	1.954	0.488
(№-3) Uzbekskaya sladkaya	0.650	0.350	0.810	0.800	2.610	0.653
Golden	0.5	0.310	0.853	0.820	2.108	0.527
Siyuma	0.250	0.180	0.450	0.910	1.603	0.401
Rukhshona	0.26	0.20	0.60	0.740	1.605	0.401
Orzu	0.3	0.24	0.74	0.68	1.735	0.434
Alyona	0.19	0.15	0.45	0.400	1.48_	0.262
No. 14-7 (Memorial)	0.571	0.474	0.82	2.16	4.10	1.024
Dustlik	0.32	0.21	0.45	0.68	1.42	0.355
Elixir	0.42	0.31	0.49	0.69	1.595	0.399
No. 15-5	0.29	0.21	0.56	0.91	1.753	0.438
No. 13-7	0.41	0.312	0.74	0.95	2.105	0.526
No. 13-2	0.31	0.24	0.39	0.785	1.493	0.373
#13-27 (Oydin)	0.454	0.362	0.882	1.69	4.048	1.012
#13-16	0.32	0.32	0.76	0.86	2.02	0.505

#13-17	0.51	0.29	0.91	0.91	2.238	0.559
No. 15-12	0.54	0.25	1.1	1.01	2.495	0.624
No. 5-11	0.45	0.301	0.65	0.89	1.954	0.488
Zolotistaya	0.25	0.25	0.51	0.8	1.623	0.406
EKF05	***0.01***	***0.0***	***0.02***	***0.04***		***0.03***
ECF%	***3.2***	***2.6***	***3.0***	***4.0***		***4.9***
V	***2.3***	***1.9***	***2.2***	***2.9***		***3.6***

During the years 2011-2015, the average productivity of one bush of the model variety "Plotnomyasaya" (st) is 0.624 kg. and it was found that 16 of the 19 studied varieties and hybrids had lower results from 65 to 360 grams compared to the standard variety.

These varieties and samples are effective in the early spring changing climate of Uzbekistan compared to the model "Plotnomyasaya" (st) variety, and the productivity was significantly higher in years with warm spring months, as a result of many years of observations. For example, the yield of one bush in the variety "Siyuma" was 250-180 g in 2012-2013, and 0.450 and 910 g in 2014-2015 (years with even spring days). The lowest productivity was observed in "Alyona" (0.262 kg), #13-2 (0.373 kg), "Dostlik" (0.355 kg), "Elixir" (0.399 kg) varieties and hybrids.

When analyzing the research conducted with golden currant varieties and hybrids, the 4-year average yield indicators were shown as follows. It is known that the climatic conditions of Uzbekistan change rapidly, and spring months sometimes come early and sometimes late. In most cases, the air temperature rises from the end of February to the beginning of March and decreases in the end of March or the beginning of April. This situation leads to the loss of part of the crop. Sometimes, when the air temperature rises at the same time (as a result of the frost not hitting the flowers), it is observed that the yield is high. We made a conclusion based on the average indicators of the results obtained during the 4 years of research due to the different arrivals of the spring months.

"Plotnomyasaya" (st) variety of golden black currant has an average yield of 19.27 q/ha, the lowest indicator is "Siyuma" (11.4 q/ha), No. 5-11 (11.4 q/ha), "Alyona" (11.8 q/ha), No. 13-2 (12.4 q/ha), "Dostlik" (13.2 q/ha), "Rukhshona" (14.6 q/ha), "Elixir" (14.9 q/ha), "Zolotistaya" (15.1 q/ha), No. 15-5 (15.3 q/ha), "Uzbekistanskaya krupnoplodnaya" (15.9 q/ha) samples, average indicators "Oltinoy" (16.2 q/ha), "Iroda"(16.8 q/ha), No. 13-7 (16.8 q/ha), No. 13-16 (17.1 q/ha), No. 15-12 (17.9 q/ha) were observed, relatively high results were observed in No. 13-27 ("Oydin") (20.7 q/ha), "Uzbekskaya sladkaya" (21.5 q/ha) and No. 14-7 ("Yadgor") (34.0 q/ha) were observed (Table 2).

Table 2. Four-year average yield indicators of golden currant varieties and hybrids (q/ha).

Varieties	2012	2013	2014	2015	Average
Plotnomyasaya (control)	17.82	9.93	36.30	26.40	19.27
Uzb ekistanskaya krupnoplodnaya	14.85	11.55	21.45	27.06	15.94
Uzbekskaya sladkaya	21.45	10.23	26.73	30.03	21.55
Golden	16.50	10.23	28.14	22.44	16.23
Siyuma	8.25	5.94	14.85	22.77	11.41
Rukhshona	8.58	6.60	19.80	30.03	14.64
Will	9.90	7.92	24.42	31.35	16.54
Alyona	6.27	4.95	14.85	25.90	11.82

No. 14-7 (Memorial)	18.84	15.64	27.06	88.77	34.05
Dustlik	10.56	6.93	14.85	28.38	13.20
Elixir	13.86	10.23	16.17	30.03	14.97
No. 15-5	9.57	6.93	18.48	33.33	15.28
No. 13-7	13.53	10.30	24.42	29.37	16.87
No. 13-2	10.23	7.92	12.87	26.40	12.44
No. 13-27 (Oydin)	14.98	11.95	29.1	27.06	20.77
No. 13-16	10.56	10.56	2508	30.03	17.08
No. 13-17	16.83	9.57	3003	22.44	16.56
No. 15-12	17.82	8.25	36.30	22.77	17.94
No. 5-11	14.85	9.93	2.145	30.03	11.45
Zolotistaya	8.25	8.25	16.80	33.5	15.15

No. 13-27 ("Oydin") , "Uzbekskaya sladkaya" and No. 14-7 ("Yadgor"), the average yield compared to the variety "Plotnomyasaya" (st) is from 1.5 to 14.7 q/ha or, 7 , 3-44.6% higher results were noted (Figure 1).

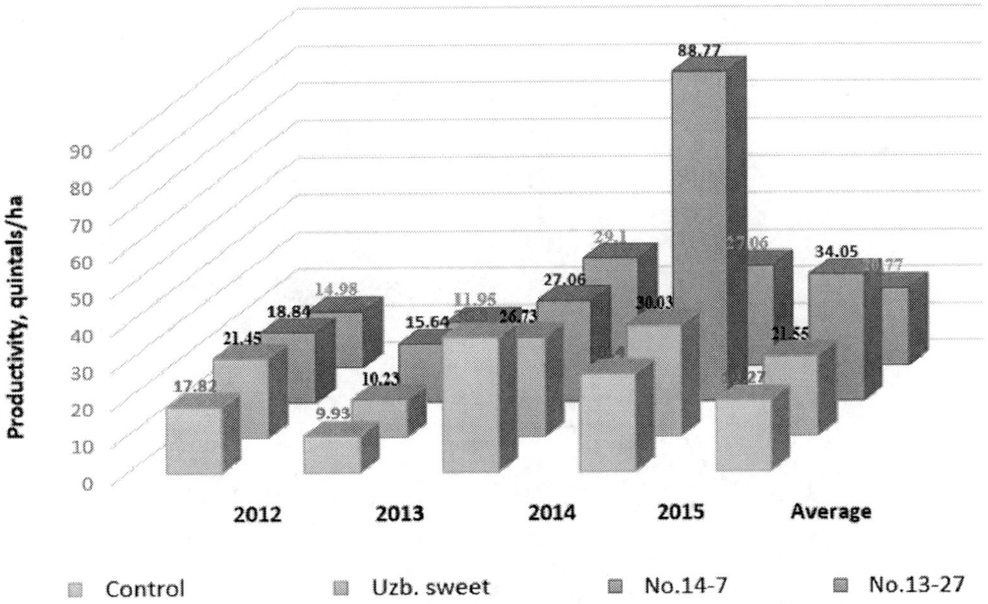

Figure 1. Productivity indicators of fruit varieties

No. 3, No. 14-7 and No. 13-27 were selected according to the complex economic characteristics. "Uzbekskaya sladkaya", "Yadgor" and "Oydin" were given the names "Uzbekskaya sladkaya", "Yadgor" and "Oydin" to the selected hybrids of golden currant.

The newly created varieties " Uzbekskaya sladkaya", "Yadgor" and "Oydin" were studied for four years (2012-2015) in comparison with the model variety " Plotnomyasaya" and the following results were obtained.

The average productivity of " Oydin" variety is 20.77 tons/ha, " Uzbekskayasladkaya" variety is 21.55 tons/ha, and " Yadgor" variety is 34.05 tons/ha, and "Oydin" variety is 1.5 It was noted that the " Uzbekskayasladkaya" variety has a high yield of 2.28 tons/ha, and the "Yadgor" variety has a high yield of up to 14.77 tons/ha (Table 3).

Table 3. Four-year average productivity indicators of golden currant varieties and hybrids (q/ha).

Varieties	2012	2013	2014	2015	Average	Difference
Plotnomyasaya (control)	17.82	9.93	36.30	26.40	19.27	0
No. 13-27 (Oydin)	14.98	11.95	29.1	27.06	20.77	1.5
No. 3 Uzb. sweet	21.45	10.23	26.73	30.03	21.55	2.28
No.14-7 (Memorial)	18.84	15.64	27.06	88.77	34.05	14.77

Another valuable economic characteristic of the golden currant is its large fruit size. The sample "Plotnomyasaya" (st) variety is 1.6 grams, and the largest one is 2 grams. Almost all of the studied varieties and hybrids are lower in average weight of one fruit (0.67-1.5 g) compared to the sample variety. got results. However, according to the weight of the largest fruit, 12 varieties and hybrids are 1.5 compared to the standard variety ("Siyuma", "Rukhshona", "Dostlik", "Elixir" varieties, No. 13-7, No. 13-2, No. 13- 27, "Oydin", "Yadgor") and 2 times (Uzbekistanskaya krupnoplodnaya, "Iroda" varieties, hybrids #13-16, No. 15-12) were noted in the research (Figure 2).

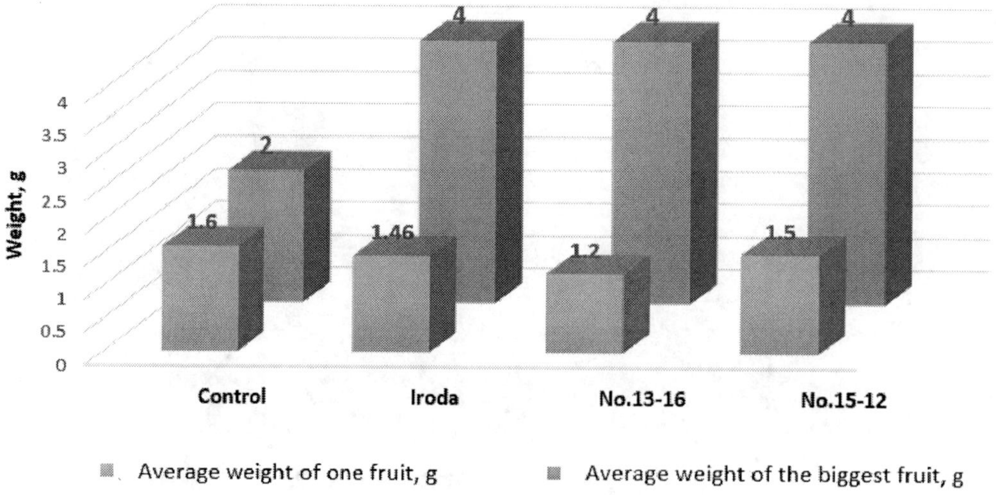

Figure 2. Varieties and hybrids selected by the size of the fruits of the black currant.

These selected specimens are the primary rare resource for the development of new large-fruited market varieties of golden currant.

One of the most valuable economic characteristics of golden currant is evaluated by its chemical composition. The fruit can be black or brown in color, round, elongated, pear-shaped or flat. According to the sources, the fruit is valuable because it contains valuable vitamins S, V, V9 , A, R, K , sugar, organic acid, aromatic substances. In our research, we analyzed the content of acid, dry matter and sugar in the fruits of 9 varieties and hybrids of golden currant (Table 4). Of the sample "Plotnomyasaya" (control) variety was 1.24% acid, 14% dry matter, and 13.4% sugar.

Table 4. Analysis of chemical active substances in golden currant fruits (2022).

Varieties name	Acid %	Dry matter	Carbohydrates
Plotnomyasaya	1.24	14.0	13.4
Uzbekistanskaya krupnoplodnaya	1.79	16.0	17.0
(№-3) Uzbekskaya sladkaya	1.08	18.0	18.3
Golden	1.72	20.0	16.9
Elixir	1.66	16.0	15.1
No. 13-27 (Oydin)	1.66	18.0	18.2
No. 14-7 (Memorial)	1.24	17.0	15.2
No. 13-2	1.60	14.0	14.5

In 6 out of 8 collection varieties and hybrids whose chemical composition was analyzed, including: "Uzbekistanskaya krupnoplodnaya", "Uzbekskaya sladkaya", "Oltinoy", "Elixir" varieties and the fruits of hybrids No. 13-27 and No. 14-7, the dry matter content was 16 up to -20%, with 2-6% higher results compared to the model. Sample No. 13-2 showed a low result compared to the model variety (Figure 3).

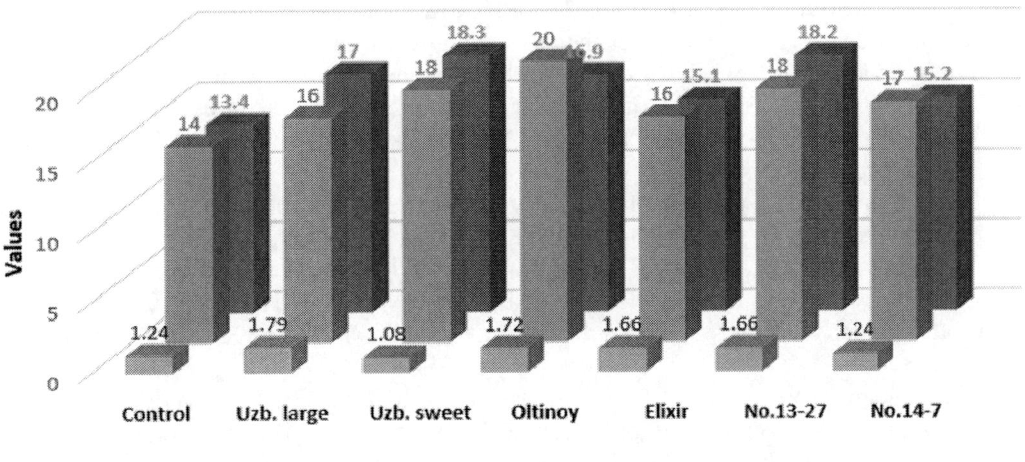

Figure 3. Analysis of chemical active substances in golden currant fruits.

It was found that all the varieties and hybrids selected for the amount of sugar in the fruit have higher results compared to the "Plotnomyasaya" variety, "Uzbekistanskaya krupnoplodnaya" variety has 17%, "Uzbekskaya sladkaya" variety has 18.3%, "Oltinoy" variety has 16.9%, 15.1% in "Elixir" variety, 18.2% in "Oydin" variety, 15.2% in "Yadgor" variety and 15.5% in No. 13-2, i.e. 1.1% to 3.9% higher than the sample was recorded.

As a result of research carried out in the nursery of the collection, 3 varieties, including No. 3 ("Uzbekskaya sladkaya"), No. 14-7 ("Yadgor"), No. 13-27 ("Oydin"), were selected for valuable farm characteristics, and to continue the selection work the control was transferred to the cultivar trial nursery.

4 Conclusions

During the years 2011-2015, compared to the "Plotnomyasaya" variety of the golden black currant (budding of buds on February 24, bud opening on March 3), only the "Uzbekskaya sladkaya" variety of the studied varieties and hybrids is the same as the model, and all other varieties and hybrids are delayed by 1-3 days observed. Budding was observed at the latest (on March 1) in "Alyona" variety and samples No. 13-2 and No. 13-27. In all other varieties and hybrids, this physiological process was recorded on February 25-27, and opening on March 2-6.

Among varieties and hybrids of golden currant, differences were also observed in the phase of entry into the flower. In the variety "Plotnomyasaya" the beginning of the pattern was on March 24, the flowering of marigolds was on March 29, and the end was on April 21, while in the varieties and hybrids "Altinoy", "Dostlik", "Elixir" the appearance of flowers was on March 21, the flowering of marigolds was on "Oltinoy" and " It was recorded in Elixir varieties on April 1, and in Dostlik varieties on March 29.

The duration of flowering in the collection nursery was 28-33 days, and the longest flowering process lasted for the "Uzbekskaya sladkaya" variety (33 days).

From the varieties and hybrids of the studied golden currant, 3 varieties, Oltinoy, No. 5-11, No. 13-7 and Rukhshona, were selected according to their early ripening characteristics, and it was found that they ripened 4-6 days earlier than the model Plotnomyasaya variety.

The hybrids of No. 3, Oltinoy, No. 14-7, No. 13-7 and No. 13-27 varieties were selected according to the duration of fruit ripening, and it was noted that the duration of fruit ripening in these hybrids was 5-8 days longer than the standard variety.

The productivity of one bush of the studied 20 varieties and hybrids is lower than the model (65-360 g) in 16 samples, the same as the model in 1 sample, only in 3 samples (No.-3 Uzbekskaya sladkaya- 0.653 kg, No. 14- 7 Yadgor - 1,024 kg and No. 13-27 Aydin - 1,012 kg) a high result was observed.

It was found that the average productivity of varieties and hybrids of No. 3, No. 14-7 and No. 13-27 was 105-164 percent higher than the standard variety.

According to the results of research carried out for four years in the nursery of the collection, varieties and hybrids No. 3, No. 14-7 and No. 13-27 were selected, and on their basis, the golden currant varieties " Uzbekskaya sladkaya", "Yadgor" and "Oydin" were created.

The average productivity of " Oydin" variety is 20.7 tons/ha, " Uzbekskaya sladkaya" variety is 21.5 tons/ha, and " Yadgor" variety is 34.0 tons/ha. "Oydin" variety is 1.5 compared to the model variety. q/ha, " Uzbekskaya sladkaya" variety - 2.28 q/ha, and "Yadgor" variety - 14.77 q/ha were noted to have high productivity.

The varieties "Uzbekistanskaya krupnoplodnaya", "Iroda" and hybrids No. 13-16, No. 15-12 were distinguished according to the sign of fruit size. These samples will serve as a unique primary source for the creation of new large-fruited, exportable cultivars of golden currant.

References

1. Abdullaev R.M., Abdullaeva H.R. Study of golden currant varieties introduced in the conditions of Uzbekistan. / Collection of materials of the republican scientific and practical conference on the topic "The importance of innovative agrotechnologies in the

prospective development of horticulture, viticulture and winemaking" (September 26, 2019). - Tashkent, 2019. - p. 12-14.
2. Abdullaev R.M., Abdullaeva H.R. Study of golden currant varieties introduced in the conditions of Uzbekistan./Collection of materials of the republican scientific and practical conference on the topic "The importance of innovative agrotechnologies in the prospective development of horticulture, viticulture and winemaking" (September 26, 2019). - Tashkent, 2019. - p. 148-151.
3. Abdullaeva H.R., Kosimov A.A. Promising varieties of Golden currant cultivated in the conditions of Uzbekistan. // European Journal of Agricultural and Rural Education (EJARE). - Spain, 2019. - Volume 2, - No. 12. – P. 112-120. Impact Factor: 7.354.
4. Abdullaev R.M., Abdullaeva H.R. Plants with berries. Monograph. Tashkent, "EFFECTNASHR" 2020.
5. Abdullaev R.M., Abdullaeva H.R., Sultanov K., Agzamkhodzhaev J., Kasimov A., Berry-bearing plants. Study manual. Tashkent, "EFFECTNASHR" 2021.
6. Abdullaeva H.R. Newly created promising varieties of golden currant. // Journal "Actual problems of modern science". - Russia, 2022. - ISSN 1680-2721. - P. 55-61.
7. Abdullaeva H.R., Amanova M.E., Kasimov A.A. Perspektivnye novye sorta zolotistoy smorodiny v to Uzbekistan. // Journal "Actual problems of modern science". - Russia, 2022. - ISSN 1680-2721. - P. 13
8. Tayjanov, K., Khojimatov, O., Gafforov, Y., Makhkamov, T., Normakhamatov, N., & Bussmann, R. W. (2021). Plants and fungi in the ethnomedicine of the medieval East-a review. Ethnobotany Research and Applications, 22, 1-20. DOI:0.32859/ERA.22.46.1-20
9. Jabeen, S., Zafar, M., Ahmad, M., Althobaiti, A. T., Ozdemir, F. A., Kutlu, M. A., ... & Majeed, S. (2023). Ultra - sculpturing of seed morphotypes in selected species of genus Salvia L. and their taxonomic significance. Plant Biology, 25(1), 96-106. DOI:10.1111/plb.13473
10. Majeed, S., Ahmad, M., Ozdemir, F. A., Demirpolat, A., Şahan, Z., Makhkamov, T., ... & Nabila. (2023). Micromorphological characterization of seeds of dicot angiosperms from the Thal desert (Pakistan). Plant Biosystems-An International Journal Dealing with all Aspects of Plant Biology, 157(2), 392-418. DOI:10.1080/11263504.2023.2165553
11. Majeed, S., Ahmad, M., Ali, A., Althobaiti, A. T., Ramadan, M. F., Kilic, O., ... & Sultana, S. (2023). Pollen micromorphology among amaranthaceous species from desert rangeland: Exine stratification and their taxonomic significance. BioMed Research International, 2023, 4967771. DOI:10.1155/2023/4967771
12. Makhkamov, T., Sotiboldiyeva, D., Mamarakhimov, O., Yuldashov, Y., & Botirova, L. (2022, May). Morphogenesis and Seasonal Developmental Rhythm Under the Conditions of Introduction of Curcuma Longa L. In International Scientific Conference on Agricultural Machinery Industry "Interagromash"" (pp. 1460-1469). Cham: Springer International Publishing. DOI:10.1007/978-3-031-21432-5_155
13. Boboev, S., Makhkamov, T., Bussmann, R. W., Zafar, M., & Yuldashev, A. (2023). Anatomical and phytochemical studies and ethnomedicinal uses of Colchicum autumnale L. Ethnobotany Research and Applications, 25, 1-9. DOI:10.32859/era.25.6.1-9
14. TKh, M., Brundu, G., Jabborov, A. M., & Gaziev, A. D. (2023). Predicting the potential distribution of Ranunculus sardous (Ranunculaceae), a new alien species in the flora of Uzbekistan and Central Asia. BioInvasions Records, 12(1), 63-77. DOI:10.3391/bir.2023.12.1.05
15. Ameen, M., Zafar, M., Ahmad, M., Ramadan, M. F., Eid, H. F., Makhkamov, T., ... & Majeed, S. (2023). Assessing the Bioenergy Potential of Novel Non-Edible Biomass

Resources via Ultrastructural Analysis of Seed Sculpturing Using Microscopic Imaging Visualization. Agronomy, 13(3), 735. DOI:10.3390/agronomy13030735

16. Noor, W., Zafar, M., Ahmad, M., Althobaiti, A. T., Ramadan, M. F., Makhkamov, T., ... & Khan, A. (2023). Petiole micromorphology in Brassicaceous taxa and its potential for accurate taxonomic identification. Flora, 303, 152280. DOI:10.1016/j.flora.2023.152280
17. Aziz, A., Ahmad, M., Zafar, M., Gaafar, A. R. Z., Hodhod, M. S., Sultana, S., ... & Chaudhay, B. (2023). Novel Copper Oxide Phyto-Nanocatalyst Utilized for the Synthesis of Sustainable Biodiesel from Citrullus colocynthis Seed Oil. Processes, 11(6), 1857. DOI:10.3390/pr11061857

Biotechnologic farming: an innovative approach to sustainable agriculture

Gulnara Djumaniyazova[1,*], *Khurshida* Narbaeva[1], *Boriy* Alikhanov[2], *Sanjar* Rakhimov[3] and *Farhad* Yuldashev[4]

[1]Research Institute of Microbiology, Tashkent, 100128, Uzbekistan
[2]International Academy of Sciences on Ecology and Life Safety, Saint-Petersbourg, 199206, Russia
[3]Innovation Ideas LLC, Tashkent, 100101, Uzbekistan
[4]Belgorod State Technological University, Belgorod, 308012, Russia

Abstract. The symbiotic relationship between agriculture and the environment has undergone a profound shift with the advent of extensive chemicalization. While this approach initially bolstered yields, it simultaneously disrupted the delicate equilibrium of nature. The repercussions have been felt across the spectrum, from weakened natural defenses in plants and animals to human health concerns. The efficacy of traditional agricultural techniques has dwindled in the face of these mounting challenges. The urgency of the situation calls for an exploration of innovative farming methods that can reconcile productivity with ecological integrity. This article presents a foray into a novel biotechnological approach to farming, a transformative endeavor that champions environmental friendliness, resource conservation, cost-effectiveness, and innovation. The crux of this endeavor is the cultivation of agricultural crops on soil marred by salinity and heavy metal degradation - a locale perilously poised on the brink of desertification in the arid expanse of Uzbekistan. This research encapsulates an intensive investigation, delving into the intricate mechanics of this newfangled bio-agrotechnology. Through painstaking experimentation and strategic implementation, the innovative methodology triumphs over the adversities posed by challenging soil conditions. The results are not only heartening but transformative. The application of this bio-agrotechnology rejuvenates saline soils, heralding a resurgence in fertility that ripples across the ecosystem. This reinvigoration is evidenced in elevated crop yields and heightened product quality, illustrating the synergistic harmony between nature and innovation.
Keywords. Saline soils, biological fertilizers, bioagrotechnology, soil fertility, crops, sustainable agriculture, food security.

1 Introduction

In recent years, agriculture, using intensive technologies, is everywhere facing the problems of land degradation, a drop in the quantity and quality of products and, as a result, a

* Corresponding author: gulnara_djumaniyazova1001@gmail.com

decrease in the economic performance of all economic activities in the agricultural sector. There is a need for a deep critical analysis of the development of intensive production methods, when even elementary environmental requirements were ignored in the pursuit of volume, quantity and quick profit [1, 2].

The ever-increasing population of the planet, the constant decline in crop yields, an annual average of 30% yield loss due to pests, plant diseases and weeds were the basis for the total chemicalization of agriculture [3]. However, such an approach provoked a global ecological crisis, putting, on the one hand, endangered the life of mankind, and on the other hand, caused serious violations in agricultural activities [4]. As a result of chemicalization, mechanization and melioration, the chemical and physical load on the components of the agricultural landscape grew exponentially [5]. A significant increase in the use of nitrogen, phosphorus fertilizers and pesticides (seed protectants, herbicides, fungicides, insecticides, chemical growth stimulants and pollinators) has led to the suppression of beneficial soil microflora - the basis of fertility and the activation of harmful microflora that destroys soil humus, as a result of [6]:

- soil properties have deteriorated - structure, water permeability, aeration;
- the number of mobile forms of nitrogen, phosphorus, potassium, macro-microelements decreased;
- the level of soil fertility is catastrophically reduced;
- pathogenic microflora develops in the soil, causing various plant diseases;
- the development of the aerial parts of plants is enhanced, root formation is inhibited;
- metabolic processes in plants are disturbed, which leads to a decrease yield and quality of agricultural products in connection with the accumulation of nitrates, nitrites, pesticide residues in it;
- the environment is polluted with toxic substances as a result of the accumulation of chemicals in the soil, groundwater, plant and livestock products, causing contamination of feed and food products;
- the imbalance of the microflora of the gastrointestinal tract of the body is disturbed, which leads to the formation of many diseases (gastritis, peptic ulcer of the stomach and duodenum, liver damage, rheumatoid arthritis, joint damage, urolithiasis, bronchial asthma, dermatitis, allergic, oncological and other diseases).

Thus, chemicalization violated self-regulation in living nature, weakened the defenses of plants, animals and humans. Old, proven agricultural technologies are no longer able to cope with these problems [7]. Mankind faced the problem of finding alternative ways of farming, maintaining its high productivity and environmental safety [8]. Under these conditions, old technologies should be replaced by biotechnology, and at the present stage only with its help it is possible to solve the environmental, energy and food problems facing humanity.

Undoubtedly, today the soil is involved in all the most important processes of the functioning of terrestrial ecosystems and the biosphere as a whole - from providing resources and space for terrestrial vegetation to maintaining the parameters of the atmosphere and hydrosphere, including the problems of greenhouse gases, the purity of surface and ground waters, and the elimination of xenobiotics [9].

An exceptional role in the formation of soil quality or "soil health" is played by soil microorganisms, which perform all the main ecosystem functions. Modern intensive technologies in agriculture, aimed at maximizing yields and quick profits, are essentially waging a costly battle with the soil microbial system - the foundation of ecosystem resilience [10].

The way out of the situation is the emphasis on " soil health " in the broadest sense and its sustainable maintenance with the possibility of an acceptable livelihood for the current and future generations. That is why soil microorganisms are the main element of a sustainable and efficient life support system.

A decrease in the composition and abundance of useful soil microflora, which is involved in the circulation of important macro- and microelements, suppresses the processes of restoring soil fertility and structure, which leads to the process of soil degradation - erosion, phosphatization, salinization and desertification . As a result, greenhouse gases such as carbon dioxide (CO_2), nitrous oxide (N_2O) and methane are released from the soil into the atmosphere, which seems to be a powerful factor in climate change on the planet.

2 Materials and methods

Plants do not feed on fertilizers applied to the soil, but on the products of their processing by soil microorganisms. Therefore, the growth, development of plants and the value of yields are determined by the soil microflora, especially those living in the rhizosphere, i.e. in the root zone of plants. Consequently, all agricultural technologies should ensure the development of the microflora of the rhizosphere, taking into account the genetic characteristics of plants.

Now all over the world there is a revision of the concept of development of agricultural production and a gradual transition from intensive technogenic methods of agricultural production to biological methods.

In recent years, in a number of developed countries, in particular in Europe, the USA, Canada, Japan, China and Russia, cheap production methods have been used that bring a high effect at relatively low costs. Main point biological farming technologies consists in applying to the soil effective microorganisms, which enrich it with readily available nutrients, make it fertile and supply plants with the necessary products of their vital activity (vitamins, enzymes, amino acids, etc.). At the same time, mineral fertilizers (only in combination with organic ones), pesticides and other chemical plant protection products are not used, the products become environmentally friendly and completely safe for humans.

Restoring degraded soils lies in the efficient and rational use of the biological potential of the soil, in optimizing the plant-microbial interaction in agrophytocenoses and improving the ecosystem as a whole.

One of the components of ecological farming is the use of new biotechnologies based on the use of biofertilizers (bacterial fertilizers, biological products, biocomposts, biofungicides, etc.). One of the urgent problems of farming in the world is the problem of soil phosphatization, which arose as a result of the annual introduction of a large amount of phosphate fertilizers into soils for agricultural crops. The utilization rate of phosphorus from fertilizers is low and amounts to only 10-20% of the applied amount, while the rest of it, reacting with soil complexes, turns into compounds that are insoluble in water and inaccessible to plants. At the same time, the phosphate regime of soils is disturbed, which leads to a deficiency of available phosphorus and as a result:

- the activity of beneficial microorganisms in the soil is reduced (especially the nitrogen cycle - nitrifiers, azotobacter, cellulose-decomposing);
- the balance of nutrient biogenic (nitrogen, potassium) and other macro-microelements (Ca, Mg, Zn) is disturbed;
- the efficiency of photosynthesis, respiration decreases, root formation, growth and development of plants;

- favorable conditions are created for the development of microscopic fungi in the soil, including phytopathogens;
- the intake of heavy metals into plants increases;
- soil fertility decreases, because there is a direct correlation between soil fertility and phosphate regime;
- the ability of soil and plants to retain water is deteriorating;
- the productivity of agricultural crops and their quality is reduced.

Improving the phosphate regime of soils and phosphorus nutrition of plants is extremely necessary due to the fact that during phosphorus starvation, the effectiveness of all other nutrients noticeably decreases - the absorption of nitrogen, potassium introduced with fertilizers, as well as macro- and microelements by plant roots and their movement to aboveground soil organs is inhibited [3].

However, the process of biological transformation of phosphorus in the soils of Uzbekistan has not been practically studied, there were no specific ways and methods of directed regulation of the processes associated with the transformation of phosphorus and its use by plants, in connection with which great damage was done to agriculture every year.

In connection with the foregoing, we were faced with the task of developing biologically balanced farming systems based on the maximum use of the biological potential of soils and obtaining environmentally friendly products on this basis.

One of the promising ways to solve the problem of soil phosphatization is the microbiological transformation of sparingly soluble phosphorus-containing compounds. It is known that some soil microorganisms are capable of secreting various organic acids and enzymes as a result of their vital activity and, being in symbiotic interactions with plant root systems and other microorganisms, they can increase the content of soluble phosphorus in the soil, which is easily assimilated by plants.

3 Results and discussion

In the laboratory of soil microbiology and biotechnology of the Institute of Microbiology of the Academy of Sciences of the Republic of Uzbekistan, we have created new environmentally friendly bacterial fertilizers and biopreparations to improve soil fertility, crop yields and quality of agricultural products [4-6].

Bacterial fertilizers of complex action based on new local strains of phosphorus mobilizing bacteria:

1) FOSSTIM-1 - for pre-sowing treatment of industrial crops (cotton, sugar beet) (patents no. IAP 02787, no. IAP 02788).

2) FOSSTIM-3 - for pre-sowing treatment of seeds and seedlings of vegetable crops both in open ground and in greenhouses, melons, legumes, potato tubers, seedlings and trees of horticultural crops and vineyards (patent No. IAP 4712).

Biopreparations of complex action based on new local strains of salt-resistant phosphorus and potassium-mobilizing rhizobacteria with polyfunctional properties.

3) RIZOKOM-1 - for pre-sowing treatment of cotton seeds when cultivating it on saline and native soils (patents No. IAP 05712, No. IAP 05713, No. IAP 05714, No. IAP 05715).

4) RIZOKOM-2 - for pre-sowing treatment of wheat seeds when cultivating it on saline and native soils.

5) SERHOSIL - for foliar feeding of all types of agricultural crops based on green microalgae (patent No. IAP 04933) (Figure 1).

https://doi.org/10.1051/e3sconf/202342101005

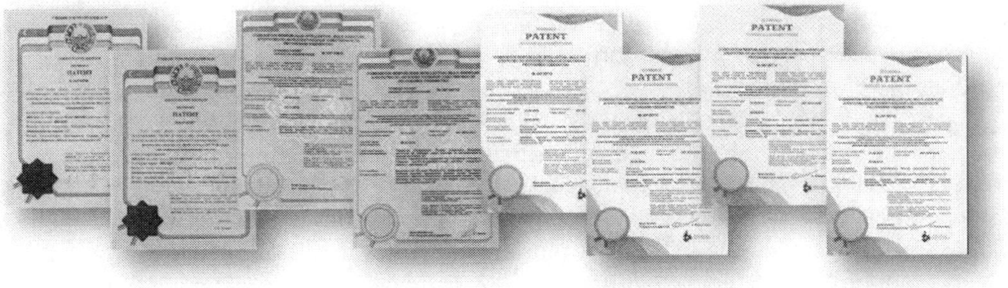

Figure 1. List of patents derived for biopreparations.

At the same time, it should be noted that biofertilizers are not toxic to animals, bees, fish, earthworms, do not contain or emit toxic substances, and do not require special precautions when working with it. They belong to the 4th hazard category - that is, to non-hazardous products.

The mechanism of action of bacterial fertilizers of the FOSSTIM series and biological preparations of the RIZOKOM series in crop production is due to an increase in the symbiosis of plants and soil microbial communities, through an increase in the number of beneficial microorganisms, in particular, a group of free-living nitrogen fixers, phosphorus mobilizing, cellulose-decomposing, nitrifying and other types of microorganisms that are functionally necessary for soils and plants. This mechanism practically steadily increases the level of phosphorus, nitrogen and potassium available to plants in the soil, crop yields, product quality indicators, and also reduces the level of nitrate content in products and the degree of plant disease by several times.

Effective phosphorus mobilizing bacteria introduced into the soil convert soil hard-to-reach phosphates into forms that are assimilated by plants and increase the absorption rate of applied mineral fertilizers, nourish plants with their dead biomass and create optimal living conditions for the rest of the soil microflora and fauna.

Useful microflora of bacterial fertilizers heals plants and strengthens their immunity. This allows plants to tolerate adverse conditions much better, including drought, frost, salt and dust storms, and sudden changes in air temperature.

The mechanism of action of the biological product SERHOSIL is explained by an increase in the photosynthetic surface of the leaves, an improvement in the processes of photosynthesis due to the additional nutrition of plants with extracellular physiologically active substances of green microalgae (amino acids, lipids, extracellular polysaccharides, vitamins, available forms of macro-microelements).

A new science-based bioagrotechnology for environmentally oriented farming, based on the integrated use of bacterial fertilizers and biological products, has been developed and tested in practice.

In many years of field experiments, comprehensive studies have been carried out and the effective influence of the new bioagrotechnology on the state of the soil microbial community, the content of organic matter, mobile forms of nitrogen, phosphorus and potassium, the intensity and direction of biological processes in soils, the yield and quality of agricultural products have been proven.

The use of new biofertilizers in the cultivation of agricultural crops increases the field germination of seeds, stimulates root formation, growth and development of plants, improves nitrogen, phosphorus and potassium nutrition of plants, increases their resistance to phytopathogens and, as a result, helps to increase productivity and product quality, allows get earlier products, increases its safety.

The use of microbiological preparations in the cultivation of agricultural crops and a decrease in the doses of mineral fertilizers leads to an increase in economic efficiency and profitability of production.

The advantages of the new bioagro technology include such factors as a reduction in the consumption of mineral and organic fertilizers by 25-50%; almost 2 times reduction in irrigation water consumption; improving the balance of the soil microbial community in favor of beneficial microflora; increasing soil fertility; normalization of alkaline pH of saline soils; improving the balance of nutrients in the soil and plant nutrition; reduction in the incidence of cotton root rot, gommosis, alternariosis and wilt; reduction in the incidence of winter wheat with root rot, rust, smut, wilting; lowering the degree of phosphatization, soil salinity, reducing soil pollution with mycotoxins and organochlorine pesticides; increasing winter hardiness and drought resistance of plants; increase in the yield of raw cotton - by 8-10 q/ha, vegetable crops (cucumbers - by 3.6-4 t/ha, tomatoes by 12 t/ha), potatoes - by 5.2-5.9 t/ha and quality of agricultural products, increasing the profitability of agricultural production: raw cotton by 52%, wheat by 45%, vegetables and potatoes by 45-90%.

New bioagrotechnologies were developed within the framework of six state applied and one international projects and passed scientific and production state tests at the head institutes of the Ministry of Agriculture of the Republic of Uzbekistan - the Uzbek Research Institute of Cotton Growing, the Uzbek Research Institute of Cotton Breeding and Seed Growing, the Uzbek Research Institute of Vegetables, Melons and Potatoes, the Gallyaaral branch of the Research Institute grain and leguminous crops on irrigated lands, on saline soils of the Syrdarya and Bukhara scientific and experimental stations of the Research Institute of Selection, Seed Production and Agricultural Technology of Cotton Growing, as well as in several farms.

Biological preparation of complex action RIZOKOM-1 in 2014-2015 was introduced on cotton on saline soils in the farm "Suvchi Mirob Dalasi" of the Mirzaabad district of the Syrdarya region. As a result of its use, the fertility of saline soils, the yield of cotton and the quality of the fiber have increased. The increase in the yield of raw cotton amounted to 6-7 q/ha, profitability increased by 20-25%. On saline soils in the Syrdarya branch of the Research Institute of Grain and Leguminous Crops, when using RIZOKOM-2 and SERHOSIL on wheat, an increase in yield of 8-9 q/ha was obtained, the quality of grain improved [6, 7].

Bacterial fertilizers of the FOSSTIM series and biopreparationsof the RIZOKOM series in 2018 were included in the List of drugs approved by the Decree of the President of the Republic of Uzbekistan dated July 14, 2018, No. PP-3855 “On additional measures to increase the efficiency of commercialization of the results of scientific and scientific and technical activities”.

Biopreparations RIZOKOM-1 and SERHOSIL were included in the List of drugs approved by the order of the government of the republic dated October 14, 2018. No. 850-f "Comprehensive program for the development and implementation of the concept of personalized agriculture in the Republic of Uzbekistan for the period 2019-2021".

Biopreparations RIZOKOM-1 and SERHOSIL in 2019-2021 were introduced on cotton in the country on a total area of 20,000 hectares. As a result, the fertility of saline soils, the yield of cotton and the quality of the fiber have increased. The increase in the yield of raw cotton on average in the republic amounted to 7-8 q/ha.

Biopreparations RIZOKOM-2 and SERHOSIL in 2019-2021 were introduced on wheat in the country on a total area of 10,000 hectares. As a result, the fertility of saline soils has improved, the cost of mineral fertilizers has decreased by 50% and the cost of irrigation water has decreased by 20-25%.

The increase in the wheat harvest on average in the republic amounted to 9-10 centners per hectare, the quality of grain improved. Meanwhile, it should be noted that soil phosphating (an increase in phosphorus content) occurs due to the low solubility of soil phosphorus compounds due to the fixation (by 80%) of the applied phosphorus fertilizers with soil calcium in tricalcium phosphate $Ca_3(RO_4)_2$ and a low coefficient (20%) assimilation of phosphorus by plants from applied phosphate fertilizers. As a result, phosphorus and calcium become unavailable for plant nutrition. Soil phosphating also leads to salinization [8].

In addition, soils are also salinized with black salt as a result of the use of potash fertilizers, which leads to the fixation of potassium in the soil with soil silicon in the form of potassium aluminosilicates ($KAlSi_3O_8$), which are insoluble in water and, therefore, potassium and silicon also become unavailable for plant nutrition. According to the degree of saturation with salts, weakly saline soils are distinguished, on which the yield of agricultural crops falls by 25%, on medium saline soils - by 50%, on saline soils - by 75% and on highly saline soils - by 100%.

Soil pollution also occurs when mineral fertilizers are used in excessive amounts. Recently, another an unfavorable aspect of the immoderate consumption of mineral fertilizers and, first of all, nitrogen fertilizers. It turned out that a large amount of nitrates reduces the oxygen content in the soil, and this contributes to an increased emission of greenhouse gases into the atmosphere - nitrous oxide (N_2O) and methane (CH_4). Nitrates also have a negative effect on the human body. Thus, when nitrates enter the human body at a concentration of more than 50 mg/l, their direct general toxic effect is noted [9]. Soil pollution also occurs when a large number of chemical pesticides (fungicides, insecticides, herbicides, etc.) are used in agriculture for seed treatment, control of diseases, pests and weeds of agricultural crops.

Pesticides are able to penetrate into plants from contaminated soil through the root system, accumulate in biomass and subsequently infect the food chain. When spraying pesticides, intoxication and death of birds and beneficial insects is observed. The long-term use of pesticides is also associated with the development of resistant (sustainable) pests and the emergence of new pests whose natural enemies have been destroyed. The surface layers of soils are easily contaminated with pesticides. Large concentrations in the soil of various chemical compounds - toxicants adversely affect the vital activity of soil organisms - microorganisms, flora and fauna.

Currently, heavy metal pollution is a major environmental problem because metal ions remain in nature due to their non-degradable nature. One of the global manifestations of soil degradation, and of the entire environment in general, is desertification. Desertification is a process of irreversible change in soil and vegetation and a decrease in biological productivity, which in extreme cases can lead to the complete destruction of the biospheric potential and the transformation of an area into a desert. On the territory subject to desertification, the physical properties of soils deteriorate, vegetation dies, groundwater becomes saline, biological productivity drops sharply, and, consequently, the ability of ecosystems to recover is undermined.

According to available data, every minute in Central Asia, 9 sq.m. of fertile land turns into a desert. The fight against salinization, degradation and desertification of soils today is one of the foundations of food security. The areas of saline soils in 13 regions of the republic presented in Fig. 1 indicate the need for an urgent solution to improve the fertility of saline soils and the productivity of agricultural crops.

For this reason, it is considered an urgent problem to increase fertility and obtain a high-quality crop on saline, degraded soils through the use of biotechnologies based on the use of new generation biological fertilizers that have an environmentally safe, highly effective effect on degraded soils with the ability to increase seed germination, accelerate the growth and development of plants, improve nourish roots and plants, reduce soil salinity and pollution, prevent desertification, and improve crop immunity and protect plants from diseases and pests.

We, a group of scientists and specialists in the field of microbiology, biotechnology and ecology, have developed an environmentally safe, resource-saving, cost-effective innovative bioagrotechnology for growing agricultural crops on saline, degraded, contaminated with heavy metals, soils close to desertification in Uzbekistan.

The new bioagrotechnology is based on the complex application of new generation bacterial fertilizers of the TERIA series (patent No. FAP 02090) and biological product SERHOSIL (Patent No. IAP 04933).

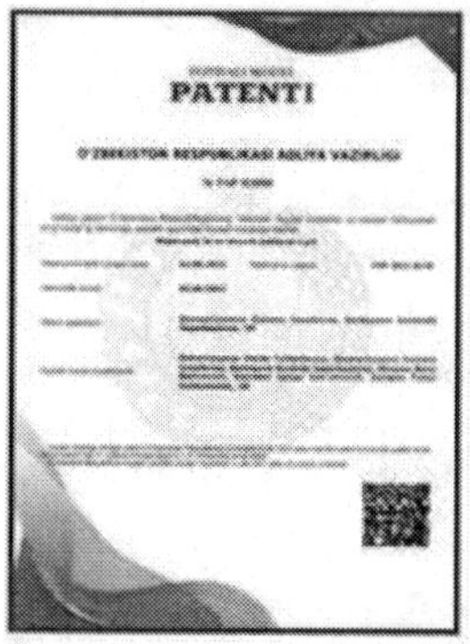

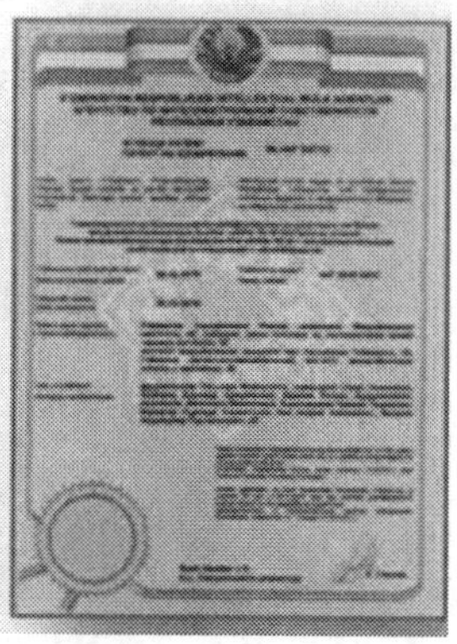

Figure 2. List of patents derived for new generation bacterial fertilizers of the TERIA series.

We have studied the following polyfunctional properties of microorganisms that are part of the bacterial fertilizer of complex action TERIA:

1. Salt tolerance. Strains from the composition of the TERIA bacterial fertilizer - *Bacillus subtilis* K-4, *Bacillus thuringiensis* K-7 and *Priestia megaterium* K-8 have high salt-tolerant activity. The influence of increasing concentrations of sulfate (Na_2SO_4 and $MgSO_4$) and chloride ($NaCl$ and $MgCl_2$) toxic salts on their growth activity. The strains showed good growth in the presence of increasing concentrations from 3% to 10% of the above toxic salts in the peptone nutrient medium.

2. Phosphate mobilizing activity. Strains soil bacteria have phosphate-mobilizing activity, as evidenced by the results of a study of the dissolution of $Ca_3(PO_4)_2$.

Strains *bacillus subtilis* K-4, *Bacillus thuringiensis* K-7 and *Priestia megaterium* K-8 showed the dissolution zones of $Ca_3(RO_4)_2$ on the Pikovskaya medium - 30(±1.5) mm, 31(±1.2) mm and 12(±1.6) mm, respectively.

The mobilization of P_2O_5 from $Ca_3(PO_4)_2$ by *Bacillus subtilis* K-4, *Bacillus thuringiensis* K-7, and *Priestia megaterium* K-8 strains was studied when the strains were cultivated on a peptone medium with a single source of phosphorus $Ca_3(PO_4)_2$: mobilization of P_2O_5 from $Ca_3(PO_4)_2$ in strains of *Bacillus subtilis* K-4, *Bacillus thuringiensis* K-7 and *Priestia megaterium* K-8 was 115±0.12 mg P_2O_5/100 ml, 128±0.15 mg P_2O_5/100 ml and 128±0.1 mg P_2O_5/100 ml respectively.

3. Potassium mobilizing activity. Strains soil bacteria *bacillus subtilis* K-4, *Bacillus thuringiensis* K-7 and *Priestia megaterium* K-8 dissolve potassium aluminosilicates $KAlSiO_4$, as evidenced by the dissolution zones of $KAlSiO_4$ strains on solid nutrient media Zak and A-27.

The strains showed the following dissolution zones on the 2 media above, respectively: *Bacillus subtilis* K-4 - 50 ± 1.6 mm and 20 ± 1.4 mm; *Bacillus thuringiensis* K-7 - 20 ± 1.5 mm and 20 ± 1.9 mm; *Priestia megaterium* K-8 - 30 ± 1.4 mm and 20 ± 1.3 mm. Thus, it has been proven that *Bacillus strains subtilis* K-4, *Bacillus thuringiensis* K-7 and *Priestia megaterium* K-8 have a high potassium mobilizing activity. The release of mobile and water-soluble potassium from potassium aluminosilicate ($KAlSiO_4$) was tested. Mobilization of mobile potassium from $KAlSiO_4$ was 1265 mg/kg in the strain *bacillus subtilis* K-4, in *Bacillus thuringiensis* K-7 and *Priestia megaterium* K-8 - 1229 mg/kg. Release of water-soluble potassium from $KAlSiO_4$ was 0.048 g/l *bacillus subtilis* K-4, in *Bacillus thuringiensis* K-7, this figure was 0.047 g/l and 0.049 g/l in the strain *Priestia megaterium* K-8.

4. Antifungal activity. The antifungal activity of strains of *Bacillus subtilis* K-4, *Bacillus thuringiensis* K-7 and *Priestia megaterium* K-8 was tested by the well method. It was found that the studied strains to some extent have antifungal activity against phytopathogenic fungi *Fusarium moniliforme, Alternaria alternata, Alternaria solani, Fusarium solani, Fusarium oxysporium, Aspergilius niger* and *Cladosporium oxysporum* (Table 1).

Table 1. Growth inhibition of phytopathogenic fungi by salt tolerant soil bacteria.

Phytopathogens agricultural crops	Zone of inhibition growth	Soil bacteria strains		
		***Bacillus subtilis* K-4**	***Bacillus thuringiensis* K-7**	***Priestia megaterium* K-8**
Fusarium moniliforme	D, mm	50±0.58	75±0.58	50±0.58
Alternaria alternata	D, mm	67±1.53	52±1.53	60±1.43
Alternaria solani	D, mm	11.3±0.1	13.4±0.1	-
Fusarium solani	D, mm	-	-	51±0.58
Fusarium oxysporum	D, mm	11.6±0.1	10.2±0.1	-
Aspergilius niger	D, mm	11.2±0.1	8.3±0.1	15.6±0.2
Cladosporium oxysporum	D, mm	13.2±0.2	14.1±0.2	-

5. Ability to decompose gypsum. Strains soil bacteria *bacillus subtilis* K-4, *Bacillus thuringiensis* K-7 and *Priestia megaterium* K-8 dissolve gypsum - CaSO4 . _The study of the effect of bacteria on the content of gypsum, pH of the medium and the survival of bacterial cells was carried out in the dynamics of deep cultivation in flasks on a liquid peptone nutrient medium (200 ml) + gypsum (6 g). Deep cultivation was carried out on a rocking chair at 220 rpm at a temperature of 28 ^{0}C for 20 days. The use of a consortium of 3 strains of soil bacteria reduced the content of gypsum in the nutrient medium by 2 times, in the control medium - without bacteria - by 0.5 times. The use of a consortium of bacteria will contribute to the dissolution of gypsum, which can positively affect the development of crops on gypsum soils.

6. Properties of accumulation of heavy metals in the soil. In the last decade, under the influence of high anthropogenic loads, agrolandscapes have been polluted with various ecotoxicants (mercury, lead, cadmium, fluorine, aluminum, arsenic), which are considered toxic heavy metals. Exceeding the norms of toxic substances leads to a violation of soil

quality, an increase in pests and plant diseases, and a decrease in the quality of agricultural products. This negatively affects the health of people and animals. Thus, the use of nitrogen fertilizers increases the acidity of soils, increases the mobility of heavy metals (aluminum, cadmium, copper, lead) and leads to the accumulation in the soil of toxic elements for plants and microorganisms, in particular fluorine. Potash fertilizers, and especially potassium chloride (with excessive application), chlorine and lead are deposited in the soil, soil acidity increases, and the activity of soil microorganisms decreases.

Strains soil bacteria *bacillus subtilis* K-4, *Bacillus thuringiensis* K-7 and *Priestia megaterium* K-8 are able to reduce the adverse effects of heavy metals entering the soil along with mineral fertilizers and pesticides, increase the fertility of saline, polluted degraded soils, reduce the consumption of heavy metals by plant organs, and increase productivity and product quality.

When using a consortium of 3 strains of bacteria, the amount of toxic heavy metals in the soil decreased compared to the control, without the use of bacteria against the background of mineral fertilizers (NPK): the amount of Zn in the soil decreased by 69.8 mg/kg, Cd - by 0.149 mg/kg, Pb - by 72.5 mg/kg, Mo - by 0.38 mg/kg, Cu - by 58.2 mg/kg, Co - by 7.8 mg/kg, Ni - by 29.9 mg/kg, As - by 2.1 mg/kg and Cr - by 18.7 mg/kg (Table 2).

Table 2. Influence of a consortium of strains of *Bacillus subtilis* K-4, *Bacillus thuringiensis* K-7 and *Priestia megaterium* K-8 on the accumulation of mobile forms of heavy metals in soils under winter wheat crops, adjacent fields of the Almalyk Mining and Metallurgical Combine of the Republic of Uzbekistan (mg/kg, maturation phase, June, n=3, average for 2021-2022).

Experience options	Zn	Cd	Pb	Mo	Cu	Co	Ni	As	Cr
Control, NPK	150,0	0,320	97,8	1,42	90,4	13,8	44,2	20,2	49,8
Experience, NPK, treatment of wheat seeds with bacterial fertilizer TERIA	80,2	0,171	25,3	1,04	32,2	6,0	14,3	18,1	31,1
MPC	55,0	0,5	32,0	2,1	30,0	5,0	14,0	20,0	30,0

Based on the complex application of bacterial fertilizer TERIA and biological product SERHOSIL we have developed a new environmentally friendly and cost-effective, resource-saving bioagro technology for growing agricultural crops on medium and highly saline soils of Uzbekistan (Table 3).

Table 3. Influence of complex application of bacterial fertilizer TERIA and biopreparation SERHOSIL on crop yields on saline degraded soils (field trials, average for 2021-2022).

Agricultural essential culture	Yield, q/ha		Yield increase	
	Control, NPK traditional sowing	Experience, NPK TERIA + SERHOSIL	q/ha, t/ha	%
Cotton	25.0±0.7	34.0±1.5*	9.0±0.05	36.0
Sugar beet, t/ha	34.5 ±1.2	75.0 ±2.6*	40.5 ±0.4	17.4
Winter wheat	3 8, 4 ±1.3	51.2±2.5*	12.8 ±1.3	13.3
Soya	22.1 ±0.6	34.1 ±1.2*	12.0 ±1.2	54.3
Potato, t/ha	41.5±0.5	46.7±1.3*	5.2±0.03	12.5
Tomatoes, t/ha	23.6±0.8	30.1 ±1.5*	6.5 ±0.4	27.5
Cucumbers, t/ha	19.0 ±1.2	22.5 ±0.6*	3.1 ±0.02	18.4

The created new bioagro technology can be effectively used in the production of environmentally friendly organic agricultural products that meet international Organic and Global standards. G.A.P (Figures 3-8 and Table 4).

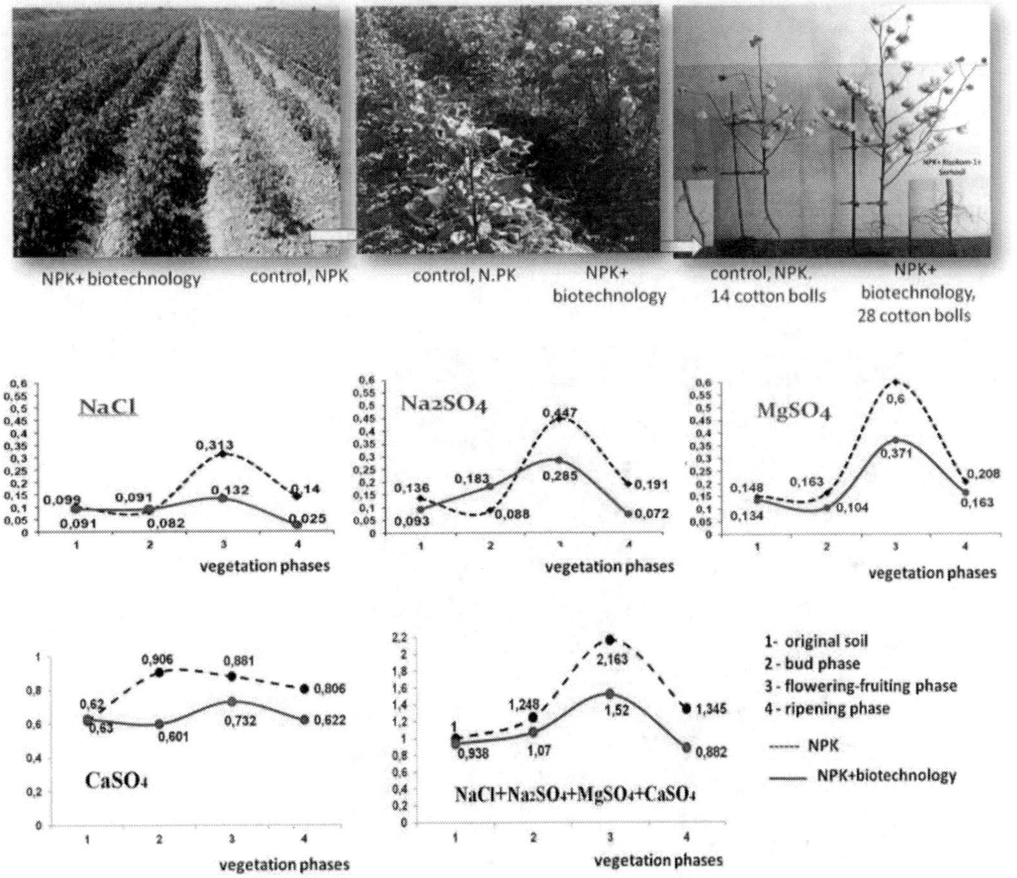

Figure 3. Influence of biotechnology on the development of cotton on highly saline soils of the Sirdarya Branch of the Institute of Cotton Growing (Uzbekistan).

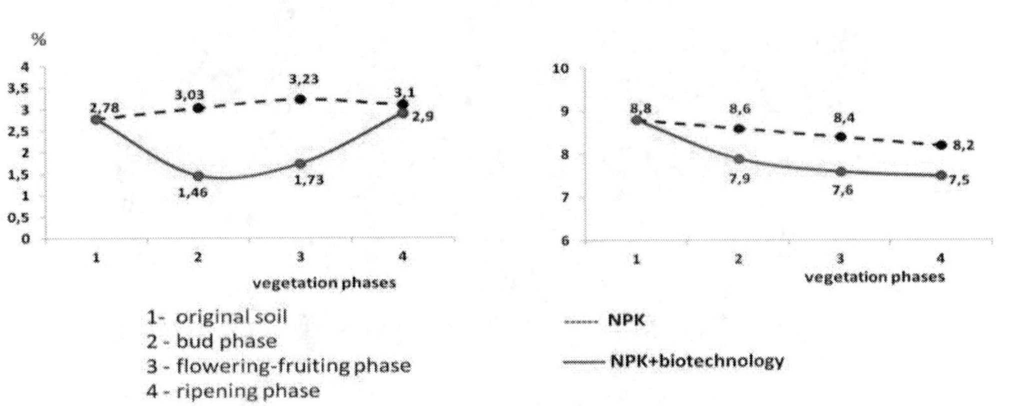

Figure 4. Influence of biotechnology on the dynamics of changesin the pH and humus content of saline soils during the growing saeason.

Table 4. Use of bioagrotechnology on highly saline soils of Karakalpakstan (Uzbekistan).

Region	Farm	Year	Cotton variety	Cotton yield, q/ha		Increase in yield, q/ha	Increase in yield, %
				control, NPK	NPK+ Teria+Serhosil		
Turtcul district	"Turtcul Agroclaster" LLC	2022	C-4727	12.0±0.6	30.0±1.5	18.0±1.0	25.0
Biruni district	"Fozilbek Komolovich"	2022	Sulton	20.0±2.7	40.0±1.7	20.0±2.1	100.0
Average				17.0±0.7	28.5±2.6	11.5±1.0	53.1

Figure 5. Use of bioagrotechnology on highly saline soils of Karakalpakstan (Uzbekistan).

Figure 6. Cotton crop of "Uztex Cluster" based cotton farms using bioagrotechnology in Khorezm (Uzbekistan).

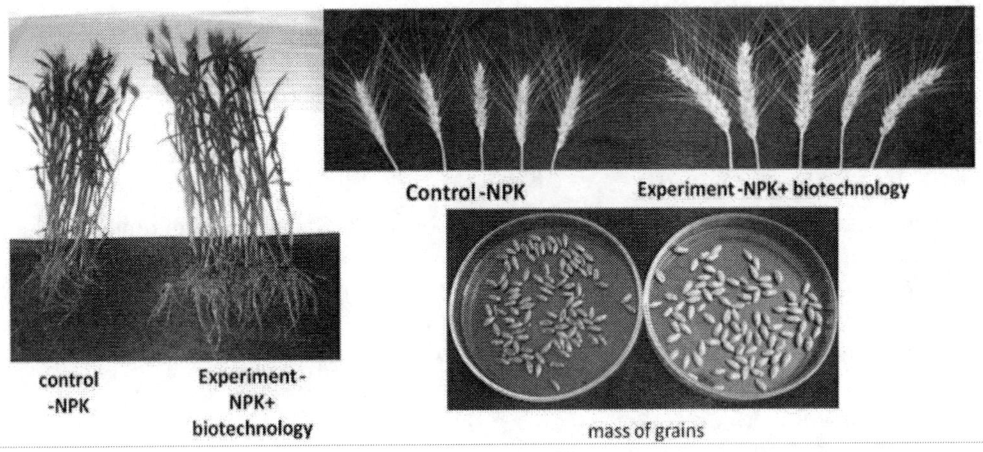

Figure 7. Influence of new biotechnology on the development of the root system of the aboveground mas of wheat (Uzbekistan).

Figure 8. Efficiency of biotechnology on winter wheat on highly saline soils of Sirdarya (Uzbekistan).

4 Conclusions

Thus, based on the conducted scientific research and field tests and implementation on saline, degraded soils in the Syrdarya, Jizzakh, Bukhara, Namangan, Andijan, Fergana, Khorezm regions and Karakalpakstan, a new bioagrotechnology on cotton on an area of 7,500 hectares and on wheat on an area of 2,500 ha, the following tangible 4 beneficial effects were identified:

1) resource saving effect:
- increase in the coefficient of digestibility of applied fertilizers;
- restoration and increase of soil fertility;
- increasing the field germination of seeds of agricultural crops on saline soils;
- reduction of seeding rates by 10-30%;
- reduction of irrigation water consumption by 30-40%.

2) economic effect:
- increase in productivity of agricultural crops by 12-54% and product quality;
- reduction of damage to agricultural crops by diseases and pests;
- increase of immunity, drought resistance and winter hardiness of plants;
- reduction of terms of ripening of agricultural products by 15-20 days;
- increasing the profitability of agricultural production up to 35-40%;
- increasing the shelf life of finished products in warehouses.

3) environmental effect:
- reduction of chemical load on the soil and agricultural crops;
- reduction of doses of chemical fungicides for dressing seeds;
- decrease in the degree of phosphatization, salinity, soil pollution with heavy metals, health improvement of soil environment.

4) social effect:
- improving the quality of agricultural products (reducing the content of nitrates in products, increasing the content of vitamins and useful macro-microelements);
- improving the health of the population.

The developed new bioagro technology may well be used to manage the fertility of saline and degraded soils and the processes of plant nutrition. The soil becomes healthy, its productivity increases significantly.

References

1. Kojevin P.A., Population ecology of soil microorganisms. Abstract of the dissertation for the degree of Doctor of Biological Sciences, 2006
2. Shablin P.A. EM-technology is a technology for improving the environment, 2011.
3. Tueva O.F. Phosphorus in plant nutrition / O.F. Tuev. M.: Nauka, 1996.
4. Wildflush I.R., Tsyganov A.R., Lapa V.V., Persikova T.F. Rational application of fertilizers, Minsk, 2002.
5. Soil salinization: a problem and solutions. Resource-saving agriculture. 1(33) 2019, (RF).
6. Narbayeva Kh.S. Salt-resistant rhizobacteria of cotton with polyfunctional properties and a biological product of complex action based on them. Abstract of the dissertation for the degree of Doctor of Biological Sciences. Tashkent. 2016
7. Babina A.E. Biopreparations of complex action RIZOKOM-1 and SERHOSIL for cotton cultivation on saline soils. Abstract of the dissertation for the degree of Doctor of Biological Sciences. Tashkent. 2019
8. Jumaniyazova G.I. Phosphormobilizing bacteria and biofertilizers based on them. Abstract of dissertation for the degree of Doctor of Biological Sciences. Tashkent. 2012
9. Duvick, D. N. (2001). Biotechnology in the 1930s: the development of hybrid maize. Nature Reviews Genetics, 2(1), 69-74.
10. Obembe, O. O., Popoola, J. O., Leelavathi, S., & Reddy, S. V. (2011). Advances in plant molecular farming. Biotechnology advances, 29(2), 210-222.

Monitoring of the current state of tugai ecosystems in the conditions of the southern Aral Sea region

Gulnaz Saparova[1,*], *Gulzar* Kutlymuratova[2], *Svetlana* Mambetullayeva[3] and *Nadira* Yuldasheva[4]

[1]Tashkent State Agrarian University, Tashkent, 100140, Uzbekistan
[2]Nukus State Pedagogical Institute, Nukus, 230100, Uzbekistan
[3]Karakalpak Research Institute of Natural Sciences, Nukus, 2301400, Uzbekistan
[4]Tashkent Financial Institute, Tashkent, 100060, Uzbekistan

Abstract. This article presents the results of monitoring the current state of the tugai in the South Aral Sea region. Over the past decades, as a result of a sharp increase in anthropogenic pressure on tugai forest ecosystems, there has been a catastrophic reduction in their areas, a violation of the ecological stability of ecosystems. The main area of distribution of tugai vegetation in the territory of Karakalpakstan is the lower reaches of the Amu Darya. Considering that tugai forests are a complex biological system with a long reproduction period, it is important to comprehensively know the influence of climatic and soil conditions on their formation and development. Tugai massifs are a unique type of ecosystems. They should be considered as a natural reserve of unique flora and fauna. During the formation and development of tugai communities, the ecological conditions of habitats change, which, in turn, affect communities. This mutual influence occurs during the entire period of their existence, being reflected in the accumulation of biological mass. In this connection, it is extremely interesting to consider the dynamics of changes in the productivity of tugai communities in the Amudarya delta. It is shown that as a result of a sharp increase in anthropogenic pressure on tugai forest ecosystems, their areas have sharply decreased, and the ecological stability of tugai ecosystems has been violated. Currently, in the tugai massifs of the lower reaches of the Amudarya, there are 55 species of tugai plants belonging to 21 families and 47 genera.
Keywords. Tugai massifs, dynamics, monitoring, anthropogenic impact, Aral Sea.

1 Introduction

The current ecological situation in the Aral Sea region is characterized by intensive processes of desertification and aridization of the territory, associated with a drop in the level of the Aral Sea and changes in hydrological conditions in the lower reaches of the Amu Darya. The

* Corresponding author: arabova_nodira@mail.ru

destabilization of the ecological situation in this region had an ambiguous effect on various ecological types of vegetation. The most significant changes occurred in the vegetation of the tugai type [1, 2].

Over the past decades, as a result of a sharp increase in anthropogenic pressure on tugai forest ecosystems, there has been a catastrophic reduction in their areas, a violation of the ecological stability of ecosystems [3, 4].

Tugai massifs are a unique type of ecosystems. They should be considered as a natural reserve of unique flora and fauna. In this regard, and also taking into account the environment-forming and water-protecting role of the tugai, they are subject to mandatory protection [5]. At the same time, the development of an effective scientifically based program of rational nature management in tugai, which includes issues of protection, restoration, environmental monitoring and management of natural processes, will become especially relevant [6].

The main area of distribution of tugai vegetation in the territory of Karakalpakstan is the lower reaches of the Amu Darya [7]. Considering that tugai forests are a complex biological system with a long reproduction period, it is important to comprehensively know the influence of climatic and soil conditions on their formation and development. Therefore, we considered it appropriate to give a brief description of the current natural conditions of the lower reaches of the Amu Darya [8].

According to climatic conditions, the region belongs to the northern region of the desert zone with a sharply continental temperature regime. Aridity is determined by the amount of precipitation, which is equal to 100 mm per year and huge insolation, leading to an excess of evaporation over precipitation by 15-20 times [9]. An important climatic factor here is frequent winds of a predominantly northeasterly direction. The climate is sharply continental. Tugai soils are close to meadow soils, where the abundant annual leaf fall, together with the grassy mass, leads to the fact that a large amount of organic matter accumulates on the surface every year [10]. Therefore, on such soils, the supply of nutrients is quite significant, and they themselves are the most fertile.

At present, the change in hydrological conditions in the lower reaches of the Amu Darya, caused by the drying up of the Aral Sea, directly affected the current state of the vegetation cover, which led to a violation of the ecological stability of phytocenoses and the development of anthropogenic desertification processes, a decrease in the biomass of vegetation and wildlife. These factors led to the transformation of the structure and functioning of ecosystems [11].

Violations affected all ecological types of vegetation presented in the Amudarya delta. The vegetation of the tugai type has undergone the most significant changes. For the period from 1978-1985. The composition and structure of tugai communities changed radically, successional changes took place [12].

During the formation and development of tugai communities, the ecological conditions of habitats change, which, in turn, affect communities. This mutual influence occurs during the entire period of their existence, being reflected in the accumulation of biological mass. In this connection, it seems extremely interesting to consider the dynamics of changes in the productivity of tugai communities in the Amudarya delta.

Tugai complexes of the 50s of the last century were characterized by well-developed turanga forests with an admixture of oleaster *(Eleagnus angistifolia)* and willow species *(Salix songarica)*, (*S. Wilhelmsiana*), as well as coastal aquatic plants . The main massifs of tugai massifs were confined to the valley and delta of the Amu Darya and were located along its main channel. Until 1961, their area was about 300 thousand hectares, and 260 thousand hectares were located in the delta [13]. In addition to the arboreal tugai forests inhabiting the fore-delta and flooded areas, on the dried-up channels grew tree-shrub vegetation composed of species of comb, licorice, karabarak, and wolfberry. Reeds and reed grass, and on the banks - kendyr, bitter, abash, azhrek, kermek, parnolistnik, annual saltworts, yantak [1].

The anthropogenic impact on the vegetation of the lower reaches of the Amu Darya, which continued in the 1980s and 1990s, led to the degradation of forest vegetation over a large area. From 300 thousand hectares of tugai areas by 1984, no more than 50 thousand hectares remained [14].

In the 1990s, the natural formation of young tugai communities on the floodplain terraces of the river ceased. Studies carried out in these years [6] revealed that the primary colonization by trees and shrubs occurs only occasionally. The species composition of tugai pioneers has been significantly reduced. Seedlings of previously common kendyr (*Apocynum scabrum*), licorice (*Glycyrrhiza glabra*), chingil (*Halimodendron halodendron*), wolfberry (*Lycium ruthenicum*). Extremely rare single specimens of willow (*Salix songarica*) and parnophyllum (*Zygophyllum oxyanum*). *Quinoa* species (Atriplex *tatarica*, *A. dimorphostegia*) and saltworts (*Salsola paulsenii, S. micranthera, S. foliosa*). Common early pioneers are reed and reed species (*Calamagrostis epigeios, C. pseudophragmites*) met occasionally.

2 Material and Methods

In 2011, on the basis of the reserve and additional territories, the Nizhne-Amudarya State Biosphere Reserve was created with an area of 68,717.8 ha, of which the protected area is 11,568.3 ha [2]. The goals of the reserve are the protection of tugai forests in the Amudarya delta, scientific research, the protection of cultural values, and the socio-economic development of the region [8].

The main tugai species, the comb (*Tamarix ramosyssima*), proved to be more resistant than the turanga (*Populus ariana*). At the same time, it was established that in the Amudarya delta, the following row of plant communities can be traced in the direction from the riverbank to the interchannel descending: turanga tugai - comb tugai - comb-karabarak tugai - annual hodgepodges. In lower, more humid places, there are reed thickets [6, 7].

Thus, with the change in the hydro regime of the Amudarya and the lowering of the Aral Sea level, the tugai phytocenoses of the Amudarya delta have changed. The cessation of floods, a decrease in the level of groundwater, an increase in their mineralization, and salinization of soils contributed to the degradation of tugai: their territory began to be occupied by drought- and salt-tolerant phytocenoses [9].

Currently, in the tugai massifs of the lower reaches of the Amudarya, there are 55 species of tugai plants belonging to 21 families and 47 genera. Of these, the most representative are the haze and Compositae families. All woody edificators belong to the families of willows (5 species) and oleasters (1 species), accounting for 11% of the tugai flora [6, 7]. About 26 species (40%) of the flora are herbaceous, perennials, which usually form the basis of the grass cover of the tugai. Annual and biennial herbs (37%), as a rule, are rare in the composition of communities and do not have a large abundance. The classification of tugai trees and shrubs in the lower reaches of the Amudarya was compiled on the basis of the typology and nomenclature of K.Z. Zakirov and P.K. Zakirov [4], which are most convenient for characterizing and deciphering succession shifts. Based on them, A. B. Bakhiev compiled a classification of the vegetation of the lower reaches of the Amudarya with the allocation of taxa of the first order - climatypes - in accordance with special hydrothermal conditions [1, 3, 4].

3 Results and discussion

In the modern vegetation cover of the tugai of Central Asia, turanga communities occupy the largest areas. In the lower reaches of the Amudarya, two turanga formations were noted: gray-leaved poplar and mixed-leaved poplar [2, 5, 6]. Since in the lower reaches of the Amu Darya *turanga sizolifolia* is very rare, and tree tugai consist mainly of poplar variegated, then in the classification only the turanga formation from variegated poplar is distinguished [3].

In recent years, due to the regional anthropogenic impact, there has been a steady reduction in the areas occupied by the comb (*Tamarix*) in both the ancient and modern Amudarya delta, mainly due to the destruction and degradation of tree and shrub tugai. Comb tugai are one of the stages of vegetation change during the transition of the floodplain hydromorphic regime [7].

Fluctuations in the total phytomass in tugai communities are largely due to changes in the environmental conditions of their habitats. So, with the deterioration of water supply, accompanied by an increase in soil salinity, the share of root mass in shrub and halophyte-shrub tugai increases to 64-67% and, on the contrary, the abundance of moisture gives a significant increase in above-ground phytomass in tree and shrub tugai, thereby confirming the fact that soil conditions are the most important environmental factors influencing the entire course of formation and development of tugai vegetation. Tugais are the most highly productive (Table 1).

Table 1. Productivity of tugai communities in the lower reaches of the Amu Darya.

Index	Wood tugai	Shrub tugai	Herbal tugai
Phytomass , t/ha	60.7	44.2	16.0
Green part, %	15.3	11.9	21.3
Wood, %	23.4	19.9	21.0
Roots, %	56.2	68.0	54.6

It can be seen from Table 1 that the average amount of phytomass in arboreal tugai of the lower reaches of the Amudarya is 60.7 t/ha, of which 56% is for the root system. Anthropogenic interference has a huge impact on the productivity of tugai.

The dynamics of the watering of the delta has led to the fact that the areas of the tugai type of vegetation in the Amudarya valley, which occupied 300 thousand hectares in the 60s, Currently do not exceed 25 thousand hectares (Figure 1). Of the 10 types of pasture communities described in the Amudarya delta in 2006, only 3 were found, with the dominance of reed, azhrek and camel's thorn [6, 7].

The exponential trend indicates a reduction in the areas of tugai vegetation in the lower reaches of the Amu Darya, in our opinion, this is primarily due to the large anthropogenic pressure on the ecosystem, as well as the processes of desertification and aridization. The rate of reduction in the areas of tugai thickets is 4 thousand hectares per year.

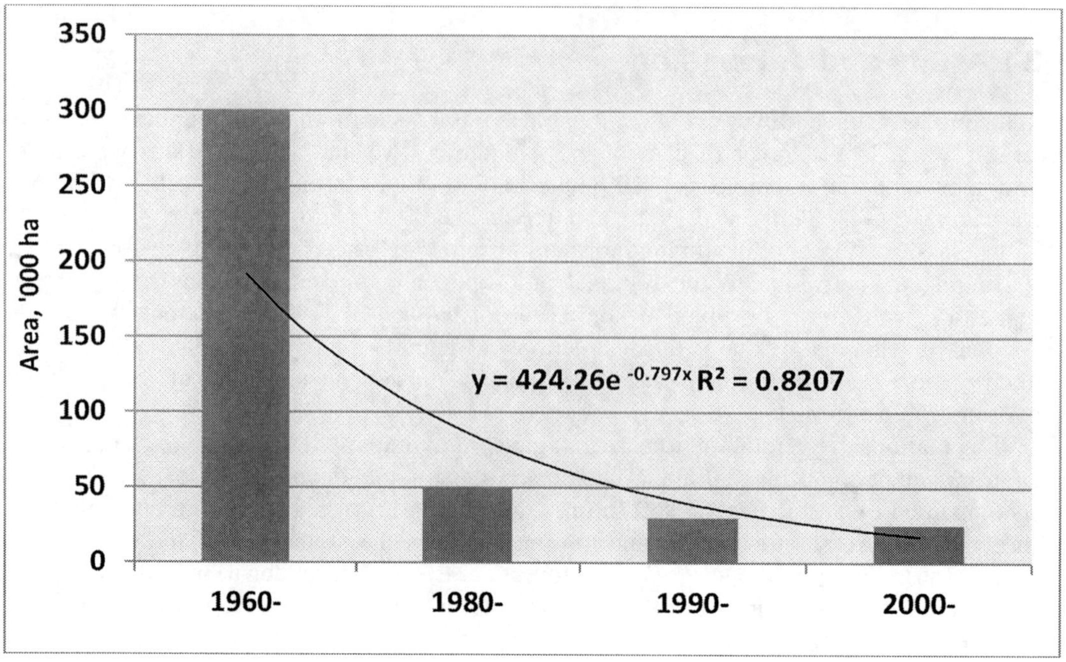

Figure 1. Dynamics of areas of tugai vegetation.

The predicted impact of climate change on the water resources of Central Asia can lead to a reduction in the water content of rivers by 20-40%, which will increase the drying of modern floodplain territories, with the activation of the solonchak formation process in soils, which in an extremely short time can lead to a complete loss of tugai in Central Asia [6, 7]. In order not to completely lose the tugai type of vegetation under the conditions of climate change, it is necessary to develop new approaches to the conservation and stabilization of these ecosystems in the absence of floods. First of all, we must not allow the complete loss of vegetation and the formation of open solonchaks in place of the former tugai alluvial soils of slightly or moderately saline.

4 Conclusions

Change in the hydrogeological regime of the territory leads to a radical change in the species composition of communities, which, in turn, affects the productivity of tugai in the conditions of the lower reaches of the Amudarya.

At present, the main direction of forestry activities in tugai communities is the protection of forests from fires and damage, reforestation , sanitary felling, thinning for the purpose of natural regeneration of the forest, reconstruction of low-value plantations, collection of medicinal plants, use of tugai plants as honey plants. A necessary condition for improving management in tugai forests is to bring the current tasks of the industry in line with the current state of forests and their potential.

References

1. Treshkin S.E., Bakhiev A.B. Development of desertification processes in the delta and lower reaches of the Amudarya and problems of rational nature management // Vestnik KKO AN RUz . No. 6. 2001. - p. 9-11.
2. Treshkin S.E. Degradation of the tugais of Central Asia and the possibility of their

restoration // Abstract of the thesis . diss . ... doc . agricultural Sciences. - Volgograd. - 2011. - 48 p.
3. Avezov A.K. Fauna of the Lower Amudarya State Biosphere Reserve // World Science. - 2022. - No. 6. - pp. 31-33.
4. Mambetullaeva S. M. Biodiversity of flora and fauna of the Lower Amudarya State Biosphere Reserve // Theory and practice of modern science. - 2018. - No. 4. - S. 389-392.
5. Tayjanov, K., Khojimatov, O., Gafforov, Y., Makhkamov, T., Normakhamatov, N., & Bussmann, R. W. (2021). Plants and fungi in the ethnomedicine of the medieval East-a review. Ethnobotany Research and Applications, 22, 1-20. DOI:0.32859/ERA.22.46.1-20
6. Jabeen, S., Zafar, M., Ahmad, M., Althobaiti, A. T., Ozdemir, F. A., Kutlu, M. A., ... & Majeed, S. (2023). Ultra-sculpturing of seed morphotypes in selected species of genus Salvia L. and their taxonomic significance. Plant Biology, 25(1), 96-106. DOI:10.1111/plb.13473
7. Majeed, S., Ahmad, M., Ozdemir, F. A., Demirpolat, A., Şahan, Z., Makhkamov, T., ... & Nabila. (2023). Micromorphological characterization of seeds of dicot angiosperms from the Thal desert (Pakistan). Plant Biosystems-An International Journal Dealing with all Aspects of Plant Biology, 157(2), 392-418. DOI:10.1080/11263504.2023.2165553
8. Majeed, S., Ahmad, M., Ali, A., Althobaiti, A. T., Ramadan, M. F., Kilic, O., ... & Sultana, S. (2023). Pollen micromorphology among amaranthaceous species from desert rangeland: Exine stratification and their taxonomic significance. BioMed Research International, 2023, 4967771. DOI:10.1155/2023/4967771
9. Makhkamov, T., Sotiboldiyeva, D., Mamarakhimov, O., Yuldashov, Y., & Botirova, L. (2022, May). Morphogenesis and Seasonal Developmental Rhythm Under the Conditions of Introduction of Curcuma Longa L. In International Scientific Conference on Agricultural Machinery Industry “Interagromash"” (pp. 1460-1469). Cham: Springer International Publishing. DOI:10.1007/978-3-031-21432-5_155
10. Boboev, S., Makhkamov, T., Bussmann, R. W., Zafar, M., & Yuldashev, A. (2023). Anatomical and phytochemical studies and ethnomedicinal uses of Colchicum autumnale L. Ethnobotany Research and Applications, 25, 1-9. DOI:10.32859/era.25.6.1-9
11. TKh, M., Brundu, G., Jabborov, A. M., & Gaziev, A. D. (2023). Predicting the potential distribution of Ranunculus sardous (Ranunculaceae), a new alien species in the flora of Uzbekistan and Central Asia. BioInvasions Records, 12(1), 63-77. DOI:10.3391/bir.2023.12.1.05
12. Ameen, M., Zafar, M., Ahmad, M., Ramadan, M. F., Eid, H. F., Makhkamov, T., ... & Majeed, S. (2023). Assessing the Bioenergy Potential of Novel Non-Edible Biomass Resources via Ultrastructural Analysis of Seed Sculpturing Using Microscopic Imaging Visualization. Agronomy, 13(3), 735. DOI:10.3390/agronomy13030735
13. Noor, W., Zafar, M., Ahmad, M., Althobaiti, A. T., Ramadan, M. F., Makhkamov, T., ... & Khan, A. (2023). Petiole micromorphology in Brassicaceous taxa and its potential for accurate taxonomic identification. Flora, 303, 152280. DOI:10.1016/j.flora.2023.152280
14. Aziz, A., Ahmad, M., Zafar, M., Gaafar, A. R. Z., Hodhod, M. S., Sultana, S., ... & Chaudhay, B. (2023). Novel Copper Oxide Phyto-Nanocatalyst Utilized for the Synthesis of Sustainable Biodiesel from Citrullus colocynthis Seed Oil. Processes, 11(6), 1857. DOI:10.3390/pr11061857

Efficiency of growing jujube (*Ziziphus jujuba* Mill) varieties that imported from China

Hao Qing[1], *Xikmat* Shaumarov[2], *Husen* Hamroyev[2] and *Musirmon* Ochilov[2,*]

[1]Institute of Horticulture, Xinjiang Academy of Agricultural Sciences, Xinjiang, 830000, China
[2]Tashkent State Agrarian University, Tashkent, 100140, Uzbekistan

Abstract. Jujube (*Ziziphus jujuba* Mill) holds a prominent position within the spectrum of tree species in Uzbekistan, primarily due to its noteworthy nutritional and medicinal attributes. With approximately 400 distinct varieties having been developed in China, the task of introducing those best suited to our local conditions is of utmost importance. This article is dedicated to the presentation of outcomes arising from the cultivation of seedlings belonging to ten imported varieties from China, facilitated through the technique of clef grafting. In the grafting process, it is recommended to employ rootstocks and scions featuring an average diameter of no less than 20.0 mm, while the discrepancy between them should not exceed 0.9 mm. Following this approach, a notable increase in grafting efficiency up to 68.8% can be achieved, concurrently contributing to the attainment of seedlings with an elevated stature, reaching up to 143.5 cm. The findings underscore the significance of meticulous grafting methodology in optimizing the success rate of introducing diverse jujube varieties. This contributes to the enrichment of the horticultural landscape, fostering the cultivation of jujube specimens that can flourish effectively in the unique Uzbekistani conditions.

Keywords. *Ziziphus jujuba*, jujube, grafting, seedling, horticulture.

1 Introduction

Climate change and population growth in recent years require to introduce in all countries of varieties of medicinal plants that are resistant to external factors, as well as having food value. According to the FAO, 44% of the world's orchards account for the share of plants introduced [1].

Today, in the world of fruit growing, special attention is paid to the cultivation of jujube seedlings, gardening and increasing the gross yield. In particular, to date, the total area of orchards in the subtropical fruit group is 410,000 hectares, with a total yield of 7.5 million tons. Jujube is grown in large quantities in countries such as China, India, South Korea, Afghanistan, Pakistan, the United States and Russia. In China, intensive jujube orchards have been established on more than 200,000 hectares [2, 3].

Today there are more than 400 varieties of this crop in China, its plantings occupy an area of 200 thousand hectares and are the third largest after apple and citrus orchards.

* Corresponding author: m_ochilov83@mail.ru

Ziziphus as a useful drought-resistant plant is grown in India, Pakistan, Afghanistan, Algeria, Israel, Egypt and the Caucasus countries. In recent decades, more and more attention has been paid to this plant in the USA, Italy, Spain and France [4].

Jujube or Chinese date (Ziziphus jujuba Mill.). It is one of the oldest main fruit trees in China. This plant was cultivated in China 4,000 years ago and created large varieties of edible local fruits [5].

Jujube varieties in China and methods of growing them in cultural conditions were studied by A.S. Koverga during a scientific expedition to China. According to him, jujube is widespread in the mountainous regions of China, especially in northern China. In China, mainly varieties and forms of one species - common Jujube (Ziziphus jujuba) are cultivated [6].

Jujube was named Ziziphus yuyuba in 1768 by the researcher Miller (Miller) on the basis of a tree that first spread in Southern Europe [7]. In Central Asia, the forms of the native jujube are mainly found in the arid subtropical region of southern Uzbekistan, Turkmenistan and Tajikistan, and are grown by amateur gardeners in their backyards. Samarkand Fruit Experimental Station In 1934, grafted jujube seedlings from California began to be grown locally [8].

Jujube - is widespread in Uzbekistan as a new, promising subtropical fruit tree. Jujube fruits are distinguished by high productivity, valuable nutritional feature, and medicinal properties. Jujube varieties are a fruit variety successfully introduced to our republic from China. In recent years, jujube has become increasingly popular as a new fruit species in the field of horticulture and medicinal plants in Uzbekistan. It is important to enrich the flora of Uzbekistan with new valuable fruit varieties and species, including the introduction of new Jujube varieties introduced from the Republic of China in the horticultural system of the our Republic. Cultural varieties and forms of folk selection of this fruit species are widely cultivated by the population in China as a valuable fruit tree [9].

In 1953, the Uzbek Research Institute of Horticulture, Viticulture and Wine-making was brought to the Samarkand Breeding Experimental Station as a test and began to grow jujube varieties created in China. As a result of many years of observations, promising varieties have been recommended for cultivation in Uzbekistan [10].

2 Materials and methods

The experimental work was carried out by the Extension center of Tashkent State Agrarian University. Promising varieties of Xiangzao, Zanhuangdazao, Tangzao, Mayazao, Junzao, Dabailing, Dongzao, Cuiwang Zao, Li Zao and Jixin Zao which are widely used in China, were whip grafted on rootstock grown from the seeds of the sour jujube (*Ziziphus acidojujuba*, *Ziziphus jujuba* Mill. var. spinosa) form of Jujube. Grafting was carried out in the last decades of March and the first decades of April. The following varieties were used in the research.

Shang jujube. The cultivar originated from Beixiang town in Yuncheng City of Shanxi Province. It is the dominant variety there and has a history of over 3 000 years. The medium-sized tree is half-spreading with a strong central leader trunk, dense branches and a semi round crown. The reddish-brown 1-year-old shoots are 30.0 cm long. The secondary branches are 25,3 cm long with less-developed thorns. The medium-sized column-shaped mother fruiting spurs can germinate 2~5 deciduous fruiting shoots. The small leaves are oval shaped and dark green. There are 48.1 flowers per deciduous fruiting shoot and 3.8 ones per inflorescence with a diameter of 6.8 mm for the zero-level flowers.

The large oval-shaped fruit weighs 19.1 g with irregular sizes. It has purplish-red thick skin and a rough surface. The greenish-white flesh is thick and hard, tight-textured, sweet, with little juice. Dried fruits have whippy flesh, tolerant to extrusion. It has an excellent quality for dried fruits. The percentage of edible part of fresh fruit is 96.8%, and SSC, TTS, TA, Vc is 28.50%, 25.51%, 0.34 and 474.00 mg per 100 g fresh fruit. The content of flavones and c AMP in mature fruit skin is 2.95 mg/g and 101.97 ug/g. The percentage of fresh fruits which can be made into dried ones is 53.0%. And the content of TTS and TA is 70.29% and 0.82%. The small stone is spindle-shaped averaging 0.61 g.

Zhuanen jujube. The cultivar originated from Zanhuang County in Hebei Province. It is the dominant variety there with a history of over 400 years and a cell chromosome of 2n=3x=36.

The large tree is half-spreading with a medium-strong central leader trunk, sparse and strong branches and a conical-shaped crown. The yellowish-brown 1-year-old shoots are 83.0 cm long. There are 7~10 secondary branches. The large column-shaped mother fruiting spurs can germinate 3~4 deciduous fruiting shoots. The thick wide leaves are oval-shaped and dark green. There are 50 flowers per deciduous fruiting shoot.

The large fruit is column-shaped or nearly obovate, averaging 18.6 g. It has a smooth surface and medium- thick red skin. The thick flesh is nearly white, tight-textured, crisp, sweet and a little sour, with medium juice and a good quality for fresh-eating, processing dried fruits and candied fruits. The percentage of edible part of fresh fruit is 96.7%, and SSC, TTS, TA and Vc is 33.30%, *29.32%*, 0.78% and 324.70 mg per 100 g fresh fruit. The content of flavones and cAMP in mature fruit skin is 13.43 mg/g and 50.58 Ug/g. The percentage of fresh fruits which can be made into dried ones is 47.8%. The content of TTS and TA in dried fruit is 66.91% and 1.93%. Dried fruits have plump and whippy flesh, tolerant to storage and transport with a better than normal quality. The small spindle-shaped stone weighs 0.62 g.

Tong jujube. The cultivar originated from Mayang, Xupu, Huayuan in west Hunan Province and in Lingling, Hengshan, Guiyang, Qiyang in southern Hunan Province. It is the dominant variety in Xupu County.

The large tree is spreading with an irregular crown and pendulous external branches. The reddish-brown 1-year-old shoots are 75.3 cm long and 1.01 cm thick. The intemodes are 5.7 cm long with gray wax on the sun-side and developed thorns. The secondary branches are 29 cm long with 7 nodes of small curvature. The column-shaped mother fruiting spurs can germinate 4 deciduous Suiting shoots which are 24.3 cm long with 17 leaves. The small green leaves are oval-shaped. There are many medium sized flowers produced, averaging 6 ones per inflorescence. It blooms in the daytime.

The small oblate fruit has a vertical and cross diameter of 2.54 cm and 2.19 cm averaging 8.0 g. It has a flat-round shoulder, a shallow and wide stalk cavity, a flat fruit apex and medium-thick red skin. The light-green or white flesh is tight- textured and sweet with medium juice. It can be used for making dried fruits with medium quality. The percentage of edible part of fresh fruit is 93.1 %, and the content of TTS, TA and Vc is 25.05%, 0.44% and 403.33 mg per 100 g fresh fruit. The percentage of fresh fruits which can be made into dried ones is 34.0%, and the content of TTS in dried fruit is 63%. The medium-sized obovate stone weighs 0.55 g, and the percentage of containing kernels is 63.3 %.

Mya jujube. The cultivar originated from Beijing, with a long cultivation history. It is planted in different districts of Beijing with a large quantity.

The medium-sized tree is half-spreading with a round crown. The grayish-brown trunk bark has wide striped fissures, easily shelled off. The reddish-brown 1-year-old shoots are 84.7 cm long and 1.24 cm thick. The intemodes are 7.7 cm long with less wax. Most thorns are degrading, gradually falling off in 1 or 2 years. The secondary branches are 31.7 cm long with 6 nodes. Young mother fruiting spurs are conical-shaped and mature ones are column-shaped. The mother fruiting spurs can germinate 5 deciduous fruiting shoots which are 22.1 *cm* long with 13 leaves. The medium-sized leaves are oval-shaped with a gradually-cuspate apex, a round or wide-cuneiform base and a sharp saw-tooth pattern on the margin. Each inflorescence has 5 flowers.

The medium-sized conical fruit has a vertical and cross diameter of 10.10 cm and 4.42 cm, averaging 10.1 g with irregular sizes. One side of the fruit is flat while the other fs half-moon-shaped. It has a flat and asymmetrical shoulder, a shallow and wide stalk cavity, a pointed-round fruit apex, a smooth surface and thin red skin. The light-green flesh is crisp, juicy, sour and sweet. It has an excellent quality for fresh-eating. The content of SSC, TTS, TA and Vc in fresh fruit is 31%, 27.91 %, 0.30% and 258.78 mg per 100 g fresh fruit. The content of flavones and cAMP in mature fruit skin is 12.27 mg/g and 8.70 Ug/g. The large spindle-shaped stone weighs 0.45 g.

Joou jujube. The cultivar originated from remote some mountains of Jiaocheng County in Shanxi Province. It is one of the ten famous jujubes in Shanxi Province with a history of over 1 000 years.

The large tree is half-spreading with a strong central leader trunk, strong and medium-dense branches and a natural-round crown. The reddish-brown 1-year-old shoots are 54.8 cm long. There are 6~7 permanent secondary branches with less developed or degraded thorns. The large conical-shaped mother fruiting spurs have a long life and can germinate 3—4 deciduous fruiting shoots, which are 15.0 cm long with 10 leaves. The large leaves are long oval-shaped and dark green. The number of flowers is medium large.

The large fruit is column-shaped in earlier growth stage and long obovate in later growth stage. It weighs 26.3 g with a regular size. The thick flesh is white or greenish white, loose-textured, crisp and sweet with medium juice and a good quality and wide applications. It can be used for fresh-eating, processing dried fruits, candied fruits and alcoholic fruits. The percentage of edible part of fresh fruit is 96.3%, and SSC, TTS, TA and Vc i s 33.00%, 28.68%, 0.45% and 430.20 mg per 100 g fresh fruit. The content of flavones and cAMP in mature fruit skin is 1.78 mg/g and 102.14 Ug/g. The percentage of fresh fruits which can be made into dried ones is 56.8%, and the content of TTS and TA in dried fruit is 71.77% and 1.58%. SSC, TTS and TA in alcoholic jujubes is 36.30%, 30.83% and 0.83%.

Debay jujube. The cultivar originated from Xiajin County in Shandong Province, and spreads in Linqing, Wucheng, Yanggu in Shandong and Xianxian County in Hebei Province. It is dispersedly planted.

The small tree has moderate vigor, spreading with a strong central leader trunk, medium-dense branches and an umbrella-shaped crown. The reddish-brown 1-year-old shoots have moderate growth vigor, 81.0 cm long with the intemodes of 8.0 cm. The secondary branches are 29.1 cm long with 6 nodes and less-developed thorns. The medium-sized mother fruiting spurs can germinate 3~4 deciduous fruiting shoot, which are 23.5 cm long with 16 leaves. The medium-sized leaves are oval-shaped and dark green. There are many small flowers with a diameter of 6 mm produced, averaging 10-11 ones per inflorescence.

The large fruit is nearly round, averaging 31 9 g with a regular size. It has a purplish-red thin skin and an unsmooth surface. The greenish-white thick flesh is loose-textured, Crisp and sweet, with medium juice and a good quality for fresh eating. The percentage of edible part of fresh fruit is 97.2%, and SSC, TTS, TA and Vc is 33.00%, 24.50%, 0.28% and 280.08 mg per 100 g fresh fruit. The small stone is spindle-shaped, averaging 0.9 g almost without kernels.

Doon jujube. The cultivar originated from and spreads in Huanghua, Yanshan in Hebei Province and Zhanhua, Zaozhuang in Shandong Province.

The tree is spreading with a semi-round crown, a strong central leader trunk and thin and dense branches. The grayish-green 1-year-old shoots are 57.4 cm long with the intemodes of 7.8 cm, with almost degraded thorns. There are 4~7 secondary branches on each 1-year-old shoot. The small mother fruiting spurs can germinate 4.3 deciduous fruiting shoots. The medium sized leaves are oval-shaped. There are many small flowers with a diameter of 5.0-5.8 mm produced.

The small fruit is nearly round with a vertical and cross diameter of 3.10 cm and 2.80 cm, averaging 11.9 g with irregular sizes. It has thin red skin and a smooth surface. The delicate flesh is crisp and juicy, with a sweet taste and a very good quality for fresh-eating. The percentage of edible part of fresh fruit is 94.1%, and the soluble solid content (SSC), monosaccharide, disaccharide, total sugar (TTS) and Vitamin C (Vc) is 37.8%, 14.80%, 19.00%, 33.80% and 292.60 mg per 100 g fresh fruit. The content of flavones and c AMP in mature fruit skin is 25,59 mg/g and 94.91 Ug/g. The oval-shaped stone has a vertical and cross diameter of 1.70 cm and 0.70 cm, averaging 0.70 g with a short apex and medium-deep veins. The percentage of containing kernels is 86.7%. Most stones contain a well-developed kernel, which can be used as breeding material.

Lee jujube. The cultivar originated from Linyi and Yuncheng in Shanxi Province with a cultivation history of over 3 000 years.

The small tree is spreading with a weak central leader trunk and a round crown. The trunk bark has deep fissures, easily shelled off. The reddish-brown 1-year-old shoots have strong germination ability and moderate growth potential, averaging 83.1 cm long with 6~8 nodes. The secondary branches are 28.3 ctti long with less-developed thorns. The small conical mother fruiting spurs can germinate 4.4 deciduous fruiting shoots. The thick leaves are oval-shaped and dark green. The number of flowers is rather small, averaging 2.2 ones per inflorescence. The medium-sized blossoms have a diameter of 7.2 mm for the zero-level flowers.

The extremely large fruit is obovate pear-shaped, with a vertical and cross diameter of 4.50 cm and 4.00 cm, averaging 31.6 g with irregular sizes. It has light-red thin skin. The rough flesh is loose-textured, crisp, juicy and sweet. It has a very good quality for fresh eating. The percentage of edible part of fresh fruit is *96.6%,* and SSC, monosaccharide, disaccharide, TTS, titratable acid (TA) and Vc is 27.90 17.00 H, 5.25$), *22.25%, 0.31%* and 292.25 mg per 100 g fresh fruit. The content of flavones and cAMP in mature fruit skin is 3.91 mg/g and 122.17 ug/g. The small spindle-shaped stone weighs 1.08 g.

Jeshine jujube. The cultivar originated from Xinzheng, Zhongmou, Xihua and the suburbs of Zhengzhou City in Henan Province. It is mainly planted in Xinzheng which is the 2nd dominant variety there. There are still some trees of over 400 years old.

The medium-sized tree is half-spreading with a round crown. The yellowish-brown 1-year-old shoots are 88.2 cm long and 1.22 cm thick, with the intemodes of 6.3 cm. The secondary branches are long, averaging 34.6 cm with 6 nodes of a small curvature and less-developed thorns of 1.2 cm long. The mother fruiting spurs are small and column-shaped. The deciduous fruiting shoots are 23.5 cm long with 14 leaves. The large green leaves are oval-shaped. There are many large flowers with a diameter of 7.6 mm produced. It blooms in the daytime.

The small fruit is chicken-heart-shaped or oval-shaped, averaging 4.9 g with a regular size. It has purplish- red thin skin and a smooth surface. The greenish-white flesh is medium thick and tight-textured, crisp and sweet, with less juice and a good quality for dried fruits. The percentage of edible part of fresh fruit is 93.8%, and SSC, TTS, TA and Vc is 37.20%, 32.68%, 0.70% and 488.90 mg per 100 g fresh fruit. Dried fruits have tight-textured and whippy flesh with strong tolerance to extrusion, storage and transport. The large stone is spindle shaped, averaging 0.48 g.

3 Results and discussion

This research was carried out by grafting 10 prospective varieties on seedlings grown from seeds of the wild offspring form Jujube that imported from China. These varieties are valuable in terms of fruit size, taste, color, appearance, exportability and composition. Grafting was carried out in the spring, and at the end of the growing season, the degree of retention of grafts, diameters of rootstocks and scions, as well as the height indicators of the grafted seedlings were determined.

Grafting regulations require that the rootstocks and scions be matched to each other in terms of dimensions. Especially, it is very important that their diameters match during the cleft grafting process. The diameter of the seedling used as a graft in the grafting process affected its retention in our research, too. In the research, scions were used that the average diameters were 11.7-20.0 mm. Accordingly, grafting work was carried out to the seedlings the sort of Shang jujube with 17.0 mm, Zhuanen jujube with 18.9 mm, Tong jujube with 15.3 mm, Mya jujube with 14.5 mm, Joou jujube with 20.8 mm, Debay jujube with 16.2 mm, with Dooon jujube 15.5 mm, Cuiwan jujube with 14.8 mm, Lee jujube with 16.9 mm and Jeshine jujube with 12.6 mm in diameter. During the grafting, dimensions of the scions differed from the dimensions of the rootstocks by an average of 0.1-3.2 mm. This allowed the grafting work to be carried out efficiently (Figure 1).

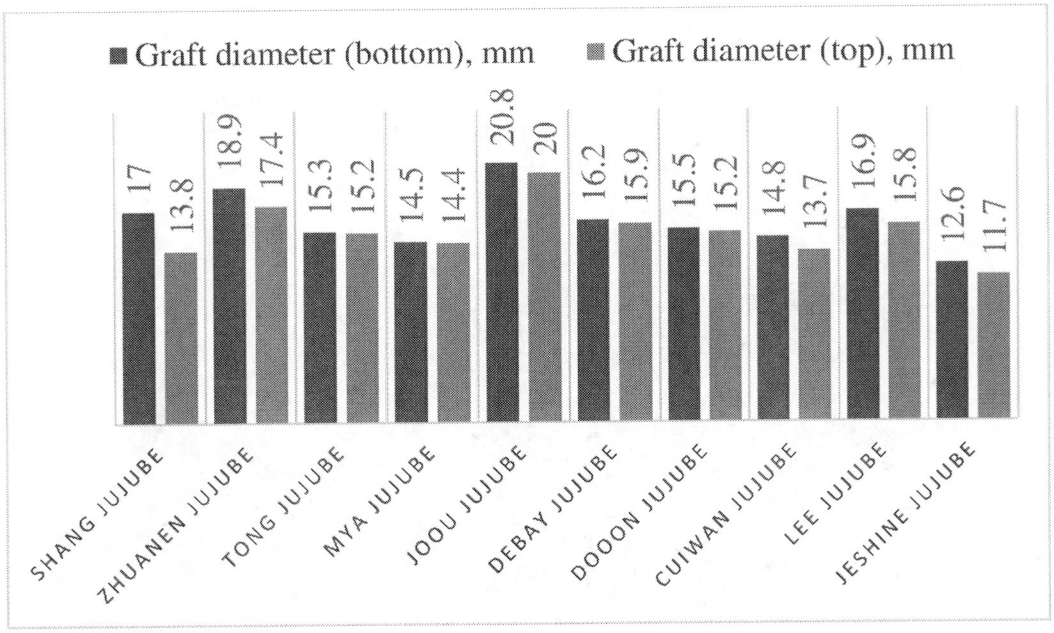

Figure 1. Dimension of rootstocks and scions in the method of cleft grafting.

The efficiency of grafting by sorts, that is the retention rate was 45.2-75%. The difference in the size of the rootstocks and scions used in grafing was 0.1-0.3 mm (Tong jujube, Debay jujube, Mya jujube ва Dooon jujube), the efficiency was low (45.2-51.5%) compared to other varieties. This difference was higher rate (68.8-75.0%) in varieties with 0.8–0.9 (Cuiwan jujube, Joou jujube and Jeshine jujube). This figure was equal to 59.9-65.8% in varieties with a difference of more than 1.0 (Lee jujube, Shang jujube and Zhuanen jujube). The indicator of the Jeshine jujube sort was 75%, while the Tong jujube was 45.2%. Compared with other varieties, we can observe that the grafting efficiency of this variety is 1.09-1.65 times higher (Figure 2).

 https://doi.org/10.1051/e3sconf/202342101007

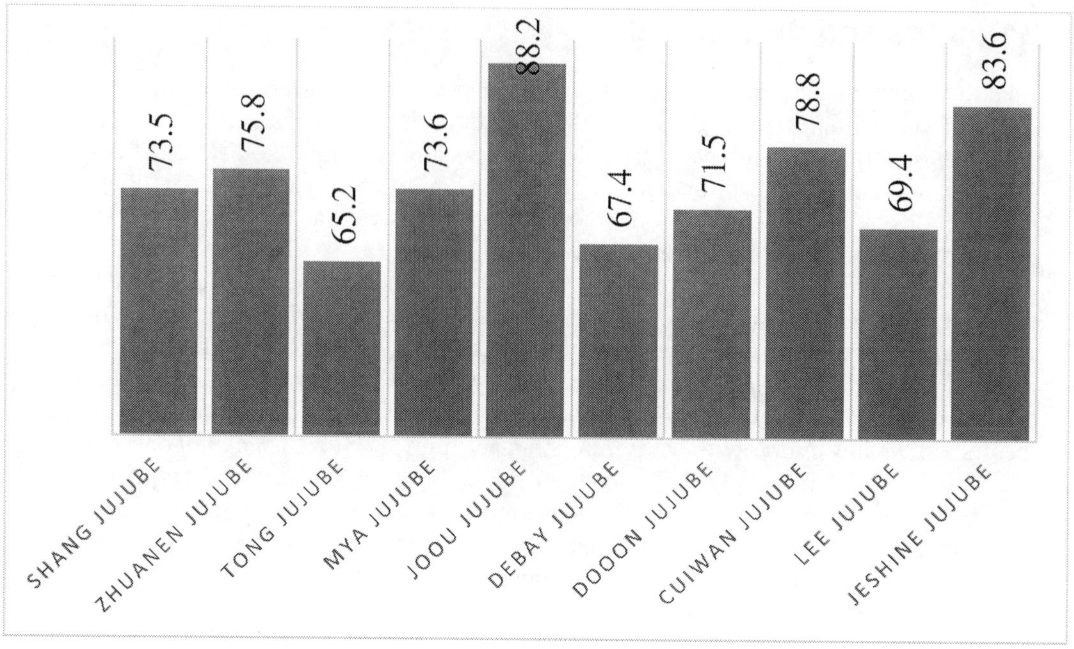

Figure 2. Retention rate of the sorts which was graft by the method of cleft grafting, %.

When the growth rates of grafted seedlings were studied, their height was equal to 82.1-143.5 cm. The highest value was observed in the Joou jujube (143.5 ± 3.50) and the lowest in the Jeshine jujube (82.1 ± 6.12) (Figure 3).

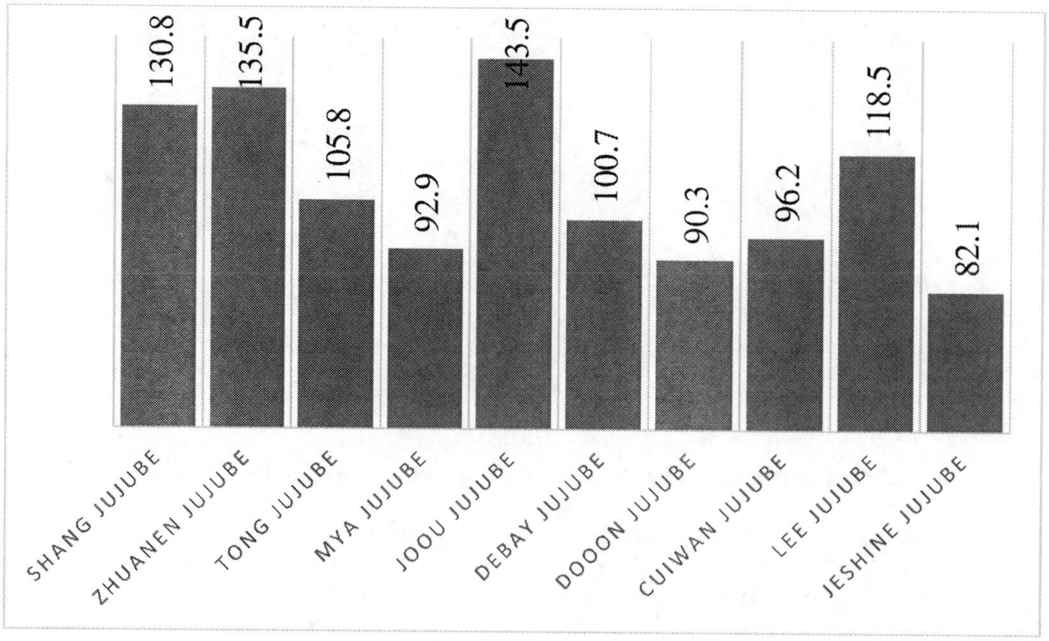

Figure 3. Height of grafted seedlings, cm.

4 Conclusions

In our research, it has been found that the role of the rootstocks and scions' diameters in the cultivation of grafted seedlings of Jujube is very important. Grafting efficiency and seedling growth rates have been also found to be good when the difference between them was 0.8-0.9 mm. In our research, the grafting efficiency of a 20.0 mm scion to a rootstock with an average diameter of 20.8 mm has been 68.8%. This ensures that the dimensions of rootstock and scions in the cultivation of Jujube seedlings are not more than 20 mm, and the difference between them does not exceed 0.8, the growth rate of seedlings reaches 143.5 cm.

References

1. Mahajan, R. T. C. M., & Chopda, M. (2009). Phyto-Pharmacology of Ziziphus jujuba Mill-A plant review. Pharmacognosy Reviews, 3(6), 320.
2. Ji, X., Peng, Q., Yuan, Y., Shen, J., Xie, X., & Wang, M. (2017). Isolation, structures and bioactivities of the polysaccharides from jujube fruit (Ziziphus jujuba Mill.): A review. Food chemistry, 227, 349-357.
3. Cheng, G., Bai, Y., Zhao, Y., Tao, J., Liu, Y., Tu, G., ... & Xu, X. (2000). Flavonoids from Ziziphus jujuba Mill var. spinosa. Tetrahedron, 56(45), 8915-8920.
4. Gao, Q. H., Wu, C. S., & Wang, M. (2013). The jujube (Ziziphus jujuba Mill.) fruit: a review of current knowledge of fruit composition and health benefits. Journal of agricultural and food chemistry, 61(14), 3351-3363.
5. Rodriguez Villanueva, J., & Rodriguez Villanueva, L. (2017). Experimental and clinical pharmacology of Ziziphus jujuba Mills. Phytotherapy research, 31(3), 347-365.
6. Zhang, H., Jiang, L., Ye, S., Ye, Y., & Ren, F. (2010). Systematic evaluation of antioxidant capacities of the ethanolic extract of different tissues of jujube (Ziziphus jujuba Mill.) from China. Food and chemical toxicology, 48(6), 1461-1465.
7. Zhao, Z., Liu, M., & Tu, P. (2008). Characterization of water soluble polysaccharides from organs of Chinese Jujube (Ziziphus jujuba Mill. cv. Dongzao). European Food Research and Technology, 226, 985-989.
8. Ji, X., Cheng, Y., Tian, J., Zhang, S., Jing, Y., & Shi, M. (2021). Structural characterization of polysaccharide from jujube (Ziziphus jujuba Mill.) fruit. Chemical and Biological Technologies in Agriculture, 8, 1-7.
9. Guil-Guerrero, J. L., Delgado, A. D., Gonzalez, M. M., & Isasa, M. T. (2004). Fatty acids and carotenes in some ber (Ziziphus jujuba Mill) varieties. Plant Foods for Human Nutrition, 59, 23-27.
10. Zhang, Q., Wang, L., Liu, Z., Zhao, Z., Zhao, J., Wang, Z., ... & Liu, M. (2020). Transcriptome and metabolome profiling unveil the mechanisms of Ziziphus jujuba Mill. peel coloration. Food chemistry, 312, 125903.

 https://doi.org/10.1051/e3sconf/202342101008

Study of vegetable beans collection in climatic conditions and selection of promising sources for breeding

Makhfurat Amanova[1,*] and *Abror* Hasanov[1]

[1]Tashkent State Agrarian University, 2, University street, Tashkent, 100140, Uzbekistan

Abstract. During the span of 2018 to 2021, a comprehensive investigation was conducted encompassing the morphobiological and economically significant attributes of 20 distinct samples of vegetable beans. These samples were meticulously gathered through collaborative expeditionary ventures spanning the landscapes of Kyrgyzstan, Kazakhstan, and Uzbekistan. The interdisciplinary efforts of scientists hailing from the Scientific Research Institute of Plant Genetic Resources and the Rural Development Administration of the Republic of Korea (RDA) culminated in an in-depth analysis of these plant varieties. The outcome of this thorough exploration yielded noteworthy results, highlighting the identification and isolation of four early maturing and five high-yielding samples of vegetable beans. The amalgamation of expertise, fieldwork, and scientific scrutiny brought forth the delineation of these promising bean varieties with traits poised to contribute significantly to agricultural productivity and resilience across the studied regions. The collaborative nature of this endeavor underscores the significance of international cooperation in advancing agricultural innovation and enhancing the cultivation of vital crops.
Keywords. Vegetable beans, collection accessions, prospective accessions, ripe, medium-ripe, high-yielding.

1 Introduction

Today, when we talk about food safety in developed countries, we mean not only the amount of products grown, but also its variety, ecological purity, and the richness of minerals useful for the human body and health [1, 2]. In this regard , mutual integration of the processes of production, processing, storage, service and sale (export), development of clusters (cooperation), ensuring food safety and increasing the volume of export is of great importance in Uzbekistan . is being given.

Therefore, along with expanding the range of vegetable and fruit crops, radical improvement of scientifically based seed breeding systems of their created varieties, wide use in production, development of resource-saving technologies of seed cultivation remain one of the important tasks of today [3].

Every year in Uzbekistan, more than one million fields freed from grain crops are

* Corresponding author: m.amanova@tdau.uz

cultivated with vegetables, sugarcane, legumes, corn for livestock and sunflower products. Among these products, mash, beans and groundnuts are mainly exported [4]. However, due to the lack of early, high-yielding and exportable varieties suitable for the climate of Uzbekistan, the average yield is 12-15 t/ha in mash, 10-14 t/ha in beans, and 25-30 t/ha in groundnut [5].

One of the urgent tasks of today is to create and introduce into production new early, high-yielding, exportable varieties for the main and repeated planting periods suitable for the local climatic conditions in growing a high yield of vegetable beans in Uzbekistan [6].

The purpose of the research is to study the varieties and samples of vegetable beans introduced from abroad in the climatic conditions of the Tashkent region according to valuable economic characteristics, to select primary sources for different directions of selection.

Tasks of the research are to study the morpho-biological characteristics and valuable economic characteristics of the collected variety samples of vegetable beans in the climatic conditions of the Tashkent region and to select primary sources for different directions of selection.

2 Materials and methods

This research was carried out in open field conditions, in which "Methodology of conducting experiments in vegetable growing, potato growing and potato growing" [7-9] conducted on the basis of methodological manuals. The statistical analysis of the data was carried out using the Microsoft Excel program according to the "Metodika polevogo opyta" method by B. A. Dospehov.

These studies were carried out in the cultivated fields of the "Educational-consultative center" of the Tashkent State Agrarian University, located in the Qibrai district of the Tashkent region in Uzbekistan. According to the information provided by the Agrochemical Analysis Center of the Tashkent region, in the arable layer (0-30 cm) of soils where field experiments are conducted from the beginning of the growing season, humus is 1.6-1.87%, nitrogen-0.135%, phosphorus-0.37%, potassium-1 , was 10%.

Tashkent region is located in the north-east of Uzbekistan, bordering the Republic of Kazakhstan to the north-west, Kyrgyzstan to the north-east, Namangan region to the east, Republic of Tajikistan to the south and Syrdarya regions to the south-west. The climate of the region is characterized by a sharp continentality, its winters are mild and wet, and summers are hot and dry.

Tashkent region has strong sunlight, the duration of sunny days during the year is 2800-2900 hours, 360-400 hours in summer months and 90-100 hours in winter months. Cloudy days occur in rare periods, they are observed mainly in the winter months, that is, from 10 to 25 days in December and January [10].

According to the data of the Qibray district "Aq Zukuk" weather station, the average annual air temperature of the region is 14-17 ^{0}C in March-April, May, June-August, September 28.2-32.4 ^{0}C, respectively and starting from September, the air temperature drops to 18.4 ^{0}C· Qibray district is located in the arid region, where a total of 250-500 years of rainfall is observed. The main part of precipitation periods corresponds to the winter-spring months. March is the busiest month. In these areas, snow piles are kept for 25-35 days.

3 Results and discussion

Samples were planted in the main crop in Kibrai district of Tashkent region in 2018-2021 and studied for morpho-biological and economic characteristics.

Bean seeds were sown on typical gray soils on April 1 at a seedling thickness of 70x15 cm (in the feeding area) and 10% of the seeds germinated on April 7 and 75% of the seeds germinated on April 10. For the studied samples of beans, the local variety "Qaraqosh", widespread in Uzbekistan, was taken as a template (St).

The stem of the local variety "Qaraqosh" is a creeper, the length of the stem is on average 40-80 cm, the technical ripening period is 45-50 days, the physiological ripening of the seeds after the formation of pods is 65-70 days, and the growing period is 120 days. did When planted for grain, the average yield is 20 ts/ha, it is also consumed during technical and biological ripening.

Important morphological characters. The inter-row irrigation method is widely used in growing beans in Uzbekistan. This, in turn, leads to a number of additional costs. The shape of the stem in bean varieties is one of the main economic characteristics, and the growth period of the creeping varieties is relatively long, and it causes certain difficulties in inter-row processing and harvesting.

The seeds of the 21 samples included in the research were planted in 3.5 m^2 experimental plots with a seedling thickness of 70x15 cm. 33 seeds were planted. Biometric measurements were taken on 10 plants during the harvest. 7 of the studied samples: Tabletka, Chipor, Kirmizi (small) (Turkiye), Rakushka (unknown), Lapatka (Kyrgyzstan), Krasnaya shapochka (Russia) and Kirmizi (Turkiye) have a creeping stem, the growth period is 100-120 days reached, the growth period was equal to 95-100 days in samples of semi-erect growing white mai (Kyrgyzstan) and Black Pearl (Bolivia) (Table 1).

Table 1. Valuable farm signs of bean collection samples (April 1).

#	Conditional name of the samples	Origin	Growth period	Shape of the stem
1	Productive	Uzbekistan	90	Steep growth
2	Ravot	Uzbekistan	75	An upright grower
3	Black eyebrow	Uzbekistan	1 2 0	A creeper
4	Tablet	Turkiye	120	A creeper
5	Cypress	Turkiye	115	A creeper
6	Rakushka	Unknown	100	A creeper
7	Cypress	Kyrgyzstan	90	An upright grower
8	Shovel	Kyrgyzstan	100	A creeper
9	Caesar	Unknown	80	An upright grower
10	Tomatnaya	Turkiye	80	An upright grower
11	Giant (white)	China	1 6 0	An upright grower
12	Small white	Kyrgyzstan	100	A semi-erect grower
13	Little Red Riding Hood	Russia	120	A creeper

14	Black prince	Kazakhstan	70	An upright grower
15	Korolevskaya 1	Kyrgyzstan	70	An upright grower
16	Black (small)	Bolivia	100	A semi-erect grower
17	Black (giant)	unknown	130	An upright grower
18	Swallow	Kyrgyzstan	80	An upright grower
19	Red (small)	Turkiye	100	A creeper
20	44444	Kazakhstan	85	An upright grower
21	Shahika	Ukraine	75	An upright grower
22	Zolnushka	Ukraine	75	An upright grower

Vegetative period lasted for 130-160 days, along with the stem growing upright. The pods of the white giant variety are very large, rough, and difficult to separate from the shell by hand. Black Giant is distinguished by its long flowering period compared to other bean varieties. The flowers are relatively large and bright red in color.

The shape of the stem of the plant is very important when growing high yields of beans in large areas. Creeping varieties are not suitable for inter-row cultivation during the growing season. It is recommended to grow such varieties in residential plots.

In order to harvest with the help of technology, it is necessary to grow varieties with upright stems, ripening at the same time, pods that are easily separated from the pods, and pods that are resistant to rotting. Bean seeds were sown in the collection nursery on April 1, and we observed that the germination of seeds was not the same in all samples.

Including 5% germination of seeds Productive, Rowat, Black brow (st), Tabletka, Chipor, Rakushka, Chipor-2, Lopata, Caesar, Tomanaya, White May, Krasnaya Shapochka, Black Prince, Korolevskaya 1, Black (small), In the samples of the varieties Swallow, Kirmizi, Shahika, Zolnushka, 5% of seeds germinated 5 days after sowing, and 75% of seeds germinated 8 days after sowing, while in Black and White giant giant samples, 5% of seeds germinated on 7-8 days, and 75% of seeds germinated on 10 and 11 days were recorded. The transition of plants to the flowering stage (10%) was observed on 28-35 days after seed germination, and 75% of plants entered flowering on 35-40 days after sowing. However, White and Black Giant cultivars were exceptions, with 10% of plants reaching 40-50 days in White Giant and 35-45 days in Black Giant, respectively.

In total, 75 days for Tomatnaya, Black Prince varieties, 80 days for Rovot, Caesar, Kirmizi, Shahika, Zolnushka varieties, 85 days for Swallow, Black Pearl, Korolevskaya 1 varieties, 90 days for Krasnaya Shapochka varieties, Chipor, Rakushka and Ravot varieties for 75% ripening of seeds. 95 days, while 100-105 days were needed for the varieties Black Brow and Tabletka, Lapatka, 150 days for White Giant, and 130 days for Black Giant (Table 2).

In the climate of Uzbekistan, it was observed in the studies that the duration of the growth period of creeping varieties is extended by 10-15 days in accordance with the external environmental factors. However, the flowering process was not recorded in the plants during this period.

When the samples of the collection of beans were studied according to valuable farm signs, 6 varieties and samples (Tomatnaya, Black Prince, Caesar, Kirmizi, Shahika and Zolnushka) were selected for early ripening. The growth period of these varieties is 75-80 days, and in the future, they will serve as a primary source for the creation of early bean varieties that can be ripened in the northern regions of Uzbekistan.

In Uzbekistan, bean plant is used in two ways, the first is when it is technically ripened with beans, and it is eaten with physiologically ripened grain in the fall and winter seasons (Table 2).

Table 2. Development stages of the samples of the studied collections.

Varietal name of samples	Origin	Irrigation days	Germination of seeds, one day after sowing		Wait for the plants to bloom		Physiological ripening of beans (from germination) days	
			10%	75%	10%	75%	10%	70%
Productive	Uzbekistan	100	5	8	30	40	80	95
Rawat (st)	Uzbekistan	85	5	8	31	35	72	80
Black eyebrow	Uzbekistan	110	5	8	32	38	75	100
Tablet	Turkiye	110	5	8	35	36	85	105
Cypress	Turkiye	100	5	8	31	39	85	95
Rakushka	Unknown	100	5	8	30	38	80	95
Cypress 2	Kyrgyzstan	100	5	8	28	35	85	95
Shovel	Kyrgyzstan	105	5	8	35	40	85	100
Caesar	Unknown	85	5	8	30	34	75	80
Tomatnaya	Turkiye	80	5	8	30	34	70	75
giant (white)	China	160	8	11	40	50	120	150
Small white	Kyrgyzstan	95	5	8	30	36	80	90
Little Red Riding Hood	Russia	95	5	8	30	34	80	90
Black prince	Kazakhstan	80	5	8	30	34	70	75
Korolevskaya 1	Kyrgyzstan	85	5	8	30	36	75	85
Black pearl	Bolivia	85	5	8	30	34	70	85
giant (black)	unknown	135	7	10	35	45	115	130
Swallow	Kyrgyzstan	90	5	8	30	40	80	85
Red (m)	Turkiye	85	5	8	30	35	75	80
Shahika	Ukraine	82	5	8	30	33	70	80
Zolnushka	Ukraine	82	5	8	30	33	70	80

During the growth period, when the pods were technically ripened, harvesting was carried out 3 times. The first time was 50 days after germination, the second time was 70 days after germination, and the third time was 90 days after germination.

Among the bean samples selected for early ripening, Caesar (0.55 kg/1 m^2), Tomatnaya (0.54 kg/1 m^2), Shahika (0.61 kg/1 m^2) and Swallow (0.51 kg/1 m^2) was found to have higher results than the standard Ravot variety in terms of productivity.

The yield of these bean varieties is up to 120-153 percent higher than the standard variety, and will serve as a primary source for the creation of early, high-yielding new varieties for future main and repeat seasons (Figure 1).

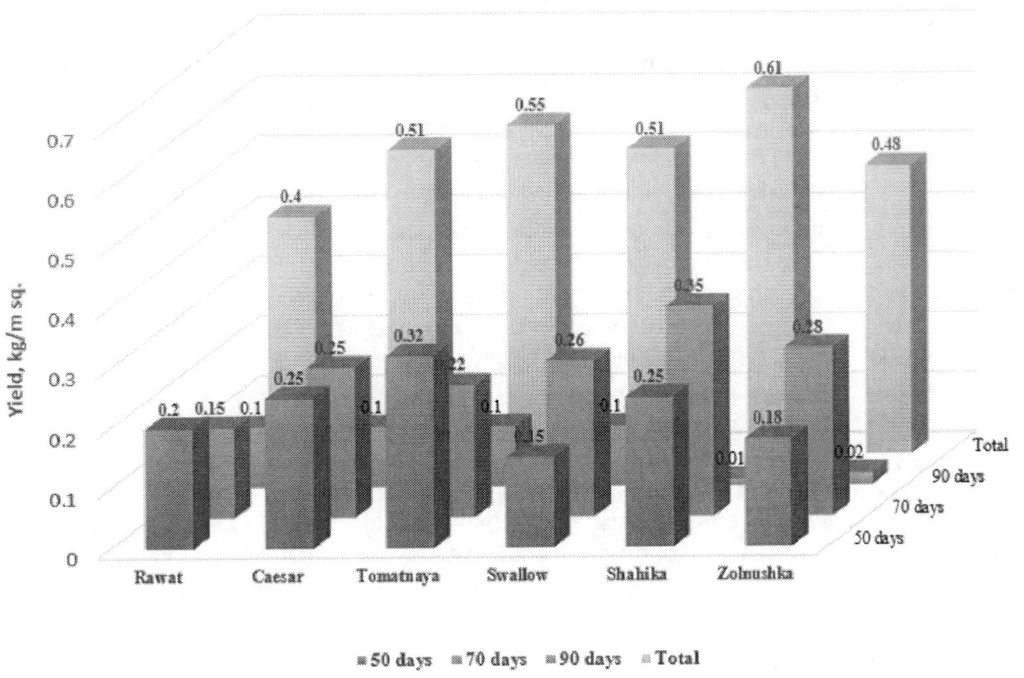

Figure 1. Yield of bean samples selected for early ripening, kg/m^2.

In total, 9 of the studied specimens with a yield of 20 grams per plant of the model Ravot variety were selected, including: Mahsuldor, Black Eye, Chipor, Caesar, Tomatnaya, White and Black Giant, Shahika and Zolnushka varieties. It was found that the productivity of one plant of these varieties and samples was on average 24-60 grams (Table 3).

Table 3. Valuable economic characteristics of bean collection samples during biological ripening of seeds.

#	Conditional name of the sample	Growth period	Productivity of one plant, g	1000 seed weight	Yield, kg/m^2
1	Productive	90	25	410	0.250
2	Rawat (st)	75	20	320	0.200
3	Black eyebrow	120	30	210	0.300
4	Tablet	120	22	350	0.220
5	Cypress	115	21	450	0.210
6	Rakushka	100	15	380	0.150
7	Cypress	90	26	410	0.260
8	Shovel	100	19	380	0.190
9	Caesar	80	27	370	0.270
10	Tomatnaya	80	24	350	0.240
11	giant (white)	160	60	980	0.600
12	Small white	100	18	250	0.180
13	Little Red Riding Hood	120	18	350	0.180
14	Black prince	70	18	310	0.180
15	Korolevskaya 1	70	18	340	0.180

16	Black pearl	100	12	275	0.120
17	Black (Giant)	130	47	850	0.470
18	Swallow	80	19	410	0.190
19	Red (small)	100	20	220	0.200
20	44444	85	19	300	0.190
21	Shahika	75	26	290	0.290
22	Zolnushka	75	25	300	0.250

In June, when the average air temperature exceeded 25 °C, it was noted that the flowering process of Lapatka, Krasnaya shapochka and Kara marvarid varieties almost stopped, and it was observed that they could not adapt well to local conditions.

From the studied samples, the following samples according to the productivity indicator, including: Black-browed 0.30 kg from kg/m^2, Cypress 0.26 kg, Caesar 0.27 kg, Shahika 0.29 kg, Zolnushka 0.25 kg, Black (giant) 0.47 kg and Giant (white) 0.60 kg were selected. This indicator was 0.20 kg/m^2 in the Rivot variety.

The seeds of the studied vegetable bean collection samples were divided into 4 groups according to their size, including small (≥ 300 g), medium large (301-450 g), large (451-600 g), very large (600 ≤).

Today, the scientific research work with Gigant (white), Shahika and tomato varieties of beans is continued, and the selection process for creating early, fruitful new varieties, suitable for planting in the main and repeated periods suitable for the climatic conditions of Uzbekistan.

4 Conclusions

In the central and southern regions of Uzbekistan, it is recommended to plant Shahika, Zolnushka, Kara Kosh, Tomatnaya, Caesars, and Mahsuldornavs in order to produce a high yield of beans in the spring and repeated crops.

The central regions of Uzbekistan, it is effective to plant Shahika, Zolnushka, Kara Kosh, Caesarlar, Tomatnaya, Mahsuldar, Aq and Kara giant varieties to produce a high yield of beans. In the northern regions of Uzbekistan, it is recommended to plant Shahika, Tomatnaya, Caesarlar and Ravot varieties in order to obtain a high-quality harvest of beans in fields free from grain.

In the northern regions of Uzbekistan, it is effective to plant Black Prince, Korolivskaya 1 and Tomatnaya varieties to produce a high yield of beans in a repeated crop.

References

1. Dospekhov B.A. Methodology polevogo opyta. - M.: Kolos, 2015. -317 p.
2. Balashov N.N., Zeman G.O. Voshchevodstvo. - Tashkent: Ukitchich, 2001. - 156 p.
3. Atabaeva H.N. Plant science. - Tashkent: Labor, 2000. -134-136 p.
4. Abdurahmonov T.-"Introduction to soil science". Educational and methodological complex. Tashkent - 2016. p. 85-96.
5. Blagoveshchensky A.V. Biochemistry and metabolism of azotsoderzhashchikh veshchestv and plant. - M.: izd. AN Uzbekistan, 2018. – 395 p.
6. Bodner G.V., Lavrinenko G.T. Zernobobovy culture. - M.: Kolos, 2007. - 246 p.
7. Control measures against the main pests of leguminous crops. Recommendation. Tashkent-2018. 18 p.
8. Zakirov K.Z., Jamolkhanov H.L. Russian-Uzbek encyclopedic dictionary of botany. I roof. - Tashkent: Teacher Press, 2013. - 233 p.

9. Kenjaev Yu, Oripov R. Characteristics of growth and development of repeated crops. "Agro ilm" - the scientific application of the agricultural journal of Uzbekistan. - Tashkent, 2007, v. 3. p. 14.
10. Kogai M.T. Cultivation of legumes on irrigated lands. - Tashkent: Uzbekistan, 2013. - 40 p.

Propagation methods of *Acorus Calamus* in Uzbekistan

Nodira Arabova[1,*], *Khasan* Akhmedov[1] and *Shukhrat* Abdurasulov[1]

[1]Tashkent State Agrarian University, 2, University street, Tashkent, 100140, Uzbekistan

Abstract. This article is devoted to methods of reproduction of *Acorus calamus* L. in the conditions of Tashkent region. According to our three-year research and analysis of the literature, it was found that the pollen grains developed in *Acorus calamus* plants are sterile. It was noted that it does not produce seeds in Tashkent conditions. Therefore, the methods of vegetative reproduction of the plant have been thoroughly studied. Biological methods of growing *Acorus calamus* have been developed in running water ponds and propagation from rhizome joints in laboratory conditions. Based on the set of the most important indicators of biological properties, the scientific basis of the "Node" method was created, and promising agrotechnical methods were developed based on experiments. In the process of studying the reproduction of plant nodes, the growth of 1-10-noded cuttings was observed in different nutrient media. It was found that the cuttings of the plant were fully grown, and the rhizomes and leaves grew well. It was determined that it is appropriate to plant the plants in 7-10 joints in different reservoirs. Planted rhizome cuttings were found to grow intensively even at 13-18 C°. For the first time, plants have been observed to grow faster in sheep manure-fed media than in cattle manure.
Keywords. *Acorus calamus*, rhizome, vegetative propagation, node method, medicinal.

1 Introduction

Medicinal plants have been widely used in folk medicine since ancient times. Continuous use of chemical preparations in highly developed modern medicine can cause certain diseases in

* Corresponding author: arabova_nodira@mail.ru

the human body. In our independent Republic of Uzbekistan, the group of higher plants consists of 4650 species, of which 700 species form the group of medicinal plants. The demand for medicinal plants is increasing in recent years, because drugs made from synthetic drugs have little effect and can cause functional changes in the body when taken for a long time. That's why studying, breeding, restoring and adequately using medicinal plants is one of the most important issues.

The great scholar Abu Ali Ibn Sina [1], who grew up in Central Asia and contributed to the development of world science - medical science, obtained 811 plain medicines from plants and animals. He revealed ways to prepare and use medicine from many medicinal plants. One such plant is Acorus calamus. Ibn Sina used ointments made from this unique plant in the treatment of liver, black and gastrointestinal diseases and as a diuretic.

According to H.H. Kholmatov and A.I. Kasimov [2, 3], the rhizome of *Acorus calamus* contains up to 5% essential oil, acorin glycoside, additives, resin, glue, up to 25% starch and other substances. It also contains terpenes such as azarone, proazulene, pinene, calamene and camphene. The leaves contain up to 150 mg % of vitamin C, essential oil and flavoring substances. In modern medicine, it has been found that under the influence of acorin glycoside in the rhizomes of *Acorus calamus*, the secretion of gastric juice increases by the reflex way, the bile secretion function of the liver and the gall bladder are strengthened, and diuresis is increased. The extract from this plant causes the loss of pain sensation and lowers arterial pressure. *Acorus calamus* preparations are used to treat stomach and duodenal ulcers (gastritis, enteritis, colitis). Tinctures have the properties of accelerating digestion and appetite. In addition, it emits a pleasant smell, improves the composition of the air and cleans it from various microbes. The original homeland of *Acorus calamus* plant is tropical countries - China, Burma, India and Japan, where it is widely distributed [4-6]. Currently, it is grown by local residents in their backyards and used as a medicinal and ornamental plant. As we mentioned above, the range of application of this plant in folk medicine and scientific medicine is wide, and the demand for it is increasing. It is clear that if the raw materials of the plant are planted and propagated in our country, it will have a great economic effect and medicinal properties will have a good effect on the people living here.

Acorus calamus decorates rivers, lakes, ponds and marshes with a beautiful appearance, and does not allow the unique microclimate of those lands to be disturbed by environmental influences. At the same time, it prevents water pollution and erosion of river and lakeside soil. Today, the poisonous water and gases coming out of cities, factories, and factories are polluting our atmosphere, and are seriously affecting the health of plants, animals, and people. Some infectious diseases are emerging. As a result of planting crops in many areas, the ecological situation is disturbed and the land is filled with salt. Waters are becoming polluted, and the ecological condition of water bodies is deteriorating several times. Nowadays, an international problem, the island problem, has arisen in our Republic, and the ecological situation is getting worse. It is known that the deterioration of the ecological situation has a serious impact on the world of plants, leading to the loss of rare and medicinal plants. In order to prevent this problem, we must follow the laws of our independent Republic and use plants sparingly and wisely. Therefore, planting and breeding of medicinal plants is the demand of the times. We need to use plants properly and expand their cultivation. That is why the development of methods of mass reproduction of ornamental, rare and medicinal plants such as *Acorus calamus*, studying their chemical composition and growing and using them on a large scale remains one of the urgent problems of today. According to the analysis of literature sources, pollen grains in *Acorus calamus* anthers were found to be sterile. It was found that the number of chromosomes is 2p-33.34 in the species of the cotton plant population spread in Armenia, and 60-70% of the pollen grains in the flowers are sterile. A defect in this reduction division caused the formation of anomalous gametes. Aneuploidy has been found to be one of the main reasons why a plant almost does not produce seeds [7, 8]. *Acorus calamus* does not produce seeds even in its natural range. According to the scientific point of view of K. Fegri and L. van der Peil [9], the species in the European populations were created as "unmatched clones".

2 Materials and methods

The object of research is *Acorus calamus* L., a perennial herbaceous plant belonging to the Acoraceae Agardh family, the Acorales ordo, the subclass of the Aridae [4].

The rhizome is creeping, slightly flattened and covered with tiny rhizomes. Leaflets grow from the rhizome, the leaves are linear or lanceolate, 60-120 cm long. The stem grows upright, is unbranched and leafless, tube-shaped on one side, and sharp on the opposite side. A spadix inflorescence develops on the plant. The length of the stem is 4-12 cm, it is covered with small yellow flowers of both sexes. **Perianth** is simple, six-leaved, blooms from May to June. The fruit is a red multi-seeded berry. The seeds usually do not ripen, this plant reproduces only from rhizomes [5, 6, 10, 11].

We used the "Node" method in the propagation of medicinal and ornamental - *Acorus calamus* L. in running water under conditions of introduction. When propagating from the joints of the plant, 1 to 10:5:1 and 12:1:1 ratios of pre-prepared soil, sand and manure (cattle manure) in nutrient mediums prepared at 5g/l of cattle manure and 5g/l of sheep manure and glass plastic trays up to 10 noded rhizomes were planted in 3 rows. In this case, the distance between the pens was 15 cm, and the distance between the rows was 20 cm. In running water conditions, 3 kinds of experiments were conducted. Scientific research work was carried out at the experimental farm of Tashkent State Agrarian University and F.N. Glass plastic trays and artificial ponds in the territory of Rusanov Botanic Garden were completed for three years.

Experiments in flowing water conditions were carried out on plants growing in the place of the Salar canal flowing through the city of Tashkent, flowing through the territory of the "Tashkentkabel" production association, and in the place where the Karasuv canal enters the Burjar canal, which exits from the territory of Sirgali district. The effects of soil and water chemical composition, temperature, nutrient medium, pH index, air temperature and amount of light on plants in the experimental areas were studied and experiments were carried out. Scientific work was carried out based on the recommendations of B.A. Dospekhov [12].

3 Results and discussion

The *Acorus calamus* is propagated in Russia and Western Europe only vegetatively - using rhizome cuttings. The herb does not produce seeds, because these regions are not suitable for it, and its pollinating insects are not found here [13, 14].

Scientific observations were made on young seedlings and older plants. The reproduction of young seedlings was studied under natural and laboratory conditions. In this case, observations were made on young seedlings planted along the Salar and Karasuv canals under natural conditions from April until the end of vegetation. Laboratory conditions during the growth period of the plant in the botanical garden, the beginning of spring growth, active growth, the appearance and shedding of leaves: the growth of the generative branch, budding, flowering and formation of the node and their shedding before fully ripening were monitored during the development period. Vegetative reproduction of the plant. It was carried out in different periods, taking into account different phases of the growing season and the number of joints. We divided the rhizome of *Acorus calamus* into joints and observed how many of them increase in number. In the first experiment, 1-3 new vegetative nodes were produced from each node. The experiment was observed from April to November, until the end of vegetation. During their first year of growth, the number of leaves was 1-6, the old leaves turn yellow and new leaves grow. At the end of the growing season, the height of the plants reached 60-130 cm. During the growth and development of *Acorus calamus*, it was observed that air temperature changed to 13-29-45 C^o, water temperature to 13-30-45 C^o, pH 5.5-7.0.

In some nodes of the *Acorus calamus*, only one stem emerged and grew well until the end of the growing season. When the length of the root is 10-15 cm, it produces 2-3 stems. When we observed the artificially propagated plants the second year, it was observed that 1-3 stems were produced, and then the development of their roots decreased. During the growing season, rotting of old roots and frequent formation of new roots was observed.

No generative organs were produced during the two-year follow-up. In order to study under natural conditions, weed plant was planted on the waterside (Salar canal water). Each bush produced 1-3 stems from the root, as a result, the plant developed well and the leaves grew to 60-130 cm. Old leaves, that is, the first leaves, turn yellow and stop growing. Observation was carried out until the end of vegetation. When the observations were carried out the second year, it was observed that the plant produces stems and starts growing through root nodes from April.

Up to 15-18 vegetative (stem) organs emerged from one rhizome, rotting of the rhizome was not observed. The fescue grew well and developed, even growing to the sides and taking the place of some aquatic plants. In the experiment, we did not allow weeds to grow, we created conditions for the direct sunlight and good development of the plant. We cleaned it of foreign plants.

At present, for the cultivation of plants, it is necessary to protect the plant by a man. In addition, the plant was also planted along the Karasuv canal. It was observed that the planted seedlings were well developed. Cultivated plants (in natural state) did not develop generative organs. Rhizomes prepared for propagation in November without buds, rooted stems; b) stem with side shoots and roots. Rhizomes prepared for propagation in November; c) rhizomes with side shoots, various nodes; g) stems with tip shoots, roots.

As a result of the observations, it was found that the rhizome of the plant in its natural state is well developed. *Acorus calamus* should be propagated in the margins of flowing water areas, by building special ponds. According to J.A. Abdullaev [15], the growth and development of plants planted from Samarkand region to Fergana valley is given. Plant cuttings are planted to a depth of 5-10 cm. After 2 months, the length of the leaves has grown to 60.7-66.8 cm, width 1.32-1.53 cm. The next year, the development of the plant continued longer, and at the end of the growing season, the number of plant leaves was up to 12-15. Summer leaves are reported to be 156.2-172.5 cm long or spring leaves are 32-47 cm long. The flowering process lasted until June. The plant has been found to reproduce only from rhizomes.

Acorus calamus was brought to Tashkent by T. Taubaev from Samarkand region in 1968 and its biology was observed. The flowering of the plant lasted from May to the end of June, and during this period the height of the leaves was up to 120 cm. The plant is mainly propagated by rhizomes [16]. The following are found in the plant communities growing in the Siyob and Koksuv rivers of Samarkand region: *Tlersia oryzoides*, *Phragmites australis*, *Alisma plantago-aquatica*, *Sagittaria trifotia*, *Typha latifolia* are rare. According to its ecological status, it is a hydrophytic plant [17].

It is recognized by many authors that the pollen grains in the anther of *Acorus calamus* are sterile. But the property of reproduction from the rhizome is very high, and its suppression of other useful plants is also reported in the literature. These data are the basis for the development of methods of propagation of *Acorus calamus* using rhizomes and to conclude that its future perspective is a plant. From a biological point of view, there is no doubt that propagation from rhizomes is useful for this species, because this condition allows the rhizome to grow in the soil for a long time, and crops can be propagated under human protection. That's why we aimed to do such things as propagation of rhizomes from nodes, study of their reproduction in different waste water basins, different nutrients, how many nodes are planted, this plant will give good results.

Acorus calamus was grown in April 2019 with 5 joints and 5% cattle feed. The cuttings were cut and broken in 3 rows, 10 pieces per row 40 cm long and 1.7 cm diameter cuttings were planted in portable trenches (the length of the container is 7 m, the width is 45 cm). Observation work started from May to September. The following table shows the biology of the cultivation of cuttings (Table 1).

Table 1. Biological parameters of transplanted rhizome cuttings (cut/broken).

Acorus calamus L.			Rhizome		Leaf		
Planting time	**Planting condition**	**Planting method**	**Length, cm**	**Diameter, cm**	**Length, cm**	**Width, cm**	**Amount, quantity**
2019 IV	Laid down	Whole	10	1.2	20	1.5	5
			8	1.1	17	1.3	4
			7	1	12	1.2	3
2019 IV	Laid down	Whole	9	0.9	19	1.4	4
			7	0.7	16	1.3	3
			6	0.8	18	1.1	5

In the spring, the leaves of the cuttings planted in the following form are 25-40 cm long, and the summer leaves are 60-70 cm long. The number of leaves increased to 8-9 in spring and 10-13 in summer. The temperature of the food medium was 15-35-450, the air temperature was 13-28-450, the light was 400-450 W/m2 FAR, pH 5.5-7. In the process of observation, it was observed that the first leaves stopped developing and formed a new rhizome. Hence, joint formation also produces a new rhizome joint during each leaf shedding process. Table 2 shows the results of determining how much the number of experimental plants increased or decreased in September.

Table 2. Biological characterization of cuttings planted by cutting and breaking rhizomes nodes.

End of experiment	**Dried leaf**	**Saved**		**Rhizome**		**New leaf**		
		New rhizome, quantity	**Bruised plant, quantity**	**Length, cm**	**Diameter, cm**	**Length, cm**	**Width, cm**	**Quantity**
2019 IX	7	3	10	35	2.4	70	1.5	13
	6	1	8	31	2.2	68	1.4	12
	5	2	9	28	2	65	1.1	10
2019 IX	4	2	8	26	2.1	66	1.2	11
	3	1	7	24	1.9	60	1.3	10
	5	-	6	20	1.7	62	1.1	9

From the data presented in Table 2, it can be seen that the number of nodes in the rhizome has increased to 9-13. The best result was observed when the rhizome of the plants was cut. 27 seedlings germinated in cuttings, 90%, and 21 seedlings germinated in broken seedlings, 70%. These observational experiments were carried out in the 2nd year as well. Spring leaves grew to 40-70 cm, summer leaves to 60-105 cm. The observation was continued until the end of vegetation. In this case, the air temperature was 13-29-48 0C, the temperature of the food environment was 13-30-49 0C, the light was 410-440 W/m2 FAR pH 5-7. The results obtained about the growth and development of the plant, how much it multiplied, are given in Table 3 below.

Table 3. Biological description of cut and broken cuttings planted in the second year of vegetation.

End of experiment	Dried leaf, grain	Saved		Rhizomes		New leaf		
		New rhizome drain	Blue-green plant, grain	Length, cm	Diameter, cm	Length, cm	Width, cm	Quantity, grain
2020 IX	11	35	-	15	1.2	104	1.5	14
	10	22	-	11	1.1	105	1.4	12
	9	29	-	10	1.0	95	1.1	10
2020 IX	4	25	-	8	0.8	98	1.1	11
	3	20	-	7	0.7	94	1.0	10
	5	14	-	5	0.6	85	0.9	9

In our experiment, it was observed that the plant rotted old rhizomes and produced new rhizomes with a length of 5-15 cm and a diameter of 0.6-1.2 cm with 14-35 nodes. The number of leaves is 9-14, their length is 85-104 cm, width is 0.9-1.5 cm. During the growing season, the old leaves of the plant die and produce new leaves. It was observed that the air temperature changed up to 14-33-480 C during the vegetation. The number of rhizomes that were cut and planted reached 86, and when they were broken, they reached 59. At the end of the experiment, the bruising of cut rhizomes increased by 409 % compared to the initial state, and the bruising of broken rhizomes increased by 315 %. In March 2021, experimental works were carried out on the basis of 4 repetitions in the Salar and Karasuv canals flowing from Tashkent and in the artificial basins in the Botanical Garden. These seedlings were observed for 3 years. A rhizome with a length of 4-10 cm was selected for propagation. The number of leaves in them was 1-3, and the length was 5-12 cm. Prepared stems were planted to a depth of 5-8 cm and observed until the end of vegetation. Tables 4, 5, 6 below show the results of 3-year follow-up.

Table 4. Biological description of cuttings propagated from rhizomes in the first year.

End of experiment	Dried leaf, pc	Saved		Rhizomes		New leaf		
		New rhizome, drain	Blue-green plant, grain	Length, cm	Diameter, cm	Length, cm	Width, cm	Quantity, grain

2021 X Botanical Garden	6	3	2	20	2.7	110	3.0	16
Salar canal	8	5	3	19	2.8	115	2.4	14
Karasuv canal	7	4	3	18	2.5	106	1.9	12

Table 5. Biological description of cuttings transplanted from rhizome nodes in the second year.

End of experiment	**Dried leaf, pc**	**Saved**		**Rhizome**		**New leaf**		
		New rhizome	**Blue-green plant, grain**	**Length, cm**	**Diameter, cm**	**Length, cm**	**Width, cm**	**Quantity, grain**
1999 X Botanical garden	10	12	5	24	2.8	116	2.8	19
Salar canal	11	15	8	25	3.0	120	3.0	20
Karasuv canal	9	13	7	23	2.4	114	2.6	18

Table 6. Biological description of cuttings transplanted from rhizome nodes in the third year.

End of experiment	**Dried leaf, pc**	**Saved**		**Rhizome**		**New leaf**		
		New rhizome	**Blue-green plant, grain**	**Length, cm**	**Diameter, cm**	**Length, cm**	**Width, cm**	**Quantity, grain**
2000 X Botanical garden	12	31	14	28	2.9	121	2.9	21
Salar canal	14	35	20	34	3.0	120	3.0	24
Karasuv canal	13	32	16	27	2.8	117	2.8	20

When we observe the planted rhizomes until October, the length of the leaves is 110-115-116 cm, the width is 3.0-2.4-1.9 cm, the number is up to 12-14-16, the number of leaves is 6-7-8, and at the end of October the rest of the leaves also stopped developing.

The rhizomes are 20-19-18 cm long, 2.5-2.7-2.8 cm in diameter, of which 2-3-3 plants have grown up to 3-5 new rhizomes from the main rhizome, leaves are 7-9-10 , the length was 25-45 cm. Air temperature changed to 13-32-47°C, water temperature to 12-30-46°C, pH-5-7. Thus, we continued to observe the growth and development of the plant in the 2nd year. Autumn rhizomes begin to produce new leaves to replace them in early spring in March. During the observation, the number of leaves and rhizomes of the plant was studied. At the end of vegetation, leaves are 116-120-114 cm long, 2.8-3-2.6 cm wide, 18-19-20 in number, rhizomes are 2.4-2.8-3 cm in diameter, 24-25-23 in length was up to cm. 5-7-8 bruised rhizomes, 12-13-15 new rhizomes. Leaf fall was 9-10-11. At the end of November, the plants began to go into a dormant period. Air temperature varied from 13-33-450, water temperature to 14-30-47^{0}C, pH-5-7.

Observations were carried out as in previous years. By the third year, the length of the leaves is 121-124-117 cm, the width is 2.9-3-2.8 cm, the number is 20-21-24, the newly formed rhizomes are 14-20-16, the rhizomes formed by sympodial branching are 31 - increased to 32-35. The length of the rhizome is 27-28-34 cm, the diameter is 2.9-3-2.8 cm, the dead leaves are 12-14-13. Thus, the number of Acorus calamus plants in the Salar Canal increased by 4-20 compared to the Botanical Garden and the Karasuv Canal. Stem joints increased to 4-17.

In addition, under the conditions of the Botanical Garden, water was given to the plant and the soil, and its growth was observed. For the experiment, in October, 120 rhizomes with 3-4 joints were planted in 4 rows, with 30 cuttings per row, 10-15 cm long, with 3-4 leaves. Planted cuttings began to grow in early spring. 5-7 new leaves have grown in place of the winter leaves. Spring leaves grew to 25-45 cm, 2-2.5 cm wide, and summer leaves grew to 60-85 cm, 2-2.7 cm wide. The plant grew 12-19-17-12 from 30 plant cuttings on April 12, 17-12-24-14 on April 24, and 17-13-24-15 on May 6 (Table 7). These numbers did not change in the next determination.

Table 7. Biological characterization of rhizome cuttings germination.

When planted, row	**When planted, rhizome**	**12 April**	**24 April**	**6 May**	**Leaf**		**Root**	
					Length, cm	**Width, cm**	**Length, cm**	**Diameter, cm**
1	30	12	17	17	85	1.6	9	1.3
2	30	19	12	13	70	1.5	7	1.1
3	30	17	24	24	73	1.4	8	1
4	30	12	14	15	68	1.2	7	0.8

The plant has 7-10 leaves, and the autumn leaves have died. 3-4 new leaves with a length of 10-14 cm were formed from the rhizome. These leaves protect the plant from cold in winter. Air temperature reached 14-34-49^{0}C, light reached 400-460 W/m^2 FAR.

According to the results of the above experiments, in order to determine the best results if the rhizome of the plant is planted into several nodes, the rhizomes are divided into 10 nodes, i.e. 1 node, 2 nodes, 3 nodes, etc. was grown on sheep and cattle manure in the environment (Table 8).

Table 8. Biological characterization of cuttings transplanted from rhizome joints into nutrient medium prepared from sheep manure.

Nodes	Saved		Rhizome		New leaf		
	When planted	When blue-green	Length, cm	Diameter, cm	Length, cm	Width, cm	Quantity, piece
1	3	-	-	-	-	-	-
2	3	-	-	-	-	-	-
3	3	1	12	1	68	2.1	10
4	3	2	24	1.1	80	2.3	11
5	3	2	25	1.3	84	2.5	13
6	3	3	27	1.4	88	2.7	19
7	3	3	31	2.4	96	2.9	22
8	3	3	32	2.5	97	2.9	24
9	3	3	35	2.7	103	3	25
10	3	3	36	2.8	108	3	26

The rhizomes of the plant with 1-2 nodes are not blue-green, the remaining joints have up to 1-2-2-3-3-3-3-3, the length of the leaves is up to 68-80-84-88-96-97-103-108 cm, the number of leaves up to 10-11-13-19-22-24-25-26, rhizome length 12-24-25-27-31-32-35-36 cm, width 1-1,1-1,3-1, It was 4-2.4-2.5-2.7-2.8 cm.

1-2-3 noded cuttings planted on cattle manure medium did not germinate. The cuttings in the next joints are blued up to 1-2-3-3-3-3-3. The number of leaves is 9-11-11-13-15-22-23 pieces, the length of the rhizome is 10-15-16-20-22-24-29 cm, the diameter is 1-1,2-1,4-2-2, It reached 1-2.4-2.7 cm. Plant growth was faster on the sheep manure medium than on the cattle manure medium (Table 9).

Table 9. Biological characterization of cuttings transplanted from rhizome joints on nutrient medium prepared from cattle manure.

Nodes	Saved		Rhizome		New leaf		
	When planted	When blue-green	Length, cm	Diameter, cm	Length, cm	Width, cm	Quantity, piece
1	3	-	-	-	-	-	-
2	3	-	-	-	-	-	-
3	3	-	-	-	-	-	-
4	3	1	10	1	65	2	9
5	3	2	15	1.2	67	2.2	11
6	3	3	16	1.4	69	2.4	11
7	3	3	20	2	73	2.6	13
8	3	3	22	2.1	78	2.8	15
9	3	3	24	2.4	85	2.9	22
10	3	3	29	2.7	94	3	23

When analyzing the composition of manure chemically, it was found that N_2-0.55, K_2O-0.15, P_2O_5-0.31 in sheep manure, N_2-0.29, K_2O-0.15, P_2O_5-0.17 in cattle manure (Table 10). So, sheep manure has much higher biogenic elements.

Table 10. Chemical analysis of cattle and sheep manure, %.

Animal manure	Composition of substances					
	N_2	K_2O	CaO	Mg	P_2O_5	H_2SO_3
Cattle	0.29	0.10	0.35	0.13	0.17	0.04
Sheep	0.55	0.15	0.46	0.15	0.31	0.14

4 Conclusions

Methods of propagation of *Acorus calamus* L. by vegetative means were developed for the first time in Uzbekistan under conditions of introduction.

Biological methods of growing *Acorus calamus* have been developed in running water ponds and propagation from joints in laboratory conditions. Based on the set of the most important indicators of biological properties, the scientific basis of the "Node" method was created, and promising agrotechnical methods were developed based on experiments. In the process of studying the reproduction of plant joints, it was found that it is possible to grow cuttings with 7-10 nodes in both nutrient media. The cuttings of the plant are fully green, and the development of the rhizome and leaves is high. It was determined that it is appropriate to plant the plants in 7-10 nodes in different reservoirs. Planted rhizome cuttings were found to grow intensively even at 13-18°C. Methods of propagation of *Acorus calamus* by vegetative means were developed for the first time in Uzbekistan under conditions of introduction.

Plants were first observed to grow faster in sheep manure media than in cattle manure. It was found that the productivity of plants grown in certain environments is high. It was determined that the demand can be met as a result of planting. It has been proven that it can be grown on an industrial scale depending on the demand. The results of the research can be used in lectures and practical training in higher educational institutions.

References

1. Abu Ali, Ibn Sina. (2016). Tib qonunlari. (3-10), 16-213.
2. Kholmatov, H.Kh., & Kasimov, A.I. (2014). Medicinal plants, 90. Khalq Press.
3. Kholmatov, H.Kh., & Ahmedov, O'.A. (2017). Pharmacognosy, 200-265.
4. Pratov, O.P., & Nabiyev, M.M. (2007). Modern system of higher plants of Uzbekistan, 46.
5. Takhtadjian, A.A. (2006). Sistema i phylogeny tsvetkovykh rasteniy, 651.
6. Takhtadjian, A.A. (2007). Sistema Magnoliofitov, 341-343.
7. Oganezova, G.G. & Barsegyan, N.A. (2001). Osobennosti generativnoy sfery aira bolotnogo (Acorus calamus L.) iz armyanskikh population vida. Flora rastit. res. Armenia, 39-41.
8. Atlas lekarstvennykh rastenii USSR. (2002). Medgiz, 701.
9. Fegri, K. van der. Paul, L. (2012). Osnovy ekologii opylenia, 377.
10. Arabova, N.Z. (2022). Morphobiology of some plants with rhizomes, Bulletin of Science and Education, 127, 13-15.
11. Arabova, N., Abdurasulov, Sh., Ismoilov, Kh, & Pulatova, M. (2021). Flowering biology of Acorus calamus L. growing in Tashkent. Int. J. of Aquatic Science, 12(3), 2900-2907.
12. Dospekhov, B.A. (2009). Methodology polevogo opyta, 208-209.
13. Kondratenko, P.T., Kur, S.D., & Rojko, F.M. (2005). Zagotovka, vyrashchivanie i obrabotka lekarstvennyx rasteniy, 35-36.
14. Korkhov, V.V., Mats, M.N., & Khamidov M.Kh. (2007). Herbal preparations and obstetrics and gynecology, 127.
15. Abdullaev, Dj.A. (2002). Rational use of water and wetland plants in the Fergana valley and national economy, 128-132.
16. Taubaev, T. (2010). Flora i rastitelnost vodemov Sredney Azii, 305-306.
17. Rychin, Yu.V. (2008). Flora hygrophytov, 118.

Morpho-biological features of growth and development of golden currant in Tashkent province, Uzbekistan

Durdona Zhanakova[1,*] and *Ulugbek* Uzakbergenov[1]

[1]Tashkent State Agrarian University, 2, University street, Tashkent, 100140, Uzbekistan

Abstract. This scientific article presents the results of a study on the growth and development of varieties of golden currant in order to identify the most promising for cultivation in the conditions of the Tashkent region. Studies have established that, according to the development of shoots in the above-ground part, the varieties Uzbekistanskaya large-fruited and Elixir differ in vigorous growth, in which the height of the bushes by the end of the second year of vegetation reached 109.7-112.5, and in other varieties this figure was in the range of 88.2-97.5 cm. Of the tested varieties of golden currant, the earliest onset of the flowering phase was observed in the Venera variety - March 23, in the rest a little later. The variety of golden currant Rukhshona had the latest start of flowering, in which this phase occurred on March 26, that is, four days later. In the Uzbekistan large-fruited variety, the completion of the flowering phase occurred five days later than in the Uzbek sweet variety. Of the varieties of golden currant, the phase of complete set of berries was observed most intensively within 30 days in the variety Venera. In the Elixir variety, the full set of berries ended two days later, and in the rest within 34-36 days.
Keywords. Berry, growth, development, shoots, flowering, set, variety, assortment, buds, bush, aerial part.

1 Introduction

Golden currant (*Ribes aureum*) holds significant botanical and horticultural importance due to its ornamental value, high nutritional content, and potential pharmaceutical applications. As a deciduous shrub native to North America, golden currant has garnered increasing attention in recent years for its diverse attributes, making it a subject of interest for researchers and horticulturists alike [1, 2].

The growth and development of golden currant encompass a multitude of morpho-biological aspects, influenced by various intrinsic and extrinsic factors [3]. Understanding the intricacies of its growth patterns, vegetative and reproductive development, as well as environmental responses, is essential for optimizing cultivation practices, enhancing its productivity, and exploring its broader potential in various industries [4-6].

* Corresponding author: d.janakova@mail.ru

This manuscript aims to provide a comprehensive exploration of the morpho-biological features of golden currant. By delving into the mechanisms governing its growth and development, we seek to shed light on the underlying physiological processes, responses to environmental stimuli, and genetic factors that govern its unique characteristics [7, 8]. Moreover, we endeavor to establish the implications of these insights for horticultural practices, sustainable agriculture, and potential applications in functional foods and medicine.

Golden currant (*Ribes aureum L.*) is a valuable fruit crop with a high quality source of biologically active substances. Regular consumption of golden currant berries by a person serves as a prevention of metabolic disorders, sclerosis, diseases of the cardiovascular system and respiratory viral infections of the respiratory system [9, 10]. Currant berries in their composition contain various organic acids, 8.16-12.29% sugar, 0.73 mg% provitamin A, pectin - 2.84%, mineral salts, tannins and many vitamins that have a beneficial effect on human life [11].

Recently, special attention has been paid to the development of fruit growing in the republic, increasing the range and volume of exports, and the gradual transfer of low-efficient extensive orchards to intensive berry orchards [12, 13]. In this direction, the range of berry crops in the republic is expanding in order to create currant berry plantations.

2 Materials and methods

The experiments were carried out according to the methodology developed in the Department of Fruit and Vegetable Growing and Viticulture. Research on the chosen topic was carried out at the training and experimental station of the "Information and Consulting Center" (Extension center) at the Tashkent State Agrarian University.

As an object of study, five varieties of currants were used - Venera, Rukhshona, Uzbekistan large-fruited, Elixir and Uzbek sweet. Varieties of golden currant on the experimental plot were planted according to the scheme of 2.5x1.0 meters [14, 15].

During the study period, the growth and development of plants were controlled by phenological observations of their passage through the main phases of vegetation, as well as by biometric measurements of the growth of the main structural parts of the aerial part. Experiments on the variety study of golden currant were carried out using the "Program and methodology for the variety study of fruit, berry and nut crops" and "Programs and methods for breeding fruit, berry and nut crops", "Method of calculations and phenological observations in experiments with fruit and berry plants", and "Methodology of accounting and observations in experiments with fruit and berry crops" [3, 5, 7].

3 Results and discussion

Currant is a perennial shrub, and this year it began to bloom in the third decade of March. In early spring, pure nitrogen and phosphorus fertilizers are applied under the bushes, row-spacing is plowed to a depth of 25-30 cm, the area around the bush is softened and should be watered during the flowering period.

Golden currant blooms for 10-30 days in April. The flowers are formed in dense, curved, cone-shaped inflorescences. The fruits are round, of different sizes, their length is preserved. The color of the fruit can be orange, brown, and black. The fruits do not ripen at the same time, ripe fruits remain on the bush for a long time, and some varieties do not ripen until August. The flowers are small, tubular, golden yellow, with a strong spicy smell reminiscent of cinnamon [2, 4, 6].

Studies have shown that the intensity of flowering and the setting of berries are directly dependent on the varietal characteristics of plants. Of the varieties of golden currant, the phase of the beginning of flowering in the variety Venera, compared with other varieties of the collection, occurred on March 23, that is, one or two days earlier than in other varieties.

The intensity of the flowering phase was also observed in correlation with the varietal characteristics of plants. For example, in the Venera variety, the mass flowering phase of 75% occurred on April 10, in the Uzbek sweet variety on April 20, that is, 10 days later, which is five days in comparison with the Uzbek large-fruited variant (Table 1).

Table 1. Intensity of flowering varieties of golden currant, 2022.

Varieties	Beginning of flowering	Bloom intensity					
		after bud break, days	25% color, date	after bud break, days	50% color, date	after bud break, days	75% color, date
Uzbekistan large-fruited	24/III	8	30/III	16	7/IV	24	15/IV
Venera	23/III	8	26/III	15	2/IV	23	10/IV
Uzbek sweet	25/III	8	30/III	19	10/IV	29	20/IV
Rukhshona	26/III	7	29/III	17	7/IV	27	17/IV
Elixir	25/III	9	28/III	19	7/IV	27	15/IV

The set of golden currant berries is one of the main factors in the productivity of varieties. Golden currant is practically a self-fertile plant (on average no more than 15%), and in order to set berries, cross-pollination is necessary. That is why, in order to obtain a good harvest, it is inapplicable to grow several varieties at once. When self-pollinated, the ovaries fall off.

From the studied varieties of golden currant in the conditions of the Tashkent region The earliest and most intense berry set was observed in the Venera variety, where the berry set phase ended by April 18. And in the Uzbek sweet variety, a later setting of berries was observed. In addition, they had a longer period of berry infestation, which ended on April 25, that is, seven days longer than that of the Venera variety (Table 2).

Table 2. Intensity of the set of berries of golden currant varieties, 2022.

Varieties	Beginning of flowering	Berry set					
		From the beginning of flowering, days	25% berry set, date	From the beginning of flowering, days	50% berry set, date	From the beginning of flowering, days	75% berry set, date
Uzbekistan large-fruited	24/III	25	18/IV	27	20/IV	31	24/IV

Venera	23/III	18	10/IV	23	15/IV	26	18/IV
Uzbek sweet	25/III	21	15/IV	26	20/IV	31	25/IV
Rukhshona	26/III	22	17/IV	26	20/IV	28	22/IV
Elixir	25/III	23	17/IV	24	18/IV	26	20/IV

Golden currant is a powerful upright tall, deciduous shrub, up to 2-3 meters tall. It has a rounded, beautiful crown. Plants can have different shapes: compact, slightly spreading and sprawling, and therefore it is often formed in the form of a small shrub-tree. In this case, the plant is well lit.

A bush of golden currant is formed from numerous basal shoots of renewal, which are formed from underground adventitious buds on a long rhizome and consists of 15-20 branches of different ages. From adventitious buds on the roots, shoots are formed. The shoots of the golden currant are thick, straight, without thorns. In terms of growth strength and longevity of branches (8-10 years), it significantly exceeds black and red currants.

Uzbekistan large-fruited varieties stood out , in which this growth parameter reached 117.5 and 119.7 cm. The intermediate position was occupied by the varieties Venera (97.5 cm) and Rukhshon (88.2 cm). C ort Uzbekskaya sweet was less tall with a height (85.5 cm) of the aerial part (Table 3).

Table 3. Growth of the vegetative part of plants of golden currant varieties, 2022.

Varieties	**The height of the main shoot, cm**	**The number of side shoots in the bush, pcs**	**Total length of lateral shoots, cm**	**Number of internodes on the central shoot, pcs**	**The length of the internodes of the central shoot, cm**
Uzbekistan large-fruited	119.7	9.5	516.3	12.0	5.0
Venera	97.5	9.0	465.0	10.5	3.4
Uzbek sweet	85.5	9.5	462.5	9.5	3.0
Rukhshona	88.2	7.25	132.3	9.5	4.9
Elixir	117.5	10.25	547.5	9.5	3.5

4 Conclusions

1. According to the development of shoots in the aerial part, variety an Uzbekistan large-fruited and Elixir, in which the height of the bushes by the end of the second year of vegetation reached 119.7-117.5 cm.

2. Of the tested varieties of golden currant, the earliest onset of the flowering phase was observed in the variety Venera - March 23.

3. The phase of mass flowering (75%), as well as its beginning, occurred the earliest and most intensively in the Venera variety, 10 days later in the Uzbek sweet variety.

4. V grade an Uzbekistan large-fruited, the completion of the phase of mass flowering (75%) occurred five days later than in the variety Uzbekskaya sweet.

5. Of the varieties of golden currant, the phase of complete set of berries was observed most intensively within 30 days in the variety Venera.

References

1. Aladina O.N. "Smorodina" - Moscow, Ed. “Niola -press” 2007.
2. Abdullaev R., Yagudina S. "Home garden berries", textbook. Tashkent, 2008.
3. Glebova E.I., Mandrykina V.I. Currant. Moscow: Rosselkhozizdat , 2014.
4. Burmistrov A.D. Berry crops. - L., 2012. - Head of Currant.
5. Sergeeva K.D. Features of variety study of currants and gooseberry.//Sat. scientific works/VNIIS , 2014. – vol. 19, 12-17.
6. Program and methodology for the study of fruit, berry and nut crops. - Eagle, 2019.
7. Jumaev, R., Gazibekov, A., Sulaymonov, O., & Sobirov, B. (2021). Representatives of Lepidoptera groups occurred in forestry and agricultural crops and their effective entomophage types. In E3S Web of Conferences (Vol. 244, p. 02020). EDP Sciences. DOI:10.1051/e3sconf/202124402020
8. Lebedeva, N., Akhmedova, Z., Kholmatov, B., & Jumaev, R. (2021). Revision of stoneflies (insecta: plecoptera) fauna in Uzbekistan. In E3S Web of Conferences (Vol. 258, p. 08030). EDP Sciences. DOI:10.1051/e3sconf/202125808030
9. Jumaev, R. (2023). Invitro rearing of parasitoids. In E3S Web of Conferences (Vol. 371, p. 01032). EDP Sciences. DOI:10.1051/e3sconf/202337101032
10. Tayjanov, K., Khojimatov, O., Gafforov, Y., Makhkamov, T., Normakhamatov, N., & Bussmann, R. W. (2021). Plants and fungi in the ethnomedicine of the medieval East-a review. Ethnobotany Research and Applications, 22, 1-20. DOI:0.32859/ERA.22.46.1-20
11. Jabeen, S., Zafar, M., Ahmad, M., Althobaiti, A. T., Ozdemir, F. A., Kutlu, M. A., ... & Majeed, S. (2023). Ultra-sculpturing of seed morphotypes in selected species of genus Salvia L. and their taxonomic significance. Plant Biology, 25(1), 96-106. DOI:10.1111/plb.13473
12. Majeed, S., Ahmad, M., Ozdemir, F. A., Demirpolat, A., Şahan, Z., Makhkamov, T., ... & Nabila. (2023). Micromorphological characterization of seeds of dicot angiosperms from the Thal desert (Pakistan). Plant Biosystems-An International Journal Dealing with all Aspects of Plant Biology, 157(2), 392-418. DOI:10.1080/11263504.2023.2165553
13. Majeed, S., Ahmad, M., Ali, A., Althobaiti, A. T., Ramadan, M. F., Kilic, O., ... & Sultana, S. (2023). Pollen micromorphology among amaranthaceous species from desert rangeland: Exine stratification and their taxonomic significance. BioMed Research International, 2023, 4967771. DOI:10.1155/2023/4967771
14. Makhkamov, T., Sotiboldiyeva, D., Mamarakhimov, O., Yuldashov, Y., & Botirova, L. (2022, May). Morphogenesis and Seasonal Developmental Rhythm Under the Conditions of Introduction of Curcuma Longa L. In International Scientific Conference on Agricultural Machinery Industry “Interagromash"” (pp. 1460-1469). Cham: Springer International Publishing. DOI:10.1007/978-3-031-21432-5_155
15. Boboev, S., Makhkamov, T., Bussmann, R. W., Zafar, M., & Yuldashev, A. (2023). Anatomical and phytochemical studies and ethnomedicinal uses of Colchicum

autumnale L. Ethnobotany Research and Applications, 25, 1-9. DOI:10.32859/era.25.6.1-9

State of the management mechanism of fruit and vegetable cluster in Uzbekistan

Uchkun Beglaev[1,*], *Suyun* Khalikov[1], *Jahangir* Rozikov[1] and *Alisher* Turdibaev[1]

[1]Tashkent State Agrarian University, 2, University street, Tashkent, 100140, Uzbekistan

Abstract. This article provides a comprehensive overview of scientific research and information concerning the development of fruit and vegetable clusters. It delves into the evolution of management structures within these clusters, considering their implications on the country's economic situation. The research also explores the overall development of the national economy and the efforts made to create a stable economic system. The cited scientific research works shed light on the dynamic growth and evolution of fruit and vegetable clusters, highlighting their importance in the economic landscape. These clusters play a crucial role in the production, processing, and distribution of fruits and vegetables, contributing significantly to the country's agricultural and economic sectors. Moreover, the article discusses how the establishment and proper management of these clusters can enhance economic growth, create job opportunities, and stimulate rural development. The research cited in the article underscores the significance of effective management structures in optimizing the performance of these clusters and maximizing their economic potential. Additionally, the article touches on the broader context of the country's economic situation and the efforts undertaken to create a stable and sustainable economic system. The development of fruit and vegetable clusters is viewed as an integral part of this economic strategy, as it fosters a diversified and robust agricultural sector, ultimately contributing to the country's overall economic stability and prosperity.

Keywords. Cluster, management, structure, system, method, digitization, network, development.

1 Introduction

As a result of the proper organization of the management method in clusters, its criteria and structure, the development of the field of storage of fruit and vegetable products, the construction of modern storage warehouses, the wide use of modern techniques and technologies in their equipment, the prevention of perishability of the grown products and the reduction of quality deterioration to farms and the service sector companies will be able to make a large amount of profit [1].

Today, farmers are forced to sell their crops in the field due to the use of various agro-technologies in order to increase the amount of fruit and vegetable products grown in farms and peasant farms. First of all , the relatively low prices of products will lead to low incomes

* Corresponding author: u.beglaev@tdau.uz

of farmers and peasants; secondly, fruits and vegetables are mainly calculated according to their size, not enough attention is paid to their quality indicators [2, 3].

The economic scientists of our country and foreign countries have been conducting scientific and research work in various directions on the problems of improving and developing the scientific and theoretical methodology of fruit and vegetable industry development, the mechanism of effective organization and management of fruit and vegetable clusters in the conditions of market economy. In particular, foreign scientists: ME Porter, JH Trienekens, RA Stanley, J. Mutura, N. Nyairo, M. Mwangi and S. Wambugu, M. Mehdi, B. Ahmad, A. Yaseen, A. Adeel, N. Sayyed [4-9] reflected in the concepts, theories and practical research of scientists on these issues.

E.B. Falkovich, D.A. Kotlyarov, T.V. Tsikhan, A.S. Khukhrin, O.A. Anichkina, G.B. Kleiner, R.M. Kachalov, V.V. Mitenev from scientists of the CIS countries [10-13], O.B. Kirik, N.N. Balashova S.A. Ernst, G.L. Azoev, V.N. Ivanov, V.V. Paltsev, Z.N. Burjalieva, Sh. Rakhimi, A.Yu. Sachkova, R.A. Karimov [14-19] and others conducted scientific research in the development and management of fruit and vegetable clusters.

Therefore, buyers who buy products try to deliver to consumers in a short period of time. For this, it is necessary to pay attention to the financial capabilities of fruit and vegetable clusters, social and political position, system participants, large and small volumes of production, regional conditions, the share of modern technologies in production, the role of the cluster in free market competition.

2 Materials and methods

Mathematical modeling, economic analysis, SWOT analysis, systematic and comparative analysis, integrated approach, questionnaire survey, systematic approach, comparison, classification methods were widely used in the research process.

Clusters established in the field of fruit and vegetables are a structure based on horizontal and vertical cooperation of legally independent enterprises and organizations. This structure includes producers of agricultural products, enterprises that store and process cultivated crops, suppliers of seeds and chemical protection agents and fertilizers to agriculture, manufacturers, suppliers, technology creators and suppliers of machines and mechanisms used in these fields, various infrastructure departments, financial organizations, research institutions, etc. will exist. Clusters in the agrarian sector can be large in terms of size and can be organized at the district or regional level according to the territory they occupy. The mechanism for organizing clusters in the fruit and vegetable sector should be as follows. (Figure 1)

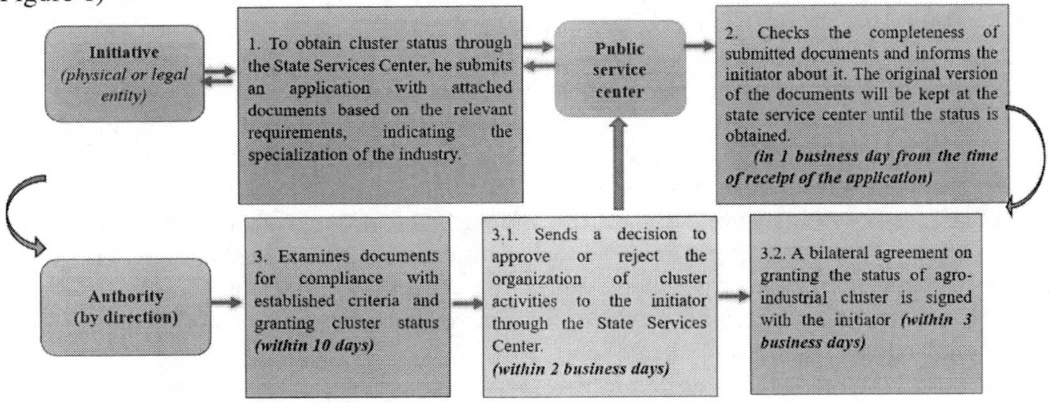

Figure 1. Mechanism of organization of fruit and vegetable clusters.

The mechanism of organization of fruit and vegetable clusters, special attention should be paid to the main criteria in the organization of clusters (Figure 2). By this means, taking into account all aspects of the organized fruit and vegetable cluster, it is possible to perform the most important tasks and to determine the causes of the problems.

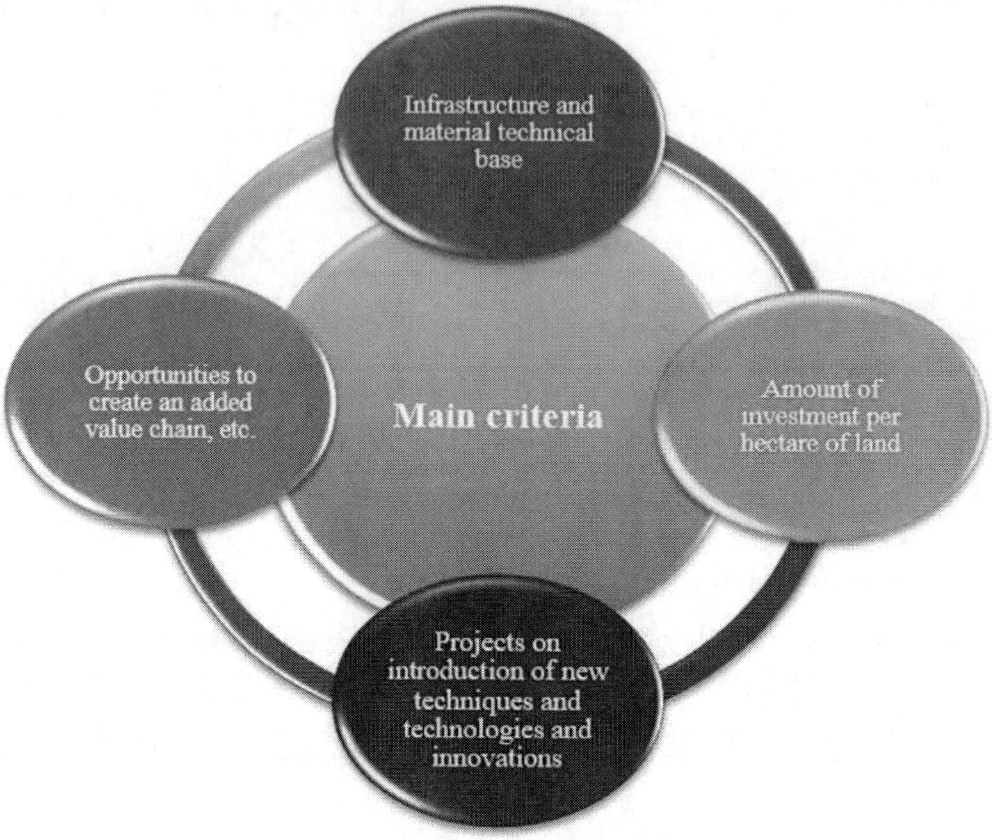

Figure 2. Main criteria for the establishment of fruit and vegetable clusters.

Fruit and vegetable clusters will work towards a common goal, and the main goal will be to achieve the efficiency of the sector and increase the export potential. However, the entities in the existing structure must have their own purpose. An important task in the organization of clusters is to effectively connect the goal of each subject with the overall goal of the cluster. Otherwise, each organization strives to realize its own goal, and because the common goal of the cluster is not realized, inefficient activities are carried out.

At the same time, it will not be possible to find some types of fruit and vegetable products in certain periods, and as a result of the increase in demand, the level of shortage will occur. It is observed that the import of chemical preparations from abroad and the preservation of seeds necessary for the cultivation of fruit and vegetable products from various diseases and the export of fruits and vegetables lead to a sharp decrease. From this point of view, there is a need to modernize the organizational structure of fruit and vegetable clusters, to maintain and process fruit and vegetable products, to achieve their constant availability throughout the year and to increase the volume of export.

Today, the establishment of an effective management system in agriculture, along with the development of clusters, completely eliminates the existing shortcomings in the field. Currently, cluster cultivation of fruits and vegetables is showing its positive results.

Significant work has been done on introducing resource-saving techniques and technologies into agricultural production, bringing industry to rural areas and creating new jobs, producing finished products with high added value, and developing infrastructure.

At the same time, effective mechanisms have not been created to coordinate the activities of fruit and vegetable clusters, eliminate systemic problems arising in their activities, and protect their interests in state management bodies.

To further increase the efficiency of the activity of fruit and vegetable clusters and further improve the management mechanism, to increase the processing capacity of fruit and vegetable products, to widely introduce scientifically based methods and intensive technologies to this process, as well as to eliminate the problems arising in the activity of fruit and vegetable clusters in a timely manner, indicates that a number of reforms should be implemented in order to establish a system of protection of their interests in the administrative bodies of state power.

3 Results and discussion

In our research work, we have developed scientifically based proposals on the organization of the "Uzbekistan Fruit and Vegetable Clusters" association of organizers of fruit and vegetable clusters and its organizational management structure (Figure 3).

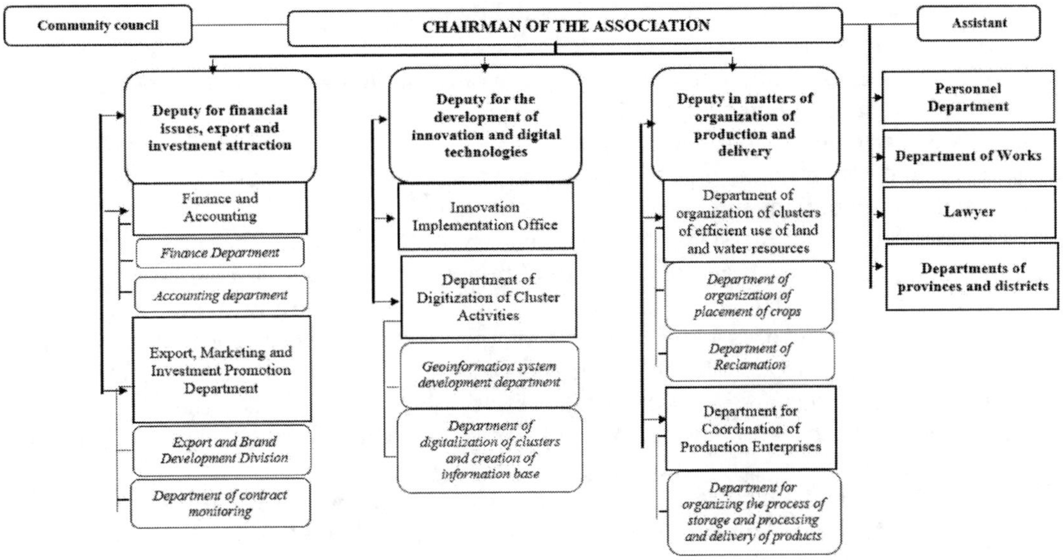

Figure 3. Management structure of the association "Fruit and vegetable clusters of Uzbekistan".

Proposed *"fruit and vegetable clusters of Uzbekistan"* The main purpose of the organization of the association: elimination and management of systemic problems in the field, development of proposals for the active introduction of marketing innovations, as well as effective functioning of fruit and vegetable clusters, achieving high final results, setting specific tasks and constant regulation of their scattered activities, Harmonization of interests of cluster participants, creation of financial opportunities for clusters connected to the system, and implementation of economic relations between the cluster and the state on the basis of modern mechanisms are considered.

It is planned to control and create conditions for the main product-producing clusters, as well as to rapidly develop the industry and unify it in one system.

The following were defined as the main tasks and activities of the Association:

- coordinating the activities of fruit and vegetable clusters and the implementation of projects by them, protecting their rights and interests, organizing effective cooperation with state bodies and organizations and economic associations;
- support the placement of fruit and vegetable clusters based on the prospects of further development of the potential of the regions, raw material base, existing infrastructure, labor resources and market conditions;
- introducing advanced agrotechnologies and resource-saving techniques and technologies into the process of growing fruit and vegetable products, as well as creating a transparent mechanism for regulating relations between fruit and vegetable clusters and farms;
- production and export of ready-made products with high added value, expansion of activities of fruit and vegetable clusters, and implementation of comprehensive measures for wide implementation of digital technologies in these processes;
- to develop proposals for the application of unified benefits and preferences for the development of fruit and vegetable clusters and the development of agrotourism as an additional sector;
- organization of marketing research in the domestic and foreign markets, participation in the development and implementation of medium and long-term programs of network development, as well as wide introduction of market mechanisms to the network;
- modernizing fruit and vegetable production, technical re-equipment, helping fruit and vegetable clusters to attract investments, including foreign direct investment, in introducing high-performance innovative technologies;
- identifying and eliminating systemic problems that hinder the rapid development and stable operation of fruit and vegetable production, as well as preparing proposals for further improvement of legislation related to the sector.

It would be desirable to unite the union and fruit and vegetable clusters in one area, study their main problems, and eliminate them.

At the same time, the state of cluster management principles, the causes of the main problems in management, and the ways to eliminate them require a separate study. Today, when studying the management structure and management methods of fruit and vegetable clusters, the absence of a clear structure in the clusters, the presence of a number of shortcomings in the application of management methods, shows that special attention should be paid to the management system of the sector. As a result of the introduction of the management structure of the association and vegetable clusters, the mechanism of development of the sector will have a collective character.

One of the main and most important stages of planning, which is considered one of the important functions in the management of fruit and vegetable clusters, is the selection of the objectives of the clusters. It is important to note that clusters have a wide range of purposes in multi-link systems.

the fruit and vegetable cluster is defined in its task. The task describes the situation of the cluster in detail and defines ways to achieve goals and strategies at each organizational stage.

Therefore, the task of the fruit and vegetable cluster should be expressed as follows:

- in terms of main services and products, markets and technologies;
- from the ratio of the external environment, which determines the working principles of the entities in the structure of clusters, to the enterprises in the structure ;
- the culture of organizations in clusters , their internal activities.

Clusters are structured according to their goals. It is impossible to create a structure that is the same for everyone. Each cluster constructs a structure based on its capacity, and each department and structural units in the structure contribute to the realization of the cluster's goal. At the same time, new structural units can be established that can contribute to the purpose of the clusters in cases where it is necessary to ensure the effective functioning of the clusters (Figure 4).

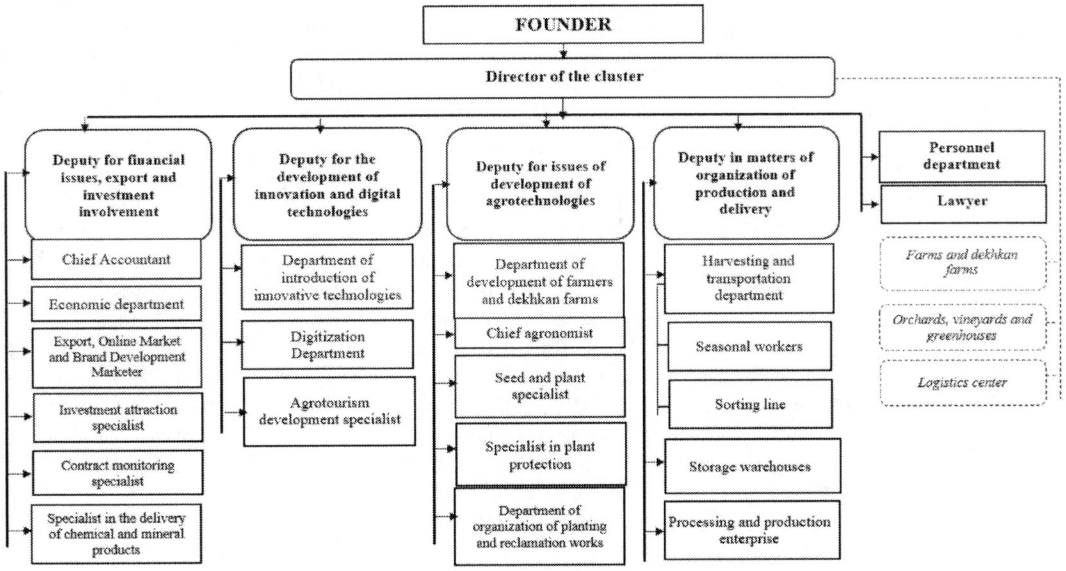

Figure 4. Proposed governance structure of fruit and vegetable cluster.

Conditions are created for elimination of clusters independently or in cooperation with the association, taking into account all the factors affecting it. When improving the management mechanism of fruit and vegetable clusters, it is necessary to pay special attention to the organizational and economic mechanism of clusters. Therefore, it is desirable to develop organizational and economic mechanisms of clusters in the following directions:

- ✓ identify all influencing factors and determine their scope;
- ✓ planning the activities of clusters for the current and long term, establishing the most proportionate proportions and relationships between all participants. To achieve their full economic and legal independence;
- ✓ to establish the correct mutual calculations between units and organizations that are part of the cluster. In this case, not to allow their rights to be violated, but rather to take measures to protect and support them;
- ✓ establishing mutually efficient distribution relationships to ensure the development of clusters in the fruit and vegetable sector. taking into account only the cluster development prospects when distributing the received funds. At the same time, taking into account the interests of economic entities within the cluster;
- ✓ proper establishment of contractual relations between entities in the fruit and vegetable cluster. Taking into account the contribution and level of development of each party.

In addition, it is necessary to pay special attention to the organizational, economic and social aspects of the management of fruit and vegetable clusters (Table 1). In order to do this, it is necessary to create a single digitized platform for clusters, through which to analyze the activities of clusters, to introduce the sale and purchase of products, delivery service, and to ensure that cooperation relations are integrated into one system.

Table 1. Peculiar aspects of the organizational and economic basis of fruit and vegetable cluster management.

ORGANIZATIONAL	*ECONOMIC*	*SOCIAL*
- to fully satisfy the population's demand for quality fruit and vegetable products; - production of various products as a result of fruit and vegetable processing and increasing the volume of export in addition to domestic consumption; - creation of new jobs by establishing industrial enterprises in rural areas and providing employment to the unemployed population; - complete regulation of economic and social relations between entities operating in the fruit and vegetable sector; - creation of a value chain in the cultivation of fruit and vegetable products and extensive use of modern management methods;	- creating an opportunity to increase economic efficiency indicators of enterprises operating in the fruit and vegetable sector; - to establish a wide use of resource-saving technologies in the production of products; - by organizing the processing of fruit and vegetable products, to achieve a profit of more than 15 times compared to wet sales; - the possibility of correct organization of monetary policy with banks and other financial organizations and timely repayment of creditor debts; -in order to increase the income of the population and establish a financial incentive mechanism; -increasing foreign exchange earnings as a result of exporting processed products, etc.	- production of products that fully meet the sanitary requirements of the population; - organization of agrotourism zones in rural areas and radical change of living standards of the population; - financial stimulation of the vulnerable strata of the population through social support, etc.

Therefore, it is ensured that each cluster's goal setting serves to complement the overall goal. If we analyze the production, storage, processing and economic indicators of fruit and vegetable clusters and farms, we pay special attention to the high share of clusters, as well as the provision of a continuous work process (Table 2).

Table 2. Analysis of economic indicators of fruit and vegetable cluster and farms operating in Bukhara region, Uzbekistan.

#	**Indicators**	**Unit of measure**	**Product cultivation**		**Difference (-/+)**
			Fruit and vegetable cluster of "Emir Fruit Industry" LLC	**"Kelajak Zamini" farm**	
1	Total harvested land area	*(ha)*	70	13	57
2	*Fruit growing*	*(ha/t)*	15.5	12.3	3.2
3	*Growing vegetables*	*(1/ha ton)*	17.6	14.7	2.9
4	Product storage capacity	*(tons)*	600	0	600
5	Product processing capacity	*(tons)*	1000	0	1000
6	Product sold wet *(average selling price per kg (fruits 10,000) (vegetables 3,000) 13,000 soums.)*	*'000 UZS*	445000	112500	332500
7	As a result of processing	*'000 UZS*	1356000	0	1356000

	and selling *(1/kg selling price on average 30,000)*				
8	Jobs created	*(number)*	43	8	35
9	The number of techniques	*(piece)*	45	2	43
10	Electricity consumption	*'000 UZS*	36000	600	35400
11	Mineral fertilizer purchase costs	*'000 UZS*	77000	8000	69000
12	Costs of purchase of chemicals	*'000 UZS*	68000	4500	63500
13	Costs of fuel lubricants	*'000 UZS*	80000	4500	75500
14	Labor costs	*'000 UZS*	1032000	76000	956000
15	Depreciation	*'000 UZS*	4700	300	4400
16	Transportation costs	*'000 UZS*	30000	100	29900
17	Other expenses	*'000 UZS*	45000	1.000	44000
18	Marketing and modernization	*'000 UZS*	17000	550	16450
19	**Total cost**	***'000 UZS***	**1299.0**	**95550**	**1203450**
20	Income	*'000 UZS*	1801000	112500	1688500
21	Net profit	*'000 UZS*	502.000	16.95	485050
22	**Profitability**	*(%)*	**38.6**	**17.7**	**20.9**

It is known that production in farms has a clearly expressed seasonal character, in which the work period does not correspond to the production period due to the uneven use of labor resources throughout the year, sometimes there is a shortage of labor, and at other times there is a surplus. In particular, fruit growing and viticulture, which require less labor during the growing season and lack it during the harvest season, suffer more. A similar situation is observed in vegetable growing, where the need for labor force and changes in its use are relatively small due to the use of methods of planting crops in one row and alternating planting, and the use of technologies for growing vegetables under film.

If we look at the comparative analysis between the cluster and farms, it can be seen that the net profit of the cluster is 40 times more than the net profit of the farm and the profitability index is 20.9% higher. In addition, the creation of more than 100 jobs on the basis of clusters is of particular importance.

When organizing mutual relations in fruit and vegetable clusters, it is necessary to take into account the situation and development of each participant. To achieve this state, the following directions should be implemented:

— assessment of the state of development of the regional fruit and vegetable network and organization taking into account the tradition of achieving prospective results.

— clearly defining the purpose of the cluster organization and aligning the goals of all participants to this main goal.

— developing a development model of each participant and the complex as a whole and setting criteria in it.

— taking into account the impact of clusters in the fruit and vegetable network on the environment and ecology. Establishing measures to increase the impact on the improvement of the environment, if necessary.

4 Conclusions

Taking into account the positive impact of clusters in the fruit and vegetable sector in the future development of the country's economy and their role in providing employment to the population and increasing the country's export potential, the state should exert its influence on the development of this sector and support it in necessary cases. It is expedient for the state to bear part of the main expenses for the establishment of fruit and vegetable clusters. the analysis of the activities of developed countries in this field shows that they are receiving a lot of attention from the state.

Therefore, in the formation of the management structure of these fruit and vegetable clusters, it allows to cover the activities of farms, farmers and LLCs in cooperation and to eliminate the existing problems with them in the field. This, in turn, is explained by the increase in the production indicators of fruit and vegetable clusters and the achievement of high economic efficiency.

of the above-mentioned fruit and vegetable cluster has been determined that the development of the fruit and vegetable sector will serve to increase varieties suitable for climatic conditions and fully satisfy the population's need for quality, cheap fruit and vegetable products.

Regarding the activity of fruit and vegetable clusters in the Bukhara region, which is being studied as the object of our research, the following conclusions were reached:

- increase the level of productivity of products grown in fruit and vegetable clusters;
- increasing the volume of fruit and vegetable production and focusing on standardization and certification processes in accordance with quality requirements;
- wide establishment of logistics transport system to deliver products grown in fruit and vegetable clusters to world markets and consumers;
- implementing the use of modern management methods and innovations in the management of fruit and vegetable clusters;
- in order to improve the development and efficiency of fruit and vegetable clusters, first of all, it is necessary to carry out a SWOT analysis of the farm.

Factors in fruit and vegetable clusters were studied, divided into internal and external factors according to the level of influence, as well as positive and negative factors according to the obtained result.

References

1. Soliev A., Kodirov Kh. Economic clustering is a factor of strategic planning. Economy: analyzes and forecasts. Tashkent-2019., #1. 104-108 p.
2. Tsikhan, T. V. Cluster theory economic development // Teoriya i praktika upravleniya. - 2003. - No. 5. - S. 40. Khukhrin A.S. The concept of agricultural development in the Russian Federation. Economics, labor, upravlenie v selskom hozyaystve. 2011. - No. 1(6). - p. 15-20
3. Rozikov J.M. Development of fruit and vegetable clusters and ways to adapt the management mechanism to market principles. Tashkent State University of Economics // Proceedings of the International Scientific and Practical Conference on "Actual tasks of the effective use of modern marketing concepts in the development of the national economy" - T., 2022 - pages 30-36.
4. Anichkina, O. A. Principle formirovania clusterov malogo i srednego biznesa v APK /O. A. Anichkina // Materialy mejdunarodnoy nauchnoy conference "Problemy sovremennoy ekonomiki". - Chelyabinsk, 2012. - 39 p.
5. Rozikov JM Prospects for the development of fruit and vegetable clusters. // electronic scientific and practical periodical publication "Economics and society" №11(102) 2022 pp. 1-13.
6. Rozikov JM Scientific basis for further improvement of fruit and vegetable cluster management mechanisms in Uzbekistan. // ISSN: Galaxy international interdisciplinary research journal (giirj) issn (E): 2347-6915Vol. 10, Issue 4, April. (2022) pp. 526-532.

7. Rozikov J.M. Urgency of organization of fruit and vegetable clusters. // "Agro ilm" agrarian economic, scientific and practical magazine. - T., 2020, No. 3. – pages 96-98. (08.00.15)
8. Mitenev V.V. i Kirik O.B. Klaster kak factor razvitiya ekonomiki/ / (Electronic new resource): http // journal vscc . ac . ru / php / joi /36/ apt 36 03. php ,
9. Balashova N. Formirovanie clustera po proizvodstvu i pererabotke nuta//APK: ekonomika i upravlenie. - M., 2004. - P.38-41.
10. Mutura, J., Nyairo, N., Mwangi, M., and Wambugu, S., 2016. Analysis of Determinants of Vertical and Horizontal Integration among Smallholder Dairy Farmers in Lower Central Kenya. International Journal of Agricultural and Food Research. Vol. 5 No. 1, pp. 1-13 ,
11. Mehdi, M., Ahmad, B., Yaseen, A., Adeel, A., Sayyed, N., 2016. A comparative study of traditional versus best practices mango value chain. Journal AgriculturaL Science. Vol.53 (3) Stanley, RA (2016). Creating new, high value innovative products from vegetables, root and tuber crops. Acta
12. Falkovich E.B., Kotlyarov D.A. Agrarian yy cluster kak innovatsionnoe napravlenie integratsii v APK stran tamojennogo soyuza i edinogo ekonomicheskogo prostranstva. Vestnik Belgorodskogo University Cooperation, Economics and Law,
13. Altukhov I. A , D evyatkina L. N. , S tanovlenie fermerstva v r ossii // A grarnya nauka . – Moscow, 2000. - #8. - S.2-3 ;.
14. Burjalieva Z.N. Organizational-economic fundamentals of innovative development of sadovodstva: na primere Respubliki Dagestan . // autoref. Makhachkala.: 2017. 20 pages;
15. Rahimi Sh. Klastery kak factor povyshenia konkurentosposobnosti regiona: na primere Sogdiyskoy oblasti Respubliki Tadjikistan . // autoref. Khujand.: 2013. 23 pages;
16. Sachkova A. Yu. Organizational and economic mechanism razvitiya ovoshchevodcheskix teplichnyx organization and usloviyakh chlenstva rossii v vto (na primere Saratovskoy oblasti) . // autoref. Saratov.: 2013. 24 str;.
17. Karimov R.A. Osobennosti formirovaniya konkurentnyx atnosheniy v usloviyax perekhoda k ustoychivomu razvitiyu selskogo hozyaystva: na materialax Republic of Tajikistan . // autoref. Khujand.: 2012. 23 str
18. Jumaev, R., Gazibekov, A., Sulaymonov, O., & Sobirov, B. (2021). Representatives of Lepidoptera groups occurred in forestry and agricultural crops and their effective entomophage types. In E3S Web of Conferences (Vol. 244, p. 02020). EDP Sciences. DOI:10.1051/e3sconf/202124402020
19. Kimsanboev, K., Rustamov, A., Jumaev, R., & Usmonov, M. (2021). Euzophera Punicaella Mooze (Lepidoptera) bioecology and development of host entomophagic equilibrium in biocenosis. In E3S Web of Conferences (Vol. 244, p. 01003). EDP Sciences. DOI:10.1051/e3sconf/202124401003

Effectiveness of the application of herbicides against pests in sunflower fields

Usman Norkulov[1,*], *Shavkat* Ahmurzayev[1], *Bakhtiyar* Nasirov[1], *Makhkam* Shodmanov[1], *Batir* Tukhtashev[1] and *Jamoliddin* Eshonkulov[1]

[1]Tashkent State Agrarian University, 2, University street, Tashkent, 100140, Uzbekistan

Abstract. This article investigates the impact of herbicides Pilot, Roundup, 36% se, and Sprut, 54% se on the growth, development, and productivity of soybean plants. The study examines how different application rates of these herbicides affect soybean crops when used to control the spread of sunflower fields. Additionally, the research includes an assessment of the economic efficiency associated with the use of these herbicides. The study aims to understand the effects of these herbicides on soybean plants, considering factors such as plant growth, development stages, and final yield. Different application rates of the herbicides are tested to determine their optimal dosages for effective sunflower field control while ensuring minimal impact on the soybean crop. Furthermore, the article presents an analysis of the economic implications of using these herbicides. This includes considerations of the cost-effectiveness and benefits associated with the use of each herbicide in soybean cultivation when compared to other weed control methods. By providing valuable information on the influence of these herbicides on soybean crops, the research aims to aid farmers and agronomists in making informed decisions about weed management strategies. The findings can contribute to the development of more efficient and sustainable approaches to control sunflower fields while maximizing soybean productivity and overall economic returns.

Keywords. Pilot, Roundup, herbicides, sunflower, yield.

1 Introduction

Currently, increasing the gross yield and quality of agricultural crops is becoming very important in meeting the needs of the population for food products in the world. 10-15% of Lalpi crop is lost due to weeds [1-3]. There are more than 3,000 types of weeds in world agriculture, and more than 40 types cause great damage. High results have been recorded in USA, Germany, China, Australia, Russia, South Korea, India, and other countries using combined agrotechnical and chemical control measures against them [4-6].

In many countries around the world, for example, in China, weed control, surface tillage and application of herbicides at planting, continuous application during the growing season, deep once every 2-3 years in the USA and India (32-35 cm) tillage and application of herbicides before planting, along with planting and during the growing season has been

* Corresponding author: 1992. jamoliddin@mail.ru

observed to give good results [7-9]. The use of herbicides is the most effective way to keep the fields where agricultural crops are grown free of weeds, it creates favorable conditions for the growth and development of crops and ensures an increase in productivity. But the chronic use of one herbicide in one field causes the proliferation of weed species that are resistant to that drug [10-14]. Based on this, in the conditions of Tashkent region in Uzbekistan, it is very important to use new herbicides.

Recently, the global climate change conditions in the world, the growth of the population of our country require the rational use of land resources [15, 16]. In recent years, special attention has been paid to the effective use of agricultural crops and food security of the population in Uzbekistan. Providing the growing population with ecologically clean and high-quality agricultural products, improving the phytosanitary status of products, preventing the entry of objects under quarantine into the republic, and sharply reducing the number of weeds in agricultural crops, including soybean fields, in Uzbekistan. in order to improve the chemical control measures, it is urgent to conduct research on the use of new herbicides, i.e., to increase the yield of vegetables as a result of updating the range of herbicides.

2 Materials and methods

In the context of the Tashkent region in Uzbekistan, this study investigates the spread, growth, and damage caused by wild plants in oil crop fields and field edges. The research aims to develop effective strategies for preventing the expansion of these wild plants into sunflower fields, field edges, ditches, and canals. To achieve this, the study focuses on implementing scientifically-grounded measures for control.

Figure 1. Phenological conditions in Jizzakh region observation activities.

A comprehensive approach is adopted, encompassing both agrotechnical and chemical control methods. The research explores soil cultivation techniques and mechanical measures that inhibit the germination, growth, development, and dissemination of sorghum seeds. By incorporating these preventive strategies, the study seeks to curtail the establishment and spread of unwanted wild plants in key agricultural areas.

Furthermore, the research identifies and tests suitable herbicides for targeted application. The timing of herbicide application is crucial to ensure maximum efficacy in controlling the unwanted wild plants while minimizing any adverse effects on the desired crops. By conducting thorough tests, the study determines the most effective herbicides and establishes appropriate application standards for their use.

Ultimately, the findings from this research contribute to the development of practical and efficient strategies for managing wild plant proliferation in oil crop fields and surrounding areas. The implementation of these methods in production can help ensure the optimal growth and yield of sunflower crops and improve overall agricultural productivity in the region.

This groundbreaking study represents the first comprehensive investigation into the types, biological characteristics, and extent of damage caused to sunflower fields by wild plants. Additionally, the research delves into the impact of these wild plants on the edges of fields. By thoroughly examining these aspects, the study aims to provide a scientifically substantiated understanding of the interplay between sunflower crops and surrounding wild plant populations.

The research endeavors to develop effective agrotechnical methods for weed control in oil crops, encompassing strategies such as crop rotation, appropriate tillage practices, and mechanical removal techniques. These methods are tailored to efficiently combat weed proliferation while preserving the health and productivity of sunflower fields.

Furthermore, the study involves the rigorous selection of effective herbicides, considering their suitability for specific weed species and their compatibility with sunflower crops. The research evaluates the optimal norms, application methods, and timing of herbicide use to ensure the most efficient and eco-friendly weed control.

By integrating these research findings into practical applications, farmers and agricultural practitioners can adopt novel and effective approaches to manage weeds in sunflower fields and maintain the health and yield of their crops. The implementation of these insights can lead to increased agricultural productivity, reduced economic losses, and enhanced sustainability in oil crop production.

3 Results and discussion

The effect of the use of Pilot herbicide at different rates against the field of sunflower planted with 5 options and 4 replications at the experimental station of Tashkent State Agrarian University was continued in 2022. The plot of the variant is 84 m^2 . Along with the planting of sunflowers in the fields, safflower seeds were sown on the soil surface to a depth of 3-4 cm, then Stomp 33% ek of 1.5 l/ha as a standard and Pilot herbicide 0.5; It was sprinkled on the surface of the soil with a hand sprayer along with sowing at the rate of 1.0-1.5 l/ha. 300 l/ha of water from standard working solution was used for each application. Other agrotechnical activities were carried out on the basis of technological processes introduced in sunflower cultivation.

As shown in Table 1, all of the Pilot herbicide rates tested in the sunflower field showed good results against boll weevil. No seed germination was observed in the initial period in all the experimental variants and in the Stomp sprayed variant. In the case where Stomp 33% ek herbicide was used at the rate of 1.5 l/ha, the spread of sedum after 30 days was 3.8 %, and after 60 days it was 8.2%, Pilot, 10% sec 0.5 In the option used at the rate of l/ha, it was observed that the spread of the fungus was 1.5% after 45 days and 4.4% after 60 days. No seed germination was observed after 45 and 60 days in pilot, 10% sec 1.0 and 1.5 L/ha rates. The prevalence of measles after 75 days was 1.6 and 1.0% in these variants, respectively.

Before harvesting the sunflower crop, it was found that in the case of Pilot, 1.0 l/ha, the spread of sorghum was 2.0%. It was determined in the experimental results that the spread of this drug was 1.1% during sowing of soybeans when this drug was used at the rate of 1.5 l/ha.

Table 1. Herbicide against the cusp (*Cuscuta*) in the sunflower field (Kibray district, Tashkent region, 2021-2022).

#	Options	Herbicide, L	Date of sprinkling herbicide	Cusp spread in % Accounting received the date			Growth of the era At the end of sunflower seeds spread, %	Sunflower yield, quintals/ha
				May 30	June 15	June 30		
1	Control (herbicide not sprinkled)	-	-	16.0	38.5	52.2	60.4	23.7
2	Stomp, 33% ke (benchmark)	1.5	April 30	-	3.8	8.2	28.4	28.2
3	Pilot, 10 % sec	0.5	April 30	-	1.5	4.4	8.4	29.9
4	Pilot, 10 % sec	1.0	April 30	-	-	1.6	2.0	32.1
5	Pilot, 10 % sec	1.5	April 30	-	-	1.0	1.1	30.8

Pilot herbicide 0.5; 2 9 .9 respectively from sunflower in the options used in the norms of 1.0 and 1.5 l/ha; 32.1 and 30.8 quintals/ha yield was obtained. 30.8 quintals/ha was obtained in the control variant without herbicide spraying.

The data obtained show that the application of Pilot 10% .ec herbicide at the rate of 1.0 l/ha has given good results. Therefore, the use of Pilot solution at the rate of 1.0 l/ha provides an opportunity to get a high yield of sunflower by effectively reducing the sorghum.

The impact of the use of Pilot herbicide at different rates against the sunflower field was carried out in the "Manas" farm of the Dostlik district of the Jizzakh region in 5 variants and 4 replications. Each option plot was taken as 84 m^2 . Sunflower seeds were sown on the calculated areas of the plots at a depth of 3-4 cm on the surface of the soil. Stomp 33% ek rate of 1.5 l/ha was taken as benchmark. In the next three options, Pilot herbicide is 0.5; At the rate of 1.0-1.5 l/ha, it was sprinkled on the soil surface with the help of a manual sprinkler. The working solution was obtained at the rate of 300 l/ha of water. All other agrotechnical activities were carried out based on technological processes introduced in sunflower cultivation.

From the data presented in Table 2, it can be seen that all tested application rates of Pilot herbicide in the sunflower field have good results. In all of the experimental options, the germination of the seeds of sorghum was not observed in the initial period of observations. In the variant where Stomp 33% ek herbicide was used at the rate of 1.5 l/ha, the spread of sorghum after 60 days was 3.25 %, and after 60 days it was 11.2%.

In the pilot, 10% sec at 0.5 l/ha, it was observed that there was no spread of weevil after 45 days and 5.50% after 60 days. No seed germination was observed after 45 days in variants sprayed with this herbicide at rates of 1.0 and 1.5 l/ha.

https://doi.org/10.1051/e3sconf/202342102004

Table 2. Pilot, the effectiveness of 10% sec herbicide against *Cuscuta* in sunflower field (Dostlik District, Jizzakh Region, 2021-2022).

#	Options	Herbicide , L	Date of sprinkling herbicide	Cusp spread in %			Growth of the era At the end of sorghum seeds spread, %	Sorghum yield, quintals/ha
				Accounting received the date				
				May 30	June 15	June 30		
1	Control (herbicide not sprinkled)	-	-	9.25	43.5	59.2	66.6	22.1
2	Stomp, 33% ke (benchmark)	1.5	April 25	-	3.25	11.2	30.5	27.0
3	pilot, 10 % sec	0.5	April 25	-	1.50	5.50	9.25	28.1
4	pilot, 10 % sec	1.0	April 25	-	-	0.75	1.75	31.1
5	pilot, 10 % sec	1.5	April 25	-	-	0.5	1.5	29.4
HCP $_{05}$ = 1.60 quintals/ha								

After only 60 days, the prevalence of sporulation in these variants was 0.75 and 0.5%, respectively. During the harvest period of sunflower, the spread of sorghum was 1.75% in the case of Pilot, 1.0 l/ha. 0.5 along with planting pilot preparation; In the variants used in the norms of 1.0 and 1.5 l/ha, the sunflower yield is 2 8 .1, respectively ; 31.1 and 29.4 quintals/ha yield was obtained. In the control variant without herbicide spraying, the yield was 22.1 quintals/ha. Therefore, the use of Pilot herbicide at the rate of 1.0 l/ha provides a high yield of sunflower by effectively reducing the weeds.

4 Conclusions

In the conditions of typical gray soils of Tashkent region, in the field of sunflower, in the variants that were sprayed with Pilot, 10% sec herbicide at the rate of 1.0 and 1.5 l/ha, no seed germination was observed after 30 and 45 days. In these variants, after 60 days, the prevalence of sporulation was 2.1 and 1.0%, respectively. During the harvesting period of sunflower, the spread of sorghum was 2.3% in the case of Pilot, 1.0 l/ha, and the biological efficiency was 96.8%.

In the conditions of Dostlik district, the spread of the psyllid during the harvesting period of the sunflower crop Pilot, 1.75% of the plants were infected with the psyllium in the option applied at the rate of 1.0 l/ha, and the efficiency was 96.0% compared to the control option. organized.

Roundup, 36% se and Sprut, 54% se herbicides were used at rates of 6.0 and 3.7 l/ha against common sedum and other weeds against common sedum and other weeds in field margins. kills 96-99% of weeds.

References

[1] Isaev SS, Juraev AA, Juraev AK, Kadyrov ZZ, Soybean care in the conditions of meadow alluvial soils of the Bukhara region, prone to salting, irrigated from ancient times. Recommendation - Bukhara-2019 –12 p

[2] Abdalova, GN, Eshonkulov, JS, Sulaymanov, SO, & Abdullayeva, FM (2021). Improvement of cotton nutrition procedure and irrigation technologies. *ACADEMICIA: An International Multidisciplinary Research Journal* , *11* (4), 720-723 .

[3] Nasirov Bakhtior Salakhiddinovich Charshanbiyev Umurok Yuldashevich, Eshankulov Jamoliddin Saporboy Ugli. "Efficiency of application of herbicides which are samurai 33% ek, zellek super 10.4% ek and triflurex 48% ek against weeds in cotton fields" *Web of Scientist: International Scientific Research Journal* 2.09 (2021): 136-139.

[4] Salakhiddinovich, Nasirov Bakhtiyor., Eshankulov Jamoliddin Saporboy ugli 2021 "Development of Irrigation Procedures for Shadow Varieties Planted After Autumn Wheat." *International conference on multidisciplinary research and innovative technologies* . Vol. 1. 2021.

[5] Burievich, TB, Olimovich, A. Eshankulov JS, Turaevich, MT 2021 Groundwater consumption and cotton productivity. *Web of Scientist: International Scientific Research Journal* , *2* (09), 130-135.

[6] Norkulov U, Izbasarov B, Tukhtashev B, Eshonkulov J., Volume : 2 Issue : 2 2022 Effects of Sardoba Reservoir Flood on Irrigated Land, International Journal of Innovative Analyzes and Emerging Technology e-ISSN : 2792-4025 40-42 p.

[7] Tukhtashev B, Norkulov U, Izbosarov B. Technology of proper use of saline soils in the conditions of Uzbekistan. E3S Web of Conferences 258, 03027 (2021)

[8] Izbasarov BE, Norkulov U, Tukhtashev, Hikmatov Sh Influence Of New Types Of Horizontal Ditches On The Growth, Development And Yield Of Winter Wheat In Saline And Groundwater Surface Soils. Influence Of New Types Of Horizontal Ditches On The Growth, Development And Yield Of Winter Wheat In Saline And Groundwater Surface Soils 2021

[9] Norkulov U, Tukhtashev B, Eshonkulov J., Volume : 2 Issue : 2 2022 Change of Mechanical Composition of Soils after Flood of Sardoba Water Reservoir, International Journal of Innovative Analyzes and Emerging Technology e-ISSN : 2792-4025 36-39 p.

[10] Ziyatov Musulman Panjiyevich, Shamsiyev Akmal Sadirdinovich, Kamilov Bakhtior Sultanovich, Abdalova Guliston Nuranovna, Abdurakhimov Shavkatjon Olimovich, Eshonkulov Jamoliddin Saporboy Ugli. PJAEE, 17(6) 2020 Effective agrotechnology of cotton feeding in different irrigation methods. Palearch's Journal Of Archeology Of Egypt/Egyptology 17(6). ISSN 1567-214x. 3415-3428 p. http://www.palarch.nl/index.php/jae/article/view/1335

[11] Eshonkulov Jamoliddin Saporboy Ugli , Kamilov Bakhtiyor Sultanovich, Shamsiyev Akmal Sadirdinovich, Nasirov Bakhtiyor Salakhiddinovich, Sheraliyev Khamidulla, Ziyatov Musulman Panjievich 2020 Appropriate irrigation procedures and cultivation agrotechnology of soybean and sunbackar varieties planted as reproductive crops. *PalArch's Journal of Archeology of Egypt/ Egyptology* , *17* (6), 3399-3414. Retrieved from https://archives.palarch.nl/index.php/jae/article/view/1333

[12] Shamsiyev Akmal Sadirdinovich, Eshonkulov Jamoliddin Saporboyugli, Sultanov Umbetali Tazabayevich 2020 Growth and development of soy and sunflower varieties. ACADEMICIA An International Multidisciplinary Research Journal

10(11):1289-1291

[13] Shamsiyev Akmal Sadirdinovich, Kamilov Bakhtiyor Sultanovich., Eshonkulov Jamoliddin Saporboyugli, Ashirov YR Agrophysical and agrochemical properties of influence of recycled soy and soil of the field 202 0 ACADEMICIA An International Multidisciplinary Research Journal August – India , 2020. – Vol. 10 . – Issue 8 . - P. _ 475-479

[14] Dusbayev IR, Nasirov BS, Ashirov YR, Eshonkulov JS, Rashidov Q 2021 Methods of planting fine liquid cotton and effects of Herbicides. 2nd International Conference on Science Technology and Educational Practices. Turkey 251-254 p.

[15] Eshonkulov Jamoliddin Saporboy, Shamsiev Akmal Sadirdinovich. Vol.5 NO. 2020 Congress (2020) Changes in water-physical properties of soil in repeated crop sunflower care. International congress on modern education and integration congress – India – Volume 5 . - P. 89-90.

[16] Eshonkulov J. Kamilov B. Effect of irrigation regimes on the fertility of soybean and sunflower cultivars planted in repeated periods . *IOP Conference Series: Earth and Environmental Science* Tom 1142, Vypusk 12023 Nomer status 0120533rd International Conference on Energetics, Civil and Agricultural Engineering, ICECAE 2022Virtual, Online13 October 2022 do 16 October 2022 Code 187394

Tolerance of cotton to certain biotic factors and fiber quality parameters of the ridges created by simple and complex hybridization

Fozilbek Toreev[1,*], *Bahriddin* Orozov[2], *Gulnoza* Shodmonova[1] and *Soyibjon* Rasulov[3]

[1]Tashkent State Agrarian University, Tashkent, 100140, Uzbekistan
[2]Scientific Research Institute of Cotton Breeding, Seeding and Cultivation Agrotechnologies, Tashkent, 100140, Uzbekistan
[3]Andijan Agricultural and Agrotechnological Institute, Andijan, 170101, Uzbekistan

Abstract. Among the strains derived from simple hybrids studied in the experiments, T-3 T-9, T-18 and strains T-43, T-53, T-57 and 60 from strains derived from complex hybrids were resistant to gommosis and wilt diseases. As a result of several years of research, it was found that the ridges obtained from complex hybrids were less affected by gommosis and wilt diseases than the ridges obtained as a result of simple hybridization. It was found that wilt disease was 3-4 times more resistant in the ridges created as a result of simple and complex hybridization compared to the model seed. Lines T-155, T-158, T-163, T-167, T-188 and T-208 from simple hybrids of cotton and lines T-212, T-215, T-220 and T-222 from complex hybrids are other It was determined that the quality of the fiber meets the requirements of the world standards in relation to the ridges and pattern type. In the studied samples, the families of ridges that do not have positive quality indicators such as fiber micronaire, yarn elasticity coefficient, fiber length, relative tensile strength and fiber uniformity are discarded, and experiments are carried out in the following years in order to incorporate new ridges that meet the requirements of the world standard with other economically valuable characteristics detached to continue.
Keywords. Cotton, fiber quality, variety, cotton diseases, fiber yield, simple and complex hybrids, ridge, micronaire, fiber length.

1 Introduction

In recent years, several decisions have been taken by the government in our country in order to develop cotton cultivation and obtain high yields [1]. This gives breeders the task of creating varieties that are quick-ripening, productive, resistant to insects and diseases, and whose fiber quality meets world standards and whose fiber yield is higher than 40%. Uzbekistan takes the leading positions in cotton production and export in the world, but it is significantly different from foreign varieties in terms of fiber productivity [2, 3]. As a result, that is, due to low productivity, our farmers and clusters receive 100 thousand tons

* Corresponding author: ftoreev@mail.ru

less fiber every year. Unfortunately, this problem has not been solved yet. Therefore, creating varieties with high fiber yield is a very urgent issue today [4, 5].

Today, the fiber output of the varieties created in Uzbekistan should not be less than 40 percent, and at the same time, the fiber quality should correspond to type IV. It will be possible to solve the problem of diversification of agricultural crops by introducing varieties with high fiber yield while maintaining the current volume of fiber production [6].

Breeding scientist R.G.Kim [3] the success of selection work on the resistance of cotton to wilt disease depends on the in-depth study of the dynamics of new reproduction of isolates (strains) of *Verticillium* fungus in different soil geographical regions, the level of resistance of cultivated varieties, the initial parental forms and hybridization [7, 8]. Studies on cultivar resistance to wilt and different isolates of Verticillium fungi show that they have different genotypic levels of resistance to the studied isolates of the fungus. They note that Omad variety shows high phenotypic wilt resistance throughout the entire growing season when infecting primary plants with 28 isolates.

In the data of F. Toreev et al. [8] based on the analysis of fiber quality indicators of F5 simple and complex hybrids, T-150 and T-195 lines from F 5 simple hybrids and T-212 and T-215 lines from complex hybrids, fibers of F8 simple and complex hybrids based on the analysis of quality indicators, it was determined that the T-3 and T-31 lines from F_8 simple hybrids and the T-46 and T-50 lines from complex hybrids meet the requirements of world standards of fiber quality compared to other simple and complex hybrids and model types.

K. Khudarganov; M.Rakhmonkulov; In the scientific works of M.Umedova and others [6, 7, 9], valuable economic traits of cotton, disease and pest tolerance, fiber quality indicators were highlighted, and it was stated that valuable economic traits were superior to the model variety in new lines of cotton.

It is known that the currently created varieties must meet the requirements of the textile industry, i.e. have high fiber quality, along with high fiber yield. Fiber quality of cotton varieties is characterized by length, hardness, thinness (metric number), length, stretchability, elasticity, luster and color.

Improving the quality of fiber has always been the main task. At the same time, there are some problems in creating varieties with sertola and high quality. First, there are negative correlations between fiber yield and quality Allambergenov et al.; According to W.R. Meredith [1, 4] there are strong negative correlations between fiber yield and fiber quality. According to Australian scientist JDClement, negative correlations were found between yield and some fiber quality traits, particularly fiber length and maturity and fiber yield were negatively correlated. Fiber maturity was found to be positively associated with yield. It was found that micronour and fiber softness were not related to yield. Genes controlling fiber quality traits show their additive effect in the F_5 -F_6 generation [10-12]. Another factor that reduces fiber quality is leaf residue, as contamination is high in deciduous plants during machine harvesting. Dramatically reducing leaf hairs reduces fiber contamination.

2 Materials and methods

The laboratory analysis of fiber quality in the HVI-900 laboratory provides numerical indicators as results. These indicators represent various aspects of fiber characteristics such as length, strength, micronaire, and other relevant properties.

To make sense of the data and draw meaningful conclusions, the numerical results are subjected to mathematical processing using statistical analysis methods. Statistical analysis involves a range of techniques to interpret and understand the data, including measures of central tendency (mean, median, mode), measures of dispersion (standard deviation, variance), and correlation analysis.

By employing statistical analysis, researchers can identify patterns, trends, and relationships within the fiber quality data. This information is vital for making informed decisions related to fiber processing, production optimization, and quality control in the textile industry.

The results of the statistical analysis can also serve as a basis for comparison with industry standards and benchmarks, allowing for the evaluation of the fiber's suitability for specific applications or markets.

The combination of laboratory analysis and statistical processing ensures a comprehensive understanding of fiber quality and assists in enhancing the efficiency and competitiveness of fiber-related industries.

3 Results and discussion

New varieties of cotton being developed should be quick-ripening, productive, high in fiber yield and quality, as well as resistant to various diseases, including wilt and gommosis. In cotton fields affected by wilt and gommosis diseases, along with yield reduction, the fiber quality also decreases sharply. In the researches, resistance to these diseases in the environment artificially infested with gommose bacteria and naturally infested with wilt fungus was studied in new lines created by simple and dark hybridization.

In this case, the number of plants in the ridges obtained from simple hybrids was from 72 to 114, and in the ridges obtained from complex hybrids from 110 to 184. It was found that 10.8% of T-18 ridges and 20.8% of T-7 ridges were infected in the ridges obtained from common hybrids with the spring form of Gommoz disease, and in the autumn form, 7.1% of T-18 ridges and 16.7% of T-7 ridges were infected. . In the ridges obtained from complex hybrids of cotton , with the spring form of gommoz from 8.7 percent (T-53), to 15.4 percent (T-46), and with the autumn form from 4.8 percent (T-53), to 10.9 percent (T - 46) was observed to be infected. The C-6524 variety, taken as a model, was damaged by 20.2 % spring form of gommosis , and 14.8 % by autumn form (Figure 1) .

When the general and severe damage by wilt disease was studied, the total damage by wilt in the rows obtained from the hybrids was from 3.6 percent (T-18), to 16.0 percent (T-30), and the severe damage by this disease was from 0.9 percent. (T-18), up to 9.3 percent (T-30) were found to be infected. As a result of multi-year selection, the total damage in the lines obtained on the basis of complex hybridization is from 3.3 percent (T-53), to 10.0 percent (T-46) damage, and the severe damage in these lines is from 0 percent (T-53), 5.4 percent (T-46) was observed to be infected. C-6524, which was taken as a model, had 43.2 percent wilt damage, and 18.9 percent in severe cases. T-3, T-9, T-18 and T-43, T-53, T-57 and 60 from complex hybrids were resistant to gommosis and wilt diseases.

As a result of several years of research, it was found that the ridges obtained from complex hybrids were less affected by gommosis and wilt diseases than the ridges obtained as a result of simple hybridization. It was found that wilt disease was 3-4 times more resistant in the ridges created as a result of simple and complex hybridization compared to the model seed. On the basis of experiments, families of ridges resistant to gommosis and wilt diseases, quick-ripening, productive, with large bags, fiber output higher than 39 percent and large seeds were isolated for the purpose of breeding and creation of initial materials.

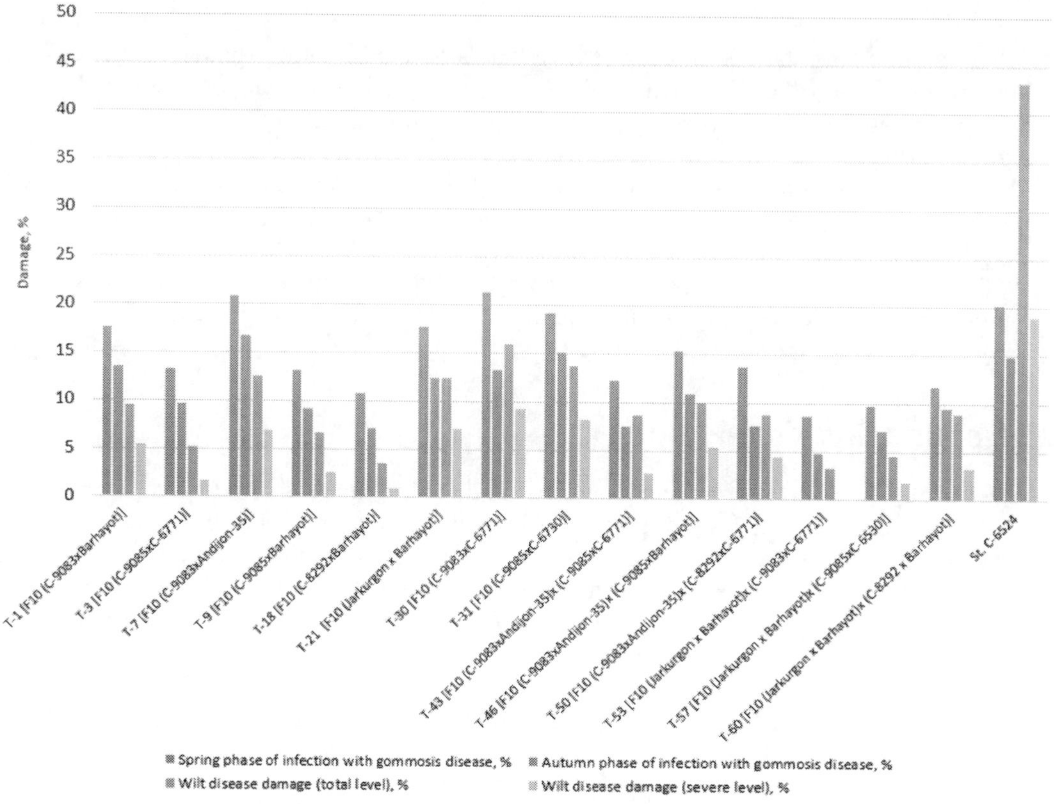

Figure 1. Tolerance of cotton ridges created by simple and complex hybridization to some biotic factors.

It is known that in all the countries that grow cotton in the world, the quality mark of the fiber is the most important in newly created varieties. In addition to the high economic characteristics of cotton, it is necessary that the fiber quality indicators meet the requirements of the world standard. In the research, the ridges created by simple and complex hybridization of medium fiber cotton with positive indicators in terms of morphoeconomic characteristics, resistant to some biotic factors, and in the samples obtained from the model C-6524 variety, fiber micronaire Mic, yarn elasticity coefficient SCI, fiber length Lend (inches) , relative tensile strength Str (g/tex) and fiber uniformity UI (%) were analyzed. Fiber quality indicators were determined in the "Sifat" certification center on a modern measuring instrument HVI.

The ridges obtained from F 6 simple hybrids ranged from 4.2 to T-188 ridges, to 5.1 to T-150 ridges, and from F 6 complex hybrids to 4.3 to T-220 ridges, to 5.2 to T-222 ridges, when these ridges were studied in terms of yarn elasticity coefficient, it was found that it was from 119 in the T-167 ridge, to 164 in the T-155 ridge, and in the ridges obtained from complex hybrids, it was from 123 in the T-220 ridge, to 162 in the T-222 ridge. The C-6524 variety, which was taken as a sample, had a microneural index of 5.0 and a yarn twist coefficient of 154. From these hybrids, the micronaire index was higher than 4.7, and the families of rows whose threadability did not meet the requirements of world fiber quality were excluded (Table 1).

From the quality indicators of the fiber, the length of the fiber was calculated from the hereditary characters and analyzed by comparison with the template variety in the lines created by simple and complex hybridization. Fiber length markers ranged from 1.07 inches in ridges derived from simple hybrids T-208 ridge [F $_6$ (L-842 x Barhayot)] to 1.20 inches in T-167 [F $_6$ (Bukhara-8 x Shodiyona)], derived from complex hybrids and in ridges from 1.05 inches T-220 ridge [F $_6$ (C-2609 x Turon) x (C-2609 x Barkhayot)], to 1.18 inches T-222 ridge [F $_6$ (C-2609 x andijon-35) x (L-842 x Barkhayot)], and in the specimen C-6524 it was 1.13 inches. According to this character, the ridges obtained from complex hybrids were slightly higher than the ridges obtained from simple hybrids and the model variety.

Table 1. Quality indicators of newly created ridges of cotton.

#	F_6 simple and complex hybrids	SCI	Mic	Land (inch)	Str (g/tex)	UI (%)
1	T-150 F_6 (C-6524xBukhara-8)	132	4.9	1.11	31.0	83.1
2	T-155 F_6 (C-6524xC-2609)	151	4.6	1.12	34.7	83.6
3	T-158 F_6 (C-6524xShodiyona)	151	4.8	1.12	35.5	83.6
4	T-163 F_6 (Bukhara-8x C-2609)	155	4.7	1.11	36.9	83.5
5	T-167 F_6 (Bukhara-8xShodiyona)	167	4.5	1.16	36.8	85.1
6	T-178 F_6 (Bukhara-8xAndijan-35)	152	4.7	1.12	37.0	82.5
7	T-188 F_6 (C-2609xTuron)	157	4.6	1.14	35.5	84.3
8	T-195 F_6 (C-2609xBarhayat)	154	4.2	1.13	37.9	81.4
9	T-201 F_6 (C-2609xAndijan-35)	135	4.8	1.12	31.9	82.5
10	T-208 F_6 (L-842xBarhayat)	147	4.4	1.14	34.4	83.5
11	T-212 F_6 [(C-6524xBukhara-8) x (C-6524x C-2609)]	155	4.4	1.11	34.6	83.3
12	T-215 F_6 [(C-6524xShodiyona)x (Bukhara-8xC-2609)]	148	4.5	1.14	33.9	83.7
13	T-218 F_6 [(Bukhara-8 x Shodiyona)x (Bukhara-8x Andijan-35)]	145	4.5	1.14	32.6	82.7
14	T-220 F_6 [(C-2609xTuron)x (C-2609xBarhayat)]	151	4.9	1.19	33.1	83.6
15	T-222 F_6 [(C-2609xAndijan-35)x (L-842xBarhayat)]	142	4.9	1.19	34.9	84.2
	St. C-6524	137	4.8	1.10	31.3	81.5

When determined by fiber uniformity and specific relative tensile strength characteristics, fiber uniformity in the ridges created by simple hybridization is from 81.4 percent for T-195 ridge, to 85.1 percent for T-167 ridge, relative relative tensile strength is 31.0 g/text for T-150 ridge Up to 37.9 g/text was observed on the T-195 ridge.

In the strands obtained from complex hybrids, the fiber homogeneity was from 82.7% in T-218 strand to 84.2% in T-222 strand, and the relative tensile strength was from 32.6

g/text to T-218 strand to 34.9 g/text to T- It was found to be on the 222 ridge. C-6524, taken as a sample, had a fiber uniformity of 81.5 percent and a specific tensile strength of 31.3 g/text.

4 Conclusions

Based on the analysis of the fiber quality indicators of the strands created as a result of simple and complex hybridization in the researches, T-155, T-158, T-163, T-167, T-188 and T-208 strands obtained from simple hybrids and T-212, T from complex hybrids. -215, T-220 and T-222 ridges were found to meet the requirements of world standards of fiber quality compared to other ridges and model types. It was observed that in these ranges there are families that meet the requirements of type IV in terms of fiber quality. The ridges that did not have positive signs in terms of fiber quality indicators were discarded, and new ridges that met the requirements of the world standard and were resistant to certain biotic factors and ridges were separated in the following years to create new varieties and initial sources by compacting most of the valuable farm characteristics.

References

1. Allambergenov T.D., Toreev F.N., Kadirova M.R., Mutalova M.K., Khakimov A.E. Inheritance and variability of fiber quality traits in cotton. Monograph, LESSON PRESS publishing house, Tashkent-2020, 124 p.
2. Ibragimov P., Orozov B., Toreev F., Shodmonova G. Creation of starting materials resistant to some biotic factors for cotton selection . Collection of the international conference dedicated to the 90th anniversary of the Tashkent State Agrarian University on the topic "Actual problems in the theory and practice of agricultural science and their solutions" . Tashkent-2020, December 14-15, pages 90-93.
3. Kim R.G., Marupov A.I. Vliyanie razlichnyx geograficheskikh isolyatov (strain) fungus Verticillium dahliae Kleb na viltoustochivost sortov i liny vida G . Hirsutum L. _ // Materials of the international scientific conference "Evolutionary and selective aspects of rapidity and adaptability of children and other cultures" devoted to the 95th anniversary of the birthday of Academician S.S. Sadykova. - Tashkent.: Science , 2005 . -S 112-113.
4. Meredith W. _ R. , Notice of release of three noncommercial stocks of Upland cotton, DES 7A-ne, DES SL-ne,and DES DK-ne .Miss.Agr.and Forest. Exp.St. _ and UDDept. Agr. Memo 1972. - P . 321-325.
5. Brubaker C. L., Brown A. D., Stewart J. M., Kilby M. J., Grace J. P. Production of fertile hybrid germplasm with diploid Australian Gossypium species for cotton improvement. Euphytica 1999;108(3): P. 199-213.
6. Khudarganov K. Variability of Agronomic Valuable Traits of Cotton Lines Depending on the Genotype and Vegetation Conditions in Different Ecological Zones. Texas Journal of Agricultural and Biological Sciences.
7. Rakhmankulov M. Efficiency of new breeding methods in creating rapid cotton varieties. PalArch's Journal of Archeology of Egyptology. Egypt 2020.
8. Toreev F., Urazov B., Shodmonova G., Allambergenov T. , Mavlonova N. Fiber quality indices of newly developed cotton lines and the test results in the enlarged and small nurseries. AIP Conference Proceedings 2432, 040023 (2022); https://doi.org/10.1063/5.0091143 . Published Online: 16 June 2022.

9. Umedova M., Rakhmankulov M., Creation of resistant specimens of cotton (Tetranychus turkestani) using marker-based selection method in Uzbekistan . E3S Web of Conferences 244 , 02010 2021
10. Jumaev, R., Gazibekov, A., Sulaymonov, O., & Sobirov, B. (2021). Representatives of Lepidoptera groups occurred in forestry and agricultural crops and their effective entomophage types. In E3S Web of Conferences (Vol. 244, p. 02020). EDP Sciences. DOI:10.1051/e3sconf/202124402020
11. Kimsanboev, K., Rustamov, A., Jumaev, R., & Usmonov, M. (2021). Euzophera Punicaella Mooze (Lepidoptera) bioecology and development of host entomophagic equilibrium in biocenosis. In E3S Web of Conferences (Vol. 244, p. 01003). EDP Sciences. DOI:10.1051/e3sconf/202124401003
12. Jumaev, R., & Rustamov, A. (2022, July). Representatives of Lepidoptera groups in the biotecenosis of Uzbekistan and their effective parasite-entomophage types. In IOP Conference Series: Earth and Environmental Science (Vol. 1068, No. 1, p. 012026). IOP Publishing. DOI:10.1088/1755-1315/1068/1/012026

Effectiveness of using herbicides against *Cuscuta* in soy field

Usman Norkulov[1], *Shavkat* Ahmurzayev[1], *Bakhtiyar* Nasirov[1], *Jamoliddin* Eshonkulov[1,*] and *Nodira* Kamalova[1]

[1]Tashkent State Agrarian University, 2, University street, Tashkent, 100140, Uzbekistan

Abstract. This research delves into the repercussions of employing pilot and zeta herbicides at varying application rates to counteract the growth of common sorghum within soybean fields. The investigation is conducted within the specific environmental context of irrigated typical gray soils prevalent in the Tashkent region. The central focus of this study is to gain a comprehensive understanding of the multifaceted impacts these herbicides exert on the growth trajectory, developmental milestones, and eventual yield of soybean plants. By assessing the influence of pilot and zeta herbicides across different dosage levels, this study contributes to elucidating the potential benefits or challenges that arise from their application in this agricultural setting. Furthermore, this research endeavors to provide critical insights into the economic efficiency of employing these herbicides, thereby offering a holistic perspective on the viability of incorporating such chemical interventions within soybean cultivation practices in the Tashkent region. The implications of this study extend beyond agronomic parameters, delving into the intricate interplay between herbicide application, crop growth, and economic considerations. The findings are poised to offer valuable guidance to farmers, researchers, and policymakers who are vested in optimizing agricultural practices while considering both the ecological and economic dimensions. Ultimately, this investigation strives to enrich the body of knowledge concerning herbicidal interventions and their potential repercussions in the realm of soybean cultivation within distinct agro-ecological contexts.
Keywords. *Sorghum, pilot, zeta, herbicide, soybean, cuscuta.*

1 Introduction

Today, increasing the gross yield and quality of agricultural crops is of great importance in meeting the needs of the population for food products in the world. However, 10-20% of the total yield is lost due to weeds [1, 2]. There are more than 3,000 types of weeds in the world's agriculture, and more than 40 types cause significant damage. High results have been achieved in USA, China, Germany, Russia, Australia, South Korea, India and other countries by using agrotechnical and chemical control measures against them [3, 4].

* Corresponding author: 1992. jamoliddin@mail.ru

In many countries around the world, including China, weed control is followed by surface tillage and herbicide application, or during the growing season, and in the USA and India, deep plowing (32-35 cm) and herbicide application every 2-3 years [5, 6]. it has been found to give good results when used before planting, along with planting and during the growing season [7].

The use of herbicides is the most effective way to keep the fields where agricultural crops are grown free from weeds, which creates favorable conditions for the development of crops and ensures an increase in productivity. However, repeated use of one herbicide in one field leads to the proliferation of weed species resistant to that drug. Based on this, in the conditions of the Tashkent region (Uzbekistan), methods of tillage processing and the use of new herbicides are considered to be very urgent issues [8-10].

In recent years, the change in the global climate conditions in the world and the growth of the population of Uzbekistan require the rational use of land resources. In recent years, special attention has been paid to the effective use of agricultural crops and food security of the population in Uzbekistan [11-14].

Providing the growing population with cheap and high-quality agricultural products, improving the phytosanitary status of products, preventing the entry of quarantined objects into Uzbekistan, and sharply reducing the number of weeds in agricultural crops, including soybean fields, in Uzbekistan [15, 16]. It is important to conduct research on the improvement of chemical control measures for the use of new herbicides, that is, to increase the yield of vegetables as a result of updating the range of herbicides.

2 Materials and methods

In the conditions of Tashkent and Jizzakh provinces of Uzbekistan, the spread, growth and damage level of sorghum in the fields of oil crops and wild plants on the edges of the fields are studied and scientifically substantiated. Effective methods of agrotechnical and chemical control measures are developed to prevent the spread of weevils in soybean and sunflower fields, field edges, ditches and canals.

Land cultivation and mechanical loss methods that prevent the germination, growth, development and spread of the seeds of sorghum, application periods of effective herbicides selected as a result of testing, and convenient standards will be determined and introduced into production.

For the first time, the types, biological characteristics, and levels of damage to wild plants in the fields of soybeans, sunflowers and field edges are studied and scientifically substantiated; effective agrotechnical methods (rotation, tillage, mechanical removal) of weed control in oil crops and wild plants on field edges will be developed; Effective herbicides are selected and convenient norms, methods and periods of their use are determined and introduced into production.

As a result of the development of effective protection measures against sorghum, the yield lost in oil crops (20-30%) is saved, as well as foreign a clean product is grown from the seeds. Laboratory and field experiments were conducted to develop enhanced control strategies targeting quarantine weeds affecting oil crops, specifically sunflower and soybeans. These strategies were devised based on rigorous scientific research and aimed at effectively managing the presence of these problematic weeds. The study encompassed both controlled laboratory settings and real-world field conditions, ensuring a comprehensive evaluation of the proposed control measures. The ultimate goal was to provide practical and evidence-based solutions for combating the spread of these quarantine weeds in oil crop cultivation, contributing to the overall sustainability and productivity of sunflower and soybean farming practices.

https://doi.org/10.1051/e3sconf/202342102006

3 Results and discussion

In the conducted experiments, the study of the effect of the use of Pilot herbicide at different rates against the soya sown field was continued in 2022 at the experimental station of the Tashkent State Agrarian University in 5 variants and 4 repetitions. Each option plot was taken as 84 m^2. Along with sowing of soybeans in the fields, sorghum seeds (200 pieces) were sown on the surface of the soil to a depth of 3-4 cm (Figure 1), then Stomp 33% ek of 1.5 l/ha as a standard and Pilot herbicide 0.5.

Figure 1. Fields affected by the projectile and surrounding areas.

Along with planting at the rate of 1.0-1.5 l/ha, it was evenly sprinkled on the soil surface with a manual spreader. 300 l/ha of water from standard working solution was used for each application. Other agrotechnical activities were carried out on the basis of technological processes introduced in soybean cultivation. From the data in Table 1, it can be seen that all application rates of Pilot herbicide tested in the soybean field against the weevil gave good results.

Table 1. Pilot, 10% sec herbicide against soybean field cusp (Cuscuta) (Kibray district, Tashkent region, 2021-2022).

#	Experiments	Herbicide, l/ha	Date of herbicide application	Prevalence of cusp, in %			Spreading of seeds before harvesting, in %	Soy harvest, q/ha
				Date of registration				
				May 30	Jun 15	Jun 30		
1	Control (no herbicide applied)	-	-	14.4	50.5	67.6	72.4	22.1
2	Stomp, 33% ke (benchmark)	1.5	Apr 30	-	4.3	15.5	30.2	28.2
3	Pilot, 10 % sec	0.5	Apr 30	-	2.2	6.2	10.4	26.1
4	Pilot, 10 % sec	1.0	Apr 30	-	-	2.1	2.3	30.3
5	Pilot, 10 % sec	1.5	Apr 30	-	-	1.1	2.1	28.5
								HCP_{05} = 1.34 q/ha; HCP_{05} = 4.99%

All the experimental options and Stomp's consumption rate obtained did not show germination of sorghum seeds in the initial period. In the variant where Stomp 33% ek herbicide was used at the rate of 1.5 l/ha, the spread of sorghum after 45 days was 4.3 %, and after 60 days it was 15.5%. In this variant, the spread of seeds before harvesting was equal to 30.2%. In the pilot, 10% sec at 0.5 l/ha rate, it was observed that the spread of weevil was 2.2% after 45 days and 6.2% after 60 days. Pilot, 10% sec at 1.0 and 1.5 l/ha rates did not show seed germination after 45 days as last year. The prevalence of measles after 60 days was 2.1% and 1.2% in these variants, respectively. Pre-harvest spread of coya was found to be 3.2% in the case of Pilot, 1.0 l/ha. In the case where this drug was used at the rate of 1.5 l/ha, the spread of this drug during soybean harvest was 2.1%.

The study of the effect of the Zeta herbicide at different rates on soybean field was conducted in the experimental station of the Tashkent State Agrarian University in 6 variants and 4 repetitions. Each option plot was taken as 84 ^{m2}. Along with sowing of soybeans in the fields, zaperechak seeds (400 pieces) were also planted to a depth of 3-4 cm on the soil surface. Zellek-super, 104 g/l ek 1.0 l/ha was taken as standard. Zeta herbicide 0.5; 0.8; After the first irrigation of soybeans at the rate of 1.0-1.2 l/ha, it was evenly sprayed with a manual sprinkler. 300 l/ha of water from standard working solution was used for each application.

Other agrotechnical activities in the field were carried out based on the technological processes introduced for soybean cultivation. From the obtained data, it can be seen that on June 5, the prevalence of sorghum in the experimental field was 15.0-18.4% (Figure 2).

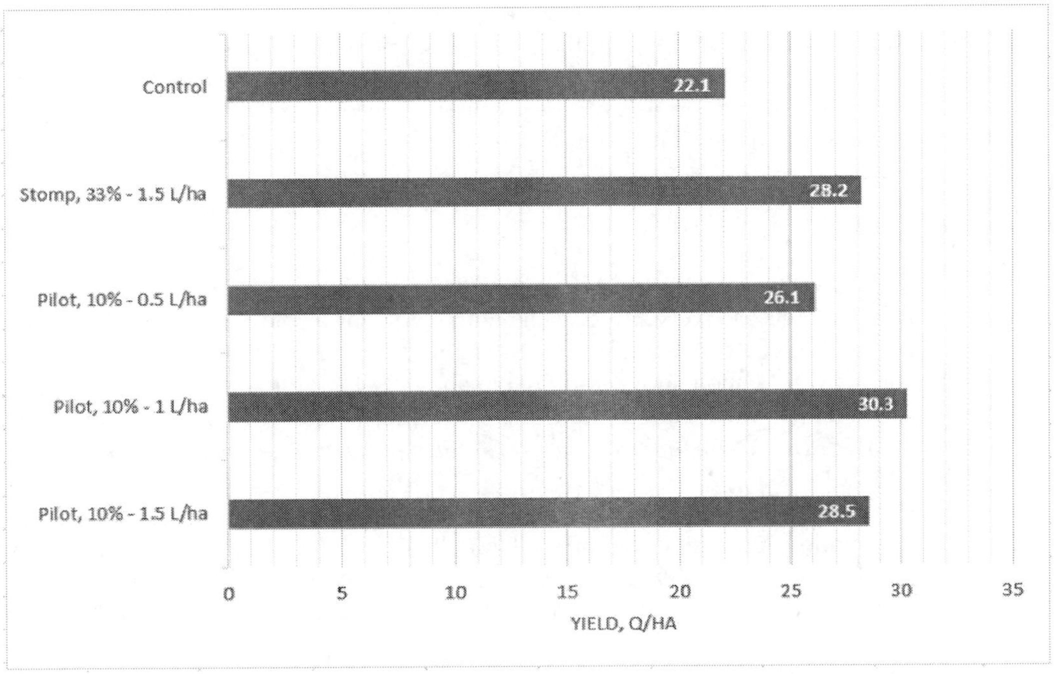

Figure 2. Soybean yield.

The first count was taken 15 days after herbicide application. In the Zellek-super sown variant, the damage of the soybean plant by aphids was 13.2%. This is an increase of 0.8% compared to the previous period. But compared to the control option, it was 10.3% less.

In the variant where Zeta herbicide was sprayed at the rate of 0.5 l/ha, 13.4% of soybean plants were infected with sorghum during the period of herbicide spraying. In the period of the first calculation, the pest infestation was 3.2% and the efficiency was equal to 81.2%. Compared to the control option, it was 82.6%. Zeta, 100 g/l herbicide at the rate of 0.8 l/ha was used before spraying the herbicide, and 16.6% of the soybean plant was infected with the sorghum. The number of plants affected by scurvy was 1.25% during the first count taken fifteen days later. The efficiency was equal to 92.4%. Damage was 22.3% less than the control variant. The effectiveness of the herbicide was 94.7% and had 15.2% of soybean plants affected by the weevil before herbicide application. After 15 days of herbicide spraying, the number of infected plants was 0.80%. In this option, the efficiency was equal to 94.7%. The efficiency compared to the control variant without herbicide spraying was 96.6%, and the number of soybean plants affected by the weevil during herbicide spraying was equal to 15.0%. 15 days after spraying the drug, the number of infected plants was 0.60%. In this option, the efficiency was equal to 96.0%.

In the second counting period, which was taken 30 days after herbicide application, the number of plants infested by the weevil increased in all options. In the control option, 48.2% of the soybean plant was damaged by the beetle. Zeta, 100 g/l at the rate of 0.8 l/ha, the damage was equal to 2.25%. This indicator was equal to 1.0% when this herbicide was used at the rate of 1.0 l/ha, and 0.75% when it was used at the rate of 1.2 l/ha.

4 Conclusions

Pilot herbicide 0.5; 1.0 and 1.5 l/ha, respectively, from the shade in the options used in the norms 26 .1; 30.3 and 28.5 q/ha yield was obtained. Soybean yield of 2 1.1 s /ha was obtained from the control variant without herbicide application. Therefore, the use of Pilot herbicide at the rate of 1.0 l/ha provides high yields of soybeans by effectively reducing the weeds.

Zeta, 100 g/l at the rate of 0.8 l/ha, the damage of the soybean plant by the weevil was equal to 3.75% during the harvest period. In this option, the efficiency compared to the initial damage was 77.4%. It was determined in the results of the experiment that the efficiency was 94.9 percent compared to the control option

References

[1] Isaev SS, Juraev AA, Juraev AK, Kadyrov ZZ, Soybean care in the conditions of meadow alluvial soils of the Bukhara region, prone to salting, irrigated from ancient times. Recommendation - Bukhara-2019 –12 p

[2] Abdalova, GN, Eshonkulov, JS, Sulaymanov, SO, & Abdullayeva, FM (2021). Improvement of cotton nutrition procedure and irrigation technologies. *ACADEMICIA: An International Multidisciplinary Research Journal* , *11* (4), 720-723 .

[3] Nasirov Bakhtior Salakhiddinovich Charshanbiyev Umurok Yuldashevich, Eshankulov Jamoliddin Saporboy Ugli. "Efficiency of application of herbicides which are samurai 33% ek, zellek super 10.4% ek and triflurex 48% ek against weeds in cotton fields" *Web of Scientist: International Scientific Research Journal* 2.09 (2021): 136-139.

[4] Salakhiddinovich, Nasirov Bakhtiyor., Eshankulov Jamoliddin Saporboy ugli 2021 "Development of Irrigation Procedures for Shadow Varieties Planted After Autumn Wheat." *International conference on multidisciplinary research and innovative technologies* . Vol. 1. 2021.

[5] Burievich, TB, Olimovich, A. Eshankulov JS, Turaevich, MT 2021 Groundwater consumption and cotton productivity. *Web of Scientist: International Scientific Research Journal* , 2 (09), 130-135.

[6] Norkulov U, Izbasarov B, Tukhtashev B, Eshonkulov J., Volume : 2 Issue : 2 2022 Effects of Sardoba Reservoir Flood on Irrigated Land, International Journal of

Innovative Analyzes and Emerging Technology e-ISSN : 2792-4025 40-42 p.

[7] Tukhtashev B, Norkulov U, Izbosarov B. Technology of proper use of saline soils in the conditions of Uzbekistan. E3S Web of Conferences 258, 03027 (2021)

[8] Izbasarov BE, Norkulov U, Tukhtashev, Hikmatov Sh Influence Of New Types Of Horizontal Ditches On The Growth, Development And Yield Of Winter Wheat In Saline And Groundwater Surface Soils. Influence Of New Types Of Horizontal Ditches On The Growth, Development And Yield Of Winter Wheat In Saline And Groundwater Surface Soils 2021

[9] Norkulov U, Tukhtashev B, Eshonkulov J., Volume : 2 Issue : 2 2022 Change of Mechanical Composition of Soils after Flood of Sardoba Water Reservoir, International Journal of Innovative Analyzes and Emerging Technology e-ISSN : 2792-4025 36-39 p.

[10] Ziyatov Musulman Panjiyevich, Shamsiyev Akmal Sadirdinovich, Kamilov Bakhtior Sultanovich, Abdalova Guliston Nuranovna, Abdurakhimov Shavkatjon Olimovich, Eshonkulov Jamoliddin Saporboy Ugli. PJAEE, 17(6) 2020 Effective agrotechnology of cotton feeding in different irrigation methods. Palearch's Journal Of Archeology Of Egypt/Egyptology 17(6). ISSN 1567-214x. 3415-3428 p. http://www.palarch.nl/index.php/jae/article/view/1335

[11] Eshonkulov Jamoliddin Saporboy Ugli , Kamilov Bakhtiyor Sultanovich, Shamsiyev Akmal Sadirdinovich, Nasirov Bakhtiyor Salakhiddinovich, Sheraliyev Khamidulla, Ziyatov Musulman Panjievich 2020 Appropriate irrigation procedures and cultivation agrotechnology of soybean and sunbackar varieties planted as reproductive crops. *PalArch's Journal of Archeology of Egypt/ Egyptology* , *17* (6), 3399-3414. Retrieved from https://archives.palarch.nl/index.php/jae/article/view/1333

[12] Shamsiyev Akmal Sadirdinovich, Eshonkulov Jamoliddin Saporboyugli, Sultanov Umbetali Tazabayevich 2020 Growth and development of soy and sunflower varieties. ACADEMICIA An International Multidisciplinary Research Journal 10(11):1289-1291

[13] Shamsiyev Akmal Sadirdinovich, Kamilov Bakhtiyor Sultanovich., Eshonkulov Jamoliddin Saporboyugli, Ashirov YR Agrophysical and agrochemical properties of influence of recycled soy and soil of the field 202 0 ACADEMICIA An International Multidisciplinary Research Journal August – India , 2020. – Vol. 10 . – Issue 8 . - P. _ 475-479

[14] Dusbayev IR, Nasirov BS, Ashirov YR, Eshonkulov JS, Rashidov Q 2021 Methods of planting fine liquid cotton and effects of Herbicides. 2nd International Conference on Science Technology and Educational Practices. Turkey 251-254 p.

[15] Eshonkulov Jamoliddin Saporboy, Shamsiev Akmal Sadirdinovich. Vol.5 NO. 2020 Congress (2020) Changes in water-physical properties of soil in repeated crop sunflower care. International congress on modern education and integration congress – India – Volume 5 . - P. 89-90.

[16] Eshonkulov J. Kamilov B. Effect of irrigation regimes on the fertility of soybean and sunflower cultivars planted in repeated periods . *IOP Conference Series: Earth and Environmental Science* Tom 1142, Vypusk 12023 Nomer status 0120533rd International Conference on Energetics, Civil and Agricultural Engineering, ICECAE 2022Virtual, Online13 October 2022 do 16 October 2022 Code 187394

Determination of the pathogenicity of the fungi *Leveillula Saxifragacearum* and *Sphaerotheca Mors Uvae* causing mildew disease

Gayrat Jumanazarov[1,*], *Ubaydulla* Rakhmanov[1], *Alisher* Omonlikov[1] and *Mukhiddin* Businov[1]

[1]Tashkent State Agrarian University, 2, University street, Tashkent, 100140, Uzbekistan

Abstract. Powdery mildew-causing fungi constitute around ten percent of the documented fungal species found in Uzbekistan. Among these fungi, a significant portion has developed adaptations to thrive in arid climatic conditions, reflecting their ability to endure dry environments. However, it's noteworthy that there also exists a substantial presence of fungi within this group that prefer moisture-rich surroundings. This variation in ecological preferences is evident not only between different fungal species but also within species belonging to the same botanical family. Such diversity in moisture requirements underscores the complex interplay between these fungi and the unique environmental conditions present in Uzbekistan. The article outlines a methodology and presents experimental data aimed at assessing the pathogenicity of two fungi, namely Leveillula saxifragacearum and Sphaerotheca mors uvae. These fungi are responsible for causing powdery mildew, a fungal disease, in the blackcurrant plant. The significance of this plant lies in its medicinal properties. The study provides insights into the mechanisms through which these fungi affect the blackcurrant plant and cause powdery mildew, which can have detrimental effects on its growth and health. By detailing the experimental approach and sharing gathered data, the article contributes to a better understanding of the interactions between these fungi and the medicinal blackcurrant plant. This research has the potential to guide efforts to manage and mitigate the impact of powdery mildew on blackcurrant cultivation, ultimately preserving and enhancing the plant's medicinal qualities.

Keywords. Pathogenicity, fungus, scab, seedling, disease, powdery mildew, spore, spread, development, *Sphaerotheca mors uvae*, *Leveillula saxifragacearum*, damage.

1 Introduction

Fungi that cause powdery mildew in plants make up one tenth of the recorded fungal species in Uzbekistan. Most of these fungi are adapted to dry climatic conditions, but there are also many moisture-loving species. Such a difference can be observed even among species belonging to the same family [1, 2].

* Corresponding author: g.jumanazarov@tdau.uz

The homeland of blackcurrant powdery mildew or spherotheca disease is North America, from where the disease entered Europe and later spread to other countries. This causative agent was considered to be an obligate parasite adapted only to kryzhovnik. But later, it was noted that this type of fungus, which causes powdery mildew, also severely damages blackcurrants. When black currant is infected with powdery mildew, its leaves are covered on both sides with white powdery mildew consisting of mycelium and conidia of the fungus. Later, the branches of the plant are completely covered with such dust [3-6].

Red currant shoots need a shorter period of time to develop than black currants. Because of this, it was observed that blackcurrant is severely damaged by powdery mildew [7, 8]. Golden black currant, which is grown as a landscape shrub, is strongly affected by powdery mildew [9, 10].

More damage from powdery mildew disease is observed when it damages the growth point of the bushes. In this case, the joint between the stem and the branches becomes short. The leaves formed in them are small, colorless, and often have a deformed appearance [11]. Diseased fruits develop slowly, if the disease is severe, they shrivel and fall off [12].

Spores of the fungus are produced in the spring and infect the growing point of blackberry shoots and young leaves. It takes 1-1.5 months for the spores to fly out of the bags, and for their germination, moisture is sufficient and the air temperature should not be lower than 15-17 °C [13, 14].

It was noted that blackberry leaves, branches, inflorescences were affected by powdery mildew caused by *Leveillula saxifragacearum* . The infected parts of blackcurrant were covered with first sparse, then dense white powders consisting of mycelia, conidiabands and conidia of the fungus. Diseased branches were behind in growth and development [15-17].

Blackberry leaves, branches and fruits are affected by powdery mildew caused by *Sphaerotheca mors - uvae* . The disease first affected the leaf bands, then the lower part of the leaf and the main veins. On the back side of such leaves, a gray mesh-like powder formed, consisting of mycelia of the disease-causing fungus [18-20]. Due to the disease, the leaves did not develop well and became colorless. On the surface of the fruits of the affected plant, dense, gray dust formed by the disease-causing fungus was visible. Fruits affected by the disease fell without development.

2 Materials and methods

To study the pathogenic properties of fungi that cause powdery mildew, it is most convenient to use cleistothecium mevatanas, which occur during the formation of their bags. The ascospores produced by these mevatanas have the potential to infect more plant species than fungal conidia. But it is not always possible to use these fruits. Because the fungal species, depending on the conditions of the area where they are studied, may produce cleistothecium fruiting bodies in very small quantities in the affected plant parts, or some species may not produce them at all. Some fungi need to be overwintered to restore fruiting bodies. Ascospores inside them lose their sterility due to rapid decay of the skin of some fruits in aqueous environment. Therefore, when determining the pathogenicity of fungi that cause powdery mildew, spores during the formation of conidia are often used. Because it is possible to multiply conidia of fungi in large quantities in natural substrates in a short period of time. To carry out the experiments, the leaves and branches of the plant infected with powdery mildew disease were collected under natural conditions and placed in wet chambers in Petri dishes.

To form a humidity chamber, Petri dishes were sterilized with filter paper and sterilized in an autoclave at 121°C for 20 minutes at a pressure of 0.5 atm. After removing the Petri dishes from the autoclave, they were moistened by pouring 1 ml of sterilized water into each of them in a laminar box. Petri dishes prepared in this way were used to conduct the above experiments.

The Petri dishes containing the parts of the plant affected by powdery mildew were kept for 1 day in a thermostat with a temperature of 24-26°C. Then, the fungal conidia that appeared on the surface of the affected plant parts inside the Petri dishes were collected by tapping them from the plant parts into new sterile Petri dishes. After the collection of conidia of powdery mildew fungi, their viability was determined in laboratory conditions. To test the viability of conidia in the laboratory, preparations were prepared from them and these preparations were observed under a microscope. If the vacuoles of conidial cells are clearly visible, such conidia have not lost their growth properties. If the cells of the conidia are filled with a granular mass, such conidia are considered to have lost their ability to germinate. For artificial infection of plants, options were selected on the basis of this criterion, that is, conidia germination of not less than 70%.

According to literary sources, the incubation period of fungi that causes powdery mildew in plants can last from 2 to 28 days, depending on the temperature. It was determined that high humidity reduces the germination of conidia and reduces the ability to infect plants [10, 13, 19]. These aspects were taken into account when artificially infecting plants with these fungi.

3 Results and discussion

In the conditions of Tashkent region, two types of powdery mildew disease were noted in currants. The causative agent of the first appearance *is Leveillula saxifragacearum* Golov. f. ribis Golov. is a type of fungus , and the appearance of the second type was recorded only once during all studies, and its causative agent is *Sphaerotheca mors uvae* (Schw.) Berk. et Curt. it was determined to be a type of fungus, and it became known that the disease in this form is called "American mildew" in scientific literature (Table 1).

Table 1. Types of fungi causing powdery mildew in currants.

The type of fungus that causes the disease	Name of the causative disease	Occurrence
Leveillula saxifragacearum Golov. f. ribis Golov.	Flour is dew	+++
Sphaerotheca mors uvae (Schw.) Berk. et Curt.		+

Experiments on the study of fungi that cause powdery mildew were carried out on seedlings grown from cuttings in flower beds. The average temperature of the laboratory during the period of conducting experiments on blackcurrant plant was 12-14°C at night and 28-30°C during the day. Experiments were conducted in three variants. In variant I, *Leveillula saxifragacearum* Golov. f. ribis Golov. of the fungus, in variant II *Sphaerotheca mors uvae* (Schw.) Berk. et Curt. Conidia of the American powdery mildew fungus were used. In option III, blackberry seedlings that were not affected by the disease were used as a control.

Currant infection with powdery mildew was carried out using the above methods. Experiments were carried out in three replicates, and four pots with two seedlings planted in each replicate were used. Disease development in blackberry seedlings artificially infested with powdery mildew fungi was observed and counted from the third day.

The first symptoms of the disease appeared in blackcurrant seedlings 6 days after infection. On the 9th day, all the seedlings in the experimental options were 100% infected with the disease (Figure 1).

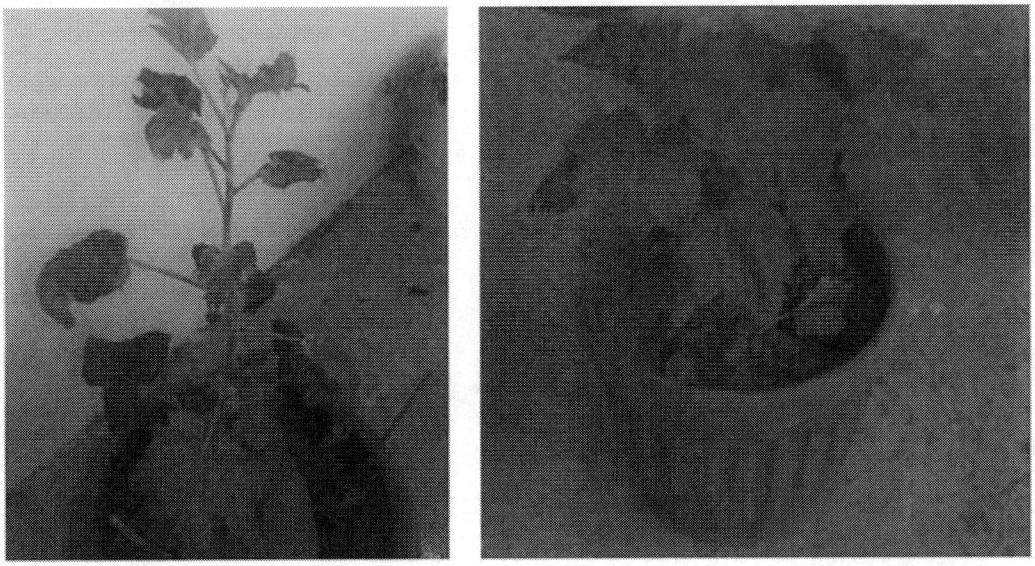

Figure 1. Blackcurrant seedlings artificially infested with powdery mildew fungus.

The development of powdery mildew was observed to be 1.8-2.0% in option I and 1.4% in option II on the 6th day, and this indicator reached 2.3-2.8% and 1, It was observed that it was 9-2.2%.

Leveillula saxifragacearum f., which causes the widespread powdery mildew disease in all farms of Tashkent region of Uzbekistan. It was found that the disease spread early and the development was accelerated in the seedlings of variant I of the experiments using ribis fungus.

Sphaerotheca mors uvae, which was recorded once in blackcurrant in the high mountain area of Parkent district of Tashkent region, the development of the disease in seedlings was relatively slow, but later its development accelerated.

Symptoms of powdery mildew diseases caused by fungi used in options I- and II- were the same as the diagnosis of the disease encountered in farms (Tables 2 and 3).

Table 2. Pathogenicity of powdery mildew fungi on currant seedlings.

#	Experiments	Number of pots in experimental iterations	Number of seedlings in one pot	Disease prevalence, %					
				Days of receipt of the bill of illness					
				3	6	9	12	15	18
1	**Option I** *Leveillula saxifragacearum* f. ribis fungus was used	4	2	-	50.0	100	100	100	100
2		4	2	-	50.0	100	100	100	100
3		4	2	-	-	50.0	100	100	100

4	**Option II** The fungus *Sphaerotheca mors uvae* was used	4	2	-	50.0	100	100	100	100
5		4	2	-	-	50.0	100	100	100
6		4	2	-	-	50.0	100	100	100
7	**Option III** Control (undamaged)	4	2	-	-	-	-	-	-
8		4	2	-	-	-	-	-	-
9		4	2	-	-	-	-	-	-

Table 3. Pathogenicity of powdery mildew fungi on currant seedlings.

#	**Experiments**	**Number of pots in experimental iterations**	**Number of seedlings in one pot**	**Disease development, % the days of receipt of the bill of illness**									
				3	**6**	**9**	**12**	**15**	**18**	**21**	**24**	**27**	**30**
1	**Option I** *Leveillula saxifragacearum* f. ribis fungus was used	4	2	-	1.8	2.5	3.1	5.4	7.6	10.5	15.3	22.6	30.8
2		4	2	-	2.0	2.8	3.2	5.6	7.8	10.9	15.7	22.8	29.5
3		4	2	-	-	2.3	2.9	4.8	7.1	8.0	12.3	18.4	25.0
4	**Option II** The fungus *Sphaerotheca mors uvae* was used	4	2	-	1.4	2.2	3.0	4.7	6.6	7.0	11.8	19.7	30.4
5		4	2	-	-	2.0	2.8	4.5	6.3	7.9	10.8	17.0	27.5
6		4	2	-	-	1.9	2.6	4.4	6.2	7.4	10.6	16.4	28.0
7	**Option III** Control (undamaged)	4	2	-	-	-	-	-	-	-	-	-	-
8		4	2	-	-	-	-	-	-	-	-	-	-
9		4	2	-	-	-	-	-	-	-	-	-	-

Although the development of the disease in blackberry seedlings artificially infested with the fungus Sphaerotheca mors uvae II was a little slower in the initial period, the disease developed rapidly, and after 45 days after infection, the leaves of all the seedlings in this variant dried and hung on the plant without shedding . All the sick seedlings turned brown.

The withering condition of the seedlings of variant I was observed on the 52nd day after infection. The leaves of the dead seedlings became brittle when plucked by hand. All such seedlings turned gray.

During the determination of the pathogenicity of blackcurrant seedlings by artificially infecting them with powdery mildew fungi, it was found that the plant will die completely within 50 days if the necessary conditions for the development of the disease exist. For the development of the fungus *Sphaerotheca mors uvae,* the environment with slightly low temperature and high humidity is considered favorable.

4 Conclusions

Blackcurrant seedlings were intentionally exposed to Leveillula saxifragacearum and Sphaerotheca mors uvae fungi. Within a span of 12 days, the disease had completely proliferated in both scenarios, encompassing a spread rate of 100%. Subsequently, in the first experimental condition (option I), over a period of 52 days, the disease displayed a developmental extent ranging from 63.5% to 64.2%.

Conversely, in the second experimental condition (option II), the disease exhibited a slightly lower development range of 61.4% to 63.9%, and this occurred over a shorter timeframe of 45 days. These findings highlight the rapid and consistent progression of the disease caused by the mentioned fungi in blackcurrant seedlings, shedding light on the differing rates of disease development between the two experimental setups.

References

1. Agabekyan M.B. Muchnistaya rosa rozy i sovershenstvovanie khimicheskikh mer borby s neyu v usloviyax Araratskoy raviny Armyanskoy SSR / / Autoref . sugar teeth - Yerevan: 2015. - 21 p.
2. Astakhov A.I. Genotype analysis sorta Bpedtopp i vliyanie spedy na ppoyavlenie istoychivosti k amepikanskoy mychnistoy pose // Dock . VASX-NIL . 2015. No. 11. - P.21-23.
3. Goleneva L. M. Evaluation of the offspring of the Scandinavian group of varieties of black currant after the reaction to the American powdery mildew (*Sphaerotheca Morse - uvae*. Schw.).-M.: 2010. -45 p.
4. Golovin P.N. Mycoflora Sredney Azii. Tom I. _ Muchnisto - Russian mushroom Sredney Azii. Vyp 1. - Tashkent: Iz- va AN UzSSR , 1999. -145 p.
5. Jumanazarov G. Kh., Zuparov M. A. Spread, vredonosnost anthracnose of currants and effectiveness of fungicides against borby s ney //innovatsionnye pokhody v sovremennoy nauke. - 2020. - S. 63-71.
6. Natalina O.B., Gorkovenko V. S., Puzanova L. A. There are many black and red currants in the Krasnodar region. 2011. -55 p.
7. Sankin L.S. Selection smorodinovo-kryzhovnikovykh hybridov na Altae//Sb. n. tr.: VNII sadovodstva. 2010. T.55. - p. 53-55.
8. Tikhonova O.A. Osobennosti biologii i selektsionnaya cennost genofon black currant v usloviyax severo-zapada Rossii.- Autoref . s. x. science SPb.: 2000. -19 p.
9. Khokhryakov M.K. Opredelitel bolezney selskohozyaystvennyx kultur. - L: Kolos, 2014. -250 p.
10. Jumanazarov G. et al. Distribution and damage of powdery mildew disease in cassis and raspberry in Tashkent province, Uzbekistan //E3S Web of Conferences. – EDP Sciences, 2021. – T. 284. - p. 03017.
11. Zuparov MA Incidence, severity and yield loss of powdery mildew disease of raspberry and efficiency of fungicides against it in Uzbekistan. International Journal of Agriculture, Environment and Bioresearch, Vol. 5, No. 05; 2020. P. 191-198.
12. Zuparov M. A. Pathogenicity of fungi causing powdery mildew disease in rosehip //Innovative investigations: theoretical and practical bases. - 2020. - P. 154.
13. Zuparov M. A. Prevalence, damage anthracnose of currant and efficiency of fungicides to control it// Innovative approaches in modern science. - 2020. - P. 63-71.
14. Allayarov A. Nosirov S. Spread and severity of fungal diseases of currant plant in the conditions of Tashkent province, Uzbekistan. IOP Conference series: Earth and Environmental Science. IOP Publishing, 2022. – No. 10. – S. 012002.

15. Jumanazarov G.Kh. The effectiveness of the use of disinfectants for rosehip seeds against Fusarium wilt of seedlings. Founders: LLC "Publishing House" Sputnik+ //actual problems of modern science. - No. 1. - P. 27-30.
16. Nazarova O., Khujaev O., Jumanazarov G. Septoria (Septoria pistaciae Desm.) pathogen infecting pistachio nuts in Uzbekistan and its biological characteristics //IOP Conference Series: Earth and Environmental Science. - IOP Publishing, 2023. - T. 1142. - no. 1. – S. 012051.
17. Khakimov, A., Salakhutdinov, I., Omolikov, A., & Utaganov, S. (2022). Traditional and current-prospective methods of agricultural plant disease detection: A review. In IOP Conference series: earth and environmental science (Vol. 951, No. 1, p. 012002). IOP Publishing.
18. Allayarov, A., Zuparov, M., Khakimov, A., & Omonlikov, A. (2021). Application of the biopreparation 'Orgamika F'against fusarium disease of cabbage and other cole vegetables. In E3S Web of Conferences (Vol. 284, p. 03011). EDP Sciences.
19. Sodikov, B., Khakimov, A., Rakhmonov, U., Omonlikov, A., Gulmatov, R., & Utaganov, S. (2022, July). Soil-borne plant pathogenic fungi biodiversity of sunflower. In IOP Conference Series: Earth and Environmental Science (Vol. 1068, No. 1, p. 012018). IOP Publishing.
20. Mutabar Khojanazarova, Rano Artikova, Gulchekhra Khalmuminova, Alisher Omonlikov . Spectrophotometric indicators of "Zamin-M" biopreparation immobilized with gypan flocculinate. E3S Web of Conf. 389 03076 (2023). DOI: 10.1051/e3sconf/202338903076

Increasing the quality and productivity of tomato fruits under the impact of a growth regulators

Rano Alimova[1,*], *Mirkosim* Sagdiev[1] and *Alisher* Omonlikov[1]

[1]Tashkent State Agrarian University, 2, University street, Tashkent, 100140, Uzbekistan

Abstract. Evaluating the impact of growth regulators is a simple, inexpensive method that is considered an environmentally safe method to improve plant yield and quality, increase growth and productivity.The obtained results of the research on the influence of the Humosteem preparation on the productivity and quality of tomato plants have been presented herein.In foreign countries, 50-80% of tomato crops are treated with such preparations. Their number has been increasing in recent years.Due to the high biological activity of this preparation, it was noted, during the research, that it had a positive effect on local and hybrid varieties of tomatoes during the entire growing season, showed high efficiency in the absorption of nutrients, growth, development and productivity of the plant. It was observed that the content of nitrites and nitrates in plant products did not exceed the state standard requirement. It was determined that the amount of nitrites and nitrates increased by 17.07% in the Yablochniy variety, by 16.5% in the Lojain F1 hybrid, and by 13.3% in the TMK-22 variety under the impact of the Humosteem preparation. Thus, it was observed that the amount of nitrates was low in tomato variety TMK-22.
Keywords. Productivity, growth regulator, “Humosteem” preparation, seed treatment, tomato quality.

1 Introduction

Cultivation of vegetable crops in the soil-climate conditions of Tashkent region consists of timely implementation of agricultural-technical measures and enhancement of economic technologies, application of highly effective methods of irrigation, fertilizing, plant protection and mechanization of technological processes.

The focus is hereby mainly on creating environmentally friendly agrotechnologies, growing products without polluting the environment and without toxic substances, applying new generation drugs to agricultural production, solving problems [1 ,2].

Evaluating the effect of growth stimulants is a simple, inexpensive method that is considered an environmentally safe method to improve plant yield and quality, increase growth and productivity. In foreign countries, 50-80% of tomato crops are treated with such preparations. Their number has been increasing in recent years [3-5].

* Corresponding author: r.alimova@tdau.uz

The advantage of growth regulators is that even if they are used in small amounts, they have a sufficient influence on the physiological and biochemical processes in plants and enable them to be controlled [5-9]. Tomato is the most consumed vegetable in the world, and its varieties are distinguished according to yield and nutritional value. Tomato fruit contains carbohydrates, organic acids, mineral salts, aromatic substances and many vitamins [10, 11].

Among the growth regulators, the biostimulant preparation "Humosteem" has successfully passed approval in practice and is widely used in agriculture [12, 13]. The composition of this preparation embodies the feature of high biological activity, affects plants throughout the growing season, increases plant growth and productivity due to better absorption of nutrients [14]. The purpose of the research was to determine the efficacy of the Humosteem stimulant in the cultivation of tomatoes in the soil-climatic conditions of the "Bur ota" farm of Kibray district of Tashkent region.

The object of the research wastomato varieties "Yablochniy" (as standard), "TMK-22" and tomato hybrid "Lojain F1". Experiments were triplicated.Experiment's layout was systematic. Drip-irrigation was used. Planting schemewas 0,90 x 0,5 m. 1 kg seed was used per ha (35 00 seedlings perha). A 1% solution of potassium permanganate (1 g of crystals per 100 ml of water) was placed into gauze bags with a label according to tomato varieties for a few minutes, and then washed in distilled water for 20-30 minutes.The working solution prepared with the growth regulator "Humosteem" at 2 g/kg was soaked for 30-40 minutes before planting the seeds. Then the seeds were dried and planted. Planting was carried out using 1.4 sections. Except the roots, other plant partsweretreated with solutionof 15 kg of Humosteem with 300 g of water per hectare during the entire vegetation period (spraying at the beginning of growth, spraying at the beginning of budding and flowering) [15]. Irrigation was in a differentiated mode, the average soil moisture was 70-75%. The average amount of water absorbed during the year was 6800-7500 m^3/ha.

2 Materials and methods

The research was conducted during 2020-2022 in the cultivated fields of the "Bur ota" farm in Kibray district of Tashkent region. The climate of this place is strongly continental, but due to irrigation and strong insolation - light and sunlight, the vegetation and development of vegetable crops is activated. For example, 2020 was characterized by drought (GTK - 0.69), 2021 was dry (GTK - 0.38) and 2022 was very dry (GTK - 0.55).

The soil includes to light brown soil subgroup, and its granulometric composition is similar to the medium and heavy clay variety,soil contains of 1.5 – 2.0% humus, 3.8 – 8.9 mg per 100 g soil hydrolyzable nitrogen, 2.7 – 3.5 mgphosphorus, 300 – 450 mg/kg exchangeable potassium, the soil solution is weakly alkaline, the amount of absorption corresponded to 26-30 mg/100 g of soil. Experimental research was carried out on the basis of "Experimental methods in vegetable growing and policing" [8] and "Methods of field experiments" [3].

Humosteem preparation is a dark brown solution with a special smell. It dissolves well in water. The exchange acidity is around pH 7–9, the total percentage of humic acids is 2.4–2.7%, it has a 4th class hazard (low hazard substance), so it is used in the cultivation of environmentally friendly vegetables.Application rate of working solution is 0,001% humic acid, for spraying to plantsat the rate 1.5 – 5 l/kg, consumption rate of Humosteem (1% humic concentration) 1.5 – 2.0 ml/kg, for spraying to plants 0.15 – 0.3 l/mg.

The phenological observations of tomato growth and development were carried out in specially designated areas by generally accepted methods. Using these methods, the beginning periods of phenological phases, the main phenological phases of growth and development were studied. The initial phase of growth included 10% of plants, total phase included 75% of plant growth.

Biometric indicators of tomato were periodically determined on the 1st day of every month. From each variant certain plants were taken to measurethe number, height, plant mass (weight), fruit diameter, size and weight. Irrigation was performed by drip irrigation method. It was in the mode of differential soil moisture in the active layer at an optimal humidity of 70-80-75% (Table 1).

Table 1. Total water absorption of tomato (average as in 2020-2022).

Atmospheric rainfall		**Irrigation rate**		**Moisture consumption from soil reserves**		**Total water absorption m³/ha**
m³/ha	%	m³/ha	%	m³/ha	%	6900.0
1390	11.3	4900	81.8	429	6.9	

It is obvious from the table that the moisture in the tomato field was 81.8% due to irrigation, and the moisture in the soil during the tomato planting period was 45 mm, which was only 6.9% of the water balance. Productivity was calculated by hand-picking, then into marketable and non-marketable product parts, and the yield structure was studied in each experimental variant during the harvesting period.

As a result of the use of Humosteem growth regulator during the vegetation period, the yield of tomatoes increased to 10.25 kg/m^2. This indicator was observed in all variants. The application of the preparation led to accumulation of sugar and dry matter from essential compounds in tomato fruits. A large amount of vitamin C was synthesized in all tomato varieties and hybrids, and their nutritional value was improved. The lowest amount of nitrate accumulation was observed in TMK-22 variety. Thus, the application of the growth regulator had a positive effect on the yield and quality of tomato.

3 Results and discussion

The effect of growth stimulant was studied when tomato was treated with it before sowing the seeds. The effectiveness of the growth regulator began to manifest in the 1st stage of the experiment of tomato ontogeny. Changes in biochemical processes and improvement of seed germination quality were observed (Table 2).

Table 2. Effect of Humosteem growth regulator on seed germination.

#	**Experimental options**	**Nuber of germinated seeds (pcs)**	**Germinability**	**Number of days from sowing to mass shoots**	**Root height, mm**	**Plant height, mm**
			Yablochny			
1	Control (soaked in water)	72	72.78	18	3.8	5.3
2	Seeds soaked in "Humosteem" preparation 1 ml/kg	82	83.8	15	4.9	5.8
			Lojain F1			
1	Control	74	75.2	18	4.3	6.2

	(soaked in water)					
2	Seeds soaked in "Humosteem" preparation 1 ml/kg	85	85.6	13	5.4	6.8
			TMK-22			
1	Control (soaked in water)	79	78.3	18	4.9	6.2
2	Seeds soaked in "Humosteem" preparation 1 ml/kg	94	94.5	12	5.6	6.9

The obtained results showed that the germination of tomato seeds in laboratory conditions was higher than those treated with water. Long primary branches and shoots were well formed in variants of application of growth regulators. Tomato productivity as a result of the application of growth regulators is presented in the accompanying table (Table 3).

Table 3. Effect of "humosteem" preparation on the productivity of tomato (in 2020-2022).

#	Experimental options	Productivity kg/m^2	Addition to control kg/m^2	Number of fruits kg/m^2	Average mass of fruits kg/m^2
		Yablochny			
1	Control (soaked in water)	87.55	-	11.2	80
2	Seeds soaked in "Humosteem" preparation 1 ml/kg	8.58	1.07	11.9	82
3	1st spray in growth phase	9.37	1.85	12.3	84
4	2nd spray in budding and flowering phases	10.15	2.65	13.3	86
		Lojain F_1			
1	Control (soaked in water)	9.4	-	14.5	65
2	Seeds soaked in "Humosteem" preparation 1 ml/kg	10.11	0.82	15.4	68
3	1st spray in growth phase	10.94	1.52	16.1	72
4	2nd spray in budding and flowering phases	11.56	2.15	16.8	78
		TMK-22			
1	Control (soaked in water)	9.8	-	13.3	85
2	Seeds soaked in "Humosteem" preparation 1 ml/kg	10.44	0.75	14.1	88
3	1st spray in growth phase	11.62	1.86	14.8	90
4	2nd spray in budding and flowering phases	12.65	2.82	15.4	93

The application of growth regulator preparation "Humosteem" had a positive effect on the quality indicators of tomato fruits (Table 4).

Table 4. Effect of "Humosteem" preparationon the quality of tomato fruits.

#	Experimental options	Dry matter, %	Amount of vitamin Cmg/kg	Total sugar content, %	Acidity, %	Nitrate content, mg/kg
Yablochny						
1	Control (soaked in water)	3.7	15.8	2.5	0.55	90.4
2	Seeds soaked in "Humosteem" preparation 1 ml/kg	4.5	15.9	2.7	0.57	95.5
3	1st spray in growth phase	5.1	16.2	2.8	0.58	98.3
4	2nd spray in budding and flowering phases	5.9	16.5	2.9	0.59	105.3
Lojain F_1						
1	Control (soaked in water)	4.5	16.0	2.8	0.53	91.3
2	Seeds soaked in "Humosteem" preparation 1 ml/kg	5.1	16.2	3.0	0.55	96.1
3	1st spray in growth phase	5.7	16.3	3.1	0.56	97.2
4	2nd spray in budding and flowering phases	6.1	16.4	3.2	0.57	106.2
TMK-22						
1	Control (soaked in water)	4.2	15.5	2.4	0.53	91.7
2	Seeds soaked in "Humosteem" preparation 1 ml/kg	4.6	15.6	2.7	0.54	94.2
3	1st spray in growth phase	5.2	15.8	2.9	0.55	98.6
4	2nd spray in budding and flowering phases	5.8	16.1	3.1	0.55	103.8

The following were taken as morphological indicators: plant height, stem thickness, number of flowers and buds. These parameters were determined in budding and flowering phases. During the application of preparation, the height of the plants increased. In all variants, compared to the control, it was observed that the increase in plant height was in Yablochny - 33.75%, in Lojain F1 hybrid - 51.68%, in TMK-22 variety - 64.16%.

The productivity of tomato was observed from 7.5 kg/m^{2}to 9.8 kg/m^{2}in the control by variants. In all variants, an increase in productivity was observed when the preparation Humosteem was used. When Humosteem growth regulator was used during at the vegetation period, the indicator ranged from 10.15 kg/m^2 to 12.65 kg/m^2 in all options.

As a result of the use of this preparation, the main compounds – sugars dry matter - accumulated in tomato fruits. In all experimental variants, the content of vitamin C in tomato fruits increased and their nutritional value improved.

It was stated that the content of nitrites and nitrates in plant products did not exceed the state standard requirements. According to this standard, the content of nitrites and vanitrates in tomato fruits should not exceed 150 mg/kg. It was determined that the amount of nitrites and nitrates increased by 17.07% in the Yablochny variety, by 16.5% in the

Lojain F1 hybrid, and by 13.3% in the TMK-22 variety under the impact of the Humosteem preparation. It was also observed that the amount of nitrates was low in tomato variety TMK-22.

As a result of the application of the growth regulator Humosteem, the taste and quality of tomato fruits improved, and it provided an opportunity to obtain high-quality raw materials for the product processing sector.

4 Conclusions

The study delved into the practical application of a growth regulator, specifically focusing on the effects of the "Humosteem" preparation. This investigation unveiled that the utilization of this growth regulator has a positive impact on the productivity and quality of tomato products cultivated in an ecologically clean manner.

Through rigorous research, the study generated valuable insights and developed recommendations that enable the efficient utilization of cultivated areas. As a result of these findings, strategies were formulated to enhance both the productivity and quality of tomatoes in the Kibray district of the Tashkent region.

This research has important implications for agricultural practices, as it offers a tangible approach to improving tomato yields and their overall quality, while also considering ecological considerations. By harnessing the positive effects of the "Humosteem" growth regulator, the study contributes to the optimization of tomato production processes in the specific geographic context of the Kibray district of Uzbekistan.

References

1. Alimova R. A. et al. Plant growth stimulants and the effect of foliar feeding on the yield of some vegetable crops //Eurasian Research Bulletin. – 2022. – T. 9. – C. 30-38.
2. Allayarov, Abdurakhman & Zuparov, Mirakbar & Khakimov, Albert & Omonlikov, Alisher. (2021). Application of the biopreparation ‘Orgamika F’ against fusarium disease of cabbage and other cole vegetables. E3S Web of Conferences. 284. 03011. 10.1051/e3sconf/202128403011.
3. Khakimov, Albert & Omonlikov, Alisher & Utaganov, Samad. (2020). Current status and prospects of the use of biofungicides against plant diseases. 13. 119-126. 10.30574/gscbps.2020.13.3.0403.
4. Khakimov, Albert & Salakhutdinov, I & Omonlikov, Alisher & Utaganov, Samad. (2022). Traditional and current-prospective methods of agricultural plant diseases detection: A review. IOP Conference Series: Earth and Environmental Science. 951. 012002. 10.1088/1755-1315/951/1/012002.
5. Sagdiev M. T., Amanova M. M., Omonlikov A. U. The influence of growth regulators on tomato productivity in the conditions of the Tashkent region //ISJ Theoretical & Applied Science, 11 (79). – 2019. – C. 241-244.
6. Sagdiev M. T., Amanova M. M., Omonlikov A. U. The influence of plant growth stimulators on tomato productivity in the conditions of Tashkent region. World Journal of Pharmaceutical and Life Sciences,(WJPLS //India. – 2020. – T. 6. – №. 8. – C. 04.
7. Sagdiev M., Omonlikov A. Influence of plant growth stimulants on tomato productivity in the conditions of the tashkent oasis //Journal of Agriculture & Horticulture. – 2023. – T. 3. – №. 4. – C. 58-63.
8. Tayjanov, K., Khojimatov, O., Gafforov, Y., Makhkamov, T., Normakhamatov, N., & Bussmann, R. W. (2021). Plants and fungi in the ethnomedicine of the medieval East-a

review. Ethnobotany Research and Applications, 22, 1-20. DOI:0.32859/ERA.22.46.1-20

9. Jabeen, S., Zafar, M., Ahmad, M., Althobaiti, A. T., Ozdemir, F. A., Kutlu, M. A., ... & Majeed, S. (2023). Ultra-sculpturing of seed morphotypes in selected species of genus Salvia L. and their taxonomic significance. Plant Biology, 25(1), 96-106. DOI:10.1111/plb.13473
10. Majeed, S., Ahmad, M., Ozdemir, F. A., Demirpolat, A., Şahan, Z., Makhkamov, T., ... & Nabila. (2023). Micromorphological characterization of seeds of dicot angiosperms from the Thal desert (Pakistan). Plant Biosystems-An International Journal Dealing with all Aspects of Plant Biology, 157(2), 392-418. DOI:10.1080/11263504.2023.2165553
11. Majeed, S., Ahmad, M., Ali, A., Althobaiti, A. T., Ramadan, M. F., Kilic, O., ... & Sultana, S. (2023). Pollen micromorphology among amaranthaceous species from desert rangeland: Exine stratification and their taxonomic significance. BioMed Research International, 2023, 4967771. DOI:10.1155/2023/4967771
12. Makhkamov, T., Sotiboldiyeva, D., Mamarakhimov, O., Yuldashov, Y., & Botirova, L. (2022, May). Morphogenesis and Seasonal Developmental Rhythm Under the Conditions of Introduction of Curcuma Longa L. In International Scientific Conference on Agricultural Machinery Industry “Interagromash"” (pp. 1460-1469). Cham: Springer International Publishing. DOI:10.1007/978-3-031-21432-5_155
13. Boboev, S., Makhkamov, T., Bussmann, R. W., Zafar, M., & Yuldashev, A. (2023). Anatomical and phytochemical studies and ethnomedicinal uses of Colchicum autumnale L. Ethnobotany Research and Applications, 25, 1-9. DOI:10.32859/era.25.6.1-9
14. TKh, M., Brundu, G., Jabborov, A. M., & Gaziev, A. D. (2023). Predicting the potential distribution of Ranunculus sardous (Ranunculaceae), a new alien species in the flora of Uzbekistan and Central Asia. BioInvasions Records, 12(1), 63-77. DOI:10.3391/bir.2023.12.1.05
15. Ameen, M., Zafar, M., Ahmad, M., Ramadan, M. F., Eid, H. F., Makhkamov, T., ... & Majeed, S. (2023). Assessing the Bioenergy Potential of Novel Non-Edible Biomass Resources via Ultrastructural Analysis of Seed Sculpturing Using Microscopic Imaging Visualization. Agronomy, 13(3), 735. DOI:10.3390/agronomy13030735

Effectiveness of the use of various biostimulants in increasing the productivity of squash crops

Mokhidil Baratova[1,*] and *Salokhiddin* Yunusov[2]

[1]Andijan Institute of Agriculture and Agrotechnologies, Andijan, 170012, Uzbekistan
[2]Tashkent State Agrarian University, Tashkent, 100140, Uzbekistan

Abstract. This article examines the problem of using natural biostimulants and creating new biologically active substances, introducing them into production, and increasing medicinal properties to increase the productivity of pumpkin crops in Uzbekistan. In this case, moisturising of pumpkin and cucumber seeds in biostimulators before planting, spraying during the growing season, and mixed (seed moisturising and spraying during the growing season) methods were used. As a solution to this problem, it was determined that it is useful to study the effect of biostimulants on pumpkin crops and to improve the quality of the product, to enrich its biochemical composition and to determine its medicinal properties. Research was conducted in 2018-2020 at Gulistan State University and Andijan Institute of agriculture and agrotechnology. In the cultivation of pumpkin and cucumber, plants were treated with Gossipren, Uchqun, Uchqun plus and Verva biostimulants, and the efficiency of Uchqun plus biostimulant at different rates was determined. When 0.01% Uchqun plus biostimulant was used in a mixed method, it was determined that the productivity of Ispanskaya 73 variety of pumpkin increased by 31.5 t/ha and that of Palov kadu 268 variety increased by 31.8 t/ha. When cucumber plants were affected by this method - Orzu F1 - 35.1 t/ha and 34.5 t/ha were obtained in Navruz variety.
Keywords. Gossipren, Uchqun, Uchqun plus, Verva, pumpkin, cucumber, Ispanskaya 73, Palov kadu 268.

1 Introduction

Squash (*Cucurbita pepo*) stands as a vital component of global agriculture, contributing to both culinary diversity and nutritional enrichment. As the demand for sustainable agricultural practices continues to grow, researchers and growers alike are exploring innovative techniques to enhance the productivity and quality of squash crops [1]. Among these strategies, the application of biostimulants has garnered considerable attention due to their potential to stimulate plant growth, improve nutrient uptake, and bolster overall crop performance [2, 3].

Biostimulants encompass a diverse range of substances, including natural compounds,

* Corresponding author: mbaratova1975@gmail.com

microbial products, and organic materials, that are specifically designed to augment plant physiological processes [4]. By influencing root development, nutrient assimilation, and stress tolerance, biostimulants offer a multifaceted approach to optimizing crop production. While biostimulants have been widely investigated in various agricultural contexts, their effectiveness in boosting the productivity of squash crops remains a subject of growing interest and scientific inquiry [5, 6].

The objective of this study is to comprehensively evaluate the impact of different biostimulants on the growth, development, and yield of squash plants. By systematically comparing the outcomes of various biostimulant treatments, we aim to discern the most promising formulations for enhancing squash crop productivity. Our investigation delves into the potential effects of biostimulants on root morphology, nutrient absorption, flowering patterns, and fruit yield. Through this research, we aspire to provide valuable insights into the utilization of biostimulants as a sustainable means to elevate squash crop performance.

As the global population continues to expand, ensuring food security through efficient and sustainable agricultural practices is paramount [7-9]. Understanding the potential of biostimulants to enhance squash crop productivity holds promise for meeting this challenge [10, 11]. The outcomes of this study may not only contribute to advancing our knowledge of biostimulant application but also offer practical solutions to farmers seeking to optimize squash production in an environmentally responsible manner.

2 Materials and methods

The research was conducted in 2018-2020 in the experimental laboratory of Gulistan State University and in the fields of Andijan Institute of Agriculture and Agro-Technology. In this case, 4 biostimulants - Gossiprin (0.01%), Uchqun (0.01%), Uchqun plus (0.01%) and Verva (5%) were added to pumpkin Ispanskaya 73, Palav kadu 268 and cucumber Navruz and Orzu F1 varieties. Effects compared. In the experiment, a variant planted with clean water was taken as a control. In addition, in the experiment, these drugs were studied in 3 different ways, i.e. moisturising the seeds for 18 hours, foliar spraying on the plant during the growth period, and mixed methods, i.e. foliar spraying with moisturising of the seeds. Experiment with 4 replicates, feeding area is 14.0 m^2 in pumpkin and 8.4 m^2 in cucumber. The area is double, 4 m long, 4.2 m wide for pumpkins and 2.1 m for cucumbers. 14 pumpkin and 20 cucumber varieties were placed in the experiment area.

The experiment was carried out on the basis of the following methodical instructions for the cultivation of pumpkin and cucumber plants: Plots and schemes of sowing in and selection, testing and primary seed production of vegetable crops. "Guidelines for the study of the world collection for assessing the physical and mechanical properties of vegetable and melon crops", International rules for seed analysis, "Wide unified classifier SEV and the international classifier SEB of the species Cucurbitaceae", "Methodology of the field experiment", "Guidelines for the use of plant growth regulators on vegetables, melons and potatoes: recommendations", Biological products in agriculture (Methodology and practice of using microorganisms in crop and feed production), "Guidelines for methods of quality control and safety of biologically active food supplements", "Methods for the analysis of minor biologically active substances in food".

"Sample technological cards for the care of agricultural crops and production of products" for the years 2016-2020 in determining the economic efficiency of the technology of growing pumpkins and cucumbers. Based on the instructions of the published collection, the technological card for growing and harvesting pumpkin and cucumber used.

3 Results and discussion

According to the results of the study, differentthe effect of biostimulants on the biometric indicators of pumpkin plants were studied. The effect of various biostimulants on the growth and development of pumpkin: stem weight and leaf area were determined (see Tables 1 and 2). The length of the main stem was 180 cm in control and 207.0-245.0 cm in options with biostimulants, stem weight was 0.250 grams in control and 261.0-291.0 grams in other options. The number of lateral branches was 4 in the control, 5-7 in other options.The number of leaves in the control was 98 pieces, and in the options biostimulant used, this indicator increased to 106-118 pieces. The best performance was observed in Uchqun plus at 0.01% amount consumption, main stem length was 245 cm, stem weight was 0.291 kg, number of side branches was 7 pieces, number of leaves was 118 pieces, leaf area of one plant was 1.931 m^2.

Table 1. Effect of different biostimulants on the biometric indicators of pumpkin variety of Ispanskaya 73.

Options	Ispanskaya 73						
	Main stem length, cm	Stem weight, g	Side horn number	# of leaves, pieces	Leaf area, m^2	Plant morbidity,%	
						Mildew	Fusarium wilt
In a method of seed moisturizing							
Control	180.0	0.250	4	98	1.729	5	1
Gossyprene -1.0%	207.0	0.261	5	106	1.755	2	0
Uchqun- 0.01%	230.0	0.278	6	110	1.770	0	0
Uchqun plus -0.01%	245.0	0.291	7	118	1.931	0	0
Verva -5.0%	225.0	0.267	5	108	1.755	1	0
In a spray method							
Control	180.0	0.250	4	98	1.720	5	2
Gossyprene -1.0%	189.0	0.251	5	105	1.780	2	1
Uchqun- 0.01%	191.0	0.252	6	108	1.901	0	0
Uchqun plus -0.01%	197.0	0.291	7	116	1.956	0	0
Verva -5.0%	189.0	0.251	5	106	1.769	1	0
In a mixed method							
Control	180.0	0.250	4	98	1.725	5	1
Gossyprene -1.0%	190.0	0.320	5	108	1.705	2	0
Uchqun 0.01%	192.0	0.340	6	113	2.157	0	0
Uchqun plus -0.01%	195.0	0.389	7	120	2.853	0	0
Verva -5.0%	190.0	0.332	5	108	1.998	1	0

Table 2. Effect of different biostimulants on the biometric indicators of pumpkin variety of Palov kadu 268.

Options	Palov kadu 268						
	Main stem length, cm	Stem weight, g	Side horn number	# of leaves, pieces	Leaf area, m^2	Plant morbidity,%	
						Mildew	Fusarium wilt
In a method of seed moisturizing							
Control	165.0	0.250	4	98	1.729	5	1
Gossyprene -1.0%	177.0	0.261	5	106	1.801	2	0
Uchqun- 0.01%	190.0	0.279	6	110	1.930	0	0
Uchqun plus -0.01%	195.0	0.291	7	118	1.996	0	0
Verva -5.0%	190.0	0.268	5	108	1.869	1	0
In a spray method							
Control	165.0	0.250	4	98	1.729	5	2
Gossyprene -1.0%	189.0	0.261	5	106	1.780	2	1
Uchqun- 0.01%	191.0	0.278	6	110	1.901	0	0
Uchqun plus -0.01%	195.0	0.290	7	118	1.950	0	0
Verva -5.0%	189.0	0.267	5	108	1.769	1	0
In a mixed method							
Control	165.0	0.250	4	98	1.729	5	1
Gossyprene -1.0%	190.0	0.320	5	106	1.755	2	0
Uchqun 0.01%	192.0	0.340	6	110	2.157	0	0
Uchqun plus -0.01%	224.0	0.389	7	118	2.850	0	0
Verva -5.0%	190.0	0.332	5	108	1.998	1	0

The Ispanskaya 73 variety has a lower rate compared to the control in the spraying method. The length of the main stem was 189 cm, the weight of the stem was 0.261 kg, the number of lateral branches was 5 pieces, the number of leaves was 106 pieces, and the leaf area of one plant was 1.755 m^2.

In the Verva variant, these parameters are main stem length 225 cm, stem weight 0.251 kg, number of side branches 5 pieces, number of leaves 106 pieces, leaf level per plant it was 1,780 and 1,769 m2. In the Uchqun variant, the length of the main stem was 191.0 cm, the weight of the stem was 0.252 kg, the number of side branches was 6 pieces, the number of leaves was 108 pieces, the leaf area of Uchqun one plant was 1.901 m^2.

The best performance was observed with Uchqun plus 0.01%, main stem length 197.0 cm, stem weight 0.291 kg, number of lateral branches 7 pieces, number of leaves 116 pieces, one plant leaf levelIt was 1,956 m^2. 5% of plants were affected by powdery mildew and 2% by fusarium wilt in control by spraying method. 2% of Gossypren with powdery mildew were affected by Fusarium wilt, and 1% of the Verva variety was infected with powdery mildew. No disease was observed in other variants.

Ispanskaya variety 73 in a mixed method, a low indicator compared to the control It was found in Gossipren and Verva variants, the length of the main stem was 190 cm, the weight of the stem was 0.320 and 0.332 kg, the number of lateral branches was 5 pieces, the number of leaves was 108 pieces, and the leaf area of Uchqun one plant was 1,705 and 1,998 m^2. In the Uchqun variant, the length of the main stem was 192.0 cm, the weight of the stem was 0.340 kg, the number of side branches was 6 pieces, the number of leaves was 113 pieces, and the leaf area of Uchqun one plant was 2.157 m^2. In Ispanskaya 73 variety, the best performance was observed at the consumption of Uchqun plus 0.01%, the length of the main stem was 195.0 cm, the weight of the stem was 0.389 kg, the number of side branches was 7 pieces, the number of leaves was 120 pieces, the leaf area of Uchqun one plant was 2,853 m^2. In the mixed method, 5% of the plants were affected by powdery mildew and 1% by fusarium wilt. 2% of the plants were infected with powdery mildew in the Gossypren variant, and 1% in the Verva variant. No disease was observed in other variants.

The length of the main stem is 165 cm, the weight of the stem is 0.250 kg, the number of side branches is 4 pcs., the number of leaves is 98 pcs.It was 1,729 m^2. A low indicator compared to the control was found in the Gossipren variant, the main stem length was 177 cm, stem weight was 0.261 kg, the number of side branches was 5 pieces, the number of leaves was 106 pieces, and the leaf area of one plant was 1.801 m^2.

The best performance was observed with Uchqun plus 0.01%, main stem length 195.0 cm, stem weight 0.291 kg, number of side branches 7 pieces, number of leaves 118 pieces, one plant leaf levelIt was 1,996 m2. 5% of the plants were affected by powdery mildew and 1% by fusarium wilt in the control method. 2% of Gossyprene and 1% of Verva plants were affected by powdery mildew. No disease was observed in other variants.

Palov kadu variety 268 showed low performance compared to the control in the spraying method The length of the main stem was 189.0 cm, the weight of the stem was 0.261 and 0.267 kg, the number of side branches was 5, the number of leaves was 106 and 108, and the leaf area of Uchqun one plant was 1,769 and 1,780 m^2. In the Uchqun variant, the length of the main stem was 191.0 cm, the weight of the stem was 0.278 kg, the number of side branches was 6 pieces, the number of leaves was 110 pieces, and the leaf area of Uchqun one plant was 1.901 m^2.

The best performance was observed with Uchqun plus 0.01% quantity consumption, main stem length 195.0 cm, stem weight 0.290 kg, number of side branches 7 pieces, number of leaves 118 pieces, one plant leaf level It was 1,950 m^2. 5% of plants were affected by powdery mildew and 2% by fusarium wilt in control by spraying method. 2% of the plants were affected by powdery mildew in gossypren, 1% by fusarium wilt. No disease was observed in other variants.

Palov kadu variety 268 showed low performance compared to control in mixed method In Gossipren and Verva variants, the length of the main stem was 190.0 cm, the weight of the stem was 0.320 and 0.332 kg, the number of side branches was 5 pieces, the number of leaves was 106 and 108 pieces, and the leaf area of Uchqun one plant was 1.705 and 1.998 m^2.

The best performance was observed with Uchqun plus 0.01% rate, main stem length 224.0 cm, stem weight 0.389 kg, number of side branches 7 pieces, number of leaves 118 pieces, one plant leaf levelIt was 2,850 m^2. In the mixed method control, 5% of plants were damaged by powdery mildew, 1% by fusarium wilt. 2% of plants were affected by powdery mildew in Gossypren, and 1% in Verva variant. No disease was observed in other variants.

In the experiment, the effect of biostimulants on the stem weight and leaf surface of Navruz cucumber plant was determined is, according to which the leaf area in one plant is under control when the seeds are sown by the moisturising method 1,620 m^2, it turned out that this indicator was slightly better in the options using Gossipren and Verva biostimulants, and the best result was obtained when using Uchqun and Uchqun plus (Tables 3 and 4).

Inferior to controlIt was found in Gossipren and Verva variant, the length of the main stem was 216.0 and 226.0 cm, the weight of the stem was 0.261 and 0.268 kg, the number of lateral branches was 5.2-5.4 pieces, the number of leaves was 106 and 108 pieces, the number of leaves per plant was 1.801 and It was 1,869 m^2.

Table 3. Effect of various biostimulants on biometric indicators of cucumber variety of Navruz.

Options	Navruz variety						
	Main stem length, cm	Stem weight, g	Side horn number	# of leaves, pieces	Leaf area, m^2	Plant morbidity,%	
						Mildew	Fusarium wilt
In a method of seed moisturizing							
Control	212	0.250	4.6	98	1.620	1.2	1
Gossyprene -1.0%	216	0.261	5.2	106	1.801	0.5	1
Uchqun-.01%	236	0.279	5.6	110	1.896	0	0
Uchqun plus -0.01%	245	0.291	5.8	118	1.931	0	0
Verva -5.0%	226	0.268	5.4	108	1.869	0	0.1
In a spray method							
Control	203	0.250	5.0	97	1.529	2.5	0.5
Gossyprene -1.0%	205	0.261	5.0	98	1.780	2.0	0.5
Uchqun- 0.01%	230	0.279	5.2	115	1.910	0	0
Uchqun plus -0.01%	232	0.291	5.2	117	1.926	0	0
Verva -5.0%	220	0.268	5.0	110	1.862	0	0
In a mixed method							
Control	203	0.250	5.0	97	1.600	2.5	0.5
Gossyprene -1.0%	205	0.261	5.0	98	1.705	2.0	0.5
Uchqun .01%	230	0.310	5.2	115	1.995	0	0
Uchqun plus -0.01%	232	0.389	5.2	117	2.157	0	0
Verva -5.0%	225	0.263	5.0	110	1.790	0	0

The best performance was observed with Uchqun plus 0.01%, main stem length 245.0 cm, stem weight 0.291 kg, number of side branches 5.8 pieces, number of leaves 118 pieces, one plant leaf levelIt was 1,931 m2. The disease resistance of the plant was also determined in the experiment. Of course, after the seeds have germinated, observations were made in 3 periods during the growth period. Powdery mildew and fusarium wilt, common in cucumber, were observed visually (visually). 1.2% of the plants were damaged by powdery mildew and 1% by fusarium wilt in the control method. 0.5% of plants were infected with powdery mildew in Gossypren, and 0.1% in Verva variant. No disease was observed in other variants.

The effect of biostimulants on stem weight and leaf area of UchqunNavruz plant was determined, according to which, when planted by spraying method, the leaf area in one plant is under control.It was found that 1,529 m^2, 1,780-1,862 m^2 were used in the options using Gossipren and Verva biostimulants, 1,910 m^2 in the Uchqun variant, and 1,926 m^2 when Uchqun plus was used.

Main stem length is lower than controlIt was determined in Gossipren and Verva variants, 205.0 and 220.0 cm, stem weight 0.261 and 0.268 kg, number of side branches 5.0-5.2 pieces, number of leaves 98 and 110 pieces, leaf area of Uchqun one plant is 1,780 and 1,862 m^2 organized. The best performance was observed with Uchqun plus 0.01%, main stem length 232.0 cm, stem weight 0.291 kg, number of lateral branches 5.2 pieces, number of leaves 117 pieces, leaf area of Uchqun one plant was 1.926 m^2. 2.5% of plants were damaged by powdery mildew and 0.5% by fusarium wilt in the control by spraying method. In Gossypren,

2.0% of plants were infected with powdery mildew, 0.5% with fusarium wilt. No disease was observed in other variants.

Table 4. Effect of various biostimulants on biometric indicators of cucumber variety of Orzu F1 hybrid.

Options	Orzu F1 hybrid						
	Main stem length, cm	Stem weight, g	Side horn number	# of leaves, pieces	Leaf area, m^2	Plant morbidity, %	
						Mildew	Fusarium wilt
In a method of seed moisturizing							
Control	212.0	0.250	4.6	98	1.729	1.2	0.5
Gossyprene - 1.0%	216.0	0.261	5.2	106	1.801	0.5	0.25
Uchqun-.01%	236.0	0.279	5.6	110	1.869	0	0
Uchqun plus - 0.01%	245.0	0.291	5.8	118	1.996	0	0
Verva -5.0%	226.0	0.268	5.4	108	1.862	0	0
In a spray method							
Control	203.0	0.240	5.0	97	1.720	2.5	0.5
Gossyprene - 1.0%	205.0	0.261	5.0	98	1.780	2.0	0.5
Uchqun- .01%	230.0	0.278	5.2	115	1.901	0	0
Uchqun plus - 0.01%	232.0	0.290	5.2	117	1.956	0	0
Verva -5.0%	225.0	0.267	5.0	110	1.769	0	0
In a mixed method							
Control	203.0	0.240	5.0	97	1.725	2.5	0.5
Gossyprene - 1.0%	205.0	0.320	5.0	98	1.705	2.0	0.5
Uchqun .01%	230.0	0.340	5.2	115	2.157	0	0
Uchqun plus - 0.01%	232.0	0.389	5.2	117	2.853	0	0
Verva -5.0%	228.0	0.332	5.0	110	1.998	0	0

The effect of biostimulants on stem weight and leaf area of Uchqun Navruz plant was determined, according to which, when planted by spraying method, the leaf area in one plant is under control. It was found that 1,529 m^2, 1,780-1,862 m^2 were used in the options using Gossipren and Verva biostimulants, 1,910 m^2 in the Uchqun variant, and 1,926 m^2 when Uchqun plus was used (Tables 3 and 4).

Main stem length is lower than controlIt was determined in Gossipren and Verva variants, 205.0 and 220.0 cm, stem weight 0.261 and 0.268 kg, number of side branches 5.0-5.2 pieces, number of leaves 98 and 110 pieces, leaf area of Uchqun one plant is 1,780 and 1,862 m^2 organized.

The best performance was observed with Uchqun plus 0.01% amount consumption, main stem length 232.0 cm, stem weight 0.291 kg, number of side branches 5.2 pieces, number of leaves 117 pieces, leaf level per plantIt was 1,926 m^2. 2.5% of plants were damaged by powdery mildew and 0.5% by fusarium wilt in the control by spraying method. In Gossypren, 2.0% of plants were infected with powdery mildew, 0.5% with fusarium wilt. No disease was observed in other variants.

The effect of biostimulants on the stem weight and leaf area of UchqunNavruz plant was determined, according to which the leaf area in one plant is under control in the mixed method.It turned out that 1,600 m^2, 1,705-1,790 m^2 in the variants using Gossipren and Verva biostimulants, 1,995 m^2 in the Uchqun variant, and 2,157 m^2 when Uchqun plus was used.

Main stem length is lower than controlIt was found in Gossipren and Verva variant, 205.0 and 225.0 cm, stem weight 0.261 and 0.263 kg, number of side branches 5.0 pieces, number of leaves 98 and 110-115 pieces, leaf area of Uchqun one plant was 1.705 and 1.790 m^2.

The best performance was observed in Uchqun plus at 0.01% consumption, main stem length 232.0 cm, stem weight 0.389 kg, number of side branches 5.2 pieces, number of leaves 117 pieces, leaf level per plant 2It was 157 m2. In the mixed method, 2.5% of plants were affected by powdery mildew and 0.5% by fusarium wilt. In Gossypren, 2.0% of plants were infected with powdery mildew, 0.5% with fusarium wilt. No disease was observed in other variants.

When the seeds of cucumber Orzu F1 hybrid were planted with biostimulants, the effect of biostimulants on plant stem length, weight and leaf area were determined, according to which, the length of the main stem in one plant was 212.0 cm in the control, 216.0-226.0 cm in the variants with Gossipren and Verva biostimulants. Cm, 236.0 cm in the Uchqun variant, and 216.0 cm when using the Uchqun plus was found to be the best.

Inferior to controlIt was found in Gossipren and Verva variant, the stem weight was 0.261 and 0.268 kg, the number of side branches was 5.2-5.4 pieces, the number of leaves was 106 and 108 pieces, the leaf area of Uchqun one plant was 1.801 and 1.862 m^2.

The best performance was observed with Uchqun plus 0.01%, main stem length 245.0 cm, stem weight 0.291 kg, number of side branches 5.8 pieces, number of leaves 118 pieces, one plant leaf levelIt was 1,931 m^2. 1.2% of the plants were damaged by powdery mildew and 0.5% by fusarium wilt in the control method. 0.5% of the plants were infected with powdery mildew and 0.25% with fusarium wilt. No disease was observed in other variants.

When biostimulants were sprayed on cucumber Orzu F1 hybrid, the length of the main stem in one plant was 203.0 cm in the control, 205.0-225.0 cm in the versions with Gossipren and Verva biostimulants, 278.0 cm in the Uchqun variant, and 232.0 cm when Uchqun plus was used. It turned out that it gave a good result.

Inferior to controlIt was found in Gossipren and Verva variant, the stem weight was 0.261 and 0.267 kg, the number of side branches was 5.0 pieces, the number of leaves was 98 and 110 pieces, and the leaf area of Uchqun one plant was 1.780 and 1.769 m^2.

The best performance was observed with Uchqun plus 0.01% rate, main stem length 232.0 cm, stem weight 0.290 kg, number of side branches 5.2 pieces, number of leaves 117 pieces, leaf level per plantIt was 1,956 m^2. 2.5% of plants were damaged by powdery mildew and 0.5% by fusarium wilt in the control by spraying method. In Gossypren, 2.0% of plants were infected with powdery mildew, 0.5% with fusarium wilt. No disease was observed in other variants.

When biostimulants are applied to cucumber Orzu F1 hybrid in a mixed method, the length of the main stem in one plant is 203.0 cm in the control, 205.0-228.0 cm in the options with Gossipren and Verva biostimulants, 230.0 cm in the Uchqun option, and 232.0 when Uchqun plus is used. It was found that cm gave the best results.

Inferior to controlIt was found in Gossipren and Verva variant, the stem weight was 0.320 and 0.332 kg, the number of side branches was 5.0 pieces, the number of leaves was 98 and 110 pieces, and the leaf area of Uchqun one plant was 1.705 and 1.998 m^2.

The best performance was observed in Uchqun plus at 0.01% consumption, main stem length 232.0 cm, stem weight 0.389 kg, number of side branches 5.2 pieces, number of leaves

117 pieces, leaf level per plant 2It was 853 m^2. In the mixed control, 2.5% of plants were damaged by powdery mildew, 0.5% by fusarium wilt. In Gossypren, 2.0% of plants were infected with powdery mildew, 0.5% with fusarium wilt. No disease was observed in other variants.

According to the results of the research, it can be concluded that in terms of the rate and methods of using different biostimulants in the cultivation of pumpkin and cucumber plants, the highest indicator in all varieties and methods is in the 0.01% version of Uchqun plus biostimulant and in the mixed method.

Data on the effect of various biostimulants on the yield of courgette crops were analyzed (Tables 5 and 6).

Table 5. Effect of various biostimulants on yield indicators of pumpkin variety of Ispanskaya 73.

Options	**Ispanskaya 73**				
	Total yield, t/ha	**Yield of 1 bush, kg**	**Commodity harvest, t\ha**	**Marketable yield, %**	**Relative to control, %**
In a method of seed moisturizing					
Control (cooling with water)	27.9	1.6	26.8	96.2	100.0
Gossyprene -1.0%	28.5	2.1	27.6	97.2	102.1
Uchqun – 0.01%	31.0	2.2	30.3	97.8	111.1
Uchqun plus -0.01%	31.1	2.2	30.8	99.1	111.5
Verva -5.0%	29.0	2.1	28.3	97.6	103.9
In a spray method					
Control (cooling with water)	27.0	1.6	26.0	96.2	100.0
Gossyprene -1.0%	27.7	2.0	26.9	97.0	102.6
Uchqun – 0.01%	30.5	2.1	29.8	97.7	113.0
Uchqun plus -0.01%	30.8	2.2	30.5	99.0	114.0
Verva -5.0%	28.1	2.0	27.4	97.5	104.1
In a mixed method					
Control (cooling with water)	28.2	1.6	27.1	96.2	100.0
Gossyprene -1.0%	28.8	2.1	28.0	97.3	102.1
Uchqun – 0.01%	31.4	2.2	30.7	97.9	111.3
Uchqun plus -0.01%	31.8	2.2	31.5	99.2	112.8
Verva -5.0%	30.5	2.1	29.7	97.7	108.2
LSD_{05}	0.5	-	0.91	-	-
Sx%	0.30	-	1.2	-	-

It provides information on the total yield of the pumpkin plant, the yield of 1 bush, the marketable yield, the percentage of marketable yield and the comparison with the control. In the experiment, when the Ispanskaya 73 variety of pumpkin was treated with biostimulants, the yield in the control variant was 27.9 t/ha, the yield of 1 bush was 1.6 kg, the yield was 26.8 t/ha, and the yield of the yield was 96.2%. . Productivity between options is 28.5-31.1 t/ha, yield of 1 bush is 2.1-2.2 kg, marketable yield is 27.6-30.8 t/ha, share of marketable yield is 97.2-99.1% organized. The best result is the Uchqun plus drug0.01%in terms of consumption, 31.1 t/ha per hectare, yield of 1 bush 2.2 kg, marketable yield 30.8 t/ha, marketable yield 99.1%, 11.5% more than control.

Table 6. Effect of various biostimulants on yield indicators of pumpkin variety of Palov kadu 268.

Options	Palov kadu 268				
	Total yield, t/ha	Yield of 1 bush, kg	Commodity harvest, t\ha	Marketable yield, %	Relative to control, %
In a method of seed moisturizing					
Control (cooling with water)	26.5	1.6	25.5	96.2	100.0
Gossyprene -1.0%	28.0	2.1	27.2	97.0	105.7
Uchqun – 0.01%	30.1	2.2	29.8	97.8	115.1
Uchqun plus -0.01%	30.5	2.2	30.2	99.0	113.6
Verva -5.0%	28.5	2.1	27.8	97.6	107.5
In a spray method					
Control (cooling with water)	26.5	1.6	25.5	96.2	100.0
Gossyprene -1.0%	27.1	2.1	26.3	97.0	102.3
Uchqun – 0.01%	29.9	2.2	29.1	97.4	112.8
Uchqun plus -0.01%	30.0	2.2	29.9	98.0	113.0
Verva -5.0%	27.8	2.1	27.1	97.4	105.0
In a mixed method					
Control (cooling with water)	26.5	1.6	25.5	96.2	100.0
Gossyprene -1.0%	27.4	2.1	26.6	97.2	103.4
Uchqun – 0.01%	30.0	2.2	29.4	97.9	113.2
Uchqun plus -0.01%	30.2	2.2	30.0	99.5	114.0
Verva -5.0%	28.5	2.1	27.9	97.9	107.5
LSD_{05}	0.5	-	0.97	-	-
Sx%	0.32	-	1.3	-	-

When treated by seed spraying with biostimulants in the Ispanskaya 73 variety, the yield in the control variant was 27.0 t/ha, the yield of 1 bush was 1.6 kg, the marketable yield was 26.0 t/ha, and the output of the marketable crop was 96.2%. The productivity of the options was 27.7-30.8 t/ha, the yield of 1 bush was 2.0-2.2 kg, the marketable yield was 26.9-30.5 t\ha, the output of the marketable crop was 97.0-99.0% . The best result is Uchqun plus biostimulator 0.01%It was observed in the amount consumption, 30.8 t/ha per hectare, the yield of 1 bush was 2.2 kg, the marketable yield was 30.5 t/ha, the output of the marketable crop was 99.0%.

When the Ispanskaya 73 variety was treated with biostimulants in a mixed method, the yield in the control variant was 28.2 t/ha, the yield of 1 bush was 1.6 kg, the marketable yield was 27.1 t/ha, and the output of the marketable crop was 96.2%.

When these indicators are analyzed among options, the productivity is 28.8-31.8 t/ha, the yield of 1 bush is 2.1-2.2 kg, the marketable yield is 28.0-31.5 t/ha, the output of the marketable crop is 97.3- It was 99.2%. The best result is Uchqun plus biostimulator 0.01%in terms of consumption, it was 31.8 t/ha per hectare, the yield of 1 bush was 2.2 kg, the marketable yield was 31.5 t/ha, the output of the marketable crop was 99.2%.

When treated with biostimulants in the 268 variety of Palov kadu, the yield in the control option was 26.5 t/ha, the yield of 1 bush was 1.6 kg, the marketable yield was 25.5 t/ha, and the output of the marketable crop was 96.2%. Productivity between options is 28.0-30.5 t/ha, yield of 1 bush is 2.1-2.2 kg, marketable yield is 27.2-30.5 t/ha, marketable yield is 97.0-99.0% organized. Also in this variety is the drug Uchqun plus 0.01%the best result was obtained in terms of quantity consumption, 30.5 t/ha per hectare, 2.2 kg of yield per bush, 30.2 t/ha of marketable yield, 99.0% of marketable yield.

When paddy pumpkin variety 268 is treated by spraying with biostimulants, the yield between options is 27.1-30.0 t/ha, the yield of 1 bush is 2.1-2.2 kg, the yield is 26.3-29.9 t/ha, the yield is 26.3-29.9 t yield was 97.0-98.0%. The best result is the Uchqun plus drug 0.01% 30.0 t/ha per hectare, yield 2.2 kg, marketable yield 29.9 t/ha, marketable yield 98.0%.

When seeds are processed in a mixed method with biostimulants in Palov kadu 268 variety, the inter-variant yield is 26.5-30.2 t/ha, the yield of 1 bush is 1.6-2.2 kg, the yield is 25.6-30.0 t/ha, marketable yield was 96.2-99.5%. The best result in this variety is the Uchqun plus drug 0.01%it was obtained in the variant where the norm was used, the yield per hectare was 30.2 t/ha, the yield of 1 bush was 2.2 kg, the marketable yield was 30.0 t/ha, the output of the marketable crop was 99.5%.

In the experiment, the effect of biostimulants on the productivity of cucumber plants was given, according to which, when Navruz variety seeds were planted with biostimulants, the total yield in the control was 27.6 t/ha, the yield of 1 bush was 1.2 kg, and the total yield was 25.7 t/ha (Tables 7 and 8). The total yield of the options was 31.0-34.0 t/ha, the yield of 1 bush was 1.35-1.5 kg, and the marketable yield was 29.3-32.8 t/ha. The best result was the total yield of 34.0 t/ha, the yield of 1 bush was 1.5 kg and the marketable yield was 32.8 t/ha at the consumption amount of 0.01% of Uchqun plus.

Table 7. Effect of biostimulants on yield indicators of cucumber variety of Navruz.

Options	**Navruz**				
	Total yield, t/ha	**Yield of 1 bush, kg**	**Commodity harvest, t/ha**	**Marketable yield, %**	**Relative to control, %**
In a method of seed moisturizing					
Control (cooling with water)	27.6	1.2	25.7	93.1	100.0
Gossyprene -1.0%	31.0	1.35	29.3	94.5	112.3
Uchqun - 0.01%	32.2	1.4	30.6	95.0	116.7
Uchqun plus -0.01%	34.0	1.5	32.8	96.6	123.2
Verva -5.0%	31.3	1.36	29.6	94.7	113.4
In a spray method					
Control (cooling with water)	27.6	1.2	25.7	93.1	27.6
Gossyprene -1.0%	32.2	1.30	30.7	95.2	116.7
Uchqun - 0.01%	33.4	1.46	32.2	96.4	121.0
Uchqun plus -0.01%	33.5	1.60	32.6	97.4	121.4
Verva -5.0%	33.3	1.40	31.7	95.2	120.7
In a mixed method					
Control (cooling with water)	27.6	1.2	25.7	93.1	27.6
Gossyprene -1.0%	32.2	1.4	30.7	95.2	116.7
Uchqun - 0.01%	33.6	1.6	32.9	97.9	121.7
Uchqun plus -0.01%	35.1	1.71	34.9	99.4	127.2
Verva -5.0%	33.1	1.45	31.5	95.3	119.9
LSD_{05}	0.5	-	0.27	-	-
Sx%	0.7	-	1.0	-	-

When applied in a mixed method, the total yield between variants was 27.6-35.1 t/ha, the yield of 1 bush was 1.2-1.71 kg, and the total yield was 25.7-34.9 t/ha. At the rate of 0.01% of Uchqun plus, the best result was determined by this method, the total yield was 35.1 t/ha, the yield of 1 bush was 1.71 kg, the marketable yield was 34.9 t/ha, and 27.2% more than the control was obtained. .

When cucumber Orzu F1 hybrid seeds were planted with biostimulants, the total yield in control was 29.9 t/ha, the yield of 1 bush was 1.3 kg, and the marketable yield was 27.7 t/ha. The total yield of the options was 32.2-34.0 t/ha, the yield of 1 bush was 1.4-1.5 kg, and the total yield was 30.7-32.8 t/ha. The best result was the total yield of 34.0 t/ha, the yield of 1 bush was 1.5 kg, the marketable yield was 32.8 t/ha, 13.7% more than the control.

When biostimulants were sprayed on cucumber Orzu F1 hybrid, the total yield between variants was 29.9-34.5 t/ha, the yield of 1 bush was 1.3-1.65 kg, and the total yield was 27.7-33.6 t/ha. The best result was the total yield of 34.5 t/ha, the yield of 1 bush was 1.65 kg, and the marketable yield was 33.6 t/ha at the consumption amount of 0.01% of Uchqun plus.

When biostimulants were applied to cucumber Orzu F1 hybrid in a mixed method, the total yield between variants was 29.9-36.8 t/ha, the yield of 1 bush was 1.3-1.75 kg, and the total yield was 27.7-36.2 t/ha . The best result was the total yield of 36.8 t/ha, the yield of 1 bush was 1.75 kg, the marketable yield was 36.2 t/ha, and 23.0% more than the control.

Table 8. Effect of biostimulants on yield indicators of cucumber variety of Orzu F1 hybrid.

Options	**Orzu F1 hybrid**				
	Total yield, t/ha	**Yield of 1 bush, kg**	**Commodity harvest, t\ha**	**Marketable yield, %**	**Relative to control, %**
Control (cooling with water)	29.9	1.3	27.7	92.8	100.0
Gossyprene -1.0%	32.2	1.4	30.7	95.2	107.7
Uchqun - 0.01%	33.6	1.46	32.4	96.4	112.4
Uchqun plus -0.01%	34.0	1.5	32.8	96.6	113.7
Verva -5.0%	33.3	1.45	31.7	95.2	111.4
Control (cooling with water)	29.9	1.3	27.7	92.8	100.0
Gossyprene -1.0%	31.0	1.35	29.3	94.5	103.7
Uchqun - 0.01%	32.2	1.4	30.6	95.0	107.7
Uchqun plus -0.01%	34.5	1.65	33.6	97.4	115.4
Verva -5.0%	32.6	1.42	30.9	94.7	109.0
Control (cooling with water)	29.9	1.3	27.7	92.8	100.0
Gossyprene -1.0%	33.3	1.4	30.7	95.2	111.4
Uchqun - 0.01%	34.5	1.46	32.8	95.2	109.7
Uchqun plus -0.01%	36.8	1.75	36.2	98.4	123.0
Verva -5.0%	33.4	1.44	32.2	96.4	107.7
LSD_{05}	0,5		0,97		
Sx%	0,34		1,1		

4 Conclusions

1. According to the amount of phytohormones in the stem and root of pumpkin crops, under the influence of 0.01% Uchqun plus biostimulator, the amount of kinetin in the above-ground part of pumpkin was 0.123 mg/ml, and in the root part was 0.041 mg/ml. The amount of indole fatty acid (IMA) obtained as a standard solution was 0.010 mg/ml in the stem, but its amount was not detected in the root.

2. The amount of kinetin in the root part of cucumber was 0.037 mg in the control, 0.131 mg in the Gossypren and Verva variant, 0.134 mg in the Uchqun variant, and 0.136 mg in the Uchqun plus, or 267.5% more than the control. The amount of indole acetic acid in the root was 0.002 mg/ml in the control variant, 0.001 mg/ml in the Gossypren, Verva, Uchqun and Uchqun plus variants, 50% more than the control.

3. The method of optimal application of various bistimulants to the biometric indicators of courgette crops was determined. In this case, the most effective mixed method, under the influence of polyprenols contained in 0.01% Uchqun plus biostimulant, has been proven to accelerate the growth and development of pumpkin crops and increase resistance to diseases. In the Ispanskaya 73 variety of pumpkin, the length of the main stem is 195.0 cm, the weight of the stem is 0.389 kg, the number of lateral branches is 7 pieces, the number of leaves is 120 pieces, the leaf area of Uchqun one plant is 2,853 m2, 2% with flour and dew in the Gossipren version, and 1 in the Verva version % plant infected. No disease was observed in other variants.

4. The length of the main stem is 224.0 cm, the weight of the stem is 0.389 kg, the number of side branches is 7 pcs., the number of leaves is 118 pcs.2,850 m2, 5% of plants were affected by powdery mildew and 1% by fusarium wilt. 2% of plants were affected by powdery mildew in Gossypren, and 1% in Verva variant. No disease was observed in other variants.

5. Cucumber Navruz variety main stem length is 245.0 cm, stem weight is 0.291 kg, number of side branches were 5.8 pieces, number of leaves were 118 pieces, one plant leaf level was 1,931 m2. In the Orzu F1 hybrid, the length of the main stem was 232.0 cm, the weight of the stem was 0.389 kg, the number of side branches was 5.2 pieces, the number of leaves was 117 pieces, and the leaf area of Uchqun one plant was 2.853 m2. In the mixed control, 2.5% of plants were damaged by powdery mildew, 0.5% by fusarium wilt. In Gossypren, 2.0% of plants were infected with powdery mildew, 0.5% with fusarium wilt. No disease was observed in other variants.

6. Uchqun plus biostimulator 0.01% in option 1, the Ispanskaya 73 variety of pumpkin produced a marketable yield of 31.5 t/ha or 12.8% more compared to the control, and the Palav pumpkin 268 variety produced a marketable yield of 30.0 t/ha or 14.0% more compared to the control. Cucumber Navruz variety yielded 34.9 t/ha, 27.2% higher than the control, Orzu hybrid yielded 36.2 t/ha, 23.0% higher than the control.

References

1. Baratova M., Khidirova N., Kosimova S.Kh. An environmentally friendly way of growing pumpkin varieties Spanish-73, XIII International Symposium on the Chemistry of Natural Compounds, Shanghai, China, October 16-19, 2019, 73 p.
2. Khidirova K., Kosimova Sh., Baratova MR "Advantages of biostimulants in growing promising pumpkin varieties". British Journal of Global Ecology and Sustainable Development Volume-10, Nov., 2022.
3. Baratova M., Khidirova N., Kosimova S.Kh. An environmentally friendly way of growing pumpkin varieties Spanish-73, XIII International Symposium on the Chemistry of Natural Compounds, Shanghai, China, October 16-19, 2019, 73 p.

4. Bolikulov F.O. Untraditional methods of growing cucumbers in the openareas. //International Journal of Recent Technology and Engineering (TM). India. 2019.- Vol. 8 – Issue 3C. - R. 586-589. (Impact Factor: 5.92 Scopus Journal).
5. Abdurakhmanova S.B., Khasanov A.P. New varieties of cucumber for the cultivation in the open area. // AJMR: Asian Journal of Multidimensional Research. - India. 2019. – Vol. 8 – Issue 10. – P. 52-58. (Impact Factor: SJIF 2018 = 6.053).
6. Akhmedov B., Yunusov S. Hybrid Properties of Cucumber Elegant F1. Middle European Scientific Bulletin, VOLUME 23 Apr 2022. R. 182-185.
7. Yunusov, S. A. Innovative technologies are growing in the open ground. Academic Research in Educational Sciences, 3(1), (2022). 341-354. https://doi.org/10.24412/2181-1385-2022-1-341-354.
8. Rhaman MS, Imran S, Rauf F, et al. 2020 Seed Priming with Phytohormones: An Effective Approach for the Mitigation of Abiotic Stress Plants (Basel) 10(1) 37.
9. Lal SK et al. 2018 Seed Priming: An Emerging Technology to Impart Abiotic Stress Tolerance in Crop Plants Advances in Seed Priming Rakshit A, Singh H (eds), Springer, Singapore.
10. Hariyanti R, Hadisaputro S, Sumarni S, Widyastuti E 2020 The Effectiveness of Cucumber Suri Juice (Cucumis Sativus) On Blood Pressure in Menopausal Hypertension STRADA Jurnal Ilmiah Kesehatan 9 1771-1778.
11. Khidyrova N.K., Shakhidoyatov Kh.M. Plant Polyprenols and their Biological Activity (review) //Chemistry of National Compounds.- New-York, Springer, 2002. -Vol.38(2). -R.107-121.

Improving elements of sesame growing agrotechnology in the climate conditions of Khorezm province, Uzbekistan

Makhfurat Amanova[1,*], *Abdumalik* Rustamov[1] and *Behzod* Rustamov[1]

[1]Tashkent State Agrarian University, 2, University street, Tashkent, 100140, Uzbekistan

Abstract. In the agricultural landscape of Uzbekistan, an area of 15-18 thousand hectares is dedicated annually to the cultivation of sesame, serving as a supplementary source of income for many. The resulting sesame harvest is predominantly distributed within domestic markets, while a portion is earmarked for international export. The average sesame seed yield in Uzbekistan stands at 4-5 tons per hectare in rain-fed conditions and 8-14 tons per hectare in irrigated settings. In light of these circumstances, the core objective of this study is to ascertain the optimal timing, seeding rate, and cultivation technique for the production of premium-grade exportable sesame goods, uniquely tailored to the climatic conditions prevalent in the Khorezm province of Uzbekistan. By delving into these variables, researchers aim to uncover strategies that maximize the quality and quantity of sesame crops in Khorezm. Such insights hold the potential to not only enhance local production and export capabilities but also contribute to the agricultural economy of the region. By striving to identify the most effective practices, this study seeks to streamline the cultivation process, boost sesame yield, and elevate the overall agricultural landscape in the Khorezm province.
Keywords. Climatic conditions, sesame, planting dates, planting scheme, planting methods.

1 Introduction

Sesame is one of the main oil-producing plants and is grown in more than forty countries of the world, including Burma, India, China, Ethiopia, Sudan, Uganda, Nigeria, Somalia, Asia Minor, Caucasus, Iran, Central Asia, Afghanistan, Far East, Japan, Southern Europe, It is widely grown in America, Mexico and Peru [1-3].

According to Internet data, the price of a ton of sesame seeds has risen sharply from $1,000 to $3,000 in the world market in the last five years [4]. The main reason for this is the prolonged drought in the last (5-7) years in the African sesame-growing countries [5].

In Uzbekistan, 15-18 thousand ha of sesame are grown annually for additional income. The main part of the grown product is sold in our domestic markets, and the rest is

* Corresponding author: makhfurat.amanova1001@mail.ru

exported. In Uzbekistan, the average productivity of sesame seeds is 4-5 t/ha in dry land and 8-14 t/ha in irrigated land [6].

In order to grow a high-quality sesame crop in Uzbekistan and increase the export potential in the world market, it is necessary to organize the primary and variety seed production of regionalized sesame varieties, to continuously provide farmers with quality seeds, and to properly introduce the agrotechnology of cultivation [7-9].

Sesame is not only the main oil-giving plant, but it is also distinguished from other oil-giving plants by its healing properties. In eastern countries, fresh sesame leaves and twigs are used to make delicious salads, rich in vitamins and minerals, white, creamy, light yellow sesame seeds, and ointments are used to treat stomach and intestinal ulcers and body burns. Tachin prepared from sesame was eaten with honey to increase tolerance [10].

Sesame has been cultivated in Central Asia for several thousand years. However, during the former Soviet Union, the Republic of Uzbekistan mainly specialized in cotton. The demand for vegetable oil is covered by cottonseed oil [11].

In Central Asia, the Central Asian branch of VIR (All-Union Institute of Plant Science), currently the Research Institute of Plant Genetic Resources, dealt only with sesame. More than 1,750 varieties of sesame introduced from more than 50 countries of the world are kept alive in this place. Z.A.Luzina, A.Cherkov, B.Amanturdiev on the systematics, biology, selection, seed production, diseases and agrotechnology of sesame plant since 1924 in this place (in the climatic conditions of Tashkent region), researches were conducted [12]. However, no research has been conducted in the saline soil conditions of Khorezm region.

In order to provide the population of Uzbekistan with a wide range of sesame products (sesame oil, tahini, kholva and seeds), the development and introduction of high-quality harvesting technology is one of the urgent issues of today. Therefore, the purpose of the study is to determine the optimal planting period, rate and method for growing high-quality exportable sesame products in the climatic conditions of Khorezm region.

2 Materials and methods

The research was carried out on the basis of "Research of the world collection of oilseeds" [10] and statistical analysis based on the method of B.A. Dospekhov [6].

The Khorezm oasis, one of the northern regions of Uzbekistan, is located in the northern part of the wide Turanian plain, which covers the plains of Central Asia. It occupied the left bank of the ancient Amudarya delta. Its eastern border is the Toshsaka plateau, and its western border is with Turkmenistan. According to its geographical location, Karakalpakstan and Khorezm region are located in the northern part of Uzbekistan. These areas fall into a very dry region, with an annual rainfall of 80-90 millimeters, and they fall mainly in winter and spring. The climate of the oasis is sharply continental. Summer is hot and dry. In July, the average air temperature is +28°C, the average absolute high temperature is +41°C, sometimes the temperature reaches +46°C.

The peculiarity of the Khorezm oasis is that it has sufficient light and heat, continental variability and dry air. The duration of sunlight is 2700–3000 hours per year, with 360–400 hours of sunlight per month in summer and 90–130 hours in winter. The variability of temperature in one day is high (10–15 ^{0}C in winter and 15–20 ^{0}C in summer).

The soil is gray soil , the groundwater is located at a depth of 1.5-1.7 meters and is highly saline. Generally, fields with agricultural crops other than rice are washed three or four times each year. The experimental field was washed twice in the winter and spring of 2022 because the predecessor plant was rice.

Agrotechnical activities carried out in the research field. Before planting, ammophos was given at 200 kg per hectare. During the growth period, 2 times cultivation, 1 time mulching and 2 times foliar feeding with micro and biofertilizers.

The climatic conditions of the Khorezm region are unique, and the method of sesame cultivation is clearly different from the regions near the Amudarya and the regions far from the river. The experimental field of 2022 was located near the shores of the Amudarya, and the groundwater was at a depth of 1.5-2 m and was highly saline.

3 Results and discussion

Sesame seeds of "Black Prince", "Tashkentsky 122", promising "Kiva" and "Urganch" varieties at different planting dates (April 30, May 10, May 20, and May 30) in 3.5 m 2 experimental plots with [4] replications planted. The plants were singled according to the feeding area of 60x15 cm.

When sesame seeds were sown on April 30, it took 10-12 days for the seeds to germinate. The arrival of cool weather in the third ten days of April prevented the seeds from germinating evenly. As a result, the differences in the development of the plants were large, 20-22 percent of the plants ripened in the first, and after 8-9 days, 75-78 percent of the plants ripened.

When sesame seeds were sown on April 30, the growing season lasted 125 days in "Black Prince", "Tashkentsky 122" and "Urganch" varieties, and 115 days in "Kiva" variety. It was 17.2 t/ha in the 122" variety, 20.9 t/ha in the "Urganch" variety and 19.1 t/ha in the "Kiva" variety (Table 1).

Table 1. Effect of planting dates on the productivity of sesame varieties planted as a main crop, q/ha.

Varieties	Black prince	Tashkentsky-122	Kiva	Urganch
April 30				
Growth period, days	125	125	115	125
Yield, q/ha	20.2	17.2	19.1	20.9
Weight of 1000 seeds, g	3.1	2.6	2.9	3.1
May 10				
Growth period, days	120	120	11 2	120
Yield, q/ha	21.3	18.4	19.3	22.9
Weight of 1000 seeds, g	3.12	2.65	2.9	3.1
May 20				
Growth period, days	132	132	125	132
Yield, q/ha	18.1	16.5	16.1	19.6
Weight of 1000 seeds, g	3.0	2.56	2.8	3.0
May 30				
Growth period, days	135	135	125	135
Yiel, q/ha	16.1	14.7	14.3	16.4
Weight of 1000 seeds, g	2.91	2.5	2.75	3.0

When the seeds were sown on May 10, they took 6-7 days to fully germinate. It took 6-7 days for the seeds to fully germinate.

The growing season lasted 120 days in "Black Prince", "Tashkentsky 122" and "Urganch" varieties, and 112 days in "Kiva" variety, and the average yield was 21.3 t/ha in "Black Prince" variety, 18, in "Tashkentsky 122" variety. 4 q/ha, 22.9 q/ha in "Urganch" variety and 19.3 q/ha in "Kiva" variety.

When sesame seeds were sown on May 20, the seeds germinated very quickly in 3-4 days. However, because the soil was dry, it was lightly watered by moistening the soil. For this, every 20 meters, a ditch was opened and water was poured.

In the field where the seeds were sown on May 20, the growing season lasted 132 days in the varieties "Black Prince", "Tashkentsky 122" and "Urganch", and 125 days in the variety "Kiva", the average yield in the variety "Black Prince" was 18.1.0 t/ha, It was 16.5 t/ha in "Tashkentsky 122" variety, 19.6 t/ha in "Urganch" variety and 16.1 t/ha in "Kiva" variety.

When sesame seeds were sown on May 30, seed water was also given and the seeds germinated in 3-4 days. However, below-normal average daily temperatures in late September and October slowed the development of tubers. In the field where the seeds were sown on May 30, the growing season lasted 135 days for the "Black Prince", "Tashkentsky 122" and "Urganch" varieties, and 125 days for the " Kiva " variety . The average productivity of the variety "Black Prince" is 16.1 t/ha, the variety "Tashkentsky 122" is 14.7 t/ha, the variety "Urganch" is 16.4 t/ha, and the variety " Kiva " is 14.8 made q/ha.

The effect of sesame planting time on seed weight was determined by weighing 1000 seeds on an electronic balance in laboratory conditions. The size of the seed depends on the characteristics of the variety, and it was observed that under the influence of the delay in the planting period, it caused a decrease in the weight of the seed, although it was a small amount (Figure 1).

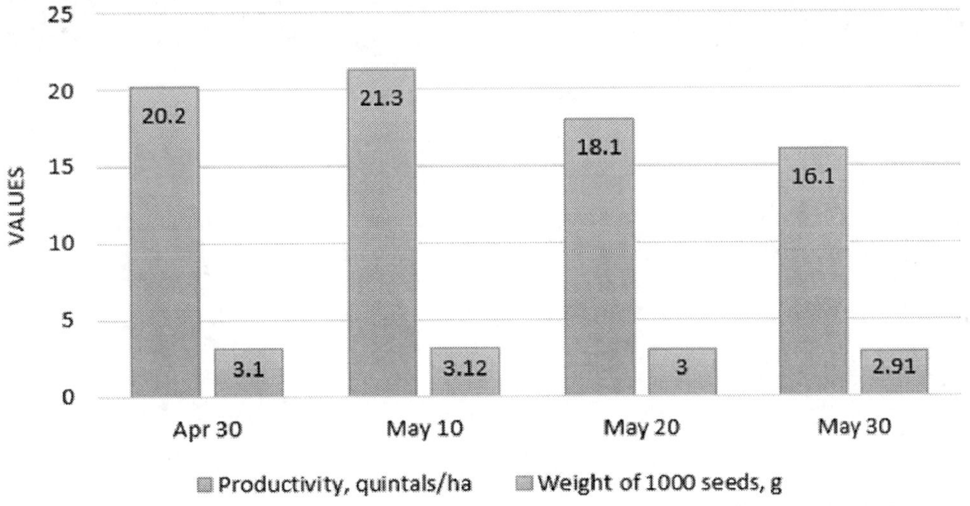

Figure 1. Dependence of productivity and weight of seeds on time periods.

Sesame varieties "Black Prince", "Tashkentsky 122" and "Urganch" are branching varieties, and plants developed well in the 60x15 cm feeding area due to the fact that one plant developed up to 3-4 side branches on average. It was observed that the yield of the " Kiva " variety was relatively low because it does not branch, and the crop is located in a spiral shape around the main stem. The non-branching feature of the Kiva variety makes it possible to plant it in double rows. By increasing the number of seedlings per unit area by 1.5-2.0 times, it is possible to increase productivity by 1.5 times.

We used the same method, that is, we planted the variety "Kiva" in double rows and increased the number of seedlings from 111,000 to 170,000 per hectare, and the average yield was 24.4 t/ha (Figure 2). In this case, the plants were 20 cm between the tapes, and the plants were left in a checkerboard pattern.

Figure 2. Kiva variety.

Another advantage of the " Kiva" variety of sesame compared to other varieties is the shorter growing period of 10-12 days. In the climatic conditions of the Khorezm region, the correct determination of planting dates is one of the most important factors in growing a high yield of sesame. Based on the results of our research, taking into account the recent short-term cold days that may occur in the Khorezm region, we determined that the optimal planting period for growing high-quality and exportable sesame in the local climate is May 1-10.

Sesame varieties were planted as a repeated crop in 10 June, 20 June, 1 July and 10 July in experimental plots of 3.5 m ^{2} with 4 replications, and the effect of sowing dates on yield was studied.

Sesame was irrigated and prepared for planting in the experimental field by side-draining before the land was prepared for planting. In June and July, the average air temperature was 28-29 $^{\circ}$C, and the seeds began to germinate within 3-4 days after sowing. By the fifth day, the number of seedlings reached 75% (Table 2).

Table 2. Effect of planting dates on the productivity of sesame varieties planted as a repeated crop, q/ha.

Varieties	Black Prince	Tashkentsky-122	Kiva	Urganch
June 10				
Growth period , days	120	120	120	120
Yield ts /ha	8.5	7.4	7.2	8.5
June 20				
Growth period , days	1 25	125	1 25	125
Yield ts /ha	5.2	5.1	4.5	5.3
July 1				
Growth period , days	110	110	110	110
Yield ts /ha	3.4	3.1	3.1	3.4
July 10				
Growth period , days	90	90	90	90
Yield ts /ha	1.4	1.4	1.4	1.45

Sesame seeds Seeds were sown on June 10, "Black Prince", "Tashkentsky 122" and "Urganch" varieties, on October 20, after the leaves were completely shed, the lower pods began to open, the crop was harvested and spread on tarpaulins for drying. Due to the cool October, it took 15-18 days for the pods to fully open.

When the seeds were sown on June 10, the average yield was 8.5 t/ha in "Black Shahzoda" and "Urganch" varieties, 7.4 t/ha in "Tashkentsky 122" variety, and 7.2 t/ha in "Kiva" varieties. If so, these indicators are 5.2 t/ha of "Black Prince" on June 20, 5.1 t/ha of "Tashkentsky 122" variety, 5.3 t/ha of "Urganch" variety and 4.5 t/ha of "Kiva" variety. was obtained (Figure 3). When sesame seeds were sown on July 1, the seeds fully germinated in 3-4 days and the plants entered the flowering stage in August. The crop had to be harvested on October 20 before physiological ripening.

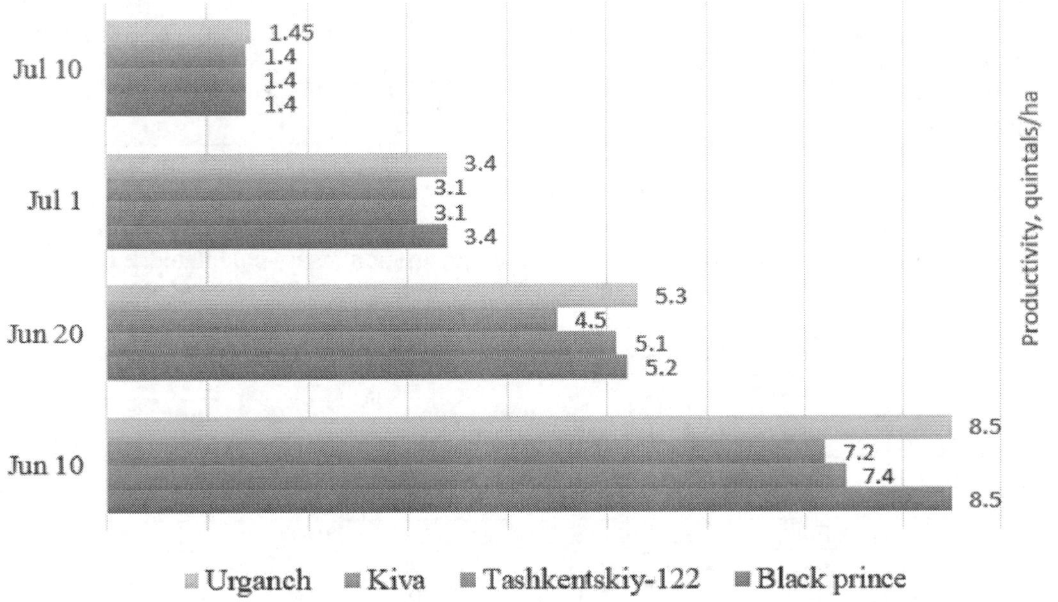

Figure 3. Productivity of different sesame varieties.

For forced germination of sesame seeds and in order to improve wind circulation during the drying period, the plants of each section were tied together and left in the open field. When the seeds were sown on July 1, the average yield was 3.4 t/ha in "Black Shahzoda" and "Urganch" varieties, and 3.1 t/ha in "Tashkentsky 122" and "Kiva" varieties.

When sesame seeds were sown on July 10, the seeds fully germinated in 4 days and the plants entered the flowering stage in August. The crop had to be harvested on October 20 before physiological ripening. For forced germination of sesame seeds (in order to improve wind circulation during the drying period), the plants of each plot were tied together and left in the open field.

When the seeds were sown on July 10, the average yield was almost the same for all varieties, including "Black Shahzoda", "Tashkentsky 122" and "Kiva" was 1.4 t/ha, while "Urganch" averaged 1.45 t/ha was harvested. Based on the results of the conducted research, it was determined that June 10 is the optimal planting date for regionalized and new promising varieties of sesame.

When determining the effect of sowing dates on the productivity of the "Black Prince" sesame variety, the highest indicator was observed in the variant planted on May 10. Total expenses (preparation of land for washing, washing, fertilizers, fertilizing, mowing, cultivation, weeding, fight against diseases and insects, harvesting by hand, cleaning) and 24% additional expenses amounted to 10,540 thousand UZS, while the net profit on April 30 it was 29,860 thousand UZS when planted, 32,060 thousand UZS on May 10, 25,660 thousand UZS on April 20 , and 21,660 thousand UZS on May 30. The highest level of profitability was observed in the first ten days of May and equaled 304% (Table 3).

Table 3. Effect of the next planting dates on the economic efficiency of the "Black Prince" sesame variety.

Indicators	Unit of measure	Apr 30	May 10	May 20	May 30
Total cost, ha	'000 UZS	10540	10540	10540	10540
Productivity, q/ha	'000 UZS	20.2	21.3	18.1	16.1
Selling price of the product, kg	'000 UZS	20.0	20.0	20.0	20.0
Gross profit	'000 UZS	40400.0	42600.0	36200.0	32200.0
Cost of the product	'000 UZS	4,207	3.99	4.70	5.28
24% overhead	'000 UZS	2,040	2,040	2,040	2,040
Net profit	'000 UZS	29860.0	32060.0	25660	21660
Rate of return	%	283	304	243	205

In general, when sesame is grown as the main crop, the net income per hectare is 21,660-32,060 thousand UZS. The economic efficiency of growing sesame in repeated periods in the climatic conditions of Khorezm region was calculated and it was scientifically proven that sowing sesame seeds after June 20 is ineffective (Table 4).

Table 4. Effect of replanting periods on economic efficiency of the cultivar "Black Prince".

Indicators	Unit of measure	Jun 10	Jun 20	Jul 1	Jul 10
Total cost, ha	'000 UZS	10540	10540	10540	10540
Productivity, q/ha	'000 UZS	8.5	5.2	3.4	1.4
Selling price of the product, kg	'000 UZS	20.0	20.0	20.0	20.0
Gross profit	'000 UZS	17000	10400	6800	2800
The cost of the product	'000 UZS	12.4	20.3	31.0	75.0
Net profit	'000 UZS	6460.0	-140	-3740	-7740
Rate of return	%	61	-1	-35	-73

4 Conclusions and recommendations

Based on the specific climatic attributes of the Khorezm region, it is advisable to initiate sesame seed sowing within the initial and subsequent ten days of May. This particular timing is strategically chosen to foster the cultivation of a robust and superior sesame harvest during the spring season. By adhering to this recommended sowing window, agricultural practitioners in the Khorezm region can harness the optimal environmental conditions, thereby increasing the likelihood of yielding a high-quality and abundant sesame crop.

In the context of the Khorezm region's climatic conditions, it is advised to exercise caution and avoid postponing sesame seed sowing beyond June 15, especially when cultivating sesame as a subsequent or repeated crop. This recommendation is underpinned by the region's specific environmental dynamics and serves as a strategic measure to ensure optimal growth and yield outcomes for sesame crops. By adhering to this timeline, agricultural practitioners can effectively mitigate potential challenges associated with adverse weather patterns and other factors that could impede successful sesame cultivation as a repeated crop.

References

1. M.E. Amanova.//Biological characteristics of sesame ecogroups and primary sources for selection//Journal of Agro-science No. 1(21), 31 p. Tashkent 2012.
2. M.E. Amanova, A.S. Rustamov //Methodological manual for studying the world collection of oilseeds// "BIOEKOSAN" educational and methodological complex of the youth of the republic. Tashkent 2010. 20 pages, number 300.
3. Dospekhov B.A. Methodology polevogo opyta. - M.: Kolos, 2015-416 p.
4. N.I.Bochkaryov, S.G.Borodin//Rekomendatsii po semenovodstvu maslichnyx kultur i efiromaslichnyx kultur// Krasnodar. 2004
5. Ya.V. Gubanov //Technicheskie kultury//Moscow. Agropromizdat 2006.
6. MEAmanova, ASRustamov //Sesame new rare sources for selection from world collection//. Proceedings of the Uzbek-Japanese symposium on ecotechnologies: Tashkent-2016.
7. M.E. Amanova, A.S. Rustamov, L.R. Allanazarova //Recommendation on agrotechnics of sesame seeding and cultivation//. Tashkent 2018
8. Boboev, S., Makhkamov, T., Bussmann, R. W., Zafar, M., & Yuldashev, A. (2023). Anatomical and phytochemical studies and ethnomedicinal uses of Colchicum autumnale L. Ethnobotany Research and Applications, 25, 1-9. DOI:10.32859/era.25.6.1-9
9. TKh, M., Brundu, G., Jabborov, A. M., & Gaziev, A. D. (2023). Predicting the potential distribution of Ranunculus sardous (Ranunculaceae), a new alien species in the flora of Uzbekistan and Central Asia. BioInvasions Records, 12(1), 63-77. DOI:10.3391/bir.2023.12.1.05
10. Ameen, M., Zafar, M., Ahmad, M., Ramadan, M. F., Eid, H. F., Makhkamov, T., ... & Majeed, S. (2023). Assessing the Bioenergy Potential of Novel Non-Edible Biomass Resources via Ultrastructural Analysis of Seed Sculpturing Using Microscopic Imaging Visualization. Agronomy, 13(3), 735. DOI:10.3390/agronomy13030735
11. Noor, W., Zafar, M., Ahmad, M., Althobaiti, A. T., Ramadan, M. F., Makhkamov, T., ... & Khan, A. (2023). Petiole micromorphology in Brassicaceous taxa and its potential for accurate taxonomic identification. Flora, 303, 152280. DOI:10.1016/j.flora.2023.152280
12. Aziz, A., Ahmad, M., Zafar, M., Gaafar, A. R. Z., Hodhod, M. S., Sultana, S., ... & Chaudhay, B. (2023). Novel Copper Oxide Phyto-Nanocatalyst Utilized for the Synthesis of Sustainable Biodiesel from Citrullus colocynthis Seed Oil. Processes, 11(6), 1857. DOI:10.3390/pr11061857

Spread, damage and control measures of powdery mildew disease of grain crops

Shaxzod Ruziev[1,*], *Muhabbat* Honkeldieva[2], *Komil* Bukhorov[2] and *Mekhriniso* Sayfiyeva[2]

[1]Research Institute of Plant Quarantine and Protection, Tashkent, 100003, Uzbekistan
[2]Tashkent State Agrarian University, Tashkent, 100140, Uzbekistan

Abstract. This article presents the outcomes of comprehensive experiments carried out within the environments of Samarkand and Kashkadarya regions. These investigations were conducted at scientific experimental stations dedicated to grain and leguminous crops, as well as on farms cultivating corn. The primary focus of this research revolves around the study of the propagation and progression of powdery mildew diseases affecting both barley and wheat plants. Additionally, the article highlights the evaluation of chemical agents' efficacy in combating these diseases. The research outcomes, culled from these trials, constitute a significant portion of the article's content. These findings provide valuable insights into the nuanced dynamics of powdery mildew diseases in the context of the local regions. Furthermore, the assessment of chemical interventions and their effectiveness in mitigating these diseases is a crucial aspect of the research. By examining the data generated from these experiments, the article contributes to the understanding of disease dynamics and potential mitigation strategies within these specific agricultural settings. This research could potentially lead to improved disease management practices, which are pivotal in safeguarding crop yields and ensuring food security within the regions under scrutiny.
Keywords. Barley, wheat, disease, powdery mildew, fungus, disease spread, disease development, fungicide, biological efficiency.

1 Introduction

Barley (*N. sativum*) and common wheat (*T. aestivum*, *T. vulgare*), a group of grain crops, belong to the cereal family, have been cultivated for centuries in world agriculture, like other agricultural crops. The main grain producing countries are Russia, Kazakhstan, China, USA, India, and Canada [1]. Since the beginning of the 1990s, in order to ensure the grain independence of the republic in Uzbekistan, the areas under cultivation have been expanded (1.2 million ha). In Uzbekistan, the total cultivated area (thousand ha) is 3785.1, of which 1789.0 are grain crops (1618.6, including wheat 1507.2). These plants contain macro and micro elements that are very important for the human body [2-4].

* Corresponding author: shaxzod_ruziev1001@mail.ru

https://doi.org/10.1051/e3sconf/202342102011

Protection of plants from harmful organisms during the growth period is one of the important measures in providing the population with quality food products. In recent years, due to the changing climatic conditions of the republic, it is observed that wheat diseases damage wheat during the growing season from year to year. It is noted that wheat powdery mildew causes a lot of damage, especially during the period of tillering, jointing, boot, flowering and grain fill of wheat [5-7].

The causative agent of wheat powdery mildew is Blumeria graminis DC.Speer's synonym (Erysiphe graminis DC.) family Erysiphaceae, order Erysiphales, Leotiomycetes class, phylum Ascomycota, kingdom Fungi, domain Eukaryota [8]. Blumeria graminis is a biologically diverse obligate parasite, highly specialized in parasitism on specific host species (special forms) and exists in numerous physiological races adapted to different cultivars of a particular host species. A special form is identified based on the ability of a given single-spore culture of Blumeria graminis to infect a specific plant species [9].

Eight specific forms of Blumeria graminis have been reported. Publication based on the host's specialty. These forms of the pathogen attack wildgrasses of the genera Dactylis, Agropyron, Poa, and Bromus, Triticum, Hor-deum, Secale, and Avena [10]. It can be assumed based on this classification that B. graminis f. sp. tritici mainly infects wheat, B. graminis f. sp. hordei infects barley, B. graminis f. sp. avenae infects oats, B. graminis f. sp. Secalis infects rye, and B. graminis f. sp. triticale infects triticale [11].

The B. graminis life cycle has two stages – ascosporal and conidial. In the ascosporal stage the fungus produces dark brown or black ascocarps (perithecia or cleistothecia or chasmothecia) 135 to 224 μm in diameter. Chasmothecia are covered with filamentous appendages and contain 8–25 asci 70–100 × 25–40 μm in size. Ascospores formed in asci in late summer or early autumn are ovoid, single-celled and 20–23 × 10–13 μm in size. After rainfall, ripe ascocarps break open and release ascospores that infect grass, volunteer cereal plants and germinating winter crops [12]. A white scurf on the infected plant is formed by mycelium, conidiophores and conidia. The pathogen overwinters as mycelium on winter cereals, volunteer plants and wild grasses and can survive until the next growing season [13].

In the spring, the growing mycelium produces conidiophores and conidia that spread infection to new plants. Conidia are colourless, ellipsoid, 24–35 × 12–17 μm in size and arranged in chains on conidiophores formed by hyphae growing on the surface of leaves. The dense mycelium of greyish colour is gradually formed [10].

Sporing is favoured by dry and warm weather [5]. B. graminis is an ectoparasite. The fungus assimilates nutrients necessary for its growth and development using haustoria penetrating epithelial cells of the host plant. Plants are infected in a wide temperature range of 5 to 30 °C and air humidity of 50–100%, but the optimal conditions are 12–20 °C and high humidity. Powdery mildew produces conidia as often as every 7 to 10 days [4].

In regions where chasmothecium are an important source of infection (e.g. southern Europe, USA), asci are released from the perithecium under favourable weather conditions and give rise to primary infections. Experimental studies have demonstrated that fungal spores can be dispersed by wind for several hundred kilometres, initiating many successive cycles of secondary infections during the growing season [10].

The first symptoms of powdery mildew of cereals and grasses develop on winter cereals in late autumn or early spring. The most severe symptoms of powdery mildew are observed on the lower leaves. Infection spreads gradually from the lower leaves to higher parts of the stem. Initially, small, fluffy white-grey pustules develop on the surface of leaves or under neath leaves [14].

Over time, the disease progresses, and under favourable weather conditions, white fluffy pustules cover increasingly larger area of leaves, sheaths, stems and ears. The scurf on infected parts of plants is white, later turning grey and farinose. Severely infected leaves become chlorotic and gradually die back. In a darkening scurf, small dark brown or black chasmothecia can be seen. The first symptoms of powdery mildew on ears occur on the inner side of the spikelets. Later, the scurf can cover the whole ear. Kernels in infected ears are poorly developed, small, and sometimes malformed [5]

Powdery mildew of wheat can cause up to 45% yield loss if initial infection occurs early and environmental conditions remain favorable for disease emergence and development during the growing season [13]. Popkova and Alexopoulos found that Blumeria graminis causes powdery mildew on the cereal crops and using haustoria (soother) for feeding. The fungi feed by sending haustoria, or root-like structures, into the epidermal (top) cells of the plant. Different formes of haustoria and dignostic sign is used to identify species of fungi [1].

Each of the many species of powdery mildew infects only one type of host plant. Of these, the monocotyledonous species Blumeria graminis is highly host specialized, and there are various specialized forms of this species. Wheat B.graminis f.sp.tritici, barley B.graminis f.sp. hordei damage.

N.G.Zaprometov, P.N.Golovin, N.I.Gaponenko in their scientific works on the study of the types of fungi that cause powdery mildew in the conditions of Uzbekistan, gave a detailed description of the fungi that cause powdery mildew [6, 7, 11].

2 Materials and methods

In addition to the detection of disease-causing fungi, our research also included the determination of their rate of damage, so we constantly calculated the degree of plant infection by pathogens.

We calculated the rate of spread of the disease according to the following formula:

$$R=P*100/N$$

where: R-Disease spread rate in %, N-Total number of plants in samples, pcs., and P-Number of diseased plants in samples, pcs.

When monitoring the spread of powdery mildew diseases in the central and southern regions of Uzbekistan in 3 regions, the widest distribution was observed in the Kashkadarya region (Table 1).

Table 1. Spread of powdery mildew diseases of grain crops in the central and southern regions of Uzbekistan.

Provinces	**District names**	**Prevalence of disease, in %**
Jizzakh	Sharof Rashidov, Pakhtakor, Gallaorol	16
Samarkand	Payariq, Ishtikhan, Akdaryo	20
Kashkadarya	Kasbi, Yakkabog, Kitab, Shakhrisabz	25

We calculated the biological effectiveness of fungicides against diseases using the following formula:

$$C = \frac{P_{k-P_0}}{P_k} * 100$$

where: C-biological efficiency, in %, Pk- disease development index in the control and Po disease development index in the experimental field (for 15, 30 or 45 days, in points).

Experiments on the use of fungicides against powdery mildew of barley were carried out in the 14 ha barley field belonging to "Kokhna Kesh Favvorasi" LLC, Kashkadarya province, Shakhrisabz district. Barley variety Amir Temur. 4 ha of the total barley area was treated with fungicides Top Krof 40% (**c.s.c**), 4 ha Titul Duo 40% (**c.s.c**), 4 ha Top-Az 10% (em.k), and 2 ha was left as a control.

During our research, Top Krof 40% (cake), Titul Duo 40% (cake) and Top-Az 10% (em.k) fungicides were tested against barley powdery mildew (Table 2).

Table 2. Damage of barley by powdery mildew (Production experience, May 9 2023, Kashkadarya province, Shakhrisabz district, "Kukhna Kesh Favvorasi" LLC).

Variants of experiment, rate of consumption of fungicides	Average damage (in points) after how many days			
	until processing	15	30	45
Top Krof 40% (c.s.c) 0.2 l/ha	4.3	1.0	2.5	4.6
Title Duo 40% (c.s.c) 0.25 l/ha	3.7	1.8	3.1	4.8
Top-Az 10% (em.k) 0.15 l/ha	3.5	2.3	3.6	5.1
Control, unprocessed	3.3	9.5	12.0	13.0
EKF $_{05}$	2.4	1.3	0.9	1.1

As can be seen from Table 2, 4.3% of powdery mildew found on barley leaves was dramatically reduced by Top Krof fungicide on days 15, 30 and 45. Other drugs tested showed similar results.

When calculating the biological efficiency (Table 3), 15 days after treatment, Titul Duo at 0.2 kg/ha showed 72.1%, Top-Az at 0.15 l/ha -70.4%. Top Krof 40% (cake) had a biological efficiency of 88.9%. From the obtained results, we can see that although all three preparations have high efficiency, Top Krof fungicide is more effective.

Table 3. Biological effectiveness of drugs belonging to the group of fungicides when used against barley powdery mildew (Production experience, 16.05.2023, Kashkadarya province, Sharisabz district, "Kokhna Kesh Favvorasi" LLC).

Variants of experiment, rate of consumption of fungicides	Biological effectiveness after how many days		
	15	30	45
Top Krof 40% (c.s.c) 0.2 l/ha	88.9	84.1	75.3
Title Duo 40% (c.s.c) 0.25 l/ha	72.1	68.0	65.1
Top-Az 10% (em.k) 0.25 l/ha	70.4	67.9	62.1
Control, unprocessed	-	-	-
EKF $_{05}$	1.6	1.8	2.4

3 Results and discussion

The central and southern regions of the Republic of Uzbekistan have a high incidence of disease, the spread and development of disease-causing agents, and the level of disease in their different soil and climate conditions. We conducted targeted monitoring in Jizzakh, Samarkand and Kashkadarya regions in order to determine the development, spread and damage of powdery mildew disease in wheat and barley fields, to prepare a herbarium from infected plants, as well as to develop scientifically based measures to combat powdery mildew disease. According to agricultural experts, wheat powdery mildew disease is dangerous and economically important after rust and scab. Among the above diseases, the appearance and spread of wheat powdery mildew is regularly observed and controlled, but in the following years cases of wheat powdery mildew damage during vegetation are regularly observed. Geographical location of our country, abundant precipitation in spring and high relative humidity of the air cause the development and spread of powdery mildew in grain fields.

The first symptoms of the disease begin with the formation of a white cotton-like powdery layer on the leaves of plants. Then the layer thickens and becomes gray or yellowish-gray, bulging pads. The disease spreads from leaves to stem leaves and spikes. Conidia of the disease-causing fungus appear on the pads (Figure 1).

A B C

Figure 1. (A and B) Wheat and barley leaves affected by powdery mildew and (C) Conidia of the fungus Blumeria graminis f.sp.hordei causing powdery mildew in barley.

Conidia are single-celled, colorless, cylindrical, and are carried by the wind to other plants. The disease spreads and develops very quickly when the air temperature is 15-20 ^{0}C and the relative humidity is 96-99%.

In unfavorable conditions, the wintering phase of the fungus, i.e. fruit bodies, appears in infected plants. Asci, and ascospores ripen inside them and damage the sprouted autumn grain crops. The fungus overwinters on spiky plants and wild grasses (Figure 2).

A B C

Figure 2. (A) Barley field affected by powdery mildew; (B) Fruiting body of the powdery mildew fungus; and, (C) Cracked state of the fruiting body of the fungus that causes powdery mildew.

According to phytopathological observations, among the above varieties, Ahmed, Navbahor, Durdona and many other varieties are affected by powdery mildew, but the most severe damage is Kesh 2016, Vekha. Recorded in Yaksart. Homer varieties. A high level of damage was noted in all 3 returns of the experimental area where this variety was planted.

Table 4 shows that when Top Krof 40% (cake) was used against wheat powdery mildew, leaf infection decreased from 20.1% to 2.5-3.5-4.2%.

Table 4. Effect of fungicides on the development of powdery mildew of wheat (small field experiment, May 8 2023, Samarkand province, Ishtikhon district, "Samarkand Branch of the Scientific Experimental Station of Cereals and Leguminous Crops").

Variants of experiment, rate of consumption of fungicides	Average damage (in points) after how many days			
	Until processing	15	30	45
Top Krof 40% (c.s.c) 0.2 l/ha	20.1	2.5	3.5	4.2
Title Duo 40% (c.s.c) 0.25 l/ha	17.6	3.5	7.6	11.6
Top-Az 10% (em.k) 0.15 l/ha	15.2	4.6	8.4	12.8
Control, unprocessed	17.1	20.3	31.2	36.3
EKF $_{05}$	-	1.1	2.4	2.1

All the above three drugs are highly effective, but it can be seen from Table 5 that Top Krof fungicide has a higher biological efficiency.

Table 5. Biological effectiveness of fungicides against powdery mildew of wheat (small field experiment, May 8 2023, Samarkand province, Ishtikhon district, "Samarkand Branch of the Scientific Experimental Station of Cereals and Leguminous Crops").

Variants of experiment, rate of consumption of fungicides	Biological effectiveness after how many days, %		
	15	30	45
Top Krof 40% (c.s.c) 0.2 l/ha	88.3	80.3	78.5
Title Duo 40% (c.s.c) 0.25 l/ha	78.6	74.9	68.0
Top-Az 10% (em.k) 0.25 l/ha	73.8	67.4	60.8
Control, unprocessed	-	-	-

The results show that after 15 days, Titul Duo has the highest efficiency with 78.6%, Top Az with 73.8%, and Top Krof fungicide with 88.3%.

In total, 52 varieties of wheat are planted on a total of 1 ha of land belonging to the "Samarkand Branch of the Scientific Experimental Station of Cereals and Leguminous Crops", Ishtikhon District, Samarkand Region, and they are as follows: Krasnadar 99, O‘zbekiston 25, Drujba, Mars-1, Davr, Asr, Velena, Yog‘du, Zimnitsa, Yuk, Grom, Aziz, Navbahor, Veha, Gurt, Taniya, Alekseich, Antonina, Chillaki, Zvezda, Andijon-2, Bezostaya-100, Andijon-4, Bobur, G‘ozg‘on, Kesh 2016, Shams, Turkiston, Yaksart, Durdona, Makuz 3, Istiqlol, Mingchinor, Yoqut 2014, Langar, Hisorak, Javohir, Hamkor, Temiryazovka 150, Vaniya, Asl chilgisi, Ahmad, Agat-donskaya, Yangi taraqqiyot, Kristella, Grand, Kovalerka, Gomer, Soberbosh, Yelanchiq, Oqdaryo fermer, Shukrona. Each variety was planted in 3 replicates in experimental plots of 0.3 ha (Table 6).

Table 6. Planting pattern of grain crops and prevalence of powdery mildew disease in different varieties.

8.Yog'du +	7. Velena +	6. Asr +	5. Davr +	4. Mars-1 +	3.Drujba +	2.O'zbekiston 25 +	1. Krasnadar +
23.Andijon-4 +	22.Bezostaya-100 ++	21.Andijon -2 +	20.Zvezda +	19.Chillaki +	18.Antonina +	17.Alekseich +	16.Taniya +
38.Hamkor qattiq +	37.Javohir qattiq +	36.Hisorak qattiq +	35.langar qattiq -	34.Yoqut - 2014 qattiq -	33.Mingchinor qattiq -	32.Istiqlol qattiq -	31.Makuz-3 qattiq -
1. Krasnadar +	52.Shukrona +	51.Oqdaryo fermer ++	50.Elanchik +	49.Soberbosh +	48.Gomer ++	47.Kovalerka +	46.Grand +
16.Taniya +	15.Gurt +	14.Veha ++	13.Navbahor +	12.Aziz +	11.Grom -	10.Yuka -	9.Zimnitsa -
31.Makuz-3 qattiq -	30.Durdona +	29.Yaksart ++	28.Turkiston ++	27.Shams +	26.Kesh-2016 +++	25.G'ozg'on -	24. Bobur -
46.Grand +	45.Kiristella qattiq +	44.Yangi taraqqiyot qattiq +	43.Agat-donskaya qattiq +	42.Ahmad +	41.Asl chilgisi +	40.Vaniya +	39.Temiryazovka-150 +
9.Zimnitsa -	8.Yog'du -	7. Velena +	6. Asr +	5. Davr +	4. Mars-1 +	3.Drujba +	2.O'zbekiston 25 +
24. Bobur -	23.Andijon-4 +	22.Bezostaya-100 ++	21.Andijon-2 +	20.Zvezda +	19.Chillaki +	18.Antonina +	17.Alekseich +
39.Temiryazovka-150 +	38.Hamkor qattiq +	37.Javohir qattiq +	36.Hisorak qattiq +	35.langar qattiq -	34.Yoqut-2014 qattiq -	33.Mingchinor qattiq -	32.Istiqlol qattiq -
2.Jasmina	1.Starshina	52.Shukrona +	51.Oqdaryo fermer ++	50.Elanchik +	49.Soberbosh +	48.Gomer ++	47.Kovalerka +

2 Jasmina	15.Gurt +	14.Veha + +	13.Navbahor +	12.Aziz +	11.Grom -	10.Yuka -	9.Zimnitsa -
3.Brigada	30.Durdona +	29.Yaksart ++	28.Turkiston +	27.Shams +	26.Kesh-2016 +++	25.G'ozg'on -	24. Bobur -
4.Farboma	45.Kiristella qattiq +	44.Yangi taraqqiyot qattiq +	43.Agat-donskaya qattiq +	42.Ahmad +	41.Asl chilgisi +	40.Vaniya +	39.Temiryazovka-150 +
Vaxshskaya javdar	8.Yog'du +	7. Velena +	6. Asr +	5. Davr +	4. Mars-1 +	3.Drujba +	2.O'zbekiston 25 +
Rossiya javdar	23.Andijon-4 +	22.Bezostaya-100 ++	21.Andijon-2 +	20.Zvezda +	19.Chillaki +	18.Antonina +	17.Alekseich +
Shalola Javdar	38.Hamkor qattiq +	37.Javohir qattiq +	36.Hisorak qattiq +	35.langar qattiq -	34.Yoqut-2014 qattiq -	33.Mingchinor qattiq -	32.Istiqlol qattiq -
Vaxshskaya javdar	1. Krasnadar	52.Shukrona +	51.Oqdaryo fermer ++	50.Elanchik +	49.Soberbosh +	48.Gomer ++	47.Kovalerka +
Rossiya javdar	16.Taniya +	15.Gurt +	14.Veha ++	13.Navbahor +	12.Aziz +	11.Grom -	10.Yuka -
Shalola Javdar	31.Makuz-3 - qattiq	30.Durdona +	29.Yaksart ++	28.Turkiston +	27.Shams +	26.Kesh-2016 +++	25.G'ozg'on -
Oqdaryo *fermer*	46.Grand +	45.Kiristella qattiq +	44.Yangi taraqqiyot qattiq +	43.Agat-donskaya qattiq +	42.Ahmad +	41.Asl chilgisi +	40.Vaniya +
2Markuz-3	*1.Soberbosh*	4.Starshina	3. Brigada	2. Pervesta-Kroshka	1.Starshina	4.Farboma	3. Brigada

Note: + light damage, ++ medium damage, +++ strong damage, - no damage

4 Conclusions

The fungicide Top Krof has exhibited noteworthy efficacy against the powdery mildew disease affecting both barley and wheat plants. With a biological efficiency rating of 88.9% for barley and 88.3% for wheat, this fungicidal treatment proves to be a reliable and potent solution for managing powdery mildew. These impressive effectiveness rates underscore the safe and reliable applicability of Top Krof in combating this fungal ailment.

By achieving such high levels of disease control, Top Krof emerges as a valuable tool for agricultural practitioners seeking effective and secure measures to address the challenges posed by powdery mildew in both barley and wheat cultivation. This research-driven solution promises to contribute significantly to the health and productivity of these crops, supporting sustainable agricultural practices and safeguarding crop yields.

References

1. Alexopolous CJ, Mims CW, Blackwell M. Introductory mycology. Fourth edition. Wiley-India, 2007, x+869 pp
2. Brown W., Cook R., Inman A., Shin H. (2002). The taxonomy of powdery mildew fungi. In: Bélanger R., Bushnell W., Dick A., Carver TLW (eds): The Powdery Mildews: A Comprehensive Treatise. St. Paul, APSPress: 13–15.
3. Daamen R. (1999): Assessment of the profile of powdery mildew and its damage function at low disease intensities in field experiments with winter wheat. Netherlands Journal of Plant Pathology, 95: 85–105
4. Draz I., Esmail S., Abou-Zeid M., Essa T. (2019): Powdery mildew susceptibility of spring wheat cultivars as a major constraint on grain yield. Annals of Agricultural Sciences, 64: 39–45.
5. Fiedorow Z., Gołębniak B., Weber Z. (2004): Choroby Roślin Rolniczych. Poznań, AR
6. Gaponenko N.I. Family Peronosporaceae Middle Asia and South Kazakhstan. – Tashkent - p. 343.
7. Gołębniak B. (2003): Wybrane zagadnienia z biologii i epidemiolii mącznaka prawizego zbóż i traw (Erysiphe graminis DC). In: Roczniki AR w Poznaniu, Rozprawy Naukowe, Zeszyt. Poznań, Akademii Rolniczej w Poznaniu.
8. Hacquard S., Kracher B., Maekawa T., Vernaldi S., Schulze-Lefert P., Ver Loren van Themaat E. (2013): Mosaic genome structure of the barley powdery mildew pathogen and conservation of transcriptional programs in divergent hosts. Proceedings of the National Academy of Sciences, USA, 110(24): e2219-28. doi:10.1073/pnas.1306807110
9. Kirk P., Cannon P., David J., Stalpers J. (2001): Ainsworth & Bisby's Dictionary of the Fungi. : 9 th edition. Wallingford. CABI Publishing
10. Legler S., Caffi T., Rossi V. (2012): A nonlinear model for temperature-dependent development of Erysiphe necator chas-mothecia on grapevine leaves. Plant Pathology, 61: 96–105.
11. Popkova K. V., Shkalikov V. A., Stroykov Yu. M., Lekomtseva S.N., Skvortsova S.N. Obshchaya phytopathology. 2nd izd. M.: "Drofa", 2005, 448 p.
12. Troch V., Audenaert K., Bekaert B., Höfte M., Haesaert G. (2012): Phylogeography and virulence structure of the powdery mildew population on its 'new' host triticale. BMC Evolutionary Biology, 12: 76. doi:10.1186/1471-2148-12-76
13. Walker A., Bouguennec A., Confais J., Morgant G., Leroux P. (2011): Evidence of host-range expansion from new powdery mildew (Blumeria graminis) infections of triticale (× Triticosecale) in France. The Journal of Plant Pathology, 60: 207–220.
14. Wicker T., Oberhaensli S., Parlange F., Buchmann J., Shatalina M., Roffler S., Ben-David R., Doležel J., Šimkova H., Schulze-Lefert P., Spanu P., Bruggmann R. .,

Amselem J., Quesneville H., Ver Loren van Themaat E., Paape T., Shimizu K., Keller B. (2013): The wheat powdery mildew genome shows the unique evolution of an obligate biotroph. Nature Genetics, 45: 1092–1096.

Prospects of hazelnut (*Corylus Avellana* L.) growing in Uzbekistan

Kamolitdin Sultonov[1,*], *Obidjon* Sattorov[1] and *Nigina* Mahmudova[1]

[1]Tashkent State Agrarian University, Tashkent, 100140, Uzbekistan

Abstract. The manuscript presents a comprehensive investigation into the propagation of hazelnut (*Corylus*) using semi-hardwood cuttings, with a particular emphasis on exploring various planting schemes to identify the most optimal approach. In an era where sustainable and efficient methods of horticultural propagation are crucial, this study contributes valuable insights to enhance hazelnut cultivation practices. The experimental trials were meticulously designed and executed during the final ten days of May, a crucial period that aligns with optimal growth conditions for hazelnut cuttings. A key innovation in this research was the utilization of a specialized fog-generating facility. With the overarching goal of determining the most advantageous planting scheme, a range of configurations was employed. Factors such as spacing, arrangement, and density were meticulously varied, allowing for a comprehensive assessment of their impact on hazelnut seedling growth. This systematic approach aimed to identify the scheme that maximizes the use of available space while ensuring optimal development and survival rates of the seedlings. The outcomes of this research represent a significant achievement in hazelnut propagation. Notably, the study demonstrated the successful cultivation of high-quality seedlings, achieving an impressive count of 42 to 45 seedlings per square meter of cultivated area. This achievement underscores the potential for efficient and productive propagation using the semi-wooden cuttings method. Furthermore, it highlights the critical role that suitable planting schemes play in harnessing the full potential of this propagation technique.

Keywords. Hazelnut, *Corylus*, tree, leaf, planting scheme, substrate, biohumus, regeneration.

1 Introduction

Corylus is a tree belonging to the birch family; a genus of deciduous shrubs. 20 species are known in the forest zones of Europe, Asia and North America. Hazelnut belongs to *Corylus* genus, *Birches* belong to *Betulaceae* family, and 10 species are found in CIS flora in different geographical conditions [1].

A common species is the hazelnut (*C. avellana*). In the Far East, Manchurian Leschina (*C. mandshurica*), different-leaved Leshchina (*C. heterophylla*) can be found. In the Caucasus, in addition to the common type, tree Leshchina, Colchida Leshchina (*C. colchica*)

* Corresponding author: k.sultonov@tdau.u

and Pontic Leshchina (*C. pontica*) species grow [2]. In the Caucasus and the Crimea, a large hazelnut or Lombard nut (*C. maxima*) is grown. Ihota forests will be built from Leshchina. The height is up to 2-5 m (sometimes 8 m). The flower is bisexual; the male ones are collected in a cluster, and the female ones are collected in a two-flowered bouquet. The fruit is a one-celled, one-nut nut [3]. It grows well in wet and fertile soils, especially in dark forests and black soils. Needs heat and humidity. The tree will bear fruit 5-6 years after planting. Up to 8 kg of nuts are obtained from one bush. Nuts contain 60-70% oil. It is propagated by cuttings from walnuts and bachki branches. Lives more than 80 years [4].

Hazelnut is a typical shrub, only some species form a tree-like form. This plant has some unisexual flowers, the stamens and stamens are located on the same branch, the stamens hang like a long earring and bloom in early spring until the leaves emerge. The arrangement of the flowers is usually simple, separated in four piles, the anthers are hairy-large (0.9-1.2x0.6 mm) and the tip of the anther is like two scythes, arranged in 8. Sawn grain (20 μm) is in the form of granules, smooth [5]. The flower is surrounded by three petals. Changchi flower buds appear in autumn, overwinter and bloom in spring. There are from one to four stamens in one panicle. Their length is from 8 mm to 35 mm, and they are cylindrical, inverted ovoid, straight, bent, rounded at the top, sharp or obtuse [6].

At the moment, there are urgent tasks before the plantation farms and homestead land owners, such as increasing the range of seedlings, increasing the volume of cultivated seedlings and reducing their cost. In the Republic of Uzbekistan, where the field of horticulture is developing, measures to master intensive types of perennial crops and advanced methods of crop care are gaining importance [7].

In order to ensure the implementation of these tasks, it is necessary to further improve the methods of increasing crop materials and methods of cultivation based on the biological characteristics of species and varieties, soil-climatic conditions of the regions, economic requirements and issues of improving production [8].

According to scientists, many factors (planting scheme, duration, state of the stem from which cuttings are taken and its part prepared for cuttings, growing substances and their concentration) affect their rooting ability when growing fruit plants from green cuttings [9]. Among these factors, the planting scheme of green cuttings is also important. Because excessive compaction leads to rotting of cuttings due to increased humidity and shading, on the contrary, increasing the distance reduces the number of seedlings per unit area [10].

Many foreign [1, 8] and local [2-6] scientists emphasized that the planting scheme is highly important in the propagation of fruit plants from green cuttings. However, the work of these scientists is on grafts of fruit plants, vines, olives, kiwi and other plants, whose biological properties are fundamentally different from those of the hazelnut plant. Therefore, choosing a planting scheme suitable for the biological characteristics of this plant when propagating the hazelnut plant from green cuttings is an important factor in ensuring the speed of its regeneration. Activation of cambium activity is observed in the basal part of the leaf during rooting of green cuttings of hazelnut plants. This, in turn, leads to the formation of callus tissue and young regenerative xylem in this zone.

2 Materials and methods

Walnut flowers are located in several small buds. They can be distinguished from the vegetative bud by its thickness and by the protuberance with a reddish blunt end (similar to a pig's snout). The buds are covered with many coins, usually there are no flowers on the lower ones, and the flowers are placed in pairs on the upper buds. Each flower bud is covered with flesh and consists of two beaks. One nut develops in one flower - the nut is monocotyledonous. Nut meat is woody to varying degrees [3].

The leaves are simple, wide, oval, often asymmetrical, unevenly distributed enlarged teeth, long pinnate, irregularly heart-shaped in the middle, sharp near the stem. The toothed, glabrous leaf is arranged in a thin hairy band (the upper part of the leaf plate is dark), the leaves are double-rowed, shedding. The edges of the leaves are also hairy, even at the tips of the teeth, this feathery hairiness is characteristic of hazelnuts. The shape and surface area of the leaf varies depending on the width of the leaf and its location on the stem. In the lower part, the leaves are small, round and oval with a blunt tip (5-10 cm^2); the largest leaves (100-300 cm2) are located in the middle, broadly inverted ovate and sharply pointed. In the upper part, it is round, egg-shaped, and the tip is slightly pointed [4].

It can be seen from the data of Figure 1 that the country of Turkey produced 600,000 tons of hazelnut products in 2010, and after 5 years, the production volume increased and produced an additional 46,000 tons of products. It is the leading country with the production of more than 19,000 tons of hazelnuts. The next place is Italy with 140,600 tons of products grown in 2020, followed by the United States of America with 64,400 tons. The Republic of Azerbaijan produced 29500 tons of hazelnuts in 2010, 32300 tons in 2015, and 49500 tons in 2020. Uzbekistan produced 1100 tons of hazelnuts in 2010, 4000 tons in 2015, and 3700 tons in 2015 [9].

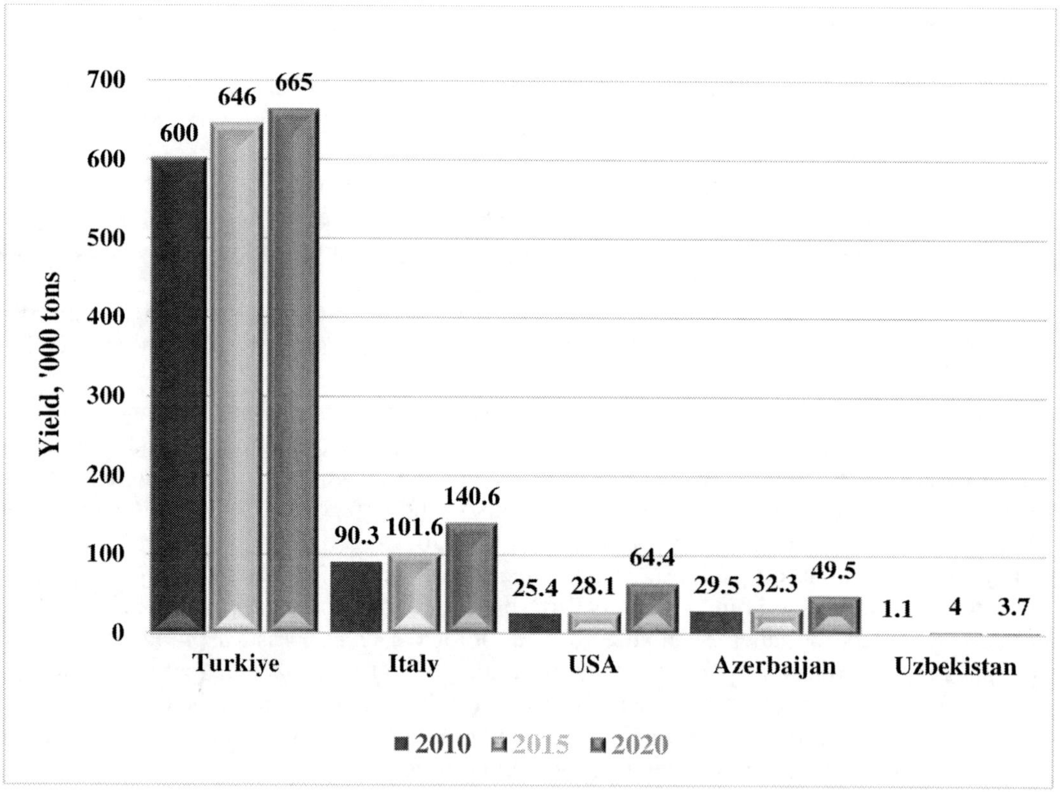

Figure 1. Hazelnut cultivation on the example of countries.

The experiment was conducted according to Ermakov's method [8]. Preparation of cuttings from hazelnut mother bushes In the third ten days of May, the experimental scheme is as follows: a microclimate controlled structure was planted from green semi-wooden cuttings in a scheme of 10x5 cm, 10x10 cm, 10x15, 10x20 cm, 10x25, 15x15 cm, 15x20 cm, and 15x25 cm. In addition, in a special substrate, the upper surface of the substrate is 8 cm of pure large-particle sand, and its lower part is planted in biohumus, which is 5 cm thick.

3 Results and discussion

Among the vegetative propagation methods, the main focus has been on the development of the technology of propagating hazelnut plants by cuttings taken from non-lignified branches with developing leaves, known in horticulture as green cuttings.

The regeneration of the green cuttings of fruit and ornamental plants is based on the process of new formation with the help of additional root and branch structural elements that stimulate the role of leaves. In the process of rooting, the leaves deliver plastic substances and hormones in the form of complex organic compounds to the shoot, and they correlate the interaction between the separate parts of the shoot. The presence of leaves in the cutting is a factor that stimulates regeneration.

When propagating hazelnuts from walnuts, it can be observed that biological signs and characteristics of the mother plant deteriorate in offspring grown from walnuts. In addition, plants grown from walnuts enter the harvest late, that is, at the age of 7-9 years. That is why hazelnut is propagated vegetatively in production conditions. When propagated in this way, the plantations begin to harvest in 4-5 years after planting seedlings. An important economic-biological feature of nut plants, especially hazelnut, is the ease of transportation of the product, the ease of keeping its nuts intact in simple facilities, and the long shelf life. This makes it possible to take care of hazel even in areas with poor roads.

The experiments on propagation of the hazelnut plant from green cuttings in different planting schemes showed that the rooting ability of green cuttings differed according to the variants. As the planting pattern expanded, the rooting capacity of cuttings increased. In accordance with the rule, the highest rooting capacity was recorded in the variant planted in the 15x25 cm scheme. In this option, the number of rooted cuttings compared to the total number of planted cuttings reached 90.4%. The least rooting was observed in the 10x5 scheme. In this experimental variant, the degree of rooting of green cuttings of the hazelnut plant did not exceed 78.32% (Figure 2).

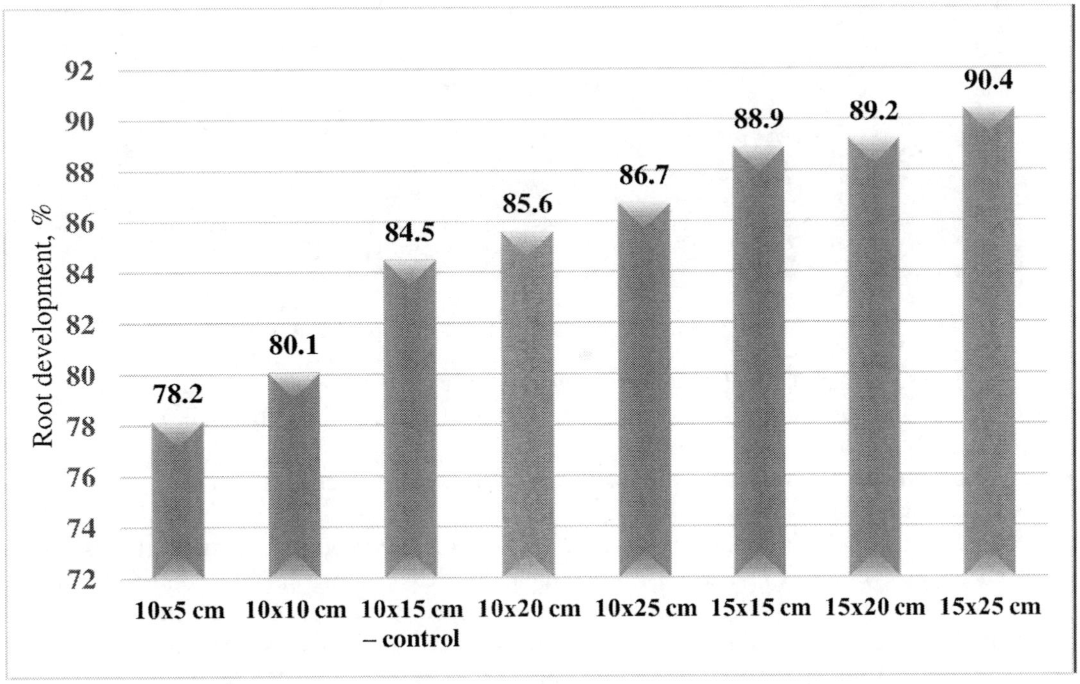

Figure 2. Green cuttings rooting of the Akademician Yablokov variety of hazelnut in different planting schemes (2021-2023).

It should be noted that the difference between the rooting capacity in the experimental options is not so high, but the thinned cuttings showed superiority in the parameters of the development of the root system. In this case, the best developed root system was noted in the cuttings of options planted in the scheme of 10x15...25 and 15x15...25 cm.

Planting scheme also had a significant effect on the development parameters of the above-ground part in green pens. In this case, the highest value of the ability to form a branch in the planted cuttings was recorded in the variant planted in the scheme of 15x25 cm. In this option, compared to the total number of planted green cuttings, the amount of them that formed a branch reached almost 97.3%. As the planting pattern became denser, the yield of shoots decreased in direct proportion. According to the rule, the least grown cuttings were recorded in the scheme of 10x5 cm. In this experimental variant, the number of sprouts did not exceed 17.4% compared to the total number of planted green cuttings (Figure 3).

Figure 3. Hazelnut cuttings planted in a special substrate.

The rest of the variants ranked between these two variants in terms of new shoot production in green cuttings and varied between 36.1 and 96.6% of the total planted cuttings, respectively, across variants.

The number of leaves formed in plants grown from green cuttings also differed depending on shoot formation. The superiority in this biometric parameter was also noted in the variant planted in the 15x25 cm scheme. It was noted that the green cuttings planted in this scheme produced an average of 18.2 leaves, which means 12 leaves more than the control. The lowest number of leaves in rooted green cuttings was recorded in the variant where the cuttings were planted in a 10x5 cm scheme. In this scheme, 2.7 fewer (6.2) leaves were formed compared to the control option (9.5). In terms of the number of leaves produced on regenerated plants, the rest of the experimental variants were intermediate in the growth sequence between these two variants and varied between 9.6-18.7 leaves according to the variants and 2.1-8.7 more leaves compared to the control.

Experimental data show that the scheme of planting green cuttings had a significant effect on the development of the conditional body of one-year seedlings. The same trend as the number of leaves was observed in the formation of this physiological indicator (Table 1).

Table 1. Developmental characteristics of the Akademician Yablokov variety of hazelnut grown from green cuttings in different planting schemes (2021-2023).

Experiments	Root system volume, cm^3	Cuttings formed by the stem, %	Number of leaves	Conditional body diameter, mm	Plant stem length, cm
10x5 cm	2.2	17.4	6.2	7.1	36.5
10x10 cm	4.1	36.1	8.43	9.0	41.6
10x15 cm – control	5.0	61.6	9.5	11.3	45.4
10x20 cm	5.8	91.2	11.6	13.4	60.1
10x25 cm	6.4	94.7	14.7	13.6	66.2
15x15 cm	6.8	95.8	16.9	14.1	68.3
15x20 cm	7.0	96.6	17.7	14.5	69.7
15x25 cm	7.1	97.3	18.2	15.2	72.6

Therefore, the thickest and well-developed plants were recorded in the variant planted in the 15x25 cm scheme. The stem of plants grown from green cuttings planted in this scheme reached an average diameter of 15.1 mm, which is 3.9 mm thicker than this physiological indicator of the control variant. In rooted green cuttings, the thinnest conditional body diameter was determined in the variant where the cuttings were planted in a scheme of 10x5 cm. The diameter of the conditional body of the plants grown in this experimental variant did not exceed 7.1 mm, and they were 4.2 mm thinner compared to this indicator (11.3) of the plants in the control variant.

The rest of the experimental variants in terms of the diameter of the conditioned body of the plants developed from the green cutting took an intermediate place between these variants and varied in the range of 13.4-14.5 mm according to the variants. The conditional trunk diameter of the plants of these experimental variants was 2.1-3.9 mm thicker compared to the control. In the planting schemes of 10x5 cm and 10x10 cm, compared to the control version, the diameter of the conditioned body of the plants was smaller by 7.1 and 9.0 mm.

It is known that in horticulture, the height of ready seedlings is also an important economic indicator. Measurement of the height of plant branches developed from green cuttings planted in different schemes showed that the highest branched plants were formed in the experimental variant planted in a 15x25 cm scheme. Plants grown in this experimental variant reached an average height of about 72.6 cm, which is 27.2 cm more than this physiological indicator of plants in the control variant (45.4 cm).

In the experiment, the lowest branched plants were determined in the variant where green cuttings were planted in a scheme of 10x5 cm. The height of plant shoots grown in this experimental option did not exceed 36.5 cm, and they were 8.9 cm lower than the height of plant shoots in the control option, and the 10x10 cm planting scheme was also 3.8 cm lower than the control option (45.4).

The remaining experimental variants in terms of the height of the plant shoots developed from green cuttings were intermediate between these variants and varied between 60.1-69.7 cm according to the variants. In these experimental variants, the shoot length of the formed plant was 14.7-27.2 cm longer, compared to the control variant.

It is worth noting that although the regeneration of plants occurs at a high rate when the planting scheme is thinned, it is important to choose the optimal option in terms of the amount of ready seedlings per unit of area.

Calculations showed that the yield of standard seedlings with the length of the main branch over 60 cm was the highest in the 10x20 cm and 15x15 cm planting schemes (42 and 45 units/m^2, respectively) compared to the total grown plants. In densely planted variants, although the number of plants that took root and grew was high, most of them (80-85%) were left for next year's cultivation. Because their height did not even reach 50 cm. On the contrary, in the sparsely planted variants, although about 95-98% of the plants sprouted and grew strongly, the number of seedlings was correspondingly low due to the fact that less cuttings were planted per unit area (Figure 4).

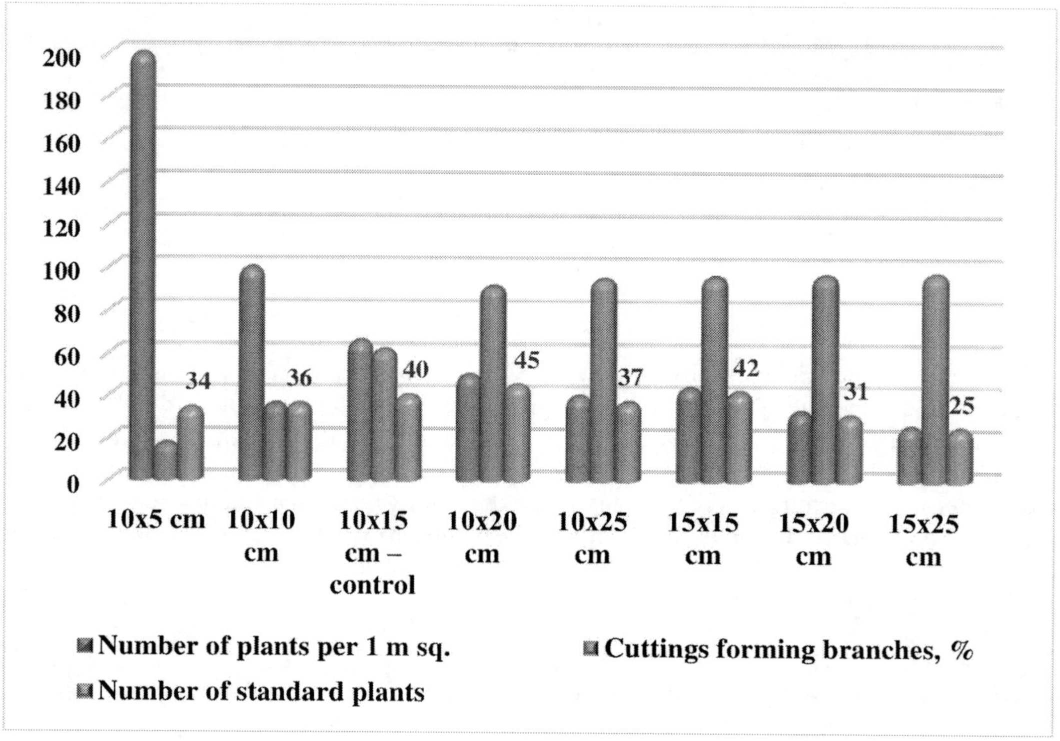

Figure 4. Effect of planting schemes on the yield of Akademician Yablokov variety seedlings of hazelnut in growing from green cuttings (2021-2023).

4 Conclusions

Based on the comprehensive analysis of the experimental findings presented above, several noteworthy conclusions can be drawn regarding the rootability and subsequent seedling yield of the Akademician Yablokov hazelnut variety. Notably, the rootability of the green cuttings exhibited a clear trend of enhancement as the spacing between the cuttings decreased. This trend culminated in a remarkable rootability rate, approaching an impressive 95%. This observation underscores the critical influence of spacing on the successful establishment of root systems in the hazelnut cuttings.

However, while the rootability rates provide valuable insights into the initial establishment of the cuttings, the ultimate goal of hazelnut cultivation lies in generating a robust and substantial yield of high-quality seedlings. In this context, it is prudent to consider both rootability and seedling yield in conjunction to determine the most optimal planting scheme.

The experimental data reveal that for achieving optimal seedling yield per unit area, specific planting schemes warrant consideration. Specifically, the planting configurations of 15x15 cm and 10x20 cm emerged as particularly favorable options. These configurations not only facilitated the successful rooting of cuttings but also translated into a highly commendable outcome in terms of seedling yield. Remarkably, under these identified planting schemes, an impressive count of 42 to 45 high-quality seedlings of standard size can be obtained from a single unit of area, measuring 1 square meter.

This convergence of favorable rootability and subsequent seedling yield serves to guide practitioners and horticulturists in making informed decisions regarding hazelnut cultivation using green cuttings of the Akademician Yablokov variety. The harmonization of rootability, seedling yield, and planting schemes not only ensures efficient resource utilization but also presents a pathway to maximize the productive potential of hazelnut propagation.

In summation, the experimental results illuminate a symbiotic relationship between spacing, rootability, and seedling yield in the context of the Akademician Yablokov hazelnut variety. The findings underline the significance of carefully tailored planting schemes to harness the full potential of this propagation method. Ultimately, the research outcomes provide invaluable insights that can drive enhanced practices in hazelnut cultivation, yielding a considerable bounty of high-quality seedlings while paving the way for sustainable horticultural endeavors.

References

1. Bulygin N.E. Dendrology. – St. Petersbourg: "Agropromizdat", 2011, 414 p.
2. Kayimov, A., Kholmurotov, M. Z., & Eshankulov, B. I. (2021, December). Justification of prospective pistachio (Pistacia vera L.) varieties and forms while creating plantations in Uzbekistan. In IOP Conference Series: Earth and Environmental Science (Vol. 939, No. 1, p. 012037). IOP Publishing.
3. Hamzayev, A. K., Eshankulov, B. I., Kholmurotov, M. Z., & Inomova, M. M. (2020, December). Study on cultivation of pistachio (Pistacia vera L.) seedlings in containers. In IOP Conference Series: Earth and Environmental Science (Vol. 614, No. 1, p. 012119). IOP Publishing.
4. Nortojiev, B. S., & Turdaliyeva, O. M. Q. (2022). LIMON KO ‘CHATLARINI QALAMCHALARIDAN KO ‘PAYTIRISHDA SUBSTRATLARNI AHAMIYATI. Academic research in educational sciences, 3(Speical Issue 1), 180-183.
5. Botirov, A., Murtazayev, A., Ochilov, B., & Rustamova, G. (2022). UZUM YETISHTIRISHNING HUDUDLAR KESIMIDAGI TAHLILI. Academic research in educational sciences, 3(Speical Issue 1), 293-297.
6. Khasanov, S., Kulmatov, R., Li, F., van Amstel, A., Bartholomeus, H., Aslanov, I., ... & Chen, G. (2023). Impact assessment of soil salinity on crop production in Uzbekistan and its global significance. Agriculture, Ecosystems & Environment, 342, 108262.
7. Odinayev, M., Buriev, K., & Eralieva, S. (2021). Analysis of mechanical properties, biochemical composition and technological parameters of grape (Vitis) raisin varieties in conditions of Uzbekistan. In E3S Web of Conferences (Vol. 284, p. 03023). EDP Sciences.
8. Ermakov B.S. Reproduction of woody and shrubby plants by green cuttings (Guidelines). - Chisinau: Shtiitsa, 2001. - pp. 21-38.
9. Azizov, A. S. & Gafurov, J. A. (2021, December). Scientific significance of using protective devices in pear storage in refrigerated warehouses (in the case of Uzbekistan).

In IOP Conference Series: Earth and Environmental Science (Vol. 939, No. 1, p. 012039). IOP Publishing.

10. Sultonov, K., Fayziev, J., Eralieva, S., Vakilova, S., & Temurov, T. (2022, May). Effect of Mineral Fertilizers on Increasing Productivity and Quality of Seedless Varieties of Grapes. In International Scientific Conference on Agricultural Machinery Industry "Interagromash"" (pp. 1930-1940). Cham: Springer International Publishing.

Effect of growth regulators on cucumber seed fertility

Salohiddinjon Yunusov[1,*], *Farkhod* Bolikulov[1], *Zhasur* Makhmanazarov[1] and *Dilshod* Yakhshiboev[1]

[1]Tashkent State Agrarian University, 2, University street, Tashkent, 100140, Uzbekistan

Abstract. This scientific article places its emphasis on investigating the impact of biostimulants and growth-enhancing preparations on the germination of cucumber seeds in the context of intensive cucumber seedling cultivation. Our study encompasses a series of experimental and practical trials meticulously designed to elucidate the effects of various growth substances on the rapid and robust germination of cucumber seeds. Throughout the course of our experiment, we conducted comprehensive assessments of the process of treating cucumber seeds with a diverse array of growth stimulants. Our goal was to unveil the precise mechanisms through which these stimulants influence and expedite the germination process. By harnessing innovative technologies and implementing the most cutting-edge methodologies available, our research seeks to delve into the intricate dynamics of cucumber seedling cultivation under intensified conditions. One of the pivotal objectives of our study was to develop a strategic framework that enables the consistent production of high-quality cucumber seedlings. This endeavor involves the integration of advanced techniques and methodologies, aligned with the overarching aim of bolstering germination rates and cultivating robust cucumber seedlings. In essence, this research endeavor amalgamates experimentation, practical application, and the utilization of modern advancements to decipher the intricate interplay between biostimulants, growth preparations, and the germination process of cucumber seeds. By providing insights into these crucial facets, our study contributes to the refinement of intensive cucumber seedling cultivation practices, ultimately enhancing agricultural productivity and the quality of cucumber yields.
Keywords. Variety, hybrid, cucumber, seed, stimulant, germination, root, temperature, humidity.

1 Introduction

In the present era, the realm of vegetable seedling cultivation within our Republic is witnessing the emergence of novel categories of growth regulators [1]. These innovative growth substances are proving to be pivotal in revolutionizing agricultural practices [2]. Their integration into our agricultural landscape offers a multitude of benefits, including heightened seed germinability, enhanced seedling resilience in the face of adverse weather

* Corresponding author: s.yunusov@tdau.uz

conditions, expedited seedling growth, amplified agricultural productivity, and the capacity to generate premium-grade, cost-effective, environmentally conscious produce [3, 4].

Our experiments have centered on the utilization of biopreparations, serving as a cornerstone in elevating seed germinability and fertility across a broad spectrum of agricultural crops [5]. This approach holds immense significance due to its widespread applicability, bolstering the resistance of seedlings against specific diseases and pests. This resilience enhancement not only contributes to improved plant health but also augments overall crop yield [6].

The integration of these innovative biopreparations constitutes a transformative force in the realm of modern agriculture [7]. By harnessing their potential, we pave the way for more robust and efficient agricultural practices that cater to both economic and environmental sustainability [8]. This shift towards growth regulators and biopreparations represents a remarkable stride towards a future where agricultural productivity is optimized, crop quality is elevated, and our environment is safeguarded. In the world, special attention is being paid to the organization of a healthy diet that provides the need for daily consumption of vegetables [9-11].

Among vegetables, cucumber is one of the main crops, and it is expedient to increase the volume of its cultivation and improve the quality of the product [12]. The total area planted with cucumbers in the world is 2 mln 496 thous. hectares, and the gross production volume is 91.3 million tons. China (65 million tons), Iran (1.981.000 tons), Russia (1.940.000 tons), Turkey (1.827.000 tons) and the United States (956.000 tons) are among the leading countries in terms of gross product volume [13-15].

2 Materials and methods

Currently, the demand to increase the production of cucumbers is increasing. Therefore, increasing its productivity in the leading cucumber-growing countries remains an urgent task. In our country, in 2022, cucumber cultivated area was more than 25.000 hectares, and the total volume of production was more than 853.100 tons. It means that it can fully meet the demand of the population of the country [15]. However, the increase in the number of population and the expansion of the volume of product exports lead to an increase in the demand for this product [7, 8]. Therefore, it is necessary to create new, high-yielding varieties of cucumber and to develop innovative technological elements of growing cucumbers in the open field.

During field experiments in research, phenological observations, biometric measurements, monitorings on the rapid germination of cucumber seeds were carried out. Observations and calculation records were carried out in accordance with generally accepted requirements. Experiments were done in 4 replications.

The research and its experiments were carried out in 2019-2022 in 100 m^2 area of protected facility of the Information and Consulting Center of Tashkent State Agrarian University. The purpose of the research is to determine the effect of biostimulants on the germination of seeds, its optimal application rate and periods in the cultivation of cucumber seedlings. The object of research is cucumber variety "Sevinch", "Fantina" G1, "Superina" F1 and "Beyt-Alfa" F1 hybrids and PLANTASTIM, Zerebra-Agro and Ifo Seed biostimulants.

3 Results and discussion

The results of research. For our experiments, we selected varieties and hybrids that are suitable for cultivation of cucumbers by the trellis method. "Sevinch" variety of cucumber created by the co-authorship and hybrids "Fantina" G1, "Superina" F1 and "Beyt-Alfa" F1 brought from foreign countries, which are suitable for the soil and climate conditions of our republic were used in the experiments. As we all know that vegetable seeds are mainly sown directly in the open field after soaked in water or in dry form. We observed that cucumber seeds are planted in the field after being soaked in water for a certain period of time, and it is noted in scientific sources that when unsoaked seeds are planted, their germination coefficient is different compared to those that have been water-soaked. Therefore, in our experiments, we studied the effect of several biostimulants on increasing seed germination by treating seeds with them and certain results were obtained. In our experiments, control option was prepared by treating cucumber seeds in normal clean water (H_2O). In other variants under study, currently recommended PLANTASTIM, Zerebra-Agro and Ifo Seed biostimulants were used.

Before sowing, the seeds were soaked in clean water and PLANTASTIM, Zerebra-Agro and Ifo Seed substances for 5 hours, then in a special heating place (heated floor), the seeds were spread on gauze or simple fabric and treated for 24 hours. If you carefully observe the processed seeds, you can notice that they are ready to germinate at the tip. After that, the seeds were sown in special cassettes or substrates prepared in cups. Germination of seeds was observed 1-2 days after sowing. After the seeds germinated fully, 2 days before transplanting in the open field, a suspension of growth regulators was prepared and applied by spraying once through the leaves.

The degree of germination of the seeds of "Sevinch" variety, "Fantina" G1, "Superina" F1 and "Beyt-Alfa" F1 hybrids under the influence of these growth preparations was determined in our experiments. It can be said that when cucumber seeds were sown in dry form without any treatment or soaked in water have lower germination rate, on the contrary when they were treated with certain biostimulants the germination rate of seeds was observed to be higher. At the same time, after germination of the seeds, intensive development of root system and soil surface part of the plant, its stem growth and leaves were observed.

The conducted experiments showed that compared to the control option, i.e. the case of cucumber seeds treated in water, the germination of the seeds treated with the preparations, the healthy development of the seedlings, and their viability in the open field were observed to be higher in our experiments. Considering above-mentioned results, it can be said that the use of modern preparations in the rapid cultivation of cucumber seedlings is expedient (Tables 1).

Table 1. Effects of types of growth regulators on the germination of cucumber seeds (in 2019-2022).

Varieties	**Type of biostimulant**	**Soaking period (hours)**	**Treatment period (hours)**	**Root formation (day)**	**Germination (day)**	**Formation of true leaves (day)**
Fantina F1	Treatment with water (*control*)	5	24	4	6	10
	ZEREBRA AGRO	5	24	3	5	9
	PLANTASTIM	5	24	2	4	8
	IFO SEED	5	24	2	4	8
	LSD_{05}	0.5	0.4	0.3	0.5	0.6

Superina F1	Treatment with water (*control*)	5	24	4	6	10
	ZEREBRA AGRO	5	24	3	5	9
	PLANTASTIM	5	24	2	4	8
	IFO SEED	5	24	2	4	8
	LSD_{05}	0.3	0.5	0.3	0.3	0.4
Beyt-Alfa F1	Treatment with water (*control*)	5	24	3-4	6	10
	ZEREBRA AGRO	5	24	2-3	5	9
	PLANTASTIM	5	24	2	3	7
	IFO SEED	5	24	2	4	8
	LSD_{05}	0.3	0.2	0.3	0.5	0.4
"Sevinch" variety	Treatment with water (*control*)	5	24	4	6	10
	ZEREBRA AGRO	5	24	3	5	9
	PLANTASTIM	5	24	2	3	7
	IFO SEED	5	24	2	4	8
	LSD_{05}	0.4	0.3	0.3	0.5	0.4

We can see in the table that when the cucumber varieties were processed, their seed germination increased. It was found that when the varieties and hybrids of cucumbers selected as control and were treated with water and other varieties treated with preparations in the same way, their germination, formation of the root system, formation of leaves differed from each other. In the control option, that is, when the seeds treated with water, the germination of seeds of varieties and hybrids showed almost the same result, as a result of the applied preparations, the germination of seeds, proliferation of roots and the above-ground part and formation of true leaves of some varieties were observed to change significantly compared to the control. In the experiments, when all varieties of cucumbers were treated with water, the seeds were soaked in water for 5 days. After that, it was spread on a cloth or gauze in a warm place and sprayed with water for 24 hours. After the tips of the water-saturated seeds were ready to germinate, they were sown individually on glass substrates. This agrotechnical measure of seed treatment was carried out in the same way in all varieties and the preparations were applied in the same process in our experiments. The result, as noted in the table, when the seeds treated with ZEREBRA AGRO preparation, compared to the control (water), the appearance of roots was observed one day earlier. It took 6 days for the seeds to germinate on the surface of the soil, and the appearance of true leaves was observed 10 days after sowing.

In the experiments, it was found that PLANTASTIM and IFO SEED preparations had a positive effect on the germination of cucumber seeds. Planting of cucumber seeds after soaking them in water mixed with PLANTASTIM biostimulant for 4-5 hours (10 ml per 1 liter of non-chlorinated water) or planting the seedlings of cucumber in cups and cassettes after treating with PLANTASTIM mixture (150 ml per 10 liters of non-chlorinated water) showed good results. At the same time, as a result of treatments on cucumber seeds, the viability and proliferation of seedlings, develop of their immunity against root rot, and vigorously formed vegetative organs, and a positive effect on high productivity were also observed during our subsequent experiments. As a result of treatment of the seeds of the cucumbers varieties and hybrids used in our experiments, it was observed that the germination of the seeds of the Fantina F1 cucumber hybrid was slightly less developed compared to other varieties and hybrids (Figure 1).

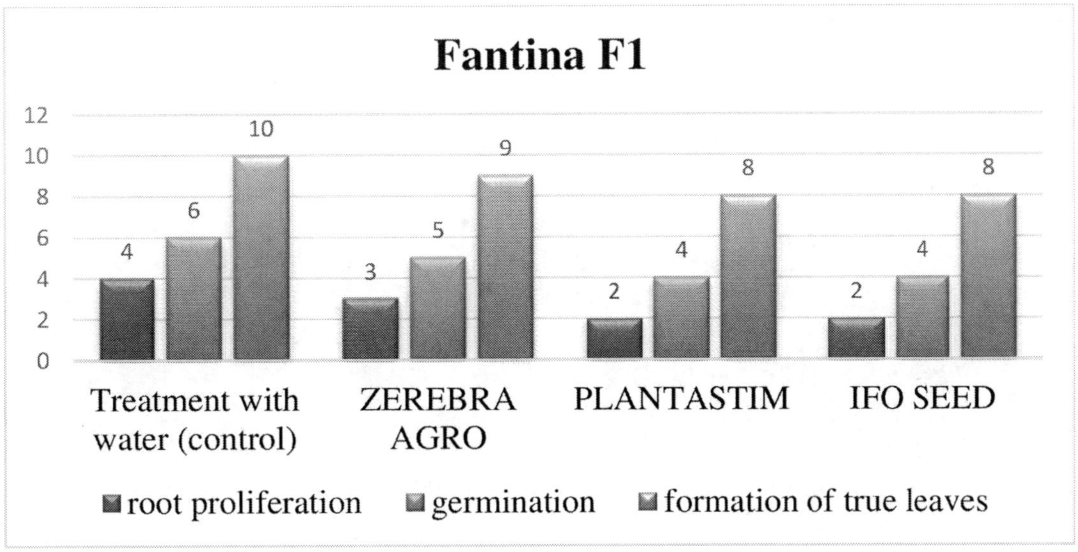

Figure 1. Rate of germinability of seeds and development of vegetative parts of Fantina F1 hybrids after treating its seeds with water, Zerebra-agro, PLANTASTIM and IFO-seed preparations.

It can be seen that when Fantina F1 hybrid seeds were treated in water, the formation of the initial root system took place on the 4th day of sowing, the germination of the seeds took place on the 6th day, and the formation of true leaves took place after 10 days.

When Fantina F1 hybrid seeds were treated with Zerebra-agro biostimulant, it was observed that the initial root system formation occurred on the 3rd day, seed germination occurred on the 5th day, and the formation of true leaves occurred after 9 days from the planting.

When the seeds were treated with the PLANTASTIM biostimulant, the formation of the first roots was observed on the 2nd day, the germination of the seeds was observed on the 3rd day, and the formation of true leaves was observed on the 8th day of the planting. When the seeds were treated with IFO SEED, their development was observed as follows: it was found that the formation of the root system began on the 2nd day, the germination of seeds began on the 3rd day, and the formation of true leaves began on the 8th day from the planting.

From this, it can be said that PLANTASTIM biostimulant and IFO SEED growth preparation showed a good effect on the germination of Fantina F1 hybrid seeds, the formation of root system and the formation of true leaves compared to the control (water).

During our experiments, the effect of the applied preparations on the germination of the seeds of the Superina F1 hybrid of cucumber was differentiated as follows: initially, as a control, when the seeds were treated in water, the result observed in the Fantina F1 hybrid was recorded. Also the effect of preparation types was observed in this variety. When Superina F1 hybrid seeds were treated with Zerebra-agro biostimulant, the formation of the initial root system was observed from the 3rd day, and the germination of the above-ground part of the seeds began from the 5th day. It was observed that the formation of tue leaves occurred after 9 days from the sowing.

When the seeds of this hybrid were treated with PLANTASTIM biostimulant, the formation of the root system and the number of roots was observed to be higher from the 2nd day, and rapid germination of the seeds was observed on the 3rd day. It was found that the formation of true leaves developed from the 8th day. The effect of IFO SEED from the used preparations was as follows when treated on seeds compared to the control. It was observed that the formation of the root system of the seeds developed on the 2nd day, the germination of the surface part of the plant on the 3rd day, and the formation of true leaves from the 8th day of the planting (Figure 2).

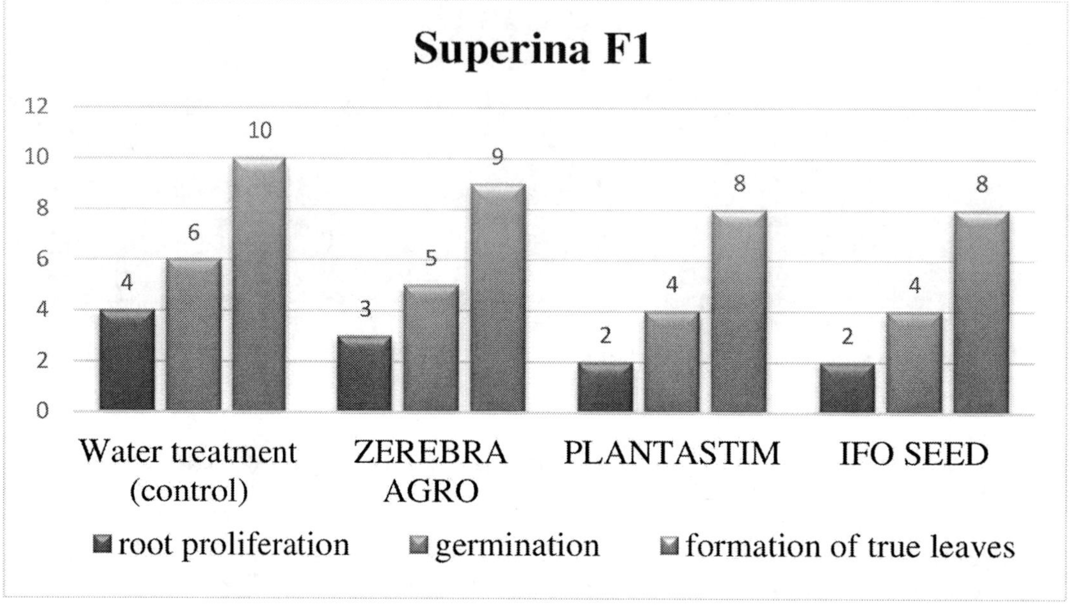

Figure 2. Rate of germinability of seeds and development of vegetative parts of Superina F1 hybrids after treating its seeds with water, Zerebra-agro, PLANTASTIM and IFO-seed preparations.

Seed germination, formation of root system and emergence of true leaves of the Superina F1 hybrid were observed to have good results with the treatment of PLANTASTIM biostimulant and IFO SEED preparation compared to control (water) and Zerebra-agro stimulant. It can be said that when vegetable seeds, including cucumber seeds, are treated with these preparations, it was found during phenological observations in our experiments that the fertility coefficient of seeds increases and, at the same time, the continued viability of seedlings and their resistance to adverse conditions are ensured.

In our experiments, it was observed that the following results were recorded when the seeds of the studied cucumber Beyt-Alfa F1 hybrid were treated with the above preparations (Figure 3). Experiments have shown that stimulators and preparations are directly important for the germination of cucumber seeds and the continued viability of seedlings and this was proven also by our phenological observations.

We observed the following effects on cucumber Beyt-Alfa F1 hybrid seeds: when the seeds were treated with water, it was found that the germination of the seeds was the same as the above varieties, but when we treated them with stimulants and preparations, the germination of the seeds and the rapid development of vegetative organs were observed. When the seeds were treated with the Zerebra-agro stimulant, the formation of their root system appeared on the 3rd day, and the initial germination of the seeds was observed on the 5th day. During our experiments, we observed that true leaves were formed 8 days after sowing the seeds.

When the seeds of this hybrid were treated with PLANTASTIM biostimulants, it was observed that the root system was strongly formed and the number of roots increased significantly compared to the roots of the above hybrids. It can be concluded that PLANTASTIM biostimulant has a positive effect on the fertility of Beyt-Alfa F1 hybrid seeds. The root system was formed from the 2nd day after the treatment of the seeds, rapid germination of the seeds was observed on the 3rd day. It was observed that the formation of true leaves developed from the 7th day. It can be seen during our experiments that this stimulant had a positive effect on the formation of leaves in the seedlings and the appearance of leaves 3 days earlier compared to the control (water).

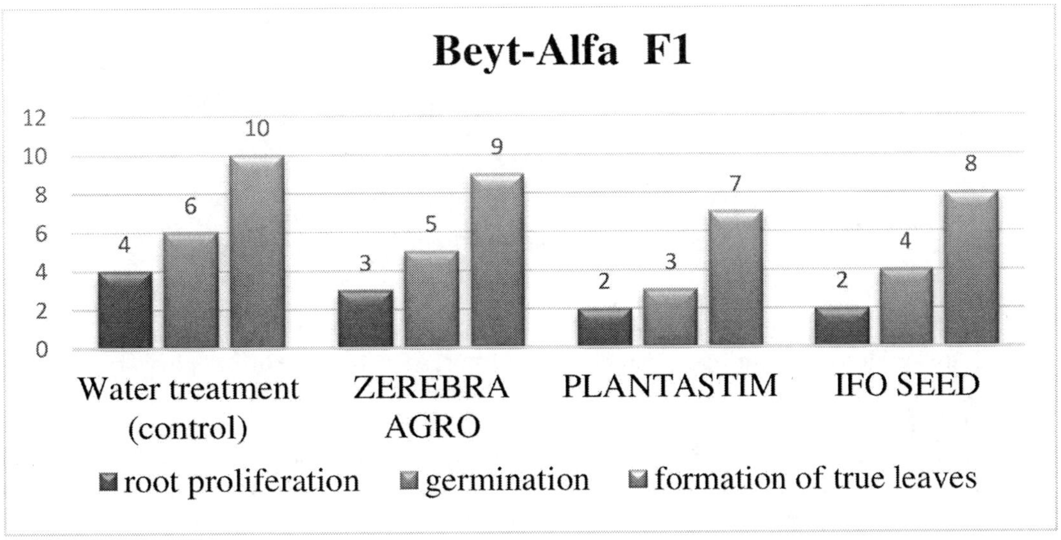

Figure 3. Rate of germinability of seeds and development of vegetative parts of Beyt-AlfaF1 hybrid after treating its seeds with water, Zerebra-agro, PLANTASTIM and IFO-seed preparations.

The Beyt-Alfa F1 hybrid seeds of cucumber showed faster germination in stimulants and preparations compared to other studied hybrids. It is under the influence of these stimulants and preparations that the biochemical processes in the seed are accelerated and the seed sprout quickly wakes up, as a result, the primary root is formed from the sprout and the seed germination is high. Therefore, in today's modern agriculture, it is an important task to quickly increase the germination of vegetable seeds and prepare healthy quality vegetable seedlings in a short period of time. Biostimulants and preparations that we use in our experiments serve as an important object in this regard. During our experiments, in addition to studying the fertility of hybrid seeds brought from foreign countries, we also tested in our research the fertility of the seeds of the Sevinch variety, created in co-authorship. It is necessary to mention that the Sevinch variety of cucumber is suitable for cultivation in the open field by the trellis method. When the seeds of this Sevinch variety were treated with stimulants and preparations, their germination was observed as follows (Figure 4).

Although the seeds of this variety showed the same results as the above hybrids when treated in water, the influence and effectiveness of stimulants and preparations were found to have better results. When the seeds were treated with the Zerebra-agro stimulant, the formation of their root system was observed on the 3rd day after sowing the seeds. The germination of seeds was observed on the 5th day, and the formation of true leaves was observed on the 8th day. During our experiments, it was found that the biochemical processes in the seeds accelerated when the seeds were treated with the PLANTASTIM stimulant. During the phenological observations, we were sure about the rapid development of seedlings and germination in the seeds compared to the control.

https://doi.org/10.1051/e3sconf/202342102013

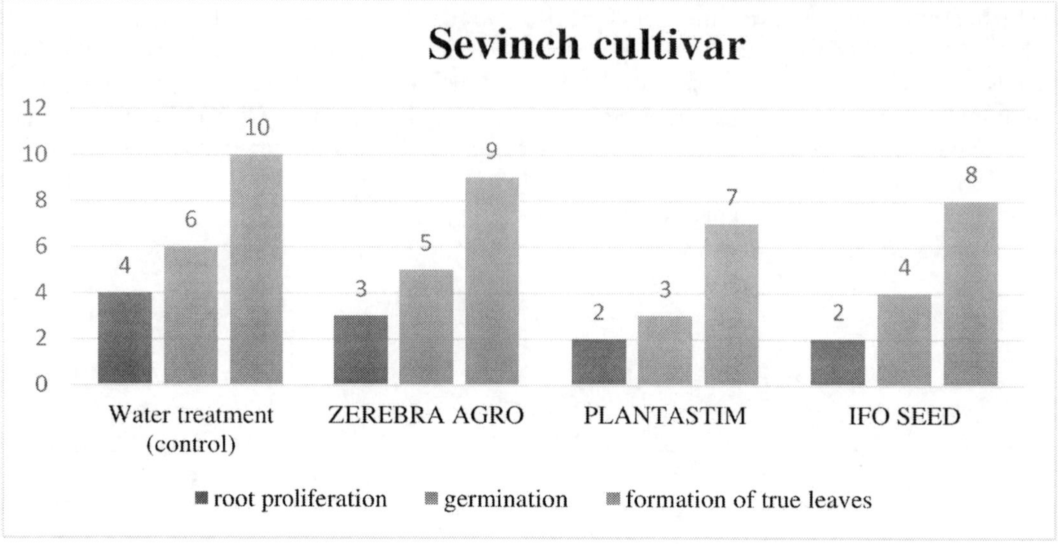

Figure 4. Rate of germinability of seeds and development of vegetative parts of Sevinch variety after treating its seeds with water, Zerebra-agro, PLANTASTIM and IFO-seed preparations.

While we observed in our experiments that PLANTASTIM stimulant had a positive effect on seed fertility in all studied hybrids, but it was observed that the PLANTASTIM stimulant had a better effect especially on the seed fertility of Beyt-Alfa F1 hybrid and Sevinch variety than other hybrids. We can say that the PLANTASTIM stimulant showed an advantage effects over other types of stimulants and preparations used in our experiments on the germination of cucumber seeds (Figure 5).

Experiments have shown that 10-day-old seedlings of Sevinch cucumber seeds treated with water are much smaller and weaker than 7-day-old seedlings of seeds treated with PLANTASTIM stimulant, as can be seen from the figure below.

a) b)

Figure 5. a) Status of 10-day-old seedlings of Sevinch variety seeds treated in water, b) Status of 7-day-old seedlings of Sevinch variety seeds treated with PLANTASTIM stimulant.

It was observed during our experiments that when cucumber seeds were treated with the PLANTASTIM stimulant, the seedlings developed 3 days earlier than the control and strong healthy seedlings were ready. All experiments were carried out in protected facilities. The effect of stimulants and preparations used during our experiments on the germination of seeds of cucumber varieties and hybrids was studied. At the same time, we have observed in our experiments that certain objects (types of substrate) and the temperature, relative humidity of the air in the facility can affect the seed germination and their norms are important for the intensive germination of seeds.

4 Conclusions

Through our meticulously designed experiments, we made a significant revelation: the applications of PLANTASTIM and IFO SEED preparations yielded highly favorable outcomes in the germination process of all cucumber varieties and hybrids studied. This breakthrough holds profound implications for the cultivation of cucumber crops.

Notably, within the realm of cucumber varieties, the Beyt-Alfa F1 hybrid and the Sevinch variety stood out as exemplars of successful seed germination under the influence of these stimulants. This observation underscores the potency of PLANTASTIM and IFO SEED in promoting optimal germination rates, with these particular cucumber variants displaying remarkable receptivity to the effects of these growth-enhancing preparations.

These findings carry great significance as they offer a practical means to enhance the germination process and potentially bolster the overall success of cucumber cultivation endeavors. By identifying specific varieties and hybrids that respond particularly well to these stimulants, we lay the foundation for tailored agricultural strategies that maximize crop yield and quality. Such discoveries further underscore the potential of innovative growth enhancers in revolutionizing agricultural practices and contributing to sustainable, high-yield food production.

References

1. Aramov M.Kh. Scientific center for selection and seed production of vegetable crops in the south of Uzbekistan. // The main directions and prospects of breeding seed production of vegetables, melons and potatoes. International Scientific and Practical Conference. - July 2-5, 2001. Abstracts. – Tashkent-Termez, 2001. (in Russian).
2. Boboev M.M. Features of cultivation of various varieties of cucumber in the conditions of irrigated lands of the Republic of Tajikistan. //Magazine. Bulletin of modern science. - Moscow, 2016. - No. 10-1 (22). (in Russian)
3. Bolotskikh A.S. Cucumber. // All about cucumber. – Kiev: Urojay. 2002. (in Russian)
4. Tayjanov, K., Khojimatov, O., Gafforov, Y., Makhkamov, T., Normakhamatov, N., & Bussmann, R. W. (2021). Plants and fungi in the ethnomedicine of the medieval East-a review. Ethnobotany Research and Applications, 22, 1-20. DOI:0.32859/ERA.22.46.1-20
5. Jabeen, S., Zafar, M., Ahmad, M., Althobaiti, A. T., Ozdemir, F. A., Kutlu, M. A., ... & Majeed, S. (2023). Ultra-sculpturing of seed morphotypes in selected species of genus Salvia L. and their taxonomic significance. Plant Biology, 25(1), 96-106. DOI:10.1111/plb.13473
6. Majeed, S., Ahmad, M., Ozdemir, F. A., Demirpolat, A., Şahan, Z., Makhkamov, T., ... & Nabila. (2023). Micromorphological characterization of seeds of dicot angiosperms from the Thal desert (Pakistan). Plant Biosystems-An International Journal Dealing with all Aspects of Plant Biology, 157(2), 392-418. DOI:10.1080/11263504.2023.2165553
7. Majeed, S., Ahmad, M., Ali, A., Althobaiti, A. T., Ramadan, M. F., Kilic, O., ... & Sultana, S. (2023). Pollen micromorphology among amaranthaceous species from desert

rangeland: Exine stratification and their taxonomic significance. BioMed Research International, 2023, 4967771. DOI:10.1155/2023/4967771

8. Makhkamov, T., Sotiboldiyeva, D., Mamarakhimov, O., Yuldashov, Y., & Botirova, L. (2022, May). Morphogenesis and Seasonal Developmental Rhythm Under the Conditions of Introduction of Curcuma Longa L. In International Scientific Conference on Agricultural Machinery Industry "Interagromash"" (pp. 1460-1469). Cham: Springer International Publishing. DOI:10.1007/978-3-031-21432-5_155
9. Boboev, S., Makhkamov, T., Bussmann, R. W., Zafar, M., & Yuldashev, A. (2023). Anatomical and phytochemical studies and ethnomedicinal uses of Colchicum autumnale L. Ethnobotany Research and Applications, 25, 1-9. DOI:10.32859/era.25.6.1-9
10. TKh, M., Brundu, G., Jabborov, A. M., & Gaziev, A. D. (2023). Predicting the potential distribution of Ranunculus sardous (Ranunculaceae), a new alien species in the flora of Uzbekistan and Central Asia. BioInvasions Records, 12(1), 63-77. DOI:10.3391/bir.2023.12.1.05
11. Ameen, M., Zafar, M., Ahmad, M., Ramadan, M. F., Eid, H. F., Makhkamov, T., ... & Majeed, S. (2023). Assessing the Bioenergy Potential of Novel Non-Edible Biomass Resources via Ultrastructural Analysis of Seed Sculpturing Using Microscopic Imaging Visualization. Agronomy, 13(3), 735. DOI:10.3390/agronomy13030735
12. Noor, W., Zafar, M., Ahmad, M., Althobaiti, A. T., Ramadan, M. F., Makhkamov, T., ... & Khan, A. (2023). Petiole micromorphology in Brassicaceous taxa and its potential for accurate taxonomic identification. Flora, 303, 152280. DOI:10.1016/j.flora.2023.152280
13. Aziz, A., Ahmad, M., Zafar, M., Gaafar, A. R. Z., Hodhod, M. S., Sultana, S., ... & Chaudhay, B. (2023). Novel Copper Oxide Phyto-Nanocatalyst Utilized for the Synthesis of Sustainable Biodiesel from Citrullus colocynthis Seed Oil. Processes, 11(6), 1857. DOI:10.3390/pr11061857
14. Yunusov, S., Sadullayev, S., Khaitboyeva, G., & Sharipova, M. (2023). Selection of high-yielding, disease-resistant, promising, and export-oriented varieties of melons. In *E3S Web of Conferences* (Vol. 389, p. 03044). EDP Sciences.
15. Egamberdiyev, S. Y., & Yunusov, A. S. (2023, March). Study on selection of high yield cherry tomatoes varieties. In *IOP Conference Series: Earth and Environmental Science* (Vol. 1142, No. 1, p. 012046). IOP Publishing.

Creation of primary resources for creation of early, high-yield varieties of sesame for the northern regions of Uzbekistan

Ruslan Kalandarov[1,*] and *Makhfurat* Amanova[1]

[1]Tashkent State Agrarian University, 2, University street, Tashkent, 100140, Uzbekistan

Abstract. This article addresses a crucial concern regarding the limited availability of raw materials and the underutilization of alternative oil-bearing plants, which hampers the establishment of a robust and competitive environment. This issue has significant implications for broadening the production spectrum of diverse oil products and fostering a sustainable, accessible supply of high-quality vegetable oils for the population. To address this imperative need for a comprehensive and affordable range of vegetable oils, the creation of novel oilseed varieties assumes paramount importance. Such varieties must possess a constellation of traits, including high oil content, remarkable yield potential, resilience against various environmental stressors, suitability for export markets, and the capability to sustainably support primary and variety seed production. The focal objective of this research initiative revolves around assembling a diverse collection of genetic resources from across the globe. Specifically, the study seeks to gather 170 world collections of sesame varieties, alongside 122 locally adapted Tashkentsky varieties that have been regionalized within Uzbekistan. By consolidating this vast array of genetic material, the research endeavors to unlock new avenues for enhancing oilseed crop productivity, bolstering quality parameters, and establishing a foundation for progressive agricultural practices. This strategic endeavor embodies a critical step towards transforming the landscape of oilseed cultivation and production in Uzbekistan, ultimately paving the way for an augmented supply of healthful vegetable oils, economic vitality, and food security for the nation's populace.
Keywords. Sesame, world collection, local varieties, early fertile, fertile, large-seeded.

1 Introduction

Sesame (*Sesamum indicum* L.) is a versatile and economically important oilseed crop that holds a significant place in global agriculture due to its rich nutritional content and diverse industrial applications [1]. Among the myriad of oil-bearing plants, sesame stands out as an invaluable source of edible oil, protein, and other bioactive compounds that contribute to

* Corresponding author: r.kalandarov@tdau.uz

human health and well-being. With its adaptability to diverse climatic conditions, this crop has found a steadfast presence in various agroecological zones, including the northern regions of Uzbekistan [2].

The economic prosperity and dietary diversity of a nation significantly hinge on the availability of high-quality, locally-produced vegetable oils [3]. In Uzbekistan, as in many other countries, sesame holds immense potential as a vital contributor to the agricultural landscape, offering an opportunity to bridge the gap between demand and supply of edible oils [4]. However, to fully harness the potential of this crop and cultivate sesame varieties tailored to the specific agroclimatic conditions of the northern regions, it becomes imperative to create a robust foundation of primary resources [5].

The creation of early maturing, high-yielding varieties of sesame that are well-adapted to the northern regions of Uzbekistan requires a comprehensive approach that amalgamates genetic diversity, breeding methodologies, and meticulous agricultural practices. The development of such varieties can revolutionize local agriculture by augmenting sesame production, enhancing yield, and ensuring food security [6].

This manuscript aims to delineate the strategic initiative undertaken to create primary resources for the development of early, high-yield sesame varieties [7]. The focus of this endeavor is on collecting, evaluating, and harnessing the genetic diversity of sesame germplasm from diverse sources, including both regionalized varieties and global collections [8]. By meticulously assembling and analyzing this wealth of genetic material, we aim to lay the groundwork for a dynamic breeding program that targets the unique challenges and opportunities presented by the northern regions of Uzbekistan [9].

In this manuscript, we will elucidate the significance of sesame as a cash crop and its potential to address the growing demand for edible oils in Uzbekistan [10]. We will outline the objectives and methodologies behind the creation of primary resources, shedding light on the collection, evaluation, and characterization of sesame germplasm. Furthermore, we will highlight the anticipated outcomes and the potential impact of our research on the agricultural landscape of the northern regions, emphasizing increased sesame production, economic prosperity, and the overarching goal of achieving sustainable food security [11].

Through a comprehensive exploration of the creation of primary resources for early, high-yield varieties of sesame, this manuscript aims to contribute to the growing body of knowledge in plant breeding, agriculture, and sustainable development [12]. Our research endeavors to open new horizons for oilseed cultivation in Uzbekistan, fostering resilience, innovation, and agricultural growth that can propel the nation towards a future of agricultural prosperity and food self-sufficiency.

2 Materials and methods

The following phenological observations were made during the growing season: initiation of germination (10%), complete germination (75%), initiation of flowering (10%), complete flowering (75%), period of tuber browning (10%) complete browning (75%) [13]. The uniformity of each sample, i.e. general flatness, the shape and height of the stem, the degree of hairiness of the stem and leaves, the color of the plant, the size of the leaf, the color of the petals and pollen grains were determined. At the beginning of the maturation period, the following measurements and calculations were carried out on 10 plants [14]. The height of the main stem (cm), the number of side branches of the 1st, 2nd and 3rd orders, the number of panicles per plant, the average length of panicles and the number of seeds per panicle [15]. In laboratory conditions, the productivity of one plant, the weight of 1000 seeds was determined.

The research carried out on seed production - the State variety test of agricultural crops [16], and the biochemical composition of the seeds of the collection samples. The method of Ermakov, N.P.Arasimovich and others [4] and the statistical analysis and dispersion analysis of the obtained results were carried out and analyzed using the method of B.A.Dospekhov [3].

Before planting, ammophos was given at 200 kg per hectare. Cultivation and foliar feeding with micro and biofertilizers twice during the growing season. 2020 - 2022 years For the first time, 170 varieties of sesame from the National Gene Fund of the Research Institute of Plant Genetic Resources were studied for valuable economic traits in the climatic conditions of Khorazm region.

These samples of sesame are from various countries around the world, including: India, China, Afghanistan, Mexico, Turkmenistan, Uzbekistan, Armenia, Israel, Germany, Iran, Egypt, Ukraine, Yugoslavia, France, Russia, Pakistan, Hungary, Bulgaria, South Korea, Tajikistan, It was imported from Venezuela, Syria and Kenya and was studied for morphobiological and valuable economic characters during the entire growth period, and primary sources were selected for different aspects of selection (earliness, yield, fertility, large seed size, tolerance to heat and salt).

3 Results and discussion

The Tashkentsky 122 variety regionalized in Uzbekistan was used as a model for the sesame samples studied in the nursery of sesame collection samples. The growing period of this variety is 130 days in the conditions of Khorezm region in Uzbekistan.

When studying the height of the main stem in the plant, it was found that the height of the Tashkentsky 122 variety was 140 cm. In 120 of the 170 samples studied, the length of the main stem was from 100 to 138 cm, in 32 samples it was 150-176 cm, and in the remaining 18 samples it was found that it was 141-149 cm.

The height of plant stems is up to 100-127 cm in samples from Uzbekistan (k-23; k-27; k-29; k-78, k-86; k-86; k-90), Turkey (k-133; k-126; k-138; k-286) up to 117-125 cm in Ukrainian samples (k-304; k-305) 118-128 cm, Yugoslavian (k-307) 110 cm, Turkmenistan (k-408; k-251) ; k-1194) 110-128 cm, France (k-712) 124 cm, Afghanistan (k-753) 126 cm, Pakistan (k-833) 124 cm, Mexico (k-849) 122 cm, Peru (k- 866) 120 cm, Hungary (k-896; k-908; k-920) 117-119 cm, Korea (k-1157; k-1362; k-1363) 122-128 cm, Tajikistan (k-1287) 122 cm, Mexico (k-1336) 123 cm, Kenya (k-1369; k-1371; k-1373) ; k-1375; k-1382) 117-122 cm, Spain (k-330) 121 cm, Venezuela (k-1435), Israel (k-41) 127 cm.

According to the VIR methodology, according to the duration of the growing season of the pea crop: vegetation period 90-100 days; the mid-vegetation period is 100-110 days; The period of mid -season vegetation is 110-120 days 120-140 days of late vegetation period is conditionally allocated. It is known that in irrigated lands, the dead water period of sesame varieties is extended by 10-15 days.

Early and mid-season samples were not distinguished from the samples involved in the research. In the middle of the evening, 5 samples were selected. These include k-78 (Uzbekistan), k-330 (Spain), k-695 (Turkey), k-866 (Peru), k-1371 (Kenya) and k-222 (Israel) samples, the growth period of these samples is 115 day and ripened 15 days earlier than the sample (Table 1).

The number of side branches in the selected samples was on average from 2 to 4, and this indicator was 4 pieces in the sample Tashkentsky-122 variety. The height to the first spikes was 28 cm in 330 k-866 (Spain), 29 cm in (Peru), k-1371 (Kenya) samples, 34 cm in k-78 (Uzbekistan) and 30 cm in k-695. It was noted that it was equal to 48 cm.

Table 1. Valuable economic characteristics of the samples selected for early ripening.

Sample	Origin	Growing period, day	Number of side branches	1st loop, cm	Number of pods	1 plant productive league, g	Weight of 1000 seeds, g
St-122	Uzbekistan	130	4	48	76	14	2.6
78	Uzbekistan	115	3	34	68	14	2.6
330	Spain	120	4	28	130	26	3.2
695	Turkey	115	2	30	64	12	3
866	Peru	115	3	29	67	14	3.0
1371	Kenya	115	3	29	67	12	3.0

When analyzing the average number of tubers per plant in samples selected for early ripening, only one sample k-330 (Spain) had 130 tubers, and the model had an advantage with 54 more tubers compared to the variety.

In all the remaining samples, the number of cocci was 64-68, with 8-12 less results compared to the standard (Figure 1).

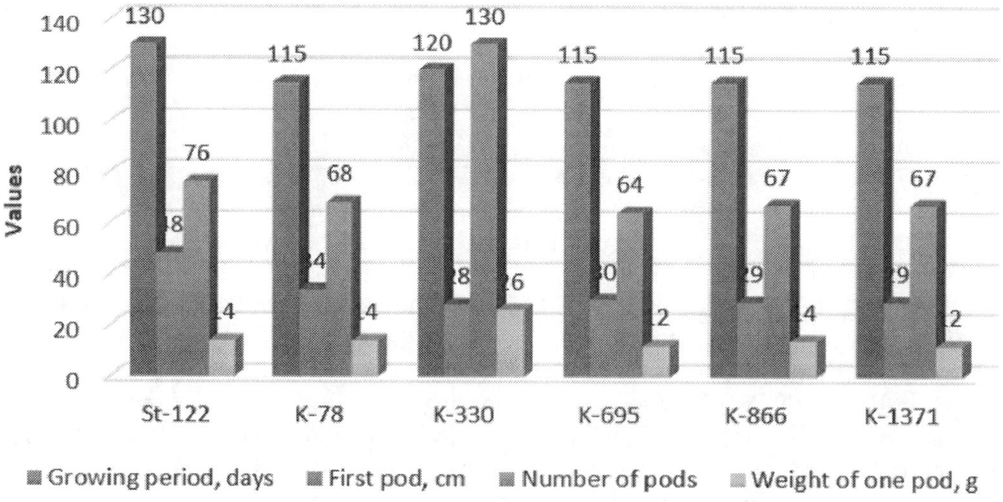

Figure 1. Valuable economic characteristics of the samples selected for early ripening.

These samples, selected for early maturity, will provide the opportunity to be used as a primary source for the creation of early varieties in the climatic conditions of the Khorezm region in the future.

Based on the results of one-year research, 18 sesame samples were selected based on the productivity of one plant. In these samples, the average productivity of one plant is up to 20-26 grams, and it was found that the sample is 6-12 grams more productive than the variety Tashkentsky-122 (Table 2).

Table 2. Main economic characteristics of the samples selected for the productivity of one plant.

Samples	Origin	Length to the 1st horn, cm	Height to the 1st tuber, cm	Number of pods	1 plant productive league, g
St-122	Uzbekistan	24	48	76	14
26	Uzbekistan	22	41	84	20
57	Uzbekistan	23	60	113	20
90	Uzbekistan	19	53	104	21
106	Armenia	11	36	105	20
170	Turkey	49	79	110	26
236	Azerbaijan	51	70	87	20
255	ETH	37	59	97	21
269	China	33	62	105	22
279	Egypt	46	68	104	23
285	Turkey	38	66	96	24
330	Spain	6	28	130	26
334	MMR	14	34	122	25
420	Turkmenistan	15	43	97	20
596	China	19	46	103	20
608	India	21	56	92	20
615	Mexico	22	41	84	20
691	Peru	23	60	113	25
723	Israel	18	53	115	23

The highest indicators of the average productivity of one plant were 26 grams observed in samples k-170 (Turkey), k-330 (Spain), and 25 grams in samples k-334 (MMR), k - 691 (Peru). In samples K-279 (Egypt), k- 723 (I s roil), this index was 23 grams, and k-269 (China) corresponded to 22 grams. K-285 (Turkey) was found to be 24 grams (Figure 2).

It was noted that 64 of the 170 studied varieties of sesame yield were higher than the standard (15-26 g), 42 were lower (5-13 g), and 64 of the remaining samples were almost the same as the standard (14 g).

It can be seen from Figure 2 that almost all of the samples selected for the productivity of one plant also achieved high results in terms of the number of tubers per plant. One of the characteristics of sesame seeds that is valued in the world market is its size. Although the large seed size of sesame plant is a characteristic of its variety, it has been confirmed several times in our previous studies that it changes under the influence of external environmental factors and applied agrotechnical measures.

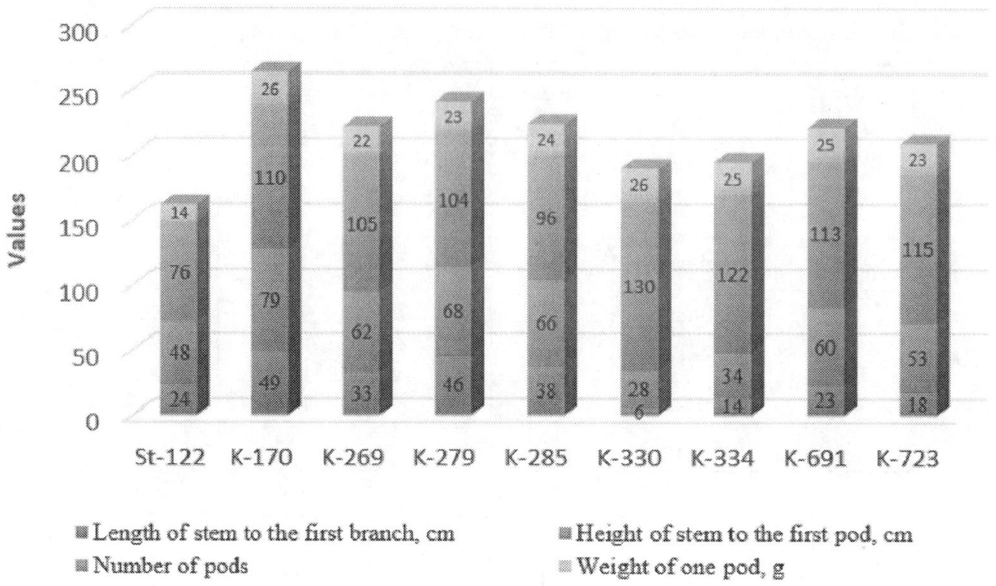

Figure 2. Main economic characteristics of the samples selected for the productivity of one plant.

From Tashkentsky 122 varieties, 1000 seeds weight is 2.6 grams, 15 of the studied samples have small seeds (2.3-2.5 g), 109 are relatively large (3-3.6 g), and 49 have a sample had almost the same results.

Among the 173 studied samples of the global sesame collection in the climatic conditions of Khorezm region, 17 samples were selected that showed high results in terms of 1000 mother seed weight as well as the productivity of one plant. The weight of 1000 seeds of these selected samples was 3.3-3.5 grams, and the model Tashkentsky achieved higher results by 0.7-0.9 g compared to 122 varieties (Table 3).

Table 3. Samples selected from the world sesame collection for their large seed size.

Sample	Origin	Number of pods	1 plant productivity, g	Weight of 1000 seeds, g
St-122	Uzbekistan	76	14	2.6
140	Turkey	87	16	3.5
217	Israel	78	16	3.3
219	Israel	85	18	3.3
236	Azerbaijan	87	20	3.3
254	Unknown	79	19	3.5
255	Unknown	97	21	3.3
278	Egypt	82	17	3.3
285	Turkey	96	24	3.3
334	MMR	122	25	3.5
507	Turkmenistan	84	17	3.3

615	Mexico	84	20	3.4
632	Vietnam	84	18	3.5
691	Peru	113	25	3.5
920	Hungary	68	14	3.4
1362	Korea	68	14	3.3
1409	Kenya	72	15	3.5

Among the studied sesame samples, 15 types of samples according to complex characteristics, including: k-90 (Uzbekistan), k-70, k-285 (Turkey), k-236 (Azerbaijan), k-255 (ETH), k-269, k-296 (Italy), k-279 (Egypt), k-330 (Spain), k-334 (MMR), k-608 (India), k -615 (Mexico), k-691 (Peru) and k-723 (Israel) samples were selected.

It was found that the samples of this selected variety were higher than the standard variety in terms of the number of pods up to 8-46 pieces, in terms of the productivity of one plant up to 7-12 grams and in terms of seed size, up to 0.4-0.9 grams (1000 seed weight) (Table 4).

Table 4. Valuable economic characteristics of sesame samples selected according to complex characteristics.

Sample	**Origin**	**Main stem length, cm**	**Number of side branches**	**Plant productivity**	**1 plant productivity, g**	**Weight of 1000 seeds, g**
St-122	Uzbekistan	140	4	76	14	2.6
90	Uzbekistan	163	4	104	21	3.2
170	Turkey	158	4	110	26	3.1
236	Azerbaijan	160	3	87	20	3.3
255	ETH	149	3	97	21	3.3
269	China	147	4	105	22	3.1
279	Egypt	166	3	104	23	3.1
285	Turkey	159	4	96	24	3.3
330	Spain	121	4	130	26	3.2
334	MMR	120	3	122	25	3.5
596	China	134	4	103	20	3.0
608	India	158	4	92	20	3
615	Mexico	131	3	84	20	3.4
691	Peru	153	5	113	25	3.5
723	Israel	139	4	115	23	3

In the future, sesame samples selected for these valuable economic traits will be presented to research institutes as a primary source for various lines of selection.

Selection works will be continued with some of the samples selected according to the complex characteristics and promising new varieties will be created for the northern regions of Uzbekistan.

4 Conclusions

In total, 170 varieties and samples of sesame introduced from different countries of the world were studied for the first time in the climatic conditions of Khorezm region according to valuable economic characteristics. On the early morning feature, including: k-78 (Uzbekistan), k-330 (Spain), k-695 (Turkey), k-866 (Peru), k-1371 (Kenya) and k-222 (I s roil) samples (ripened 15 days earlier than the sample). 18 by the productivity of one plant (k-170; k -285 (Turkey), k-330 (Spain), k-334 (MMR), k-691 (Peru), k-279 (Egypt), k-723 (I s rail) k-26; k-57; k-90 and k-269 (China), k-106 (Armenia), k-236 (O Azerbaijan), k-420 (Turkmenistan), k-596 (China), k-608 (India), k-615 (Mexico) samples were selected. It was found that the productivity of one plant of these samples is up to 8-12 grams higher than the model Tashkentsky-122 variety. 30 by seed size, including k-140; k- 170 ; k- 176 k-285 Turkey (3.3-3.5 g), k-217; k - 218 ; k-219 Israel (3.3 g), k-236 Azerbaijan (3.3 g), k-254, k-255 (3.-3.5 g), k-278 Egypt (3.3 g), k- 334 MMR (3.5 g), k-507 Turkmenistan (3.3 g), k-615 Mexico (3.4 g), k-632 Vietnam (3.5 g), k-691 Peru (3.5 g), k-920 Hungary (3.5 g), k-1362 J. Korea (3.3 g) and k-1409 Kenya (3.5 g) samples were selected. The weight of 1000 seeds of these selected samples was 3.3-3.5 grams, and it was found that it was 0.7-0.9 grams higher than the sample Tashkentsky 122 variety.

There are 15 types of sesame samples according to complex characteristics, k-90 (Uzbekistan), k-70, k-285 (Turkey), k-236 (Azerbaijan), k-255 (ETH), k- 269, k-296 (Italy), k-279 (Egypt), k-330 (Spain), k-334 (MMR), k-608 (India), k -615 (Mexico), k-691 (Peru) and k-723 (Israel) samples were selected that the samples of this selected variety compared to the standard variety have higher results in terms of the number of pods up to 8-46 pieces, in terms of the productivity of one plant up to 7-12 grams and in terms of seed size up to 0.4-0.9 grams (1000 seed weight).

References

1. M.E. Amanova.//Biological characteristics of sesame ecogroups and primary sources for selection// Journal of Agro-science No. 1(21), 31 p. Tashkent 2012.
2. M.E. Amanova, A.S. Rustamov //Methodological manual for studying the world collection of oilseeds// "BIOEKOSAN" educational and methodological complex of the youth of Uzbekistan. Tashkent 2010. 20 pages, number 300.
3. Dospekhov B.A. Methodology polevogo opyta. - M.: Kolos, 2015-416 p.
4. N.I. Bochkaryov, S.G. Borodin //Recommendations for semen production of maslichnyx culture and efiromaslichnyx culture// Krasnodar. 2004.
5. M.E.Amanova, A.S.Rustamov //Sesame new rare sources for selection from world collection//. Proceedings of the Uzbek -Japanese symposium on ecotechnologies. Tashkent-2016.
6. M.E. Amanova, A.S. Rustamov // For evaluation and selection of sesame samples allocate primary resources //. Agro-ilm 5 (55), 2018. p. 28.
7. Tayjanov, K., Khojimatov, O., Gafforov, Y., Makhkamov, T., Normakhamatov, N., & Bussmann, R. W. (2021). Plants and fungi in the ethnomedicine of the medieval East-a review. Ethnobotany Research and Applications, 22, 1-20. DOI:0.32859/ERA.22.46.1-20
8. Jabeen, S., Zafar, M., Ahmad, M., Althobaiti, A. T., Ozdemir, F. A., Kutlu, M. A., ... &

Majeed, S. (2023). Ultra-sculpturing of seed morphotypes in selected species of genus Salvia L. and their taxonomic significance. Plant Biology, 25(1), 96-106. DOI:10.1111/plb.13473

9. Majeed, S., Ahmad, M., Ozdemir, F. A., Demirpolat, A., Şahan, Z., Makhkamov, T., ... & Nabila. (2023). Micromorphological characterization of seeds of dicot angiosperms from the Thal desert (Pakistan). Plant Biosystems-An International Journal Dealing with all Aspects of Plant Biology, 157(2), 392-418. DOI:10.1080/11263504.2023.2165553
10. Majeed, S., Ahmad, M., Ali, A., Althobaiti, A. T., Ramadan, M. F., Kilic, O., ... & Sultana, S. (2023). Pollen micromorphology among amaranthaceous species from desert rangeland: Exine stratification and their taxonomic significance. BioMed Research International, 2023, 4967771. DOI:10.1155/2023/4967771
11. Makhkamov, T., Sotiboldiyeva, D., Mamarakhimov, O., Yuldashov, Y., & Botirova, L. (2022, May). Morphogenesis and Seasonal Developmental Rhythm Under the Conditions of Introduction of Curcuma Longa L. In International Scientific Conference on Agricultural Machinery Industry “Interagromash"” (pp. 1460-1469). Cham: Springer International Publishing. DOI:10.1007/978-3-031-21432-5_155
12. Boboev, S., Makhkamov, T., Bussmann, R. W., Zafar, M., & Yuldashev, A. (2023). Anatomical and phytochemical studies and ethnomedicinal uses of Colchicum autumnale L. Ethnobotany Research and Applications, 25, 1-9. DOI:10.32859/era.25.6.1-9
13. TKh, M., Brundu, G., Jabborov, A. M., & Gaziev, A. D. (2023). Predicting the potential distribution of Ranunculus sardous (Ranunculaceae), a new alien species in the flora of Uzbekistan and Central Asia. BioInvasions Records, 12(1), 63-77. DOI:10.3391/bir.2023.12.1.05
14. Ameen, M., Zafar, M., Ahmad, M., Ramadan, M. F., Eid, H. F., Makhkamov, T., ... & Majeed, S. (2023). Assessing the Bioenergy Potential of Novel Non-Edible Biomass Resources via Ultrastructural Analysis of Seed Sculpturing Using Microscopic Imaging Visualization. Agronomy, 13(3), 735. DOI:10.3390/agronomy13030735
15. Noor, W., Zafar, M., Ahmad, M., Althobaiti, A. T., Ramadan, M. F., Makhkamov, T., ... & Khan, A. (2023). Petiole micromorphology in Brassicaceous taxa and its potential for accurate taxonomic identification. Flora, 303, 152280. DOI:10.1016/j.flora.2023.152280
16. Aziz, A., Ahmad, M., Zafar, M., Gaafar, A. R. Z., Hodhod, M. S., Sultana, S., ... & Chaudhay, B. (2023). Novel Copper Oxide Phyto-Nanocatalyst Utilized for the Synthesis of Sustainable Biodiesel from Citrullus colocynthis Seed Oil. Processes, 11(6), 1857. DOI:10.3390/pr11061857

Development characteristics of underground pests of root vegetable and potato crops

Azamjon Khudoykulov[1,*], *Azimjon* Anorbaev[2], *Malika* Norova[1] and *Jabbor* Abdiev[1]

[1]Tashkent State Agrarian University, Tashkent, 100140, Uzbekistan
[2]Research Institute of Plant Quarantine and Protection, Tashkent, 100003, Uzbekistan

Abstract. The present article delves into a comprehensive examination of the species composition of pests prevalent within fields cultivating root vegetables and potatoes. Through systematic scrutiny, the study reveals the presence of pest representatives spanning three distinct classes, six genera, eleven families, and twenty-nine species within a potato field subject to repeated cultivation. Among these discerned species, a detailed analysis is focused on sixteen varieties of subterranean pests originating from four distinct genera. Concurrently, the research encompasses a decade-long assessment of the prevalence of subterranean autumn and exclamatory moths, facilitated by the use of pheromone traps. Over this span, observations indicate that the occurrence of autumn moths reached a count of 4.4 units, with spring moths totaling 1.1 units. Additionally, the study delves into the developmental attributes of root vegetables and potato crops, garnering insights into their growth patterns and characteristics. The culmination of this research brings about tangible and applicable insights. The acquired findings not only contribute to the knowledge regarding the pests affecting these agricultural sectors but also furnish practical recommendations and suggestions for enhancing production practices. The article not only underlines the significance of understanding the intricacies of pest dynamics but also reinforces the need for informed agricultural management practices to mitigate their impact and bolster crop productivity.
Keywords. Root vegetables, potatoes, underground pests, fall moth, exclamation moth.

1 Introduction

Due to the harmful effects of pests and diseases on agricultural crops, 30-40% of crops are lost every year [1]. Despite the timely control of these pests, the development of integrated control measures against them remains relevant today [2]. Today, not only in Europe, but also in the countries of Central Asia, it is urgent to ensure food safety by developing environmentally safe measures to fight against rodent and sucking pests of root vegetables and potatoes [3].

According to the long-term research of scientists, it was found that the population density of underground pests is not uniform in all areas where root vegetables and potatoes are planted, and the population density is relatively high in the areas located far from residential

* Corresponding author: a.khudoykulov@tdau.uz

areas [4]. In particular, the population density of root-gnawing moths from underground pests also depends on the plant species, and it was observed that the fertility of female breeds doubles in nectar-rich plant species [5]. 51 types of pests belonging to 15 families were recorded in crops belonging to the Solanum family. Of them, 40 types of pests were found to damage tomatoes and eggplants, and 48 types were found to damage potatoes [6].

Fall moth (*Agrotis segetum* Den. et Schiff) and exclamation moth (*A. exlamationis* Den. et Schiff.) 2-5 generations per year, these pests damage potatoes, tomatoes, eggplants and other crops. In some years, the yield of potatoes decreases by 7-12% due to the damage of root-gnawing moths [7].

Heterodera marioni Cornu., *Poeciloscytus cagnatus* Fieb., *Tetranychus urticae* Koch., *Gryllotalpa gryllotalpa* L., *Agriotes meticulosus* Cond., *Clon cerambycinus* Sem., which causes swelling of root vegetables and potato crops in the conditions of Uzbekistan, gnaws and damages the roots and crops of potatoes around ditches and ponds and in wet areas [8].

Taking into account the above, studies were conducted in 2021-2022 in order to systematically analyze the distribution and damage of omnivorous underground pests in the repeatedly planted root vegetable and potato fields in the Samarkand and Tashkent regions of Uzbekistan [9, 10].

The studies were conducted at the "Farangiz Nurli Zamini" farm, Jomboy District, Samarkand Region, Uzbekistan, and the educational experimental farms of Tashkent State Agrarian University, Qibray District, Tashkent Region, Uzbekistan.

2 Materials and methods

For the experimental endeavor, specific fields were designated, focusing on a plot designated for root vegetables and potatoes, following the cultivation of grains. The research commenced by extracting soil samples from the designated area. This extraction involved digging soil samples from layers spanning 5, 10, 15, 20, and 25 centimeters, employing a diagonal (checkerboard) pattern with a grid spacing of 25 x 25 centimeters. Subsequently, the gathered soil samples underwent a meticulous process involving passing them through a fine wire sieve, aimed at refining the extracted material [6].

With the collected data at hand, the average quantity of subterranean pests present within a designated area of 1 square meter was meticulously tallied. This calculation was conducted for both the interior of the field and its peripheral surroundings. Furthermore, the species density was meticulously quantified within the potato field that followed varying previous crops, alongside the field that had undergone winter grain cultivation.

This systematic approach to soil analysis and pest density determination establishes a solid foundation for comprehending the intricate interplay between soil conditions, crop rotation, and the prevalence of underground pests. The methodology ensures a meticulous examination of factors contributing to pest dynamics, ultimately enabling a more nuanced understanding of pest-related challenges faced within the context of root vegetable and potato cultivation.

3 Results and discussion

In the course of research, the species composition of pests distributed in the repeatedly planted potato fields after the experimental grain was systematically analyzed. According to him, representatives of pests belonging to 3 classes, 6 genera, 11 families and 29 species were identified in the repeatedly planted potato field. The majority of species belongs to the family of moths (*Noctuidae*), which comprised 10 species.

Notably, 2 types of pests belonging to the *Gelechiidae* family of representatives of the

Lepidoptera family were identified. In the course of research, 3 species of the orthoptera family belonging to the *Aphididae* family and 2 species from the *Aleyrodidae* family were identified. 4 species of representatives of the Elateridae family of the Coleoptera order were identified in repeatedly planted potato fields. 1 species of members of the *Chrysomelidae* family, 3 species of the *Scarabagidae* family, 1 species of the *Orthoptera* family, 2 species of the *Acariphormes* family, and 4 species of the *Tylenhida* family were identified (Table 1).

Table 1. Incidence of pests in root vegetables and potato fields (Samarkand and Tashkent regions, Uzbekistan in 2021-2022).

#	Name of pests	Incidence
	I. Class: *Insecta: Lepidoptera*. Family: *Noctuidae*	
1	*Agrotis segetum* Den.et Schiff	+++
2	*A. obesa*. Bd.	++
3	*A.exclamationis*. L	++
4	*A.ipsilon*. Rtt.	++
5	*Autographa gamma*. L	+
6	*Mamestra suase* Schiff	+
7	*Ochropleura flammarta* Schiff	+
8	*Pusia chrvsitis*. L	+
9	*Agrotis C – nigrum* L	++
10	*Agrotis conspicua* Hb.	++
	Lepidoptera*. Family: *Gelechiidae	
11	*Phthorimaea operculella* Zell	++
12	*Tuta absoluta*	++
	Homoptera*. Family: *Aphididae	
13	*Aphis craccivora* Koch	++
14	*Aphis gossypii* Glov	++
15	*Myzodes persicae* Sulz	+++
	Homoptera*. Family: *Aleyrodidae	
16	*Trialeurodes vaporarioum* West	+++
17	*Bemisia tabaci* Genn.	+
	Coleoptera*. Family: *Elateridae	
18	*Agrotis meticulosus* Cond	+++
19	*Clon cerambycinus* Sem	++
20	*Blaps halophila* F.W.	++
21	*Dailognatha nasute* Men	+
	Coleoptera*. Family: *Chrysomelidae	
22	*Leptinotarsa decemlineata* Say	+++
	Coleoptera*. Family: *Scarabagidae	
23	*Melonotha afflicta* Ball	++
24	*Polyphilla adspersa* Motsch	++
25	*Melonotha melonotha,M. Hypocastani*	++

Orthoptera.* Family: *Gryllotalpidae		
26	*Gryllotalpa gryllotalpa* L.	++
II. Class: *Arachnoidea: Acariphormes.* Family: *Tetranychidae*		
27	*Tetranychus urticae* Koch	++
Acariphormes.* Family: *Eriophyidae		
28	*Aculops Lycopersici* Massee	++
III. Class:. *Nematoda: Tylenhida.* Family: *Meloidogynidae*		
29	*Meloidogyne incognita* Cofoid et White.	+
Incidence rate: +++ - high, ++ - moderate, + - low		

Based on the analysis of underground pests of potatoes, the distribution and damage of underground pests in the researched areas was systematically analyzed. According to it, representatives of 2 classes, 4 genera, 5 families and 16 species of underground pests damaging repeatedly planted potatoes were identified. In terms of species, the dominance belongs to the family of moths (*Noctuidae*), which made up 7 species.

3 species of representatives of the *Scarabagidae* family were identified. According to the results of the conducted research and analysis, 1 species of representatives of the *Gryllotalpidae* family of the semicolloptera family was identified. 1 species of the *Meloidogynidae* family was identified from representatives of the *Nematoda* class (Table 2).

Table 2. Incidence of underground pests in potato fields (Samarkand and Tashkent region in Uzbekistan, 2021-2022).

#	Name of pests	Incidence
1	*Agrotis segetum* Den.et Schiff	+++
2	*A.exclamationis.* L	+++
3	*Agrotis C - nigrum* L	++
4	*Agrotis conspicua* Hb.	++
5	*A. obesa.* Bd.	++
6	*A.ipsilon* Rtt.	++
7	*Ochopleura flammarta* Schiff	+
8	*Agrotis meticulosus* Cond	+++
9	*Clon cerambycinus* Sem	++
10	*Blaps halophila* F.W.	++
11	*Dailognatha nasute* Men.	+
12	*Melonotha afflicta* Ball	++
13	*Polyphilla adspersa* Motsch	++
14	*Melonotha melonotha* M. *hypocastani*	++
15	*Gryllotalpa gryllotalpa* L.	++
16	*Meloidogyne incognita* Cofoid et White.	+
Incidence rate: +++ - high, ++ - moderate, + - low		

Based on the systematic analysis of the species composition of underground pests of potatoes, the following conclusions were reached. Based on the analysis of potato pests, 29 types of pests of 6 categories were analyzed.

Based on the systematic analysis of underground pests in potato fields, 16 types of underground pests of 4 categories were identified. In the course of our research, in order to study the damage and development phases of the eggs and larvae of subterranean moth butterflies in the potato crop, research was conducted on 1.5 hectares of repeatedly planted and potato crop fields of the farm "Maqsudjon Porloq Zamini", Urgut District, Samarkand Region.

According to him, after sorting the tubers of the "Santa" variety, the potatoes were planted in the experimental field. In a potato field planted on 15 June 2021 and fully matured on 01 July, fall and exclamation moth sex pheromone traps were installed and the flight of pest butterflies was recorded. The results showed that in the newly sprouted potato field, the number of fall moths on the traps was on average 9.0 per trap on July 5, and 0.5 on the 10th day of the counting day.

Therefore, it was observed that the flight dynamics of fall moths decreased in the second half of August. A similar situation was observed in traps with exclamation moth sex pheromones. While the number of exclamation moths was 2.1 on day 1, it was observed that it decreased slightly to 0.1 on day 10 of our observations (Table 3).

Table 3. Landing of fall and exclamation moth butterflies on traps with sex pheromone (Samarkand region Urgut district "Maqsudjon porloq zamini" farm, "Santa variety" potato, planting date June 15 2021-2022).

Pests	Average number of moths caught in 1 trap on control days										Total in 10 days	Average per day
	July 5	July 6	July 7	July 8	July 9	July 10	July 11	July 12	July 13	July 14		
Fall moth	9.0	7.0	6.5	5.2	4.2	4.0	3.5	2.0	1.2	0.5	43.1	4.3
Excl. moth	2.1	2.0	1.6	1.5	1.2	1.0	0.7	0.5	0.3	0.1	11.0	1.1

Observations were continued in order to determine the appearance and prevalence of autumn and exclamation moths in fully germinated seedlings in the potato field. According to him, the average number of fall moths per 1 m2 in the potato field was 0.3 pieces in the first decade of July, 0.5 pieces in the 2nd decade, and 0.8 pieces in the 3rd decade. These indicators are proportionally 0.2 for exclamation moth worms; It was 0.7 and 0.8 units (Table 4).

The average number of fall moth eggs per 1m2 was 0.4 pieces in the first decade of July, 0.7 pieces in the 2nd decade, and 1.2 pieces in the 3rd decade.

Table 4. Amount of fall moth in the potato field (Maqsudjon Porloq Zamini farm, Urgut District, Samarkand Region, Potatoes "Santa variety", planting date June 15 2021-2022).

Pilot fields (1 m^2)	July					
	Decade 1		Decade 2		Decade 3	
	Worm	Egg	Worm	Egg	Worm	Egg
#1	0.4	0.6	0.8	1.0	1.2	1.5
#2	0.2	0.4	0.1	0.6	0.7	1.1
#3	0.1	0.4	0.1	0.4	0.6	1.1

#4	0.5	0.8	0.1	0.5	0.8	1.2
#5	0.2	0.4	0.7	1.1	0.8	1.1
#6	0.3	0.4	0.8	1.2	0.9	1.3
#7	0.2	0.3	0.6	0.9	0.6	0.9
Average	0.3	0.4	0.5	0.7	0.8	1.2

It can be observed that the number of worms of the pest has increased. In order to monitor the development of the number of moths in other areas, researches were conducted in the potato field planted after the grain of the Faradis Thresher farm, Upper Chirchik District, Tashkent Region. Potato seeds were sown on July 16, which coincided with the germination period of potato development (Table 5).

Table 5. Number of fall moths in the potato field (Faradis threshing farm, Upper Chirchik district, Tashkent region, July 16 2021-2022).

Pilot fields ($1\ m^2$)	**August**					
	Decade 1		**Decade 2**		**Decade 3**	
	Worm	**Egg**	**Worm**	**Egg**	**Worm**	**Egg**
#1	0.9	1.2	1.5	2.0	2.5	3.1
#2	1.1	1.4	2.0	2.2	2.1	2.4
#3	0.7	1.0	0.8	1.2	1.5	1.8
#4	0.6	1.0	2.1	2.5	1.4	1.7
#5	1.0	1.3	1.5	1.9	2.8	3.3
#6	0.6	0.9	1.9	2.3	2.9	3.1
#7	0.9	1.1	1.8	2.3	3.2	3.5
Average	0.8	1.3	1.6	2.5	2.3	2.7

According to Table 5, 7 samples per diagonal of the cultivated area were taken in 3 returns. According to the observed samples, the average number of fall moths was 0.8 pieces per 1 m2 in the first ten days of August, 1.6 pieces in the second decade, and 2.3 pieces in the third decade of August. The average number of fall moth eggs per 1m2 was 1.3 eggs in the 1st ten days of July, 2.5 eggs in the 2nd ten days, and 2.7 eggs in the 3rd ten days of July.

4 Conclusions

Based on the findings garnered from the comprehensive research, a significant observation emerges regarding the disparity in the presence of rootworms between two distinct potato planting timeframes. Specifically, the relatively lower number of rootworms detected in the potatoes planted on June 15 as opposed to those planted on July 15 can be attributed to a key factor.

The study discerns that this variation in rootworm population can be attributed to the synchronicity between the germination period of the potato plants and the prevalence of moth butterflies. This simultaneous occurrence led to a consequential increase in the population of pests, notably reflected in the augmented pest density within subsequently planted crops.

This discovery underscores the intricate relationship between the timing of potato cultivation and the presence of pests, particularly moth butterflies. The study illuminates how such ecological nuances can significantly impact pest dynamics and ultimately influence the overall effectiveness of cultivation practices. This nuanced insight has profound implications for optimizing planting schedules and pest management strategies within repeated crop cycles, contributing to enhanced crop yield and quality.

References

1. Khudoykulov A., Anorbaev A., Kholikova G. Bioecological properties of underground pests in cultivation of potatoes planted after grain. E3S Web of Conferences 244, 02033 (2021) https://doi.org/10.1051/e3sconf/202124402033. EMMFT-2020.
2. Isaev S., I. Begmatov, G. Goziev, S. Khasanov, IOP Publishing 883 (1), 012080 (2020).
3. Isaev S. K., R. U. Rakhmonov, S. S. Tadjiev, G. I. Goziev, S. Z. Khasanov, IOP Publishing 614 (1), 012147 (2020).
4. Khujaev SH., Alimuhammedov S., Cotton pests and combatting with them, 37-45 (Ukituvchi Press, Tashkent, 2011).
5. Caballero P., E. Vargas-Osuna, C. Santiago-Alvarez, Journal of Applied Entomology 115 (1-5), 59-64 (2011).
6. Ebsjerg P., Journal of Applied Entomology 105 (1-5), 295-302 (1998).
7. Khudoikulov A., Muminova R., Ubaydullaev S. Type structure, meeting rate and interaction of Entomophages of underground pests in vegetable and potato agrobiotsenosis. Solid State Technology Volume: 63 Issue: 4 Publication Year: 2020. P. 297-302.
8. Bais, H. P., Loyola-Vargas, V. M., Flores, H. E., & Vivanco, J. M. (2001). Root-specific metabolism: the biology and biochemistry of underground organs. In Vitro Cellular & Developmental Biology-Plant, 37, 730-741.
9. Capinera, J. (2020). Handbook of vegetable pests. Academic press.
10. Reddy, P. P. (2015). Plant protection in tropical root and tuber crops (No. 11591). New Delhi, India:: Springer India.

Adaptation of breeding goats for the milk purposes to climate conditions

Bakhtiyor Davronov[1,*], *Izatulla* Eshmatov[1] and *Shakhnoza* Eshmatova[2]

[1]Research Institute of Animal Husbandry and Poultry, Tashkent, 100140, Uzbekistan
[2]Samarkand University of Veterinary and Medicine, Samarkand, 140102, Uzbekistan

Abstract. This article presents valuable opinions on the importation of purebred young goats from foreign countries and its impact on various aspects of dairy goat breeding. The importation of purebred young goats plays a significant role in the establishment of farm subjects specializing in dairy goat breeding. One aspect highlighted in the article is the importance of introducing purebred young goats to improve the genetic pool of local dairy goat populations. By bringing in high-quality genetics from foreign countries, dairy goat breeders can enhance the productivity, milk yield, and overall performance of their herds. This can lead to increased milk production and improved milk quality, benefiting the dairy industry as a whole. Furthermore, the article discusses the significance of adapting imported dairy goats to the local natural climatic conditions. Importing goats from foreign countries may expose them to different climates, which can impact their health and productivity. Proper adaptation measures are essential to ensure the well-being of these animals and their successful integration into the local environment. Moreover, the article explores clinical signs and morphological characteristics of blood in mother goats during the process of adaptation to the natural climatic conditions. Monitoring and understanding these physiological changes are crucial for ensuring the well-being and health of the mother goats, as well as the successful rearing of their offspring.
Keywords. Selection, import, breeding, crossbreeding, hybrid, farm, purebred, buck, mother goat, clinical indicators.

1 Introduction

Along with the rapid growth of the world's population, their demand for food products is also increasing day by day. Today, goat breeding is a rapidly developing and promising branch of world animal husbandry. According to the UN Food and Agriculture Organization (FAO), goats are raised in more than 200 countries of the world [1, 2].

Goat breeding is considered one of the oldest branches of animal husbandry in Central Asian countries, 3,681,000 goats are raised in the Republic of Uzbekistan, 23,268,000 in the Republic of Kazakhstan, and 2,300,000 in the Republic of Turkmenistan. In these countries,

* Corresponding author: b.davronov1010@gmail.com

goat breeds such as Zaanen, White Russian, Alpine, buttermilk breeds such as Nubian, Toggenburg, wool-oriented breeds such as Angora, Soviet wool breed, Tivit-oriented breeds such as Pridon, Mountain-Altai, Kashmir, Orenburg, local goat breeds, meat breeds such as damara, bursk, pafuri are being bred [3-6]. It is necessary to adapt the dairy goat breeds brought to our country from abroad to the climatic conditions of our country, to organize feeding on a balanced ration using local foods, to organize the selection and mating of breeds during breeding, to pay special attention to the use of breeding goats of the improving breed category during mating [7-10]. It is possible to breed goats that give a lot of milk in the farms of the republic by paying attention to the economic utility and selection characteristics of the breed in order to bring out the milk productivity at the level of the requirements of the breed standard of goats [11-13].

According to the 30th and 32nd goals of the "Development Strategy" approved by the Decree of the President of the Republic of Uzbekistan dated January 28, 2022 "On the Development Strategy of New Uzbekistan for 2022-2026" No. PF-60 the task of increasing the income of farmers by at least 2 times, bringing the annual growth of agriculture to at least 5%, expanding the livestock feed base and increasing the volume of production by 1.5-2 times [14].

The Decree of the President of the Republic of Uzbekistan No. PF-6198 dated April 1, 2021 establishes the tasks of creating productive genotypes of Jaidari, "Merino" sheep and "Angor" goats, maintaining and increasing the gene pool of meat and wool sheep, and the gene pool of meat and wool sheep in the Ohangaron district of Tashkent region . on the basis of the analysis determines the relevance of developing selection methods for creating productive genotypes of Jaidari sheep and "Angor" goats [15-17].

Keeping imported purebred dairy goats in the selection of dairy goats, keeping and feeding them in accordance with the technology of dairy goat breeding while adapting them to local climatic conditions, as well as taking care of them to increase milk productivity, as well as forming farms specializing in dairy goat breeding , are among the most pressing issues in the industry today [18-20].

The purpose of the study is to adapt imported Zaanen, Alpine, Anglo-Nubian, Murciano Grenada and Toggenburg goats to local climatic conditions and study of their physiological and clinical signs and morphological characteristics of blood in dairy goat breeding.

2 Materials and methods

The research focused on dairy purebred goats, specifically Zaanen, Alpine, Toggenburg, and Anglo-Nubian bucks, along with female goats and their offspring. These animals were chosen as the research materials due to their relevance in the dairy industry and their genetic purity [21].

The study employed a multidisciplinary approach, utilizing zootechnical, biological, and statistical methods. Zootechnical methods were used to evaluate and manage the breeding and production aspects of the goats, considering factors such as genetics, reproduction, nutrition, and health [22-24].

Biological methods were applied to analyze various biological processes within the goats, including metabolic functions, growth patterns, milk production, and overall physiological responses. Statistical methods were employed to analyze and interpret the data collected during the research, providing valuable insights into the relationships and patterns observed among the different goat breeds, their characteristics, and their offspring [25-27].

By combining these diverse methodologies, the research aimed to gain a comprehensive understanding of the performance, productivity, and overall potential of dairy purebred goats from different breeds. The results of this study could contribute to the improvement and optimization of dairy goat breeding programs and enhance the sustainability and efficiency of dairy production systems [28].

Zaanen is a dairy goat breed that is bred in the farms of "Agro Zangiota" Yangiyol district of Tashkent region, "Alisher Khrovador" of Qibray district, "Naslli Kumush Kurka" of Sirdarya district of Syrdaryo region, "INZIM TEX" LLC, "Bakht khokkhov khrovovi" in Altinkol district of Andijan region ., was based on the data of scientific research conducted on Alpine, Toggenburg, Anglo-Nubian buck and mother goats and their offspring, as well as local goats [29, 30].

In total, 150 female goats and 3 progeny goats from farm herds, i.e. breeding goats numbered #532903, #656808 and #837122 were selected for research.

3 Results and discussion

Dairy goat breeding is a new direction in Uzbekistan, and dairy goat farms are being formed by importing dairy breeds from European countries. In the breeding of dairy goats, the use of methods of purebred breeding and crossbreeding of imported Zaanen, Alpine, Toggenburg, Anglo-Nubian goats and their offspring with high milk productivity, especially purebred breeding in breeds and the founders of the new system within the breed. The main content of selection work is scientific justification of milk productivity indicators of offspring and selection and selection of high-productivity selection groups from their offspring , creation of new system and family generations from them. Also, in the effective use of the foals of dairy breeds brought from abroad in selection work, the use of cross-breeding methods to improve the productivity of interbreeds and local goats, it makes it possible to increase the milk productivity of goats raised in local conditions.

Today, purebred breeding of Zaanen, Alpine, Toggenburg, Nubian goats imported from abroad, preservation and reproduction of their offspring as purebreds, and the use of innovative methods of care and feeding bring higher economic results than dairy goat breeding.

Effective use of imported goat breeds in selection work, raising goats as a pure breed, while maintaining their head number, obtaining many offspring, ensuring that their productivity signs are not evident, developing the scientific and practical basis of selection is important in the dairy goat industry.

Goats are attracted by landowners, farmers and farms due to their biological characteristics, i.e. quick adaptation to different natural climatic conditions, lack of choice of food, their rapid growth and early adulthood, rapid reproduction and high fertility.

In the biology of goats, the period of physiological maturity and sexual maturity of dairy goats lasts more than 48 hours. The duration of the sexual period varies from 4 to 26 days, their sexual period is on average 17-19 days . The duration of the estrous period in goats is 150 days on average . In the natural climatic conditions of Uzbekistan, dairy goats are bred, like other goats, in the autumn season, in September - November, for 30-45 days , naturally or artificially. However, in accordance with the technology of dairy goat breeding, such technological processes as the organization of regular milk production from them and the production of dairy products ready for consumption through the processing of milk products are taken into account.

Effective use of female goats in the production of dairy goats is of great importance. At least 75% of female goats and adult female goats in herds are fathered in the first and second ten days of the breeding season. Goats usually give birth to one or two, rarely triplets or quadruplets, or 1-3% of goats in the herd give birth to twins. The duration of use of goats in farm production depends on the conditions of economic development of the farm, and the duration of use of dairy goats is 7-8 years on average. Goats with certain genetic traits have high productivity and a strong constitution, and use them for 8-10 years in farm conditions.

Adaptation of dairy goats imported from Europe to the natural climatic conditions of Uzbekistan, as well as the rational use of highly productive goats and female goats in selection work, forming a dairy goat breeding network is one of the urgent tasks in the network.

Goats are widely distributed all over the globe due to their fast and good adaptability to different climatic conditions. Goats are kept and bred in all regions of the world, even from countries located in northern latitudes to countries and islands located in southern latitudes [1, 2]; It is emphasized by [3, 4, 6-8, 16, 17]. Of the total number of goats in the world, 60.4% of goats are raised in Asia and 33.7% in Africa. In the wide spread of goats, their adaptation to different natural climatic conditions and the effective use of methods of selection and selection in the direction of productivity are of great importance in the development of the field.

The appearance of the external structure of dairy goats, i.e., the exterior, has a large size, the structure of their jaws and teeth changes with age, the length of life, the period of pregnancy, the period of throttling and throwing, the lactation period, the skin-wool cover and other dairy characteristics from goats of other breeds and Differences in breeds with their own positive biological characteristics are felt.

Villagers, landowners, peasants and farms have been raising goats for many centuries for various productivity purposes. However, scientific selection methods were not used in breeding goats, and the number of goats and productivity directions were regularly increased depending on the adaptation and viability of the goats in a natural way.

No separate statistical information on their productivity was recorded, but information on the number of sheep and goats and production indicators was regularly recorded. About 2.0 percent of the milk and 12 percent of the meat produced in Uzbekistan is attributed to the goat industry.

On average, 180-200 kg of milk, 13-15 kg of meat, 2.0 kg of skin, and up to 1.0 tons of organic fertilizer are obtained from goats reared in local conditions and are widely used in agriculture and industry. Textiles from goat wool and tweed are widely used in light industry. Leather products and other valuable leather raw materials are made from goat skin. By-products obtained from slaughtering goats - blood, bones, horns, wool, etc. are used for industrial purposes. Also, goat litter and manure waste are widely used in greenhouses as organic fertilizers to increase land productivity in agriculture and mining.

Goat's milk has been widely used by humans as a medicinal dietary food since ancient times. In folk medicine, the healing properties of goat's milk have been tested by various experiments, it improves the functioning of the stomach and stomach, i.e. it helps to relieve constipation, cough, spitting up blood, headache and upset stomach, temptation and memory loss, swelling of the tonsils, sores in the throat and bladder, rickets, liver, and pulmonary tuberculosis. , excessive weight loss, allergies, eczema, itching, asthma patients, joint pain relief, gallbladder stones, and goat's milk is consumed only in boiled form, because Bursceles disease can be transmitted to humans.

The smell and taste properties of goat milk are related to the conditions of keeping and feeding the dairy goat. The types of fodder given to goats are inextricably linked with the quality of coarse hay, silage, hay, cottonseed meal, grain fodder and the quality of the ingredients in the fodder.

Goats have a well-developed digestive system, the stomach consists of 64% connective tissue, which allows for good digestion of ingested feed [5, 9, 10, 11, 17, 19]. Goat's milk is sweet and delicious. Yoghurt, suzma, wormwood, cheese, yogurt and other dairy products are made from its milk, and suzma whey is used for liver disease and its prevention. Goat's milk mixed with sugar cleanses and beautifies the complexion. In different nations, various food products and dishes are prepared from goat's milk and meat.

Goat's milk contains nicotinic acid, which is useful for cardiovascular and nervous diseases. According to the chemical composition of goats' milk, it is closer to women's milk [12-16, 19, 20]. Goat's milk contains protein 3 g or 12 kcal, fat 4.2 g or 38 kcal and carbohydrates 4.5 g or 18 kcal, or their ratio in milk is 18%; 57% and 27% respectively. The energy of 100 g of goat's milk is equal to 66.7 kcal.

Goat's milk contains valuable macro and microelements such as calcium, magnesium, iron, manganese, potassium. Also, goat's milk is rich in fat and proteins. Goat's milk is quickly and easily digested by the human body, and the fat grains of its milk are twice as small as the fat grains of cow's milk. That's why goat's milk is used as a dietary food for young babies and elderly people. When comparing the composition of goat's milk with the composition of cow's milk, the superiority of substances in goat's milk is observed. Table 1 below shows the ratio of substances in cow's and goat's milk.

Table 1. Chemical composition of cow and goat milk, %.

Indicators	Cow's milk	Goat's milk
Dry matter	12.7	13.7
Milk fat	3.8	4.4
Milk protein	3.5	4.1
sh.j: casein	2.8	3.3
Milk sugar	4.7	4.4
Milk minerals, mg	0.7	0.8

According to the data presented above, dry matter in the milk of dairy goats is 1.0% higher than cow's milk, milk fat and milk protein are 0.6%, including casein is 0.5% higher than cow's milk. Although goat's milk fat is 0.6% superior to cow's milk fat, it has been proven in medical science that the fat content of goat's milk is quickly digested by the human body due to the smallness of the granular particles.

When importing dairy breeds, it is important to compare their adaptation to the natural climatic conditions of Uzbekistan with the natural climatic conditions of European countries, as well as to take into account the production at the level of the demand of the breed standard, as well as to analyze the biological and physiological characteristics of dairy goat breeds in depth and to study their viability in a new environment. knowing the result is one of the most important signs in keeping goats in farms.

The import of young goats for wool and milk from foreign countries to Uzbekistan began in 2015. The total number of goats imported to the republic from European countries in 2015-2020 was 8,659, including the Republic of Karakalpakstan 706, Andijan region 690, Bukhara region 103, Jizzakh region 520, Kashkadarya region 1065, Navoi region 773, Namangan region 921 head and Tashkent region imported 198 head of young goats. Business entrepreneurs continue to import purebred goats. In 2020 alone, 3,983 Zaanen, Murcia, Grenada, Toggenburg, Anglo-Nubian, Orlov, Angora goats were imported from Germany, Spain, Russia, Kyrgyzstan, and Tajikistan. indicates that he is waking up.

When observing the situation of adaptation of imported dairy goats to different natural climatic conditions of Uzbekistan, it was observed that it was a little difficult to adapt to the conditions of keeping and feeding them. Because there is a significant difference between the humid climate of European countries and the dry hot climate of Uzbekistan. In order to solve these problems, clinical conditions and morphological characteristics of blood in goats brought from abroad were observed during morning, afternoon and evening.

During the study, clinical changes in the growth and development of goats during the change of seasons in hot climates were monitored. Information about this is presented in Table 2.

Table 2. Clinical indicators of breeding foals by season.

Groups	**In the morning**			**During the day**			**In the evening**		
	Body temperature, °C	**Heart beat (minutes)**	**Breathing (minutes)**	**Body temperature, °C**	**Heart beat (minutes)**	**Breathing (minutes)**	**Body temperature, °C**	**Heart beat (minutes)**	**Breathing (minutes)**
	M ± m	M ± m	M ± m	M ± m	M ± m	M ± m	M ± m	M ± m	M ± m
In winter									
#532903	38.2	62.6	27.4	38.7	63.3	29.2	39.0	63.1	28.5
#656808	38.4	62.5	27.5	38.5	63.5	29.4	39.1	63.1	28.4
#837122	38.5	62.7	27.5	38.0	63.7	29.3	39.0	63.3	28.6
In the spring									
#532903	38.4	62.8	27.5	38.8	63.6	29.4	38.7	63.3	28.6
#656808	38.6	62.7	27.4	38.6	63.7	29.6	38.6	63.2	28.6
#837122	38.6	62.9	27.6	38.4	63.9	29.5	38.8	63.5	28.7
In summer									
#532903	39.1	63.0	31.3	39.4	63.4	32.8	39.1	63.5	31.8
#656808	39.3	62.7	30.7	39.5	63.3	32.5	39.2	63.3	31.2
#837122	39.4	63.0	30.8	39.5	63.5	32.6	39.3	63.6	31.2
In autumn									
#532903	39.1	63.0	31.3	39.4	63.4	32.8	39.1	63.5	31.8
#656808	39.3	62.7	30.7	39.5	63.3	32.5	39.2	63.3	31.2
#837122	39.4	63.0	30.8	39.5	63.5	32.6	39.3	63.6	31.2

The body temperature, heart rate, and breathing parameters of the research horses were studied according to the seasons of the year. In the winter, spring, summer and autumn months of the year, the body temperature in the soles was 0.4-0.7 °C higher than the morning temperature , and it was observed that the temperature decreased again in the evening. In the summer, the body temperature of takas reached an average of 39.2 ° C in groups, while in spring this indicator reached an average of 38.8 ° C, while in winter and autumn the body temperature of takas was 39.3 °C or air temperature. As it decreased, a partial decrease in the body temperature of the takas was also observed.

During the study, the analysis of the clinical condition of the horses showed that the change in weather temperature in the seasons also affected their heart rate. In the summer, the heart rate of the bucks was on average 63.3-63.5 beats per group, while this indicator was on average 63.3-63.9 beats in the spring, and 63.3-63.4 beats in the winter and autumn months.

Although there were differences in the breathing of horses due to the seasons and their age, it was at the level of physiological norm, i.e., the breathing rate of horses was on average 31.3 min. min./hour, 31.7 min. in winter and autumn months. organized. This situation indicates that with the continuation of growth processes in takas, significant changes in physiological conditions were observed.

Thus, it can be concluded from the data of the research carried out on the seasons in the conditions of Uzbekistan that the change of seasons does not have a negative effect on the growth of takas imported from abroad, the physiological state of the organism is at the standard level.

When the body temperature of female goats was studied according to the seasons, in the summer months, the daytime temperature was 0.4-0.7 °C higher than the morning temperature, and as the air temperature decreased again in the evening, a decrease in the body temperature of goats was observed. The body temperature of female goats was 39.2 ° C on average by group in summer and 38.8 ° C in spring , or with the increase in air temperature, an increase and decrease in body temperature was also observed (Table 3).

Table 3. Clinical performance of female goats, (p = 5).

Group	in the morning			during the day			in the evening		
	Body temperature, ° C	Heart beat (minutes)	Breathing (minutes)	Body temperature, ° C	Heart beat (minutes)	Breathing (minutes)	Body temperature, ° C	Heart beat (minutes)	Breathing (minutes)
	M ± m	M ± m	M ± m	M ± m	M ± m	M ± m	M ± m	M ± m	M ± m
In winter									
I-	38.7 ±0.04	69.36 ±0.05	28.53± 0.02	39.43 ±0.05	78.53 ±0.02	32.33 ±0.09	38.53 ±0.02	71.73 ±0.02	31.63 ±0.06
II	38.63 ±0.02	69.06± 0.02	28.46± 0.06	39.6 ±0.04	78.4 ±0.06	32.4 ±0.07	38.5 ±0.05	71.53± 0.06	31.3± 0.27
III	38.6 ±0.04	69.16± 0.02	28.33± 0.02	39.36 ±0.02	78.3± 0.02	32.2± 0.04	38.46 ±0.02	71.53 ± 0.02	31.5± 0.04
In the spring									
I	38.7 ±0.04	69.36 ±0.05	28.53± 0.02	39.43 ±0.05	78.53 ±0.02	32.33 ±0.09	38.53 ±0.02	71.73 ±0.02	31.63 ±0.06
II	38.63 ±0.02	69.06± 0.02	28.46± 0.06	39.6 ±0.04	78.4 ±0.06	32.4 ±0.07	38.5 ±0.05	71.53± 0.06	31.3± 0.27
III	38.6 ±0.04	69.16± 0.02	28.33± 0.02	39.36 ±0.02	78.3± 0.02	32.2± 0.04	38.46 ±0.02	71.53 ±0.02	31.5± 0.04
In summer									
I	38.8± 0.04	69.46± 0.05	28.63 ± 0.02	39.43 ±0.02	78.60 ±0.04	32.40 ±0.07	38.56± 0.02	72.43 ±0.28	31.66 ±0.06

II	38.73± 0.02	69.2± 0.20	28.50± 0.07	39.63 ±0.02	78.6 ±0.07	32.46 ±0.06	38.6± 0.05	71.66 ±0.05	31.36 ±0.29
III	38.7± 0.04	69.3± 0.04	28.40± 0.04	39.4 ±0.07	78.4± 0.04	32.26± 0.05	38.50 ±0.04	71.66± 0.02	31.5± 0.02
In autumn									
I	38.53± 0.02	69.46 ±0.02	28.60± 0.04	39.43 ±0.02	78.50 ±0.04	32.36 ±0.10	38.43± 0.05	72.33 ±0.28	31.60 ±0.04
II	38.73 ±0.06	69.43± 0.05	28.60± 0.04	39.4± 0.04	78.5 ±0.04	32.46 ±0.10	38.56 ±0.02	72.40 ±0.23	31.63 ±0.02
III	38.7 ±0.04	69.3± 0.04	28.40± 0.04	39.46 ±0.06	78.4± 0.04	32.26± 0.05	38.50 ±0.04	71.56 ± 0.02	31.56± 0.02

During the study, in the analysis of the clinical condition of the mother goats, changes in the weather also affected their heart rate.

Female goats had an average heart rate of 78.3-78.5 in winter and spring. In the summer, the heart rate was on average 78.4-78.6 beats per group, and respiration also increased, but it was at the level of physiological norm in goats, that is, the breathing rate of mother goats was on average 32.2-78, respectively, on the group in summer. It was 6 min./time, and in winter, spring and autumn, this indicator was 32.3-32.46 min./time.

When the body temperature of young female goats was monitored in winter, spring, summer and autumn, it was observed that the daytime temperature was 0.4-0.7 ° C higher than the morning temperature in the summer months, and the air temperature decreased again in the evening. Body temperature of young females averaged 39.2 °C in summer , 38.8 °C in spring, 38.9 ° C in autumn, and 39.0 ° C in winter organized, or with an increase in air temperature, an increase in body temperature of goats was also observed (Table 4).

Table 4. Clinical parameters of young female goats (p = 5).

Groups	in the morning			during the day			in the evening		
	Body temperature, °C	Heart beat (minutes)	Breathing (minutes)	Body temperature, °C	Heart beat (minutes)	Breathing (minutes)	Body temperature, °C	Heart beat (minutes)	Breathing (minutes)
	M ± m	M ± m	M ± m	M ± m	M ± m	M ± m	M ± m	M ± m	M ± m
In winter									
I	38.6 ±0.04	69.1 ±0.05	28.33± 0.02	39.3 ±0.07	78.33 ±0.02	32.2 ±0.04	38.46 ±0.02	71.53 ±0.02	31.50 ±0.04
II	38.46 ±0.02	68.93± 0.2	28.20± 0.06	39.4 ±0.04	78.3 ±0.04	31.86 ±0.22	38.36 ±0.05	71.36± 0.05	31.1± 0.27
III	38.63 ±0.02	69.23± 0.06	28.30± 0.02	39.3 ±0.07	78.36± 0.02	32.20 ±0.04	38.46 ±0.02	71.53 ± 0.02	31.53± 0.02
In the spring									
I	38.6 ±0.04	69.1 ±0.05	28.33± 0.02	39.3 ±0.07	78.33 ±0.02	32.2 ±0.04	38.46 ±0.02	71.53 ±0.02	31.50 ±0.04
II	38.46 ±0.02	68.93± 0.2	28.20± 0.06	39.4 ±0.04	78.3 ±0.04	31.86 ±0.22	38.36 ±0.05	71.36± 0.05	31.1± 0.27

III	38.63 ±0.02	69.0 ±0.06	28.30± 0.02	39.3 ±0.07	78.36± 0.02	32.20 ±0.04	38.46 ±0.02	71.53 ± 0.02	31.53± 0.02
In summer									
I	38.63± 0.02	69.23± 0.6	28.36 ± 0.05	39.36 ±0.10	78.4 ±0.04	32.3 ±0.04	38.56± 0.05	71.70 ±0.04	31.56 ±0.05
II	38.53± 0.02	69.03± 0.2	28.33± 0.09	39.43 ±0.02	78.3 ±0.04	31.93 ±0.22	38.40± 0.04	71.43 ±0.02	31.13 ±0.28
III	38.6± 0.04	69.26± 0.9	28.36± 0.02	39.4 ±0.07	78.43 ±0.05	32.23 ±0.02	38.5 ±0.04	71.66± 0.05	31.53± 0.02
In autumn									
I	38.56± 0.02	69.26 ±0.2	28.33± 0.05	39.4 ±0.08	78.40 ±0.08	32.33 ±0.02	38.53± 0.02	71.60 ±0.04	31.50 ±0.04
II	38.5 ±0.04	68.96± 0.2	28.33± 0.05	39.26± 0.06	78.2 ±0.04	31.9 ±0.23	38.56 ±0.02	72.40 ±0.23	31.63 ±0.02
III	38.53 ±0.02	68.9± 0.19	28.26± 0.02	39.36 ±0.11	78.33 ±0.05	32.13 ±0.02	38.50 ±0.04	71.56 ± 0.02	31.56± 0.02

The clinical status of young goats during the study are presented in Table 4. Seasonal changes in weather also affected their heart rate. The average heart rate of goats in winter and spring was 69.8 - 69.9 beats. Heart rate in summer was 78.4 beats per group and 70.1 beats in autumn. In goats, the breathing rate in summer also increased, but in young goats it was at the level of physiological norm, i.e., the breathing rate of mother goats was 32.3 min. respectively, averaged 31.6 -31.5 - 31.6 min./time.

The body temperature of young goats is observed to increase during lunch compared to the morning and evening temperatures in summer, but no negative effects on the body were observed.

Goats, like other agricultural animals, adapt their beneficial economic traits to the technological factors introduced in regions with natural climatic conditions, and it is important to study the ratio of elements in the blood during growth and development. The composition of the blood of goats varies inextricably with age, sex, breed, seasons, technology of preservation, care and feeding, as well as the physiological state of the body. Blood performs many complex tasks in the body, namely, it supplies the tissues of the body with nutrients, controls the metabolism, and maintains the water balance in the body. It carries out thermo-regulation and transportation of substances in the body . Therefore, it is important to study the blood of research goats and its composition.

The results obtained in the study showed that in the winter and spring seasons, the blood parameters of the researched goats and female goats remained within the physiological norm. Table 5 below shows the morphological composition of the blood of the founders of the system as progenitors of dairy zaanen foals.

Table 5. Chemical composition of the blood of thoroughbred stallions.

Systems	**Hemoglobin, g/%**	**Erythrocyte, ml/mm^3**	**leukocyte, '000/mm^3**
# 532903	9.73	8.83	8.70
#656808	9.67	8.20	8.50
#837122	9.69	8.44	8.60

When the hematological parameters of the blood of horses were studied in the spring months, their age, changes in summer and autumn seasons, and feeding standards were taken into account and their correlation was observed. The average amount of hemoglobin in the blood of the research horses was equal to 9.67-9.73 g/%.

The indicators of the increase in air temperature and the effect of the organism on the external environment were as follows. The number of erythrocytes in the case No. 532903 increased to 8.83 ml/mm^3 , in the case No. 656808 it was 8.20 ml/mm^3 , and in the case No. 837122 it was 8.44. was ml/mm ^{3} · On the other hand, the number of leukocytes in the blood in the spring and autumn seasons was 8.70-8.71 thousand/mm^3 in the group #-532903 , and 8.50-8.52 thousand/mm^3 in the experimental group #-656808 and #-837122 It was 8.60-8.61 thousand/mm^3. The results obtained in the study showed that all blood parameters were within physiological limits in winter and spring.

When the hematological indicators of the blood of mother goats were studied in the spring months, their age, changes in summer and autumn seasons, and feeding norms were taken into account and their correlation was observed. The morphological composition of the blood of mother goats is presented in Table 6 below.

Table 6. Chemical indicators of the blood of female goats, (p = 5).

K - indicators	Hemoglobin, g/%		Erythrocyte, ml/mm^3		leukocyte, thousand/mm^3	
	M±m	Sv, %	M±m	Sv, %	M±m	Sv, %
	Group I		Group II		Group III	
In winter, +26 °C						
Erythrocyte, ml/mm ^{3}	12.63±0.09	0.2	12.7±0.04	0.1	12.61±0.04	0.01
Letscocite, thousand/mm^3	9.02±0.00	0.02	9.05±0.00	0.01	9.04±0.00	0.01
Hemoglobin, g/%,	9.83 ±0.06	0.15	9.91±0.00	0.02	9.86±0.04	0.09
In spring, +26℃						
Erythrocyte, ml/mm ^{3}	12.63±0.09	0.2	1 2 .7±0.04	0.1	12.62±0.04	0.01
Letscocite, thousand/mm ^{3}	9.02±0.00	0.02	9.05±0.00	0.01	8.04±0.00	0.01
Hemoglobin, g/%,	9.83 ±0.06	0.15	9.91±0.00	0.02	9.86 ±0.04	0.09
In summer, +39.6						
Erythrocyte, ml/mm ^{3}	12.63±0.13	0.30	12.76±0.06	0.15	12.66±0.05	0.11
Letscocite, thousand/mm ^{3}	9.02±0.06	0.01	9.05±0.00	0.01	10.04±0.00	0.00
Hemoglobin, g/%,	9.85±0.06	0.14	9.92 ±0.00	0.02	10.6 7±0.04	0.08
In autumn, +36.6						
Erythrocyte, ml/mm ^{3}	12.6±0.01	0.2	12.7±0.08	0.2	12.7±0.04	0.1
Letscocite, thousand/mm ^{3}	9.03±0.00	0.00	9.05±0.00	0.01	10.05±0.00	0.01
Hemoglobin, g/%,	9.84 ±0.06	0.13	9.92 ±0.00	0.01	10.87±0.04	0.04

In the analysis of hematological parameters of the blood of mother goats, it was observed that the amount of hemoglobin was the same between the groups.

It determines the increase or increase in air temperature and the increase of the body's influence on the external environment. No differences were observed between groups in spring and autumn. The number of erythrocytes was 12.6-12.63 ml/mm^3 in the groups. The number of leukocytes in the body of female goats was 12.76-9.04 thousand/mm^3, hemoglobin content was 9.84-10.87 g/percent.

4 Conclusions

The amount of elements in the morphological composition of the blood mentioned above is maintained at the same rate in all seasons and does not affect the body.

In the research, the clinical and morphological indicators of the body of bucks and female goats in the spring and autumn seasons are controlled and compared, and the change of seasons in the conditions of Uzbekistan does not have a negative effect on the body of bucks and female goats.

The change of seasons in the natural climatic conditions of Uzbekistan for goats imported from abroad does not have a negative effect on the organism of breeding bucks, mother goats and young goats, and their physiological condition is maintained at a normal level.

In order to effectively use imported breeds in breeding, first of all, to preserve them, to preserve and feed young goats in accordance with the technologies used in dairy goat farming, to properly organize milking in dairy goats, and to ensure compliance with the technologies of keeping goats separately from their mothers, will serve the rise of dairy goat farms.

References

1. Berezhnaya A.V. Status of dairy industry in mire. J. Molochnaya promyshlennost. M. 2002, No. 2, S. 4-8.
2. Brynchugin V.V. Produktivnost i tekhnologicheskie svoystva moloka zaanenskoy , alpiiskoy i nubiyskoy porod. Abstract. Disser. M. 2012, S.21.
3. Erokhin A. I., Karasev E.A., Shuvarikov A. S., Erokhin S. A., Dinamika proizvodstva milk ovets i koz v mire i v Rossii. J. Ovtsy, kozy, sherstyanoe delo. M. 2015, No. 2, S.27-28.
4. Zabelina M.V., Sivkov. A. I., Dynamics of calcium and phosphorus in the blood of kozlikov russkoy porody, poluchennyx ot kozamatok s raznym irovnem molochnoy productivity. J. Ovtsy, kozy, sherstyanoe delo. M. 2012, No. 1, S. 76-70.
5. Ismailov M. i dr. - Breeding work in dairy goat breeding T. J. "Zooveterinaria" No. 6. 2016, S.27-29.
6. Ismailov M.Sh. and others. Experiences in forming breeding groups of dairy goats. T.J., "Zooveterinaria", 2017, No. 7, S.39
7. Davronov B., Khalikov Sh., Ruziev RI, Proceedings of the republican scientific-practical conference on the topic "Prospects of Livestock Development in the Republic" of dairy goats imported from abroad 99-102.b (November 25, 2019, Tashkent).
8. Davronov. BX, Eshmatov .IY, Ruziev. RI, Milk productivity of Alpine and Toggenburg goats imported from abroad., My Science Conference 2021 International Scientific and Practical Conference "Experimental And Theoretical Research In Modern Science", Tashkent 2021. pp. 24-35.
9. Kosimov M. A. - Kozovodstvo. Dushanbe 2005, S. 69.
10. Masterskikh D.G. Khozyaystvenno-poleznye priznaki, sostav I technological properties of milk in women's births at different ages. J. Ovtsy, kozy, sherstyanoe delo. M. 2004, No. 3, S. 39-40.
11. Melnikov. I. V., "Razvedenie i vyrashchivanie koz", Moscow, 2012 g.S.23.
12. Mysik A.T. Sovremennoe sostoyanie proizvodstva i potrebleniya produktsii jivotnovodstva v mire. J. Zootechnics . M. 2008, No. 1. P.41-44.
13. Novopashina S. I., Sannikov M.Yu., Kojanov T.V., Opyt formirovania

Liny pri sozdanii novogo tipa zaanenskikh koz. J. Ovtsy, kozy sherstyanoe delo. M 2014, No. 3, S.6-7.

14. Novopashina S. I., Sannikov M.Yu., Kulinich V.A., Kizilova E.I., Kondrashina I.V., Produktivnost zaanenskikh koz pri raznykh sezonakh kozleniya . J. Ovtsy , kozy sherstyanoe delo, M. 2015 g, No. 4, S.37-39.
15. Novopashina S.I. _ . Organization of zootechnical work in dairy farms. Material seminar Agro Farm 2015, Yoshkar-ola . S.35.
16. Toshchev V.K. Molochnaya productivity svoystva vymyani koz, razvodymyx v Respublike Mariy El. J. Zootechnics . M. 2007, No. 12, S. 20-22.
17. Toshchev , V.K. The creation of dairy goats in the Republic of Mariy El pri introduction of goat breeding . J. Ovtsy , kozy, sherstyanoe delo.M. 2006, No. 3, S. 59-60.
18. Eshmatov I.Ya., Norinov U.A., Sattarov N.E., Borotov A.N., Eshmatova M.I., Eshmatova Sh.I., "Selection-breeding works in cattle breeding". Ministry of Higher and Secondary Education of the Republic of Uzbekistan, Tashkent Institute of Irrigation and Agricultural Mechanization Engineering, International scientific-practical conference on "Science, education and innovation, problems and prospects for the agro-industrial complex". 1st set. pp. 59-63. November 22-23, 2019, Tashkent.
19. Eshmatov I.Ya, Sattarov N.E., Eshmatova M.I., Eshmatova Sh.I., Borotov A.N. "Care of young goats during lactation". Ministry of Higher and Secondary Education of the Republic of Uzbekistan, Tashkent Institute of Irrigation and Agricultural Mechanization Engineering, International scientific-practical conference on "Science, education and innovation, problems and prospects for the agro-industrial complex". 1st set. pp. 64-67. November 22-23, 2019, Tashkent.
20. Eshmatov I.Ya. "Domestic dairy goat herds and indices. State Committee for the Development of Veterinary and Animal Husbandry of the Republic of Uzbekistan, Research Institute of Animal Husbandry and Poultry, Proceedings of the republican scientific-practical conference on the topic "Prospects for the development of animal husbandry in the Republic", pp. 100-103, Tashkent-2019, November 25.
21. Eshmatov I.Ya. "Coefficients of connection of domestic goats' milk yield with exterior" State Committee of Veterinary and Animal Husbandry Development of the Republic of Uzbekistan, Research Institute of Animal Husbandry and Poultry, Proceedings of the republican scientific-practical conference on the topic "Prospects for the development of animal husbandry in the Republic", pp. 104-106, Tashkent-2019 ., November 25.
22. Eshmatova Sh.I., Eshmatova M.I., Eshmatov I.Ya. "Changes in exterior index indicators during the aging of dairy hybrid goats" State Committee for Veterinary and Livestock Development of the Republic of Uzbekistan, Research Institute of Livestock and Poultry, Proceedings of the republican scientific-practical conference on the topic "Prospects for the development of livestock in the Republic", pp. 106-108, Tashkent-2019 y., November 25.
23. Eshmatova M.I., Eshmatov I.Ya. "Efficiency of local dairy goats" Republic of Uzbekistan Veterinary and Livestock Development State Committee, Research Institute of Livestock and Poultry, Proceedings of the republican scientific-practical conference on "Prospects of Livestock Development in the Republic", pp. 108-110, Tashkent-2019, 25 november
24. I. Eshmatov [1], Sh Eshmatova [1], M Eshmatova [1], B Davronov [1], Kh Donayev [2], U Soatov [2], and Ch toshpulatov [3], Adaptation of imported goats to the natural climate conditions of Uzbekistan ., E3S Web of Conferences Scopus 258, 04030 1-7 r (2021)
25. IY Eshmatov , AM Svankulova , BH Davronov , Fundamentals of Creating New Drives and Lines from The Gene Fund of Angor Goats., International Journal of Academic

Research in Business Arts and Science., (IJARBAS. COM) Volume: 3 Issue 5 Page 75-81 YAER: 2021.

26. Khalikov. Sh., Ruziev.R., Khidirov. K., Davronov. B., Feeding technologies of Alpine and Toggenburg goats in the direction of milk brought from abroad, feeding during milking periods., Tashkent. Agro Science #3 2020. Pages 50-51.
27. Ruziev.R., Eshmatov. I., Davronov. B., Khalikov. Sh., Feeding technology of dairy goats in the dairy complex., Agriculture and Water Management of Uzbekistan., Tashkent. Special issue 2019. pp. 52-53.
28. Khidirov Karim Imomnazarovich Ruziev Merciful Istamovich Davronov Bakhtior Hasanovich Alp and toggenburg goats breeding, feeding technology during milking and chemical composition of goat mouth milk An International Multidisciplinary Research Journal ACADEMICIA India 8 August 2020 495-502 p
29. Kholikov Sherzod Abduhamidovich ., Davronov Bakhtiyor., Ruziev Rahmon Istamovich ., Khidirov Karim Imomnazarovich ., Technology for feeding goats in the period of lactation and cases., Asian Journal of Multidimensional Research AJMR India. July 7. 2020. 130-134 p.
30. Eshmatova M.I., Eshmatova Sh.I., Eshmatov I.Ya. "The main genetic traits passed from generation to generation in the breeding of dairy goats" State Committee for Veterinary and Livestock Development of the Republic of Uzbekistan, Research Institute of Livestock and Poultry, Proceedings of the republican scientific-practical conference on the topic "Prospects for the development of livestock in the Republic", pp. 111-113, Tashkent - November 25, 2019.

Impact of services provided in Uzbekistan on the added value of dekhkan farms

Shokhsuvor Azizov[1,*]

[1]Tashkent State Agrarian University, 2, University street, Tashkent, 100140, Uzbekistan

Abstract. This scientific article focuses on agricultural development, with special consideration given to the specialization of crop types for farmers, dekhkan farms, and landowners. The research emphasizes the significance of providing plant protection products and mineral fertilizers to enhance crop productivity and quality. Moreover, the study highlights the importance of expanding zoo-veterinary centers and artificial insemination centers for cattle, aiming to improve animal health and breeding practices. In the pursuit of agricultural advancement, it is crucial to cater to the specific needs of different agricultural entities, including individual farmers, dekhkan farms (small agricultural cooperatives), and landowners. By tailoring crop specialization to suit their respective capacities and preferences, the study aims to optimize agricultural output and foster sustainable growth. Furthermore, the research emphasizes the crucial role of plant protection products and mineral fertilizers in ensuring healthy and high-yielding crops. By promoting the use of appropriate agrochemicals and fertilizers, the study aims to enhance agricultural efficiency and contribute to food security. In addition to crop-related measures, the article underscores the need to strengthen livestock farming. Expanding zoo-veterinary centers can bolster animal health and welfare, while artificial insemination centers for cattle can lead to improved breeding outcomes and the development of superior livestock breeds. By combining these efforts, the research aims to enrich the range and quality of services offered to agricultural stakeholders. Through these interventions, the study seeks to foster agricultural progress, enhance agricultural productivity, and contribute to the overall economic development of the agricultural sector.
Keywords. Dekhkan farms, added value, service services, transportation, storage, processing, sales, packaging, technical service.

1 Introduction

It should be noted that in the next three years, specialization of agricultural industries in our republic based on regional conditions, establishment of high-yield crops, perennial orchards and vineyards, placement of their new varieties, diversification of production activities of economic entities and systematic implementation of these processes promising decrees and

* Corresponding author: sh.azizov@tdau.uz

decisions have been adopted by the leadership and government of our country regarding the formation and effective operation of service service entities related to [1-3].

However, according to the analysis, the services provided to agricultural enterprises, especially dekhkan farms and homestead land owners, do not fully meet their requirements in terms of volume, prices, and quality [4-6]. Therefore, in our opinion, it is appropriate to create a coherent and efficient system of service provision.

It is especially important to provide services related to the system of transportation, storage, processing and sale of products in order to ensure the continuity and efficiency of the production of agricultural products, which are directly related to the natural and climatic conditions, soil type and fertility of the land, and their delivery to consumers at the required level and quality plays a role [7-9]. According to the analysis, today in our country, the system of providing all kinds of services to farms producing agricultural products, including dekhkan farms, has been established by the entities that are part of the chain of added value creation [10].

It should be noted separately that in accordance with the programs of harmonious formation of this system, its modernization and diversification, in all regions and districts, taking into account their conditions and specialization in terms of crop types [11], the supply of plant protection products and mineral fertilizers, the expansion of zoo-veterinary centers and artificial insemination centers for cattle, increasing the types and quality of services provided, reducing their prices, and the rapid development of services such as the sale of high-quality goods through auctions creates the basis for a stable increase in the volume of goods and added value in dekhkan farms.

2 Materials and methods

Abstract-logical thinking, monographic observation, statistical-economic and comparative analysis, statistical grouping, social survey and other similar methods were used in the article. It is desirable to consistently implement the following measures in order to create a coherent and efficient system of servis services. Including:

- organization of infrastructure entities in places where they are highly effective and there is a need for their operation.
- provision of services in an acceptable period, quantity and quality for each agricultural enterprise, including dekhkan farms, on the basis of diversification of activities of service providers, regardless of territorial location and the structural structure of networks.
- involving dekhkan farms in the system of working on the basis of the contract with the relevant service provider, as well as ensuring its reliability, the guarantee of the fulfillment of the tasks agreed on by the clauses within the specified periods.
- organization and improvement of the system of introduction of methods of economic stimulation and preferential economic mechanisms for entities providing timely and quality services to dekhkan farms.
- organization of compact agricultural cooperatives in rural areas where dekhkan farms and landowners are mutually integrated, capable of providing all services related to their production activities, protecting their rights and having the authority to enter into various relations with other subjects on their behalf.
- in the structural structure of cooperatives of dekhkan farms and landowners, to establish a financial and accounting center equipped with information and communication tools connected to the Internet, with the possibility of constant exchange of information with other entities, and in it to collect and analyze information specific to each founding entity, to create added value implementation of the functions of accounting for their respective shares in the

added value created in the activities of entities participating in the chain links, as well as in the technological stages after product production, and so on.

Especially in the future, in order to sustainably increase the efficiency of the added value chain in dekhkan farms. Therefore, it is appropriate to dwell on the analysis of the following main types of services and directions, which create the basis for increasing the added value of these dekhkan farms, but are not developed at the level of current requirements (Figure 1).

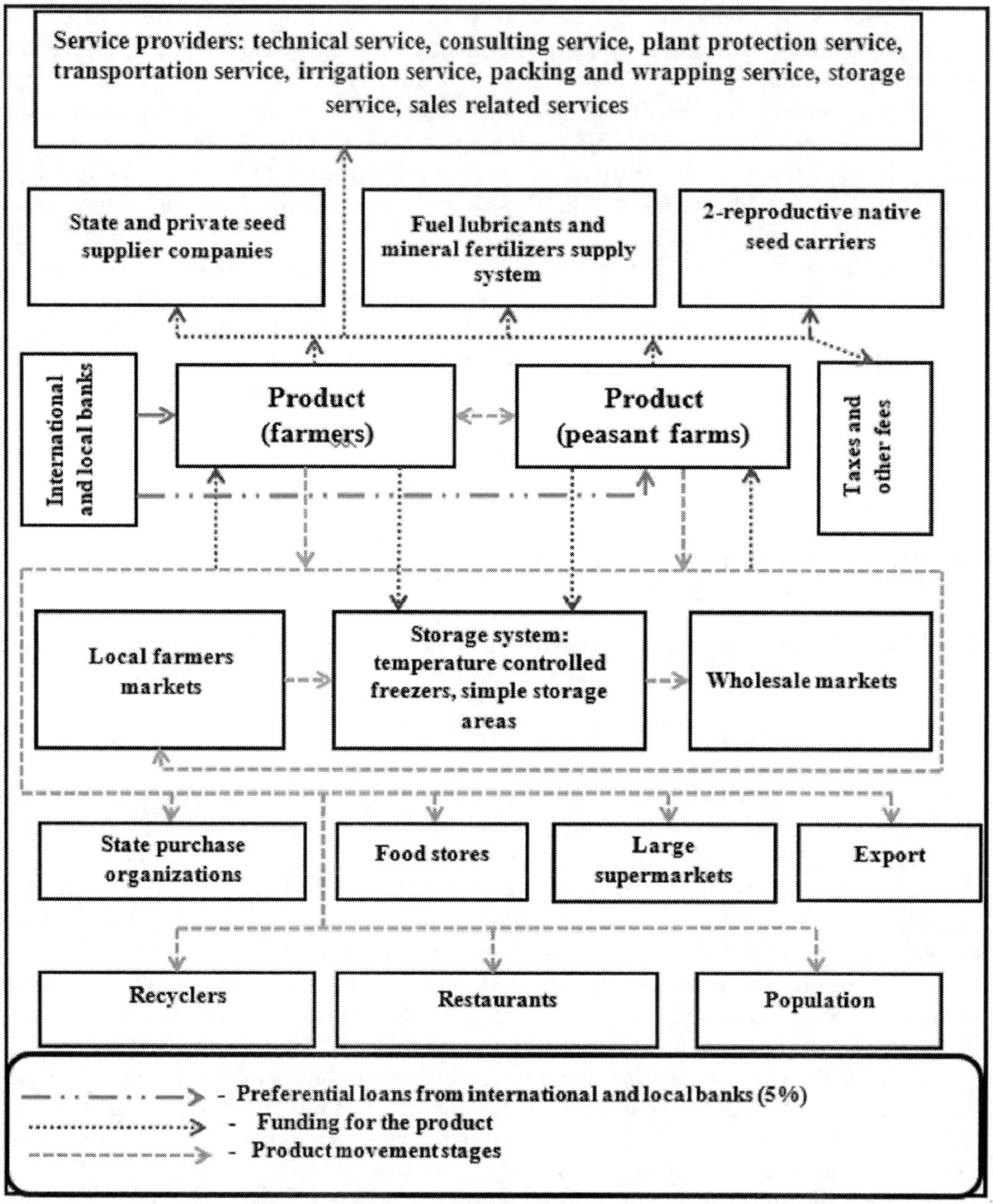

Figure 1. Stages and entities of the value added chain in the processes of production and delivery of agricultural products to consumers.

First of all, there is no integrated system for on-site preparation of farm products, and the services provided in this direction are carried out by informal intermediaries who are interested in purchasing at low prices, and in most cases, seasonal activities. This situation is considered to be one of the service lines that naturally leads to a decrease in the amount of added value created on farms.

Second, farmers, especially those who do not have legal status, face difficulties in using credit lines to expand and diversify production. As a result, due to financial weakness, they cannot fully use their existing production potential because they cannot purchase the necessary resources (quality seeds, seedlings, mineral fertilizers, plant protection products, etc.) in time. Therefore, in order to eliminate such negative situations, in our opinion, it is appropriate to support them by providing subsidies, subvention mechanisms, as well as incentives in tax and price systems.

Thirdly, as one of the important services, in our opinion, it is necessary to organize relations between farmers and trade organizations through futures contracts in a mutually beneficial manner. This system creates an opportunity for them to have financial opportunities before the production of products, to get the maximum added value based on the guaranteed sale of their goods in the agreed period, and to use it in the expansion of subsequent production activities.

Fourthly, in the future, it is important to form and develop commercial skills in the farmers' heads and workers for their effective operation in the conditions of market relations. For this reason, they apply marketing and consulting services in their activities to get acquainted with the main directions and tasks of various infrastructure entities, and for this purpose it is appropriate to establish a system of education in business schools and training courses in the regions.

3 Results and discussion

The increase in the number of farms in our country and their role and share in gross product production, in turn, requires the rapid development of infrastructure entities serving them. In particular, the analysis of the dynamics of the development of infrastructure entities operating in Kashkadarya region shows that in 2015 there were 184 alternative machine tractor parks (MMTP) in the region, and in 2021 there were 196 (increased by 12). The number of branches providing services also amounted to 198 in 2021, which decreased by 61 compared to 2015, and the number of organizations (agrofirms) purchasing agricultural products increased by 2 compared to 2015, and made 55 in 2021 (Table 1).

Based on the data of table 1, it should be noted that in 2021, compared to 2015, at the regional level, the number of branches selling mineral fertilizers, ATMs, and providing veterinary services decreased by 28.6 percent, 90.9 percent, and 23.6 percent, respectively. It is natural that this situation causes great problems, among other things, for a large number of dekhkan farms. However, in the course of the research, it was also revealed that currently infrastructure entities are mainly serving cotton and grain producing farms. Because in our republic there is a guaranteed market only for cotton and grain growers, that is, agricultural enterprises sell the cotton and grain products grown to specific processing enterprises. However, due to the incomplete formation of specific market segments for other types of products, for example, farmers are selling their products, as mentioned earlier, mainly to various informal intermediary entrepreneurs at low prices.

It is known that it is important to use vehicles in agriculture, to increase the fleet of special transport equipment for different types of products and to use them efficiently. However, the research shows that the existing problems in the service sector are currently leading to an increase in the cost of products. According to calculations, it is permissible to admit that the share of transportation costs in the cost of grain products is 26 percent, and 15 percent in the cost of corn.

Table 1. Dynamics of changes in the number of infrastructure entities serving agricultural sectors in Kashkadarya region, Uzbekistan.

Name of entities by types of services	Years:				2021 Compared to 2015, percent
	2015	2019	2020	2021	
Alternative tractor fleets	184	184	184	196	106.5
Associations of water consumers	151	151	152	151	100.0
Fuel and lubricant sales outlets	185	185	185	343	185.4
Mineral fertilizer sales outlets	126	126	126	90	71.4
Minibanks	121	1326	121	11	9.1
Information and consulting services	43	43	46	x	x
Zooveterinary branches	259	259	259	198	76.4
Transport service stations	6	7	8	6	100.0
Tare, container and packaging material supply stations	15	22	25	28	186.7
Agricultural companies	12	12	12	13	108.3
Departments for purchasing agricultural products	53	53	53	55	103.8

Source: Calculated based on the data of the General Directorate of Economy of Kashkadarya Region, Uzbekistan

Also, if the information and consulting service centers of the market infrastructure work in places with low efficiency, in some places, even many farmers do not know that this organization works. The main reason for this is the lack of adequate provision of qualified personnel, special means of communication, and propaganda work. It is natural that this is a partial obstacle to the application of innovative achievements in their production activities.

Value-added opportunities on farms depend on many factors, which determine the intensity and amount of value-added creation. These factors include the following:

- the presence of infrastructure facilities in the territorial section for the provision of services in the directions related to agricultural activities of dekhkan farms;

- compatibility of the service quality and prices of infrastructure objects related to the provision of services in the directions related to the agricultural network to the ability of dekhkan farms to pay;

- it is possible to purchase technical equipment that allows dekhkan farms to carry out activities related to storage, processing, and transportation of agricultural products;

- the existence of regulatory and legal frameworks related to the development and promotion of the system of storage, processing and sale of agricultural products;

- production, transportation, storage, processing, packaging and the effect of such factors as increasing knowledge related to sales, the degree of creation of opportunities to use advice in these directions is significant.

At the same time, under the influence of the above factors, the possibilities of creating added value in the system of production of various agricultural products (fruits, vegetables, pulse crops, potatoes, meat, milk production) are formed in different ways.

Considering the above, it is important to know the coefficients of added value creation for the listed types of products for dekhkan farms. In this case, it is recommended to determine the "Coefficient of added value creation by type of crop in dekhkan farms " through a comparative assessment of the possibilities of creating added value within the framework of agricultural products grown on dekhkan farms in separate regions.

Calculation of coefficients will have a regional character. Because the factors affecting the formation of added value have different levels of influence in different regions, and as a result, the coefficients change.

Coefficients are calculated in relation to the main products grown by dekhkan farms (in terms of agricultural products grown in the republic, the weight of farms is high). It is recommended to determine these coefficients as follows (Table 2).

Table 2. Coefficients of added value creation by types of crops in dekhkan farms.

#	The main types of agricultural products grown on dekhkan farms	***The average selling price of a unit of agricultural products within the relevant area and the coefficients determined on this basis (soums/kg as of 2021)***						
						Export		
		Sell fresh	**Selling with storage**	**Initial processing**	**Deep processing**	**fresh**	**Recycled**	**Coefficient of added value creation (K)**
		a	*b*	*c*	*d*	*e*	*f*	$K = \frac{f, d, e}{a}$
1	**Fruits**	5000	10000	12000	15000	20000	25000	**5.5**
	Coefficient of added value creation	***0.0***	***2.0***	***2.4***	***3.0***	***4.0***	***5.5***	
2	**Vegetables**	3000	4500	6500	8000	15000	28000	**9.3**
	Coefficient of added value creation	***0.0***	***1.5***	***2.2***	***2.7***	***5.0***	***9.3***	
3	**Potatoes**	2000	4000	4500	6000	0.0	0,0	**3.0**
	Coefficient of added value creation	***0.0***	***2.0***	***2.3***	***3.0***	***0.0***	***0.0***	
4	**Field**	2000	4000	5000	0.0	8000	0,0	**4.0**
	Coefficient of added value creation	***0.0***	***2.0***	***2.5***	***0.0***	***4.0***	***0.0***	
5	**Grapes**	3000	6000	9000	14000	16000	20000	**6.7**
	Coefficient of added value creation	***0.0***	***2.0***	***3.0***	***4.7***	***5.3***	***6.7***	
6	**Meat**	45000	0	60000	100000	0	0	**2.2**

Coefficient of added value creation		***0.0***	***0.0***	***1.3***	***2.2***	***0.0***	***0.0***	
7	**Milk**	3000	0	5000	8000	0	0	**2.7**
Coefficient of added value creation		***0.0***	***0.0***	***1.7***	***2.7***	***0.0***	***0.0***	

As can be seen from the table data, at the level of today's technical capabilities, knowledge and acquired practical experience, as well as the average prices of agricultural products, vegetable crops have a leading position in the agricultural production system and are evaluated with the highest coefficient of 9.3.

The cultivation of grapes focused on the possibility of its export has a coefficient of 6.7 and takes the second place. Along with this, opportunities to create added value on farms the lowest indicator is observed in meat and milk production. In this regard, it is worth noting that, first of all, due to the specific characteristics of livestock products, the extremely low capacity of processing and selling livestock products in dekhkan farms has a great impact.

In our opinion, the above-mentioned directions related to the development of the service system to increase the economic and production efficiency of the activities of dekhkan farms and landowners in the future will be the basis for solving the following problems, in particular:

- requires extensive use of innovative products and developments to specialize, diversify the activities of service infrastructure entities and increase their labor productivity, and as a result, creates the basis for an increase in the quality and efficiency of services;
- expansion of the scope of services of infrastructure entities, as well as agricultural cooperatives at the expense of dekhkan farms and homestead landowners, the establishment of mutual relations on the basis of a contract will have a positive effect on further increasing the income of participants;
- existing and newly established infrastructure entities specialize in serving not only farmers, but also dekhkan farms, homestead owners and other household entities, thus expanding the value-added opportunities for these small commodity producers.

4 Conclusions

In order to increase the added value in dekhkan farms, it is advisable to solve the following organizational-structural, economic and social-legal problems:

- harmonious organization of animal husbandry and farming in dekhkan farms, introduction of crop rotation system and periodic supply of profitable new plant varieties (seeds, seedlings) and purebred livestock;
- diversification of agricultural production activities taking into account internal and external consumption needs, and in the process of introducing measures to increase the knowledge and skills of family members;
- the establishment of agricultural cooperatives that include all types of service systems in the cooperation (foundation) of farms in regions or specific rural areas with a dense population, because they effectively participate in relations with other subjects on a large scale compared to one farm, added value have opportunities to increase creation;
- in our opinion, it is reasonable to organize cooperatives on the basis of multi-branch farms that regularly operate effectively in a certain area, export their products not only to domestic, but also to foreign market segments, and have great opportunities for creating added value, due to the contributions made by dekhkan farms and landowners. ;

- it is necessary to organize entities that are adapted to carry out appropriate agrotechnical activities on small plots of land of dekhkan farms and create a competitive environment in the supply system.

References

1. Husanov R.H., M. Kasimov. Scientific and practical foundations of dekhkan farms. Tashkent.: "Cholpon". 2000 y.-p-103.
2. Khasandjanov K.A., Mukhtorov A. X, Khamdamov S.D. Problems of development of personal subsidiary plots in the conditions of transition to market relations. Samarkand.:- 2015
3. P. Alimov. The role of private subsidiary dekhkan farms in food program implementation. T, Labor, 2008.-b-75 p.
4. N. Yusupova. Makroekonomicheskie voprosy razgosudarstvleniya i privatizatsii. Tashkent.:- "FAN", AKRUz, 2020.-p- 248.
5. S. Usmanov, M. Mominov, P. Alimov. Ways of development of homestead economy. T, 2018.
6. M. Kasimov. Issues of development of the activities of dekhkan farms in the conditions of the transition to the market economy. Iqt. science. Dissertation for obtaining the candidate's scientific degree, T.:- 2003 - 26-27 p.
7. Rudenko I. Value Chains for Rural and Regional Development: The Case of Cotton, Wheat, Fruit and Vegetable Value Chains in the Lower Reaches of the Amu Darya River, Uzbekistan. Doktorin der Wirtschaftswissenschaften - Doctor rerum politicarum - genehmigte Dissertation von. Leibniz Universitat 2008.
8. Lebedeva, N., Akhmedova, Z., Kholmatov, B., & Jumaev, R. (2021). Revision of stoneflies (insecta: plecoptera) fauna in Uzbekistan. In E3S Web of Conferences (Vol. 258, p. 08030). EDP Sciences. DOI:10.1051/e3sconf/202125808030
9. Jumaev, R., & Rustamov, A. (2022, July). Representatives of Lepidoptera groups in the biotecenosis of Uzbekistan and their effective parasite-entomophage types. In IOP Conference Series: Earth and Environmental Science (Vol. 1068, No. 1, p. 012026). IOP Publishing. DOI:10.1088/1755-1315/1068/1/012026
10. Jumaev, R. (2023). Invitro rearing of parasitoids. In E3S Web of Conferences (Vol. 371, p. 01032). EDP Sciences. DOI:10.1051/e3sconf/202337101032
11. Kimsanboev, K., Rustamov, A., Jumaev, R., & Usmonov, M. (2021). Euzophera Punicaella Mooze (Lepidoptera) bioecology and development of host entomophagic equilibrium in biocenosis. In E3S Web of Conferences (Vol. 244, p. 01003). EDP Sciences. DOI:10.1051/e3sconf/202124401003

Biochemical composition of soybean foreign varieties and samples (under typical gray soil conditions of Uzbekistan)

Nargiza Khudoyberdieva[1,*] and *Murad* Rakhmankulov[1]

[1]Tashkent State Agrarian University, 2, University street, Tashkent, 100140, Uzbekistan

Abstract. Soybean stands out as a highly prized agricultural crop due to its exceptional nutritional attributes. The proteins found in soybean possess a distinctive profile, featuring essential non-exchangeable amino acids that closely resemble those present in animal proteins. This composition renders soybean proteins a valuable and comprehensive source of vital nutrients. This article offers insights into the biochemical characteristics of imported soybean varieties and samples grown under typical gray soil conditions in Uzbekistan. Specifically, it examines various parameters to assess the quality and composition of these leguminous crops. The study presents data on crucial components such as protein and oil content, both of which constitute the primary constituents of these crops. Additionally, the research delves into the moisture content of the seeds, which contributes to understanding their overall quality. Furthermore, the article provides information about the origin of the soybean varieties under investigation, shedding light on their respective countries of origin. By analyzing these biochemical indicators and considering the seeds' moisture content and source, the study contributes to a comprehensive evaluation of the suitability, adaptability, and potential economic value of these foreign soybean varieties within the specific agroecological context of typical gray soils in Uzbekistan.
Keywords. Soybean, country of origin, variety, sample, biochemical composition, protein, oil, moisture.

1 Introduction

Soybean is one of the most valuable agricultural crops, its proteins are characterized by the presence of essential non-exchangeable amino acids similar to animal proteins, and the oils contained in it are superior to olive oil (86-95% absorption rate by the body) [1, 2].

The daily protein requirement of the body is on average 110 grams [3, 4]. But now in many countries the problem of protein deficiency is increasing. Expanding the use of soybean and its processed products will almost solve this problem [5, 6].

The biochemical composition of soybean seeds depends on the biological characteristics of the species, the weather and the correct agrotechnical measures used in each phase of the

* Corresponding author: nargiza_khudoyberdieva1001@mail.ru

plants. Obtaining a crop with high quality content is the main task of soybean cultivation [7, 8].

Soy protein constitutes a rich reservoir of essential components critical for human nutrition. This includes a robust presence of crucial amino acids, essential vitamins, vital macro-minerals, isoflavones, and phospholipids. The comprehensive nutritional composition of soy protein underscores its potential to contribute significantly to a balanced and healthful die [9, 10]t.

One of the remarkable aspects of soy protein is its affordability. When compared to animal protein sources, soy protein emerges as a cost-effective alternative, available at a fraction of the price. This cost-effectiveness makes it an attractive option for both consumers and food producers seeking to optimize their resources [11].

By planting varieties of soybeans with high-quality nutritional content and implementing appropriate agrotechnical practices in a timely manner, substantial yields can be attained. With the correct approach, it's feasible to obtain a range of 300 to 1200 kilograms of pure, top-tier protein per hectare of land sown with these high-quality soybean varieties [12]. This potential for high protein yield further underscores the significance of soybean cultivation as a viable solution to address nutritional needs and economic considerations alike.

Soybean oil meets the FAO/WHO standard in terms of biological value and quality, its composition is dominated by the most valuable unsaturated fatty acids (up to 87%) [9]. Soybean oil, in its natural and processed form, is used in food and medicine, as well as in the manufacture of a wide range of non-food products (soap, paints, plastics, biodiesel, etc.).

2 Materials and methods

Researches were carried out in the fields of the experimental farm of the Scientific Research Institute of Plant Genetic Resources. The research was carried out based on the following methodological manuals: "Methods for studying the world collection of legumes" [4], "Metodika polevogo opyta" and "Metodika gosudarstvennogo sortoispytaniya selskohozyaystvennyx kultur" by B.A. Dospehov were used [6]. The biochemical composition of the varieties and samples was determined in the German PFEUFFER-Granolyser apparatus.

Soya varieties Sparta, Arletta, Duar, Chara, Selekta-302, Selekta-201, Selena (Russia), Sava, Favorit (Serbia), Apollo (USA) and K-111, K-114, K-183 (China), K-113 (Ukraine), K-115, K-127 (Moldova), K-2, K-17, K-19, K-118, K-120, K-122, K-126, K-129, K-138, K-142 (USA), K-119 (Romania), K-124, K-152 (Russia), K-130 (France), K-141 (Germany), K-22 (Yugoslavia), Samples of varieties K-181 (Syria), K-182 (Korea), K-200, K-201 (Turkiye) and variety "orzu" (Uzbekistan) were studied.

3 Results and discussion

The experiments were conducted at the experimental field of the Scientific Research Institute of Plant Genetic Resources. The experimental farm of the scientific production department of the institute is located in the Qibrai district of the Tashkent region, 16 km from the city of Tashkent (at 41°2' north latitude and 69°2' east longitude). The natural climate of the Tashkent region is sharply continental and arid, rich in heat and sunlight throughout the year.

The climate of Tashkent region is characterized by a lot of light, rapid temperature changes during the day and seasons, dry and hot summers, and irregular winters. Duration of sunny days is 2800-2900 hours per year (360-400 hours in summer and 90-100 hours in winter). The average annual air temperature is +13...+14 ^{0}C [1].

The soil of the experimental field in Tashkent region is a typical gray soil that has been irrigated since ancient times. The parent rock of the soil is unevenly formed, and underground (seepage) water is located at a depth of 18-20 meters. According to M.A. Pankov, P.N. Besedin, P.Suchkov, one third of Central Asian soil consists of typical gray soils. Typical gray soils are characterized by low humus content and carbonation (but higher humus content than pale gray soils). Typical gray in soils rotted amount is 1.5-2.5%, nitrogen 0.08-0.1%, phosphorus 0.2-0.3 % is enough Phosphorous compounds solubility feature of soil high carbonation because of it is worth saying level no, that's why for of phosphorus quantity in the soil many p if not , him plant by appropriation level very is low [8] . Typical gray soils are heavy loamy and loamy in terms of mechanical composition. Also, the soil is characterized by low bulk density and high porosity. In these soils, biological processes are accelerated and the content of saturated cations is high. The process of mineralization of organic substances in the soil is fast, while the mobility of nitrogen is high, and the mobility of phosphorus is slow. The mobility of potassium is average compared to the mobility of nitrogen and phosphorus. Another agronomic property of this soil is that the amount of total nitrogen in the soil depends on the amount of humus in the soil. The amount of total nitrogen in the soil varies from 0.05% to 0.15%. Typical gray soils are very favorable for nitrification. The main part of nitrogen is found in the soil in the form of nitrates, and the nitrogen absorbed by the plant is in the same form. In most cases, the amount of total phosphorus is greater than the amount of total nitrogen. The amount of total phosphorus in the upper layers of the soil is 0.1%-0.2%. In general, the soil has a lot of remains of roots and other parts of the plant, soil compaction is low, and humus is present only in the plowed part of the soil.

One of the main directions of the selection process is to ensure that the quality of the varieties is high. The seeds with high quality content mean exactly the biochemical composition of the seeds.

The main biochemical component of soybeans is protein. Among all crops in the world, soybean is the plant with the highest protein index. According to various sources, the amount of protein in the seeds of this crop is on average 38-42% and can reach 50%. Soy proteins are diverse in structure and function. Soy is rich in essential amino acids, especially lysine (2.7%), which is found in small amounts in grain proteins. Most of the soy protein (about 70%) is 7S-globulins (β-conglycinins) and 11S-globulins (glycinins), which are absorbed by mammals . Since a significant portion of soy protein is water-soluble protein, obtaining plant protein from soy is the most efficient.

We can see in the table that the homeland of soybean varieties studied in the experiment is Russia, Serbia, USA. It is known that the protein content of the varieties is 33.5-47% in the specific climatic conditions of these countries, while this indicator was 33-40.7% in the typical gray soils of our country. If we give a separate explanation for each variety, Sparta of Russian origin increased from 42% to 40.7%, Arletta from 42% to 37%, Duar from 47% to 35.8%, Chara from 40% to 36.5%, Selekta-302 from 36.2 % to 36%, Selekta-201 from 33.5% to 33%, Selena from 36% to 36.1%, Apollo from the USA from 39% to 33.1%, varieties of Serbian selection Favorit from 40% to 38.1%, Sava from 35% we can see that it has changed from to 34.4 %.

The oil content of the studied varieties was 21.5-26.7% in the homeland of the variety, and 22.2-25.6% in the soil-climatic conditions of our republic. Sparta of Russian origin from 23% to 22.2%, Arletta from 23% to 23.7%, Duar from 26.7% to 24.7%, Chara from 22% to 23.7%, Selekta-302 from 25% to 24.7%, Selekta-201 from 22.4 We can see that from % to 25.6%, Selena from 23% to 23.4%, Apollo from the USA from 22% to 23.8%,

Serbian selection varieties Favorit from 21.5% to 23.2%, Sava from 22% to 24.1%. The moisture content of the seeds was found to be on average from 6.7 to 8.5% (Table 1).

Table 1. Moisture content of the seeds.

#	Country of origin	Varietal name	Protein %	Oil %	In the typical gray soils of Uzbekistan		
					Protein %	Oil %	Moisture Content %
1	Russia	Sparta	42	23	40.7	22.2	7.9
2	Russia	Arlette	42	23	37	23.7	7.6
3	Russia	Duar	47	26.7	35.8	24.7	7.3
4	Russia	Remedy	40	22	36.5	23.7	7
5	USA	Apollo	39	22	33.1	23.8	7.4
6	Serbia	A favorite	40	21.5	38.1	23.2	8.1
7	Serbia	Sava	35	22	34.4	24.1	8.5
8	Russia	Selecta-302	36.2	25	36.0	24.7	8.4
9	Russia	Selection-201	33.5	22.4	33	25.6	8.2
10	Russia	Selena	36	23	36.1	23.4	6.7

Soy foreign samples were studied in comparison with the standard variety Orzu. The obtained results are given in Table 2. The protein content of Orzu standard variety seeds was 29.8%, while the oil content was 22.7%. The moisture content of the seeds was 7.5% in the Orzu variety. Taking into account that the protein content in the samples of the world collection of soy is between 27.4% and 38.2%, compared to the standard K-141; K-126; K-127; K-152; K-124; K-183; K-182; K-181; K-129; K-138; K-113; K-114; K-119; K-118; K-120; K-115; K-19; K-22; K-17; The protein content of K-201 samples showed a high result. K-126 by oil content; K-152; K-122; K-181; K-129; K-111; K-113; K-114; K-120; K-115; K-19; K-17; K-200; K-201 samples had a higher index compared to the standard variety (Table 2).

Table 2. Soy foreign samples in comparison with the standard variety Orzu.

#	Sample catalog	Country of origin	Protein content (%)	Oil content (%)	Moisture Content (%)
1	K-141	Germany	37.4	20.3	7.6
2	K-126	USA	36.9	23.6	7.9
3	K-127	Moldova	34.5	22.2	7.2
4	K-152	Russia	29.8	24.3	7.1
5	K-122	USA	29.7	23.2	8.8
6	K-124	Russia	30.0	22.5	7.7
7	K-183	China	38.2	19.6	6.7
8	K-182	Korea	37.7	19.2	7.0
9	K-181	Syria	35.4	24.0	7.6
10	K-129	USA	31.8	23.9	7.7

11	K-138	USA	29.9	22.2	7.8
12	K-111	China	28.0	26.2	8.8
13	K-113	Ukraine	29.7	23.8	7.0
14	K-114	China	32.8	23.9	8.2
15	K-2	USA	29.2	21.9	7.9
16	K-119	Romania	35.4	22.2	7.8
17	K-118	USA	34.4	21.2	8.2
18	K-120	USA	36.8	24.3	7.4
19	K-115	Moldova	33.2	24.2	8.2
20	K-19	USA	33.5	24.3	8.6
21	K-22	Yugoslavia	33.7	22.5	8.7
22	K-17	USA	30.6	25.2	8.8
23	K-200	Turkiye	27.4	26.5	7.9
24	K-201	Turkiye	34.4	23.9	8.6

Based on experience, we inform that:

- Soy is a plant with a high biochemical composition
- The studied foreign soybean varieties were found to have a very high protein content compared to the standard. In the conditions of the typical gray soils of our country, it was found that the protein content of the Sparta variety, native to Russia, is 40.7%.
- Among foreign varieties, Duar and Selekta-201, which are native to Russia, have the highest oil content, 24.7%, respectively; It was 25.6%.
- Among the soybean samples, the ones with higher protein index compared to the standard were: K-183 (China), K-182 (Korea).
- Among the samples, K-111 (China), K-17 (USA), K-200 (Turkiye) samples had a higher oil content than the standard variety.
- The average moisture content of soybean seeds was 6.7-9. The moisture content of the seeds corresponded to the standard.

4 Conclusions

Drawing from the comprehensive array of findings we have gathered, we have arrived at several overarching conclusions. To effectively address the protein deficit within our country, a prudent approach would involve the development of soybean varieties specifically engineered to possess elevated levels of both protein and oil content. This strategic enhancement could potentially serve as a valuable means to augment our national protein supply.

Furthermore, we recommend the careful selection of existing soybean varieties that exhibit exceptionally favorable biochemical compositions. These chosen samples can serve as the foundational building blocks for subsequent selective breeding processes. By starting with these superior genetic materials, we can expedite the development of new soybean strains that are not only optimized for elevated protein and oil content but also well-suited for the unique agroecological conditions present within our country.

Ultimately, these proposed strategies hold the potential to contribute significantly to our efforts in addressing protein scarcity and fostering more sustainable agricultural practices.

References

1. Anikeev S.P., 2018, Soybean varieties, Moscow Nauka Press, 200 p.

2. Belyshkina , M. E. Problema proizvodstva rastitelnogo belka i rol zernovykh bobovykh kultur v ee reshenii / M. E. Belyshkina // Prirodoobustroystvo . - 2018. - Vypusk 2. - S. 65–73.
3. Golovina, E. V. Vliyanie pogodnykh usloviy na produktsionny protsess u sortov soi severnogo ecotipa / E. V. Golovina, V. I. Zotikov // Selskohozyaystvennaya biology. - 2013. - T. 48, No. 6. - pp. 112–118
4. Methods of studying the world collection of legumes (Plant Genetic Resources ITI 2017)
5. Kochegura, A. V. Selection soi na povyshenie pishchevoy i kormovoy tsennosti semyan/A. V. Kochegura, S. V. Zelentsov//Puti povysheniya i stabilizatsii vysokokachestvennogo zerna. - Krasnodar, 2002. - p. 25–32
6. Methodological recommendations 2.3.1.24.32-08. Normal physiological needs and energy and food for different groups of people in the Russian Federation. - Moscow: Izdatelstvo standartov, 2008. - p. 6–7
7. Pankov M.A. Pochvy Tajikistan.-Tashkent: 2005. p. 27-30
8. Zaitsev, N. I. Perspektivy i napravleniya selektsii soi v Rossii v usloviyakh realizatsii natsionalnoy strategii importozameshcheniya / N. I. Zaitsev, N. I. Bochkarev, S. V. Zelentsov // Mass culture. - 2016. - vol . 2 (166). - p. 3–11
9. Jumaev, R., Gazibekov, A., Sulaymonov, O., & Sobirov, B. (2021). Representatives of Lepidoptera groups occurred in forestry and agricultural crops and their effective entomophage types. In E3S Web of Conferences (Vol. 244, p. 02020). EDP Sciences. DOI:10.1051/e3sconf/202124402020
10. Kimsanboev, K., Rustamov, A., Jumaev, R., & Usmonov, M. (2021). Euzophera Punicaella Mooze (Lepidoptera) bioecology and development of host entomophagic equilibrium in biocenosis. In E3S Web of Conferences (Vol. 244, p. 01003). EDP Sciences. DOI:10.1051/e3sconf/202124401003
11. Lebedeva, N., Akhmedova, Z., Kholmatov, B., & Jumaev, R. (2021). Revision of stoneflies (insecta: plecoptera) fauna in Uzbekistan. In E3S Web of Conferences (Vol. 258, p. 08030). EDP Sciences. DOI:10.1051/e3sconf/202125808030
12. Jumaev, R., & Rustamov, A. (2022, July). Representatives of Lepidoptera groups in the biotecenosis of Uzbekistan and their effective parasite-entomophage types. In IOP Conference Series: Earth and Environmental Science (Vol. 1068, No. 1, p. 012026). IOP Publishing. DOI:10.1088/1755-1315/1068/1/012026

Geographical distribution of wheat fusariosis in the central and south regions of Uzbekistan

Nodira Khaytbayeva[1,*] and *Qalandar* Bababekov[2]

[1]Tashkent State Agrarian University, Tashkent, 100140, Uzbekistan
[2]Scientific Research Institute of Quarantine and Plant Protection, Tashkent, 100003, Uzbekistan

Abstract. The article offers a comprehensive exploration of the geographical distribution of wheat fusarium, focusing on both central and select southern regions within Uzbekistan. The study delves into the symptoms of fusarium disease and elucidates the extent of its impact on crop productivity. Furthermore, it investigates the spread of this disease across diverse soil and climatic conditions within Uzbekistan, spanning the period from 2021 to 2023. The research outcomes underscore critical findings related to fusarium-infected plants. Notably, infected plants exhibit an absence of a budding phase, suggesting a disruption in their growth cycle. Experimental evidence confirms the formation of only one spike from a single plant, emphasizing the negative effects of the disease on plant reproductive structures. The article additionally examines the quantitative aspects of wheat production in relation to fusarium. Findings reveal that fields plagued by the prevalence of fusarium exhibit a lower spike formation rate, with approximately 250-300 spikes forming per square meter. This figure contrasts with healthier plants, which yield a more substantial output, exceeding 400 spikes per square meter. By offering insights into the geographical spread, symptomology, and productivity repercussions of wheat fusarium, this article advances our understanding of the disease's impact on agricultural systems. The presented results provide valuable information for practitioners seeking effective management strategies against fusarium, ultimately contributing to the maintenance of healthy wheat crops and sustained agricultural productivity.

Keywords. *Fusarium, F. gramenearum, F. culmorum, F. avenaceun, F. incartanum,* plant, wheat, spike.

1 Introduction

The main food products are made from wheat flour [1]. Therefore wheat is grown on large areas in almost all regions of our republic. In the conditions of Uzbekistan, mainly soft wheat is grown [2]. Depending on the soil and climate conditions of Uzbekistan in winter wheat is planted at different times. In order to obtain a high and quality harvest from wheat, the purity of seed grains, treatment with seeders during seed storage, and timely application of planting and agrotechnical measures in the specified periods of each region are important measures [3]. But when these measures are not used in time, there are cases of decreased productivity

* Corresponding author: khaytbayevanodira@mail.ru

due to diseases in plants. In recent years, the fusarium disease in wheat causes damage to the plant from germination to the end of the growing season, causing a sharp decrease in productivity [4]. The presence of 4 types of fungi belonging to the genus Fusarium that cause diseases in wheat was determined during the research [5].

The Fusarium disease encompasses various species, including *F. graminearum*, *F. culmorum*, *F. avenaceum*, and *F. incarnatum*. Within this context, optimal temperature and humidity levels play a pivotal role in the onset and progression of Fusarium-related plant diseases [6-8]. Notably, the interplay between these environmental factors significantly influences the development of the disease [9].

The research reveals that Fusarium disease tends to thrive under specific conditions. Notably, the disease becomes more pronounced when environmental conditions feature high humidity, typically around 70%, accompanied by temperatures ranging between 24 to 27 degrees Celsius [10]. This particular combination of temperature and humidity appears to foster the growth and proliferation of the Fusarium pathogen [11].

It is important to note that the extent of Fusarium disease impact is influenced by climatic factors, especially rainfall patterns. In years marked by lower rainfall, the disease's effects tend to be less noticeable [12]. This observation underscores the intricate relationship between environmental conditions and the prevalence of Fusarium disease [13].

These findings offer valuable insights into the environmental prerequisites for Fusarium disease development, enhancing our understanding of its epidemiology and aiding in the formulation of targeted disease management strategies.

2 Materials and methods

The correct application of special methods is important for the isolation of fungi from plant parts. To clean the seed from external mycoflora, it is possible to use sulema dissolved in a ratio of 1:1000, as well as formalin solution dissolved in a ratio of 1:300 (for 30 minutes), 1% bromine water (for a few seconds), 2% potassium solution with manganese (for 15 minutes). It is recommended to keep the seed in the prepared solution for the indicated period, and then rinse it several times in sterilized water. When cleaning the seed from external infection is not effective, we recommend to use the method of flaming the seed under investigation using dena turat or technical alcohol for sterilization. Antibiotics (streptomycin) were used to ensure that the isolated fungi were free of bacteria.

In order to isolate fungi, moist chambers prepared in Petri dishes were used. For this purpose, 1 circles made of filter paper are placed in sterilized Petri dishes at 121^0C under a pressure of 1 atm and soaked in sterilized water. The seed was placed in moistened chambers in Petri dishes and stored in desiccators. The temperature in the thermostat should not exceed $27\text{-}30^0$C, the humidity should be 70-80%. After 2-3 days, the growth and development of the fungi growing in it began to be observed under a small lens of a microscope. Some pieces of mycelia, conidia of emerging fungi are planted in the agar nutrient medium in a test tube using a microbiological hook.

The use of moist chambers is the most effective method for isolating fungi parasitizing inside plant seeds.

3 Results and discussion

In the researches, the spread of fusarium disease of wheat in different soil and climatic conditions of the republic and damage to productivity were studied. The experiment was conducted mainly in Tashkent, Syrdarya, Jizzakh, Samarkand, Bukhara and Kashkadarya regions. During the research, the prevalence and damage of Fusarium disease in wheat was studied in the period from 2021 to 2023 (Table 1).

Table 1. Number of plants per 1 m^2 in fields infected with *Fusarium.*

Region	Districts	1 m^2 of healthy plants	1 m^2 plant infected with fusarium	Reduction in the number of plants due to fusarium wilt
Samarkand	Pastdarg'om	430	323	107
	Oqdaryo	390	276	114
	Payariq	412	304	108
Jizzakh	Sharof Rashidov	398	301	97
	Paxtakor	395	289	97
	G'allaorol	350	246	104
Syrdarya	Sardoba	387	283	89
	Guliston	385	275	110
Tashkent	Qibray	403	304	99
	Quyi Chirchiq	391	299	92
Kashkadarya	Yakkabog'	361	244	117
	Kasbi	398	309	89
	SHahrisabz	329	218	111
	Kitob	401	306	95
Bukhara	Buxoro	298	201	97
	Jondor	323	243	120
	Romiton	306	205	101
	Peshku	367	273	94

We are in 6 regions and 18 districts in the central and some southern regions of the republic. We observed the wheat fields of 33 farms and studied the spread of the disease (2021-2023). The disease spread the most in dry wheat crops in Yakkabog and Shahrisabz of Kashkadarya region, Jondor district of Bukhara region, Gallaorol district of Jizzakh region. In the fields where fusarium disease did not occur, the number of spikes per 1 m^2 was the maximum of 430, while in the places where the disease was widespread, the number of spikes per 1 m^2 was 89. In the fields heavily affected by fusarium disease, the number of wheat stalks and the formation of spikes are sharply reduced, as a result of the rotting of the lateral roots of plants infected with fusarium, the plant does not germinate and only one spike is formed from one plant. Therefore, the productivity of plants infected with fusarium disease is sharply reduced (Figure 1).

In the research, the geographical spread of the fusarium disease of wheat in different soil and climatic conditions was studied. In the data in this diagram, in 2021-2023, the disease level of plants during the season of Fusarium disease was studied on a total area of 325 ha in Tashkent, Syrdarya, Jizzakh, Samarkand, Kashkadarya and Bukhara regions. The obtained results show that in 2023, compared to 2021 and 2022 years, the spread of fusarium disease was found in the spring season. Fusarium infection of this wheat is especially common in Kashkadarya and Bukhara regions. In these fields, during the vegetation period of early spring wheat, there are many cases of diseased plants not budding and plants turning yellow compared to healthy plants.

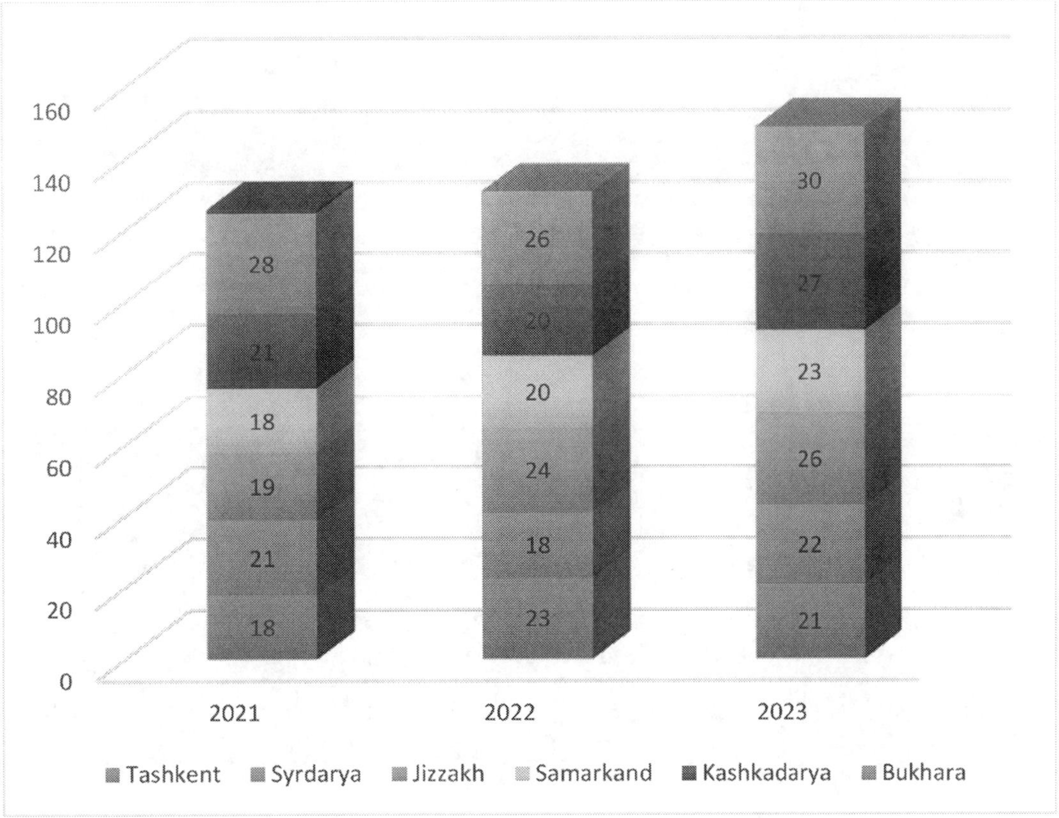

Figure 1. Geographic distribution of wheat fusarium disease in the conditions of Uzbekistan (2021 - 2023).

Wheat plants were brought from fields with disease symptoms and analyzed in the laboratory. In order to identify the pathogenic species of fusarium, which cause diseases in wheat, the roots, stems and leaves of the plant were planted in artificial nutrient media, and pure cultures of fungi belonging to the genus fusarium were isolated and identified.

These pictures show the condition of wheat infected with fusarium during the growing season in the central and southern regions of the Republic. Analysis of fusarium-infected plants in the laboratory conditions of plants affected by fusarium in wheat fields turning yellow and forming a foci, lack of a flowering phase, and plants drying before wheat ripens during the milk ripening period. we can see that the fungi belonging to the fusarium family have evolved (Figure 2).

Manifestation of wheat fusariosis in field conditions

Focal formation of fusarium disease in field conditions

Plants infected with fusarium during the vegetation period

Healthy and fusarium infected plants

Plants infected with fusarium during milk ripening

Plants infected with fusarium during milk ripening

Figure 2. Geographic distribution of wheat fusarium disease in the conditions of Uzbekistan (2021 - 2023).

4 Conclusions

In conclusion, the impact of fusarium disease on wheat cultivation is significant, potentially leading to a substantial yield reduction of up to 50%. To mitigate the detrimental effects of this disease, implementing proactive agrotechnical measures is imperative. This includes adhering to region-specific planting schedules to optimize crop health. Moreover, treating seeds with effective seeders before sowing is crucial to establish a strong foundation for disease resistance.

Furthermore, it is advised to maintain vigilance during the growing season, particularly when symptoms of fusarium disease become apparent. In such instances, timely intervention is key. One recommended approach involves the application of biostimulants to the affected plants. This strategic use of biostimulants can bolster plant health, enhancing their resilience against fusarium disease and potentially mitigating its impact.

By following these recommendations and adopting a comprehensive approach to disease management, farmers can proactively safeguard their wheat crops from the ravages of fusarium disease. Ultimately, the prudent implementation of these measures can contribute to maintaining healthy yields and sustaining the productivity of wheat cultivation in the face of this significant agricultural challenge.

Rerefences

1. Atabayeva H.N., Khudaikulov J.B. "Plantology". "Fan and Technologies" Publishing House, Tashkent: 2018.
2. Atabayeva H.N., Azizov B.M. "Wheat". Monograph, T. ToshdAU, 2008, P-10.5
3. Gagkayeva T.Y., Gavrilova O.P., Levitin M.M., Novozhilov K.V. Fusarium zernovix culture. Applied magazine "Zashita i karantin rastenii", 2011, No. 5, p-70-120.
4. Khaitbayeva N., Sheraliyev A. Fusarium diseases of wheat in saline soils of the Republic of Karakalpakstan and measures to combat them. Monograph ToshdAU, Tashkent, 2020, 130 pages
5. Leslie J.F., Summerell B.A. 2006. The *Fusarium* Laboratory Manual. Ames, Iowa, USA, Blackwell Publishing, 2006, xii + 388 pp.
6. Safarov A.A., Turdiyeva D.T., Hasanov B.A. Fusarium root rot of wheat. Agrochemical protection and plant quarantine, 2020, No. 5, pp. 95-100.
7. Khasanov B.A., Safarov A.A., Turdiyeva D.T. Fuzariozniye corneviye and biological journal, 2020, No. 5, p. 21-32.
8. N.S.Khaytboyeva. Q.B.Bobabekov, N.R.Tillyaxodjayeva, U/X.Raximov The Advantagesof Wheat's Biological Ways of Fighting against Endophit Fungus International Journal of Environmental and Agriculture Research. Volume-8, Issue-11, November 2022, pp 9-12.
9. Boboev, S., Makhkamov, T., Bussmann, R. W., Zafar, M., & Yuldashev, A. (2023). Anatomical and phytochemical studies and ethnomedicinal uses of Colchicum autumnale L. Ethnobotany Research and Applications, 25, 1-9. DOI:10.32859/era.25.6.1-9
10. TKh, M., Brundu, G., Jabborov, A. M., & Gaziev, A. D. (2023). Predicting the potential distribution of Ranunculus sardous (Ranunculaceae), a new alien species in the flora of Uzbekistan and Central Asia. BioInvasions Records, 12(1), 63-77. DOI:10.3391/bir.2023.12.1.05
11. Ameen, M., Zafar, M., Ahmad, M., Ramadan, M. F., Eid, H. F., Makhkamov, T., ... & Majeed, S. (2023). Assessing the Bioenergy Potential of Novel Non-Edible Biomass Resources via Ultrastructural Analysis of Seed Sculpturing Using Microscopic Imaging Visualization. Agronomy, 13(3), 735. DOI:10.3390/agronomy13030735

12. Noor, W., Zafar, M., Ahmad, M., Althobaiti, A. T., Ramadan, M. F., Makhkamov, T., ... & Khan, A. (2023). Petiole micromorphology in Brassicaceous taxa and its potential for accurate taxonomic identification. Flora, 303, 152280. DOI:10.1016/j.flora.2023.152280
13. Aziz, A., Ahmad, M., Zafar, M., Gaafar, A. R. Z., Hodhod, M. S., Sultana, S., ... & Chaudhay, B. (2023). Novel Copper Oxide Phyto-Nanocatalyst Utilized for the Synthesis of Sustainable Biodiesel from Citrullus colocynthis Seed Oil. Processes, 11(6), 1857. DOI:10.3390/pr11061857

Changes in indicators of semen productivity of bulls in the conditions of Uzbekistan depending on age

Sherali Allayorov[1,*] and *Mukhayo* Dosmukhamuedova[1]

[1]Tashkent State Agrarian University, 2, University street, Tashkent, 100140, Uzbekistan

Abstract. The article presents a comprehensive exploration of scientific and practical research focused on assessing alterations in both quantitative and qualitative aspects of sperm production in imported breeding bulls used as sperm donors, in comparison to the local Bushuev breed raised in Uzbekistan. The investigation spans from the age of 3 years, signifying the attainment of full physiological maturation in these bulls. Central to the research are the results detailing shifts in key sperm production indicators. The article goes beyond quantitative measures and delves into various facets of sperm productivity. Alongside these findings, the article addresses pertinent factors related to breed-specific traits. These encompass aspects such as the proportion of sperm culling, the yield of sperm doses, the duration of the bulls' tenure in the primary herd, and other relevant considerations. By conducting this comprehensive study, the article provides a deeper understanding of how sperm production evolves in both imported breeding bulls and the native Bushuev breed after reaching full physiological maturity at the age of 3 years. The analysis of quantitative and qualitative aspects, coupled with insights into breed-specific characteristics, contributes to informed decision-making in the realm of cattle breeding and management practices. Ultimately, this research enhances our understanding of reproductive efficiency and guides efforts to optimize breeding programs for sustainable livestock production.
Keywords. Cattle, breeding bulls, breed, semen, spermatozoon, ejaculate, mobility, concentration, sperm dose.

1 Introduction

It is known from the data on the production of seed products from breeding bulls in cattle breeding that, in accordance with general biological laws, most of the breeding bulls lose their sexual activity 2-3 years after being physiological, and the indicators of seed productivity decrease in terms of quantity and quality. However, this process takes place differently in imported bulls of different breeds due to their specific adaptive characteristics [1-3]. Researching this situation in the conditions of Uzbekistan and, according to its results, organizing seed production with a separate approach to each breed, carrying out large-scale selection and breeding works in the country with the help of artificial

* Corresponding author: sherallayar@gmail.com

insemination, improving the herd of cattle in a short time, and ultimately, milk, meat and other livestock products increasing production indicators is an important and urgent issue . For this purpose, the use of domestic breeding bulls brought to Uzbekistan from foreign countries, especially European countries with sharply different climates, as sperm donors at the state enterprise "Oznaslchilik" , their seed productivity at different ages was studied.

It is known that physiologically trained young breeding bulls give young bulls larger and better quality semen. This condition is associated with an increase in body weight as a result of physiological processes in the body, and the complete formation of sexual organs [1-3].

However, it is also known that the highest indicators of the total number of ejaculates and spermatozoa appear at different ages of bulls of different breeds. According to most scientists who have studied the process of spermatogenesis, it is appropriate to divide spermatogenesis into 3 periods:

- the first, the period of the formation of the first sexual function (from 18-20 months to 4 years), characterized by an increase in the volume of the ejaculate and the concentration of spermatozoa;
- the second, the period characterized by the maintenance of the same level of stability of the quantitative and qualitative indicators of sperm productivity during a certain period (4-10 years);
- the third period, due to the decline of sexual function (after 10 years), a decrease in the quantity and quality of sperm production is observed [4].

However, different researchers have different opinions about the optimal age for obtaining semen from breeding bulls. For example, Yu.V. According to Anbaza (2018), quality seed output from 2-3-year-old breeding bulls is less than that of later ages, which means that their sexual function is still being formed [5].

According to other scientists, the activity of spermatozoa reaches its maximum level at the age of 2-3 years of bulls, and the level of fertilization reaches a high level at the age of 3-4 and remains until the age of 10-12 years, and the concentration of spermatozoa is high at the age of 7-10 years, regardless of the breed, usually begins to decrease . For example, the number of sperm in the ejaculate increases up to 9 years in Simmental bulls, and up to 10 years in Holstein bulls [6-8].

According to Khokhrin and Davletkildina, sperm volume increases until the age of 6, and then decreases until the age of 9-10 [9]. According to other scientists, bulls of the black-olive breed produced 3.5-4 ml of semen in one spurt at the age of 1-2 years, 4.5-5.5 ml at the age of 4-6 years, 8 and 4.5-5 ml in young adults. however, as their age increased, the number of spermatozoa in the ejaculate decreased by 4–6% [10]. Chetvertakova E.V. in the Krasnoyarsk region of Russia. as a result of the research conducted by, the seed productivity of breeding bulls began to decrease after the age of 6 [11].

This opinion is also expressed by other scientists, who associate the increase of seed productivity indicators up to 5-6 years of breeding bulls with their physiological development [12]. As a result of the observations of some scientists, the rate of use of breeding bulls can be increased up to 31.4% at the age of 3 years [13].

The increase in the volume of the ejaculate of breeding bulls is the greatest in their 3-5 years, this change is 25-30% more than in the previous age period, and it remains the same for a long time [14].

It should be noted that, according to the scientists who have studied the use of breeding bulls in quantity and quality and as long as possible, the use of breeding bulls depends not only on their age, it also depends on their health, usage regime and maintenance conditions [15, 16]. The purpose of the study is to determine the age of optimal use of sperm donor breeding bulls under the conditions of Uzbekistan.

2 Materials and methods

The object of the research is breeding bulls belonging to Holstein, Angler, Bushuev, Schwitz and Simmental breeds kept at the state enterprise "Uzhnaslchilik " . In this case, from the time when the bulls were physiologically complete, that is, from the age of 2 to the age of 6, the feeding, storage, seed collection regimes were carried out in bulls that were cared for under the same conditions . During the experiment, according to the results of the evaluation of seed productivity, measures were taken to remove the herd, and the duration of productivity was also recorded during the experiment. The analyzes were compared to bulls belonging to the local Bushuev breed.

Such as semen productivity of breeding spermatozoa bulls: ejaculate volume (ml), sperm concentration (billion/ml) and total number of sperms (billion/ejac) were determined.

The qualitative and quantitative evaluation of sperm products, the International Council for Standardization, Metrology and Certification (MGS) of the Commonwealth of Nations was approved on 01.07.2015 "Instruments. Freshly obtained, undiluted semen of bulls", GOST - 23745-2014 - in accordance with clause 6, clauses 6.3 and 6.4 of the technical conditions of the international standard: ejaculate volume (ml), concentration of spermatozoa (billion/ml) and number of spermatozoa (billion /chin) was detected. Automatic " AndroVision " automatic "AndroVision" equipped with CASA system of the German " MINITUB" corporation , which includes a measuring tube for determining sperm volume, sperm concentration and number of sperm in sperm, and electronic microscope and counting camera for evaluating motility equipment was used .

The data obtained as a result of the research were processed using Mf Office, EXCEL software biometric processed.

3 Results and discussion

The smallest volume of semen obtained per bull during 4 years was 74.4 ± 0.18 ml in the group of bulls belonging to the Angler breed compared to the control group. R > 0.99 was less, while Simmental bulls had 145.4±0.14 ml (R > 0.95), and Schwitz bulls had more 33.4±0.11 ml (R > 0.95).

Table 1. Volume of semen according to the cattle age.

Indicators	Breeds	Age			
		III	IV	V	VI
Number of bulls by breed	Holstein	5	5	4	2
	Angler	5	5	4	3
	Bushuev	5	5	5	4
	Schwitz	5	5	4	3
	Simmental	5	5	4	2
Average total sperm from 1 head, ml	Holstein	451±0.14	481.6±0.16	410.8±0.19	258.4±0.14
	Angler	360±0.21	405.6±0.14	336±0.22	250.8±0.25
	Bushuev	350±0.26	395.2±0.21	374.4±0.28	307.2±0.30

	Schwitz	364±0.18	405±0.20	374.4±0.18	316.8±0.13
	Simmental	421.2±0.20	459.2±0.18	375±0.16	316.8±0.14
Average concentration, billion ml	Holstein	1.08±0.077	1.22±0.052	1.12±0.034	0.84±0.070
	Angler	1.2±0.052	1.60±0.048	1.36±0.028	1.3±0.036
	Bushuev	1.2±0.025	1.25±0.032	1.18±0.035	0.86±0.045
	Schwitz	1.10±0.045	1.32±0.028	1.22±0.028	0.9±0.072
	Simmental	1.06±0.056	1.32±0.036	1.22±0.045	0.9±0.027
The percentage of invalid sperm,%	Holstein	26,20	22.60	25,30	26,20
	Angler	22,27	19,21	21.51	22,27
	Bushuev	21,22	18.31	20.49	21,22
	Schwitz	21.75	18.76	21.00	21.75
	Simmental	22.53	19.44	21.76	22.53
Total sperm motility, score	Holstein	7.2±0.323	7.4±0.245	7.3±0.125	7.2±0.223
	Angler	7.7±0.220	7.9±0.198	7.9±0.184	7.8±0.230
	Bushuev	7.9±0.260	8.1±0.258	8.0±0.225	8.0±0.266
	Schwitz	7.4±0.321	7.6±0.330	7.5±0.254	7.4±0.210
	Simmental	7.6±0.168	7.8±0.180	7.7±0.265	7.7±0.212
On average, spermatozoa prepared from 1 head, thousand	Holstein	29.9±0.2	37.8±24.1	28.6±16.8	13.3±17.1
	Angler	27.9±10.2	43.6±17.8	29.8±9.8	21.1±14.1
	Bushuev	27.5±21.3	33.6±26.8	29.2±22.8	17.3±26.8
	Schwitz	26.1±10.4	36.1±9.8	30.07±13.5	18.5±11.8
	Simmental	28.8±19.5	40.6±15.7	29.8±16.5	18.4±22.8

The concentration of spermatozoa in 1 ml of semen compared to the control group was 0.243 ±0.002 billion pieces (R>0.95) in Angler bulls, 0.013 ±0.002 billion pieces (R>0.98) in Schwitz bulls , 0.002 ±0.001 billion in Simmental bulls. pieces (R>0.9 5) was more, and in Holstein bulls it was 0.06 ±0.002 billion pieces (R>0.9 8) less.

In terms of total sperm motility, the control group outperformed all imported bulls by 0.7 points over the Holstein breed, 0.5 points over the Swiss breed, 0.3 points over the Simmental breed, and 0.15 points over the Angler breed. At the same time, the percentage of unfit sperm from the total volume of sperm in the control group was 21% per year on average for 4 years, compared to them in Holstein bulls 4.7%, Simmental - 1.3%, Angler breed 1.0 % and 0.5% more in the bulls of the Swiss breed, the highest rejection rate was 26% in the Holstein breed, followed by Simmental -21.6%, Angler - 21.3% and 20.8% in the Swiss breed bulls. organized.

Depending on the above indicators, the total sperm doses received from 1 bull during 4 years were 14.8 thousand more than the control group from Angler bulls, 9.9 thousand from

Simmental bulls, 2.02 thousand from Holstein bulls and 3.1 thousand doses from Swiss bulls. The seed is taken.

When the results of Bushuev bulls in the control group were analyzed separately, the number of prepared spermatozoa was lower than the general average level for all indicators.

In the main herd used for seed production, breeding bulls of the local Bushuev breed undoubtedly have an advantage in terms of preservation, 80% of them, 60% of the Angler and Schwitz breeds, and only 40% of the Holstein and Simmental breeds are the main herd from which products are obtained up to 6 years of age. preserved in the composition.

Table 2. Volume of semen according to the cattle age of Bushuev breed.

Indicators	Breed groups	Age				
		III	IV	V	VI	Total
Average number of ejaculates from 1 head	Holstein	82	86	79	68	315.0
	Angler	72	78	70	66	286.0
	Bushuev	70	76	72	64	282.0
	Schwitz	70	75	72	66	283.0
	Simmental	78	82	75	66	301.0
Average total sperm from 1 head, ml	Holstein	451.0	481.6	410.8	258.4	1601.8
	Angler	360.0	405.6	336.0	250.8	1352.4
	Bushuev	350.0	395.2	374.4	307.2	1426.8
	Schwitz	364.0	405.0	374.4	316.8	1460.2
	Simmental	421.2	459.2	375.0	316.8	1572.2

4 Conclusions and recommendations

1. In the conditions of Uzbekistan, for the use of breeding bulls brought from abroad, the optimal period of economic use of bulls is the period of their 3-4 years of age, during which it is possible to prepare seed of maximum size and quality from them;

2. Seed productivity of breeding bulls brought to Uzbekistan from abroad, mainly from Europe, fully meets the requirements in terms of quantity and quality, however, the duration of obtaining high-quality products from them is relatively short, that is, for various reasons, it was observed that they leave the main herd before the age of 6 years. In particular, breeding bulls belonging to highly cultured breeds such as Holstein and Simmental in the experiment are observed to leave the main herd until the age of 6 years.

3. Breeding bulls of the Bushuev breed do not lag behind imported breeding bulls in most indicators of semen productivity, however, due to their relatively small live weight, fewer spermatozoa were prepared from them in exchange for the volume and number of ejaculates;

4. The quality of the seed product processed by the method of cryopreservation of breeding bulls belonging to the Bushuev breed does not differ from the quality of the seeds of other widely used breeds, and the use of the modern cryopreservation method can achieve high results in preserving the genetic resources of this breed and increasing its population ;

5. Depending on the urgency of the issue of increasing the population of Bushuev cattle, in the conditions of Uzbekistan, in contrast to the breeding bulls brought from abroad, it is possible to prepare high-quality seed from these local breeding bulls even after their 6 years of age, based on a light loading of each bull according to a separate plan;

References

1. Brito, LF Effect of age and genetic group on characteristics of the scrotum, testes and testicular vascular cones, and on sperm production and semen quality in AI bulls in Brazil / LF Brito, AE Silva, LH Rodrigues LH, FV Vieira et al. // Theriogenology. – 2021. Vol. 58. - p. - 1175-1186.
2. Fuerst-Waltl, B. Effects of age and environmental factors on semen production and semen quality of Austrian simmental bulls / B. Fuerst-Waltl, H. Schwarzenbacher, C. Perner, J. Sölkner // Animal Reproduction Science . - 2006. - Vol. 95. - p. 27–37.
3. Murphy, EM Influence of bull age, ejaculate number, and season of collection on semen production and sperm motility parameters in Holstein Friesian bulls in a commercial artificial insemination center / EM Murphy, A.K. Kelly, CO Meara, B. Eivers // Journal of Animal Science. - 2018. - Vol. 96 (6). - p. 2408-2418.
4. Markushin A.P. The term of use of agricultural products. - M.: Rosselkhozizdat, 2013. 157 p.
5. Anbaza, Yu.V. Vliyanie vozrasta na nativnuyu spermoproduktsiyu bykov-spermodonorov / Yu.V. Anbaza // Nauchnoe obespechenie jivotnovodstva Sibiri. Materialy II mejdunar. scientist and practitioner. conference. - Krasnoyarsk. - 2018. - p. 50-54.
6. Siratsky I.Z. Zavisimost kachestva spermoproduktsii ot porody i vozrasta bykov-proizvoditeley // Molochnoe i myasnoe skotovodstvo. – 2002. – No. 6. - p. 28.
7. Siratsky I.Z. Izmenchivost i nasledstvennaya obuslovlennost vosproizvoditelnoy kompetnost bykov-proizvoditeley simmentalskoy porody// Tsitologiya i genetika. - 1993. - v. 27. – #6. – P.45–52.
8. Mymrin V.S. Vosproizvoditelnye kachestva bykov holshtinskoy porody // Zootechniya. – 2011. – No. 3. - p. 55–57.
9. Khokhrin, S.N. , D.V. Davletkildina / Vliyanie vozrasta na kachestvo spermy bykov raznyx genotipov // Rol molodyx uchenyx v reshenii aktualnyx zadach APK. - 2017. - p. 180-183.
10. Basovsky N.Z., Zavertyaev B.P. Selection of cattle with the ability to produce. - M.: Rosselkhozizdat, 2005. - 144 p.
11. Chevertakova, E.V. Ekologo-geneticheskie aspekty realizatsii reproduktivenogo potensiala bykov-spermodonorov. Monograph / E.V. Chetvertakova. – Krasnoyarsk: [p. i.], 2009. – 188 p.
12. Gavrikov, A.M. Metodicheskie ukazaniya po iskusstvennomu osemeneniyu korov i telok / A.M. Gavrikov, V.I. Lebedev, V.P. Belonojkin, T.E. Taradaynik and dr. - M.: Shcherbinskaya Typography, 2009. - 40 p.
13. Zaripov, F.R. Otsenka bykov po vosproizvoditelnoy obscenity / F.R. Zaripov // Uchenye zapiski Kazanskogo GAVM. - 2011. - T. 208. - p. 402-406.
14. Sharygina, L.N. (2010) Sravnitelnaya charakteristicsika bykov ayrshirskoy and holshtinskoy porod po spermoproduktsii. Abstract diss. candidate s.x. Nauk / L.N. Sharygina. - M.: [p. i.], 20015.
15. Ilev F.V. Plemennye kachestva selskohozyaystvennyx jivotnyx.– Chisinau: Kortya Moldovenya SKE, 2021. – 102 p.

16. Volkova S.V., Alifanov V.V., Alifanov S.V. Vliyanie vozrasta bykov i vremeni goda na kachestvo spermy//Sovremennye problemy nauki i obrazovaniya. – 2008. – No. 6. - S.5. 3. – 23 p.

Live weight and exterior characteristics of lambs obtained from inbreeding F2 hybrid sheep of Hisar and native breeds

Khonimoy Berdiyeva[1,*] and *Nuraddin* Ruziboyev[1]

[1]Research Institute of Livestock and Poultry Farming, Tashkent, 100034, Uzbekistan

Abstract. In this article, live weight and body size parameters of hybrid lambs obtained from inbreeding of F2 hybrid sheep with Hisar and native breed genotypes were higher than those of pure native breed lambs. These obtained results showed that improvement of selection traits of native breed sheep using Hisar breed rams is of great scientific and practical value. The live weight of the hybrid rams obtained from inbreeding was 5.62 kg at birth, and the females were 5.17 kg, respectively. In group I, the live weight was 0.45 kg, or 8.7%, and 0.34 kg, or 7.2% higher than the live weight of purebred heifers. However, the average live weight of experimental hybrid lambs at birth was 5.37 kg. It was found that the average live weight of pure native lambs was 0.41 kg or 8.3% higher. It should be noted that in experimental group II, at the age of 12 months, crossbred lambs were higher in body size than pure native lambs. In this case, hybrid frames in experimental group II compared to those of equal frames in control group I, respectively, 1.46 cm or 2.07%; 2.4 cm or 3.43%; 0.61 cm or 0.97%; 0.35 cm or 2.15%; 0.45 cm or 2.16%; 1.2 cm or 1.51% and can be seen to be 0.03 cm higher.

Keywords. Sheep, lamb, breed, hybrid, hisar, native, live weight, ramcha, selection.

1 Introduction

Sheep breeding is one of the leading sectors in the livestock sector, 30% of the total meat supplied in Uzbekistan is obtained from sheep breeding, which requires the development of this sector on a more scientific basis [1]. For this purpose, it is important to take care of the sheep at a reasonable level, to feed them with full value, and to further improve the breeding work. From sheep, human food (meat and fat), industrial wool, skin, fur, and medical products are obtained [2].

The following sheep breeds are bred in Uzbekistan: native, Karakol, Hisar, semi-fine wool-yielding meat-serjun and partially Adilboy sheep breeds. Breeding of these sheep is carried out in the following regions - in the mountain, subalpine and semi-desert regions, mainly meat-fatty breeds are bred, and in the desert and semi-desert regions, the Karakol sheep breed is bred [3, 4].

* Corresponding author: kh.berdiyeva1001@mail.com

The native sheep breed is a breed created in our country through folk selection for several thousand years, it is well adapted to breeding in the mountainous and sub-mountainous regions of Uzbekistan, and it is considered one of the leading breeds in terms of head number. In Uzbekistan today, the number of native and native breed hybrids is about 10 million [5]. Today, in the Kashkadarya, Surkhandarya, and Bukhara regions of Uzbekistan, pure native, as well as Karakol and native hybrids, pure native, Hisar and native hybrids are being bred in Tashkent, Samarkand, Syrdarya, Jizzakh, and Vadi regions.

Native breed sheep are distinguished from other sheep breeds by being well adapted to the conditions of the outdoor environment (rainy, snowy and wet and cold) [6]. Sheep of this breed have the characteristics of bringing out their high genetic potential in terms of productivity under good grazing conditions [7].

According to the opinions of many authors, native breed sheep are called Uzbek national sheep, and these sheep are sheep that are well adapted to the climatic conditions of mountain and sub-mountain regions [8]. The native breed of Tailfat sheep is described as being distinguished by its extreme endurance, solidity of constitution, and good quickness. Sheep are kept in pastures throughout the year. In terms of meat and fat productivity, they have a lower index than the Hisar breed. One of the distinctive features of sheep of this breed is the low legs and elongated body [9]. Most sheep have long foreheads, heads and ears. Most of the sheep are black, but sometimes brown, orange and gray sheep are also found. The live weight of cows is up to 75 kg, rams - 100 kg and more. Wool is sheared once a year, and the average wool shearing is used for blankets - 2.0-3.0 kg, rams - 3.0-4.0 kg, coarse wool, coarse cloth, felt and other purposes [10].

Native breed sheep are distinguished from other sheep breeds by being well adapted to the conditions of the outdoor environment (rainy, snowy and wet and cold). Sheep of this breed have characteristics of revealing their high genetic potential in terms of productivity in good pasture conditions [11].

Native breed sheep belong to aboriginal breeds that have been created by folk selection in the mountains, highlands and hills for thousands of years, and its Persian translation means native, and they are intensively bred in the northern regions of Tajikistan [12].

In the Central Asian countries, the contribution of the Hisar breed of sheep to the improvement of the selection characteristics of tailfat sheep, to the creation of new types and breeds is considered to be very large. For example, using Hisar sheep, Tajik meat-fatty-Serjun sheep breed was created in Tajikistan, Ordabin breed in Kazakhstan, Aktyubinsky sheep with semi-fine wool, Aivyakol breed and Merino breed in Kyrgyzstan. In Uzbekistan-Usturt meat-fat-wool sheep, therefore, the Andijan type of sheep with similar characteristics was created [13].

Using the genetic potential of Hisar and native sheep, creating their productive genotypes, genetically improving productivity characteristics, establishing productive breeding flocks of the breed, raising high-yielding sheep in terms of meat productivity, developing methods for improving the selection characteristics of the offspring obtained from them is important scientific and practical. is important. Also, it is of urgent importance to crossbreed the native breed using Hisar breed rams and to improve the selection characteristics through continuous selection of the offspring obtained from them.

One of the most important selection characteristics of sheep is the live weight at birth, which affects the future growth and development of sheep, that is, lambs born with a high live weight retain a high advantage under optimal storage and feeding conditions [11].

The daily feed ration of native breed rams fed for 75 days is 1.0 kg of cotton seed meal, 1.0 kg of grass hay, 0.5 concentrate feed and 15 g of table salt, the meat and fat yield of lambs fed with fat in a stable barn, showed that it is much higher compared to tenkurs fed on pastures, that is, their slaughter yield is 52.95% and the amount of meat is 26.03%.

According to the selection results, the live weight of the best sheep compared to the general flock was 7-10 kg (12-18%) higher in ewes and 28 kg (33%) in rams. The authors noted the importance of studying the level of nutrition, breeding conditions and meat and fat productivity in the development of ways and methods of further improvement of breeding work in Hisar sheep.

The live weight of ewes with 1/2 x native ewes of Hisar breed is 4.22 kg (6.8%) higher than that of pure native ewes, while the live weight of mature native ewes is 11.66 kg (21.2%) was higher. These results are important for the improvement of the breeding characteristics of native breed sheep by their genotype selection. Similar conclusions were reached by [4, 7, 8].

Crossbred lambs have good meat qualities under pasture conditions, large body, strong constitution, well-developed body structure, good growth, satisfactory fatness, sufficient milk yield and high meat product coverage.

Tailfat is a genetic trait of this sheep breed and is considered one of the main "observable" indicators and serves as a natural nutrient reserve in unfavorable feeding conditions, mainly during severe cold periods of winter. At the moment, Edilboy sheep have the highest meat and fat productivity among other sheep breeds, and coarse wool serves to improve sheep in terms of meat-fat productivity [1, 2, 9].

2 Materials and methods

The practical phase of the study unfolded across the span of 2021 to 2023, unfolding within the premises of the distinguished "Kholto'raev Oybek KHM" breeding farm, an enclave dedicated to the cultivation of sheep in the heart of the Ohangaron district within the Tashkent region. The overarching objective of this phase was to intimately probe the dynamics within the realm of lamb rearing, a pivotal component of the sheep breeding domain.

The scope of the research encompassed two distinct groups of lambs, each with a distinctive lineage. Group I was comprised of lambs of pure native breed ancestry, showcasing an unadulterated genealogy. In parallel, Group II showcased a more complex lineage, involving hybrid lambs, a synthesis of F2 joint breeds—3/4 of the distinguished hisar breed harmoniously entwined with 1/4 of the indigenous native breed.

The meticulous selection process ensured a well-balanced representation of 46 lambs from the pristine native breed heritage for Group I. Correspondingly, Group II featured 49 hybrid lambs, each embodying the intricate genetic dance between hisar and native breeds in their F2 makeup.

This empirical treasure trove of data was meticulously subjected to the rigors of variational statistics, anchored by the scholarly insights of E.K. Merkureva, circa 1970. The statistical toolkit revealed a wealth of information, rendering the average indicators (X) of each group, fortified by their corresponding errors (Sx), in a symphony of numerical precision. The essence of variability (Cv) was distilled, offering a snapshot of the subtle dynamics that shape the realm of sheep breeding.

Further fortifying the validity of the findings, the criteria of comparative indicators were unwrapped with the discerning lens of reliability (td,P). This analytical framework lent credence to the conclusions drawn from the study, rendering a nuanced understanding of the contrasting attributes that set Group I and Group II apart.

In essence, the practical segment of the research, a span extending across two years, heralded a deep dive into the realm of lamb rearing, illuminating the juxtaposition between purebred native lambs and their hybrid counterparts. Guided by the beacon of statistical rigor, this exploration furnished a comprehensive understanding of the dynamics within this intricate world, a testament to the dedication of scholars to uncover the intricate tapestry woven within the domain of sheep breeding.

3 Results and discussion

Live weight indicators of animals are one of the most important growth indicators of the organism. The live weight of sheep at birth affects their future development and is one of the most important selection characteristics and is one of the indicators that determine the main efficiency. The growth and development of different sheep breeds are different, and they show different growth rates even under the same conditions, depending on the animal breed, fertility characteristics, direction of productivity, adaptability to natural and climatic conditions, and breeding periods.

The live weight of lambs, the total body size, is manifested by the enlargement of individual organs, which is called the growth of the body of the lambs and undergoes quantitative changes. Farm animals differ in live weight, size and growth of body parts. In the practice of zootechnics, the increase in the live weight of animals and the growth of body parts are determined based on the measurement of the body dimensions obtained from the body parts.

The development of the organism refers to the totality of morphological and physiological changes that occur in the animal organism throughout its life. One of the most important growth indicators of organicism is the live weight of the animal.

Growth and development in different breeds of sheep are different, depending on the animal breed, fertility characteristics, direction of production, adaptability to natural and climatic conditions, and methods of inbreeding, they show different growth rates even under the same storage and feeding conditions.

Taking this into account, we analyzed the live weight indicators of pure native breed and hybrid lambs by periods, the results of which are presented in Table 1.

Table 1. Live weight indicators of purebred and hybrid lambs by periods, kg.

Breed		Gender						Average live weight of lambs		
		Ram			Ewe					
		n	$\overline{X} \pm S\bar{x}$	C_v,%	n	$\overline{X} \pm S\bar{x}$	C_v,%	n	$\overline{X} \pm S\bar{x}$	C_v,%
Newborn										
I	Lambs of pure native breed	24	5.17±0.06	5.88	22	4.74±0.13	12.58	46	4.96±0.08	7.23
II	Hybrids from inbreeding	26	5.62±0.10***	9.47	23	5.08±0.07*	6.79	49	5.37±0.07***	6.30
1 month										
I	Lambs of pure native breed	24	17.05±0.27	7.72	22	16.46±0.33	9.40	46	16.77±0.24	6.85

 https://doi.org/10.1051/e3sconf/202342103006

II	Hybrids from inbreeding	26	18.64± 0.32 ***	8.74	23	17.7± 0.38*	10.36	49	18.20± 0.24***	6.41
3 months										
I	Lambs of pure native breed	24	33.06± 0.55	8.11	22	31.93+ 0.60	8.75	46	32.52± 0.42	6.19
II	Hybrids from inbreeding	26	35.60± 0.40 ***	5.72	23	34.03± 0.58*	8.16	49	34.86± 0.41***	5.79
5 months										
I	Lambs of pure native breed	24	41.26± 0.63	7.46	22	39.50+ 0.93	7.11	46	40.42± 0.47	5.56
II	Hybrids from inbreeding	26	44.52± 0.55 ***	6.28	23	41.65± 0.63*	7.22	49	43.17± 0.50***	5.71
8 months										
I	Lambs of pure native breed	24	45.50± 0.51	5.44	22	43.0± 0.47	5.11	46	44.30± 0.39	4.26
II	Hybrids from inbreeding	26	48.84± 0.46 ***	4.79	23	45.16± 0.62 **	6.59	49	447.11± 0.38***	3.99
12 months										
I	Lambs of pure native breed	24	51.40± 0.95	9.02	22	47.2± 0.52	5.21	46	49.39± 0.56	5.44
II	Hybrids from inbreeding	26	55.25± 1.04 **	9.67	23	50.0± 0.53 ***	5.09	49	52.78± 0.58***	5.37
18 months										

I	Lambs of pure native breed	24	64.80± 1.10	8.33	22	57.31± 0.87	7.14	46	61.22± 0.49	3.81
II	Hybrids from inbreeding	26	70.42± 1.09 ***	7.90	23	61.80± 0.76 ***	5.94	49	66.37± 0.55***	4.06
Note: * P>0.95; ** P>0.99; *** P>0.999										

As can be seen from the table, the average live weight at birth of hybrid rams obtained from inbreeding in group II was 5.62 kg, and the weight of females at the corresponding age was 5.17 kg, which is 0.45 kg or 8.7% compared to the live weight of equals born from pure native breeds in group I. (P>0.999) and was higher by 0.34 kg or 7.2% (P>0.95). At the same time, the average live weight of experimental hybrid lambs at birth was 5.37 kg, which was 0.41 kg or 8.3% (P>0.999) higher than the average live weight of pure native lambs.

At the age of one month, crossbred lambs achieved a higher live weight than pure native breed lambs. For example, at the age of one month, crossbred rams-18.64 kg and females-17.7 kg compared to pure native breed rams 1.59 kg or 8.9% (P>0.999) and 1.24 kg or 7, It was higher by 5% (P>0.95). It should be noted that in our research, one-month-old purebred lambs in group I had an average live weight of 16.77 kg, which was 1.43 kg or 8.5% (P>0.999) less than the average live weight of the equivalents in group II. were

In experimental groups, at the age of 3 months of lambs, the live weight of hybrid rams was 35.60 kg, 2.54 kg or 7.7% (P>0.999) than that of pure native breed lambs, in female lambs these indicators were 34.03 kg, compared to pure native breed lambs. 2.1 kg or 6.6% (P>0.95) was higher. In studies, regardless of the sex of lambs, a higher result in terms of average live weight at the age of three months was observed in hybrids. For example, it was found that the average live weight of hybrid lambs at 3 months of age was 34.86 kg, which was 2.34 kg or 7.2% (P>0.999) higher than the average live weight of pure native breed lambs.

At 5 months of age, group II crossbred rams had a live weight of 44.52 kg and female crossbreds 41.65 kg, 3.26 or 7.9% (P>0.999) compared to group I equivalent rams and 2.15 kg compared to female lambs. or 5.4% (P>0.95) was higher. When analyzing the average live weight of lambs in both groups in these months, it was noted that the average live weight of crossbred lambs was 43.17 kg, which was 2.75 kg or 6.8% (P>0.999) higher than the average live weight of purebred equine lambs.

In our research, when we studied the live weight of lambs at the age of 8 months, we can see that hybrids obtained from inbreeding have a higher result. For example, at the age of 8 months, the live weight of crossbred lambs was 48.84 kg, 3.34 kg (7.3%, P>0.999) more than that of pure native breed lambs, and 45.16 kg in female lambs, 2.16 more than that of pure native breed lambs. kg (5.0%, P>0.99) was higher.

The average live weight of lambs in all groups, regardless of gender and breed, was determined and analyzed. According to his results, lambs of pure native breed at the age of 8 months in group I had an average live weight of 44.30 kg, which was characterized by an average live weight of 2.81 kg (6.3%, P>0.999) less than the average live weight of equals in experimental group II.

A higher advantage in live weight was observed in lambs from inbreeding, where crossbred rams had an average live weight of 55.25 kg at 12 months of age, compared to pure

 https://doi.org/10.1051/e3sconf/202342103006

native-bred equal rams of 3.85 kg (7.5%, P>0.99), females belonging to 50.0 kg, 2.8 kg (5.9%, P>0.999) higher than pure native breed females. It should be noted that when we determined the average live weight indicators of lambs in groups, it was found that the live weight of hybrid lambs at the age of 12 months was higher. For example, the average live weight of crossbred lambs was found to be 52.78 kg, 3.39 kg (6.8%, P>0.999) higher than the average live weight of pure native lambs.

Even at the age of 1.5 years, crossbred lambs had a higher live weight than pure native breed lambs. For example, at 1.5 years of age, crossbred rams were 70.42 kg and females were 61.8 kg, compared to pure native rams by 5.62 kg or 8.7% (P>0.999) and 4.49 kg or 7 was higher by .8% (P>0.999). In all experimental groups, regardless of the breed and breed of lambs, according to the results of determining their live weight at the age of 18 months, the highest indicators were shown in hybrids. In particular, the average live weight of crossbred lambs was 66.37 kg, which was 5.15 kg or 8.4% (P>0.999) higher than that of pure native breed lambs.

We studied the body dimensions of purebred and crossbred lambs, the results of which are presented in Table 2.

Table 2. Exterior characteristics of purebred and hybrid lambs

Genotype	**Gender**	**n**	**Loin height**	**Rump height**	**Body oblique length**	**Chest width**	**Chest depth**	**Chest circumference**	**Poll circumference**
Lambs of pure native breed	**1 month**								
	Ram	7	49.54 ±0.26	50.52 ±0.25	42.57 +0.24	14.04 +0.11	17.75 ±0.11	54.04 ±0.32	7.11± 0.055
	Ewe	8	49.11 ±0.25	50.21 ±0.26	42.21 +0.21	13.81 +0.11	17.21 ±0.11	53.81 ±0.32	7.07± 0.037
	5 months								
	Ram	6	67.31 ±0.28	68.11 ±0.39	58.33 +0.43	15.61 +0.25	19.57 ±0.17	76.02 ±0.35	7.42± 0.084
	Ewe	8	66.52 ±0.29	67.21 ±0.21	57.56 +0.20	15.32 +0.06	18.17 ±0.07	74.82 ±0.33	7.33± 0.077
	8 months								
	Ram	8	69.12 ±0.31	70.01 ±0.35	59.75 +0.20	15.81 +0.17	19.61 +0.12	77.62 ±0.41	7.61± 0.077
	Ewe	9	68.65 ±0.31	69.46 ±0.31	59.31 +0.19	15.71 +0.20	19.45 ±0.26	77.11 ±0.34	7.56± 0.060
	12 months								
	Ram	8	70.35 ±0.37	70.01 ±0.37	62.61 +0.48	16.26 +0.24	20.75 ±0.31	79.15 ±0.47	7.71± 0.058
	Ewe	8	69.42 ±0.34	70.41 ±0.34	60.05 +0.49	16.01 + 0.18	19.71 ±0.30	78.65 ±0.49	7.66± 0.045
Hybrids from inbreeding	**1 month**								
	Ram	8	50.52 ±0.29 **	51.41 ±0.27 *	44.01 ±0.26 ***	14.41 +0.10 ***	18.15 ±0.10 ***	55.5± 0.326 ***	7.15± 0.042
	Ewe	10	50.04 ±0.23	51.11 ±0.23 **	43.81 ±0.16 ***	14.01 +0.08 *	17.45 ±0.08 *	54.61 ±0.25 *	7.11± 0.039
	5 months								

	Ram	6	68.81 ±0.29 **	69.53 ±0.31 **	59.45 ±0.42 *	15.81 +0.29	19.65 ±0.17	77.42 ±0.35 **	7.51± 0.074
	Ewe	10	67.51 ±0.25 *	68.25 ±0.27 **	58.45 ±0.28 **	15.66 +0.06	18.31 ±0.08	76.32 ±0.39 **	7.40± 0.069
	8 months								
	Ram	8	70.11 ±0.29 *	70.92 ±0.29 *	60.63 ±0.22 **	16.26 +0.17 *	19.92 ±0.13	78.68 ±0.47	7.67± 0.073
	Ewe	9	69.48 ±0.31 *	70.35 ±0.31 *	60.07 ±0.20 **	15.91 + 0.21 *	19.71 ±0.26	78.21 ±0.45	7.61± 0.068
	12 months								
	Ram	8	71.81 ±0.38 **	72.41 ±0.40 **	63.22 ±0.42 **	16.61 +0.24 **	21.2± 0.287	80.35 ±0.47	7.74± 0.042
	Ewe	8	70.55 ±0.37 **	71.51 ±0.36 **	61.65 ±0.45 **	16.68 + 0.22 **	20.48 ±0.33	80.01 ±0.51	7.70± 0.052
Note: * P>0.95; ** P>0.99; *** P>0.999									

The analysis of the table data in Table 2 shows that the body size of one-month-old crossbred lambs obtained from inbreeding was found to be higher than the body size of lambs born from pure native breed lambs. For example, group II hybrid frames have a loin height of 50.52 cm, rump height of 51.41 cm, body oblique length of 44.01 cm, chest circumference of 14.41 cm, chest depth of 18.15 cm, chest circumference of 55.5 cm and poll girth is 7.15 cm and 0.98 cm (P>0.99) compared to pure native rams; 0.89 cm (P>0.95); 1.44 cm (P>0.999); 0.37 cm; 0.40 cm; 1.46 cm (P>0.999) and 0.04 cm higher. However, at this age, Ewe crossbred lambs have loin height of 50.04 cm, rump height of 51.11 cm, body oblique length of 43.81 cm, chest circumference of 14.01 cm, chest depth of 17.45 cm, chest circumference of 54.61 cm and poll circumference is 7.11 cm, 0.93 cm compared to pure native breed Ewes respectively (P>0.99); 0.9 cm (P>0.99); 1.6 cm (P>0.999); 0.20 cm; 0.24 cm; 0.8 cm (P>0.95) and 0.04 cm higher. F2 lambs of experimental group II had higher body dimensions at the age of 5 months compared to their counterparts born from purebred lambs. In particular, at the age of 5 months, the loin height of the rams of the F2 breed is 68.81 cm, the rump height is 69.53 cm, the oblique length of the body is 59.45 cm, the chest width is 15.8 cm, the chest depth is 19.65 cm, and the chest circumference is 77.42 cm and poll circumference is 7.61 cm, compared to the body dimensions of pure native breed rams in group I, respectively, loin height-1.5 cm (P>0.999), rump height-1.42 (P>0.99) cm, body oblique length-1.12 cm (P>0.95), chest width-0.2 cm, chest depth-0.08 cm, chest circumference-1.4 cm (P>0.99) and poll circumference -0 .09 cm higher. In experimental group II, crossbred Ewes at this age had loin height of 66.52 cm, rump height of 67.21 cm, body oblique length of 57.56 cm, chest width of 15.32 cm, chest depth of 18.17 cm, chest circumference of 74.82 cm and the poll circumference is 7.33 cm, compared to the body dimensions of pure native Ewe equals of group I, respectively, loin height-0.99 cm (P>0.95), rump height-1.04 cm (P>0.99) , body oblique length-0.89 cm (P>0.99), chest width-0.34 cm, chest depth-0.14 cm, chest circumference-1.5 cm (P>0.999) and poll circumference-0 .07 cm higher.

At the age of 8 months, rams born from inbreeding crossbreeds have a loin height of 70.11 cm, a rump height of 70.92 cm, a lean body length of 60.63 cm, a chest width of 16.26 cm, a chest depth of 19.92 cm, a chest circumference of 78.68 cm and poll circumference is 7.67 cm, loin height-0.99 cm or 1.43% (P>0.95) of these indicators of purebred rams, respectively; rump height-0.91 cm or 1.30% (P>0.95); oblique body length-0.88 cm or 1.47% (P>0.99); chest width- 0.45 cm or 2.84% (P>0.95); chest depth - 0.31 cm or 1.58%; breast circumference-1.06 cm or 1.36% and poll circumference was found to be 0.06 cm higher.

In the table data, when we determined the body dimensions of crossbred Ewe lambs of this age and group, they were characterized by a higher result than that of purebred lambs of this age. For example, the loin height of hybrid Ewes is 69.48 cm on average, compared to that of pure Ewes by 0.83 cm or 1.21% (P>0.95); respectively, rump height 70.35-0.89 cm or 1.28% (P>0.95); oblique body length-60.07 cm-0.76 cm or 1.28% (P>0.99); chest width-15.91 cm-0.2 cm or 1.27%; chest depth-19.71 cm-0.26 cm or 1.33%; chest circumference-78.21 cm-1.1 cm or 1.42% (P>0.95) and poll circumference was found to be 0.05 cm higher.

At the age of 12 months, the loin height of the hybrid rams of the F2 breed in experimental group II is 71.81 cm, and the height of the pure native breed rams is 70.35 cm, respectively, the rump height is 72.41 cm; oblique length of the body 63.22 cm; chest width 16.61 cm; chest depth 21.2 cm; chest circumference was 80.35 cm and poll circumference was equal to 7.74 cm, respectively 1.46 cm or 2.07% (P>0.99) compared to that of the control group. 2.4 cm or 3.43%; 0.61 cm or 0.97%; 0.35 cm or 2.15%; 0.45 cm or 2.16%; 1.2 cm or 1.51% (P>0.99) and 0.03 cm higher.

It should be noted that in experimental group II, at the age of 12 months, cross-bred Ewe lambs were higher in body size than pure native-bred equine Ewe lambs. For example, crossbred Ewes have loin height of 70.55 cm, rump height of 71.51 cm, body oblique length of 61.65 cm, chest width of 16.68 cm, chest depth of 20.48 cm, chest circumference of 80.01 cm and poll circumference of 7 .70 cm, compared to the body size of pure native Ewe equals of group I, respectively, loin height-1.13 cm or 1.62% (P>0.99); rump height-1.1 cm or 1.56% (P>0.95); oblique length of the body-1.6 cm or 2.66% cm (P>0.99); chest width-0.67 cm or 4.18% (P>0.99); chest depth-0.77 cm or 3.9%; chest circumference was 1.36 cm or 1.73% (P>0.95) and poll circumference was 0.04 cm higher.

4 Conclusions

Thus, the results obtained from the research were observed in lambs obtained from inbreeding of crossbred sheep with high F2 joint at the age of one month, 3, 5, 8, 12 and 18 months at birth, irrespective of the sex and age of the lambs, and the performance indicators of these lambs in terms of live weight, indicates that they depend on the genotype.

The results of measuring the body size of lambs in the research and their analysis show that in all groups, regardless of age and sex, lambs born from inbreeding of hybrid lambs of the F2 breed achieved high results in terms of body size. It can be concluded that regardless of age and sex, external characteristics of lambs depend on their genotype.

References

1. Kim, Y. S., Tseveen, K., Batsukh, B., Seong, J., & Kong, H. S. (2020). Origin-related study of genetic diversity and heteroplasmy of Mongolian sheep (Ovis arie) using mitochondrial DNA. Journal of Animal Reproduction and Biotechnology, 35(2), 198-206.
2. Mehetre, S. S., Rajput, H. J., & Shinde, G. C. (2003). Genetic analysis for seed cotton yield and its components in Gossypium hirsutum. Journal of Cotton Research and Development, 17, 2.

3. Withoeft, J. A., Da Costa, L. S., Marian, L., Baumbach, L. F., Do Canto Olegário, J., Miletti, L. C., ... & Casagrande, R. A. (2022). Microcephaly and hydrocephalus in a sheep fetus infected with Neospora caninum in Southern Brazil–Short communication. Acta Veterinaria Hungarica.
4. Márquez-Quiroz, C., López-Espinosa, S. T., Sánchez-Chávez, E., García-Bañuelos, M. L., la Cruz-Lázaro, D., & Reyes-Carrillo, J. L. (2014). Effect of vermicompost tea on yield and nitrate reductase enzyme activity in saladette tomato. Journal of soil science and plant nutrition, 14(1), 223-231.
5. Saxena, K. B., Kumar, R. V., & Rao, P. V. (2002). Pigeonpea nutrition and its improvement. Journal of Crop Production, 5(1-2), 227-260.
6. Choudhary, A., Yadav, S. R., & Parewa, H. P. (2019). Effect of wool waste in combination with farm yard manure and fertilizer on soil properties in aridisol of Bikaner, Rajasthan, India. Journal of Environmental Biology, 40(5), 1067-1072.
7. Rather, M. A., Shanaz, S., Ganai, N., & Hamadani, A. (2020). Status of farm animal genetic resources of Jammu and Kashmir-A Review. International Journal of Livestock Research, 10(4), 27.
8. Bhat, P. N., & Yadav, M. P. (2018). Animal Husbandry Research, Education and Development. Scientific Publishers.
9. Pandita, S., Verma, A., Kumar, R., Chander, Y., Yadav, D., Barua, S., & Kumar, N. (2023). miRNA profiling of lumpy skin disease virus infected primary lamb testicle cells.
10. Purohit, G. N. (2013). Fetal Complications of Gestation in the Buffalo: Etiology, Antenatal Diagnosis and Management. Bubaline Theriogenology A, 5712.
11. Kaur, R., Shilpa, S., Kumar, K., & Sharma, N. (2016). Genomics in agriculture. CABI Reviews, (2015), 1-25.
12. Arya, R. L., Arya, S., Arya, R., & Kumar, J. (2015). Fundamentals of Agriculture (ICAR-NET, JRF, SRF, CSIR-NET, UPSC & IFS). Scientific Publishers.
13. de Souza, J. C., Malhado, C. H., Ramos, A. A., Mota, M. F., Jorge, A. M., de Freitas, J. A., ... & Lamberson, W. R. (2010). Age at first calving and age of dam effect on daily gain (milking phase) and weaning weight in Brazilian Water Buffalo. Revista Veterinaria, 21(1).

Land use and land cover change dynamics of Uzbekistan: a review

Yunus Karimov[1,2], *Ilhomjon* Musaev[3,*], *Sakhiba* Mirzababayeva[4], *Zebuniso* Abobakirova[4], *Shodiljon* Umarov[4] and *Zarnigor* Mirzaeva[4]

[1]Xinjiang Institute of Ecology and Geography, Chinese Academy of Sciences, South Beijing Road, 818, Urumqi, Xinjiang, China
[2]Uzbek-Finnish Pedagogical Institute, 166, Spitamen Avenue, 230044, Samarkand, Uzbekistan
[3]"TIIAME" National Research University, 39, Kori Niyozi street, Tashkent, 100000, Uzbekistan
[4]Fergana Polytechnical Institute, 86, Fergana Ave, Fergana, 150107, Uzbekistan

Abstract. Land use and land cover change (LULCC) is a dynamic process that shapes landscapes, ecosystems, and human societies. In the context of Uzbekistan, a country in Central Asia, LULCC has been driven by a complex interplay of socio-economic, political, and environmental factors. This review provides a comprehensive analysis of the historical trajectories, drivers, and implications of LULCC in Uzbekistan. Through the synthesis of diverse data sources including remote sensing imagery, government reports, and peer-reviewed literature, we trace the evolution of land use patterns over time. Agricultural expansion and industrial development have reshaped land cover, impacting biodiversity and ecosystem services. We discuss the intricate web of drivers behind these changes, encompassing demographic shifts, policy alterations, and economic priorities. Moreover, we explore the socio-economic consequences of LULCC, addressing both opportunities and challenges posed by urbanization and agricultural intensification. The interactions between LULCC and climate patterns are examined, underlining the need for climate-conscious land use planning. Drawing on these insights, we emphasize the importance of sustainable land management and evidence-based policy formulation to mitigate negative impacts and maximize benefits. This review underscores the need for continued research, adaptive strategies, and collaborative efforts to address the complex and evolving dynamics of LULCC in Uzbekistan.
Keywords. Land use, land cover change, Uzbekistan, urbanization, agriculture, industrialization, socio-economic drivers, environmental impacts.

1 Introduction

The intricate relationship between human societies and their surrounding environment has led to the transformation of landscapes at an unprecedented pace [1]. Land use and land cover change (LULCC) have become central components of global environmental change, profoundly impacting ecosystems, biodiversity, and overall sustainability [2]. Within this

* Corresponding author: ilkhommus@mail.ru

context, Uzbekistan, a landlocked country in Central Asia, stands as a compelling case study due to its rich history, diverse ecosystems, and complex socio-economic dynamics [3].

Over the past few decades, Uzbekistan has undergone significant LULCC, driven by a myriad of factors ranging from population growth and urbanization to agricultural expansion and industrial development [4, 5]. These changes have not only altered the physical appearance of the land but have also contributed to intricate shifts in hydrological regimes, climate patterns, and ecological balances [6, 7]. Understanding the patterns, drivers, and consequences of LULCC in Uzbekistan is not only crucial for academic research but also holds profound implications for policymakers and stakeholders engaged in sustainable land management and development [8, 9].

This manuscript delves into the multifaceted dimensions of LULCC within the context of Uzbekistan. By synthesizing existing literature, empirical studies, and remote sensing data, this review aims to provide a comprehensive overview of the historical trajectories of LULCC in the country. Furthermore, it endeavors to identify the key drivers that have propelled these changes, while also shedding light on the socio-economic and environmental implications that have emerged as a result.

Through the exploration of various methodologies employed to monitor and analyze LULCC, this review also seeks to highlight the advancements in remote sensing technology and geospatial analysis that have enabled researchers to gain insights into the dynamic changes occurring on the land. By doing so, this manuscript strives to contribute to the growing body of knowledge concerning LULCC dynamics in Central Asia and beyond.

In the subsequent sections, we will delve into the historical context of land use in Uzbekistan, trace the evolution of land cover patterns, dissect the primary drivers of change, and discuss the implications for ecosystems and communities. By elucidating the complex interplay between human activities and natural systems, this review aspires to inform sustainable land management strategies, policy formulation, and future research directions. Ultimately, a nuanced understanding of the LULCC dynamics in Uzbekistan will not only enrich academic discourse but will also aid in the pursuit of a more ecologically resilient and prosperous future for the country.

2 Materials and methods

Data Collection and Compilation

To comprehensively review the land use and land cover change (LULCC) dynamics of Uzbekistan, a systematic approach was adopted to collect and compile relevant data from a variety of sources [10]. The process involved the acquisition of peer-reviewed journal articles, government reports, remote sensing datasets, and historical records [10, 11]. A thorough search was conducted using academic databases such as PubMed, Web of Science, and Google Scholar, utilizing keywords related to Uzbekistan, land use, land cover change, and associated topics. Additionally, remote sensing imagery from sources like Landsat and Sentinel satellites was procured to analyze spatial patterns of LULCC over time.

Data Synthesis and Analysis

The collected data were organized and synthesized to trace the historical trajectories of LULCC in Uzbekistan (Table 1). Different time periods were delineated to capture significant shifts in land use and cover patterns. Land use categories, including urban, agricultural, forested, and natural areas, were identified and quantified based on available datasets. Changes in these categories were tracked using GIS and remote sensing tools, enabling the assessment of spatial and temporal dynamics.

Table 1. Land use and land cover categories.

Category	Description
Urban	Urban and built-up areas
Agricultural	Croplands, farmlands, and agricultural activities
Forested	Natural and planted forests
Grassland	Natural grasslands and pasturelands
Wetland	Wetlands, marshes, and water bodies
Bare Land	Barren or sparsely vegetated land

Drivers of LULCC

To comprehend the driving forces behind LULCC in Uzbekistan, a qualitative analysis was performed. Key drivers, such as population growth, urbanization, agricultural practices, policy changes, and industrialization, were identified through a comprehensive review of literature and reports. These drivers were categorized based on their socio-economic, political, and environmental origins (Table 2).

Table 2. Key drivers of LULCC in Uzbekistan.

Driver	Description
Population Growth	Increased demand for housing, infrastructure, and resources
Urbanization	Expansion of cities and urban infrastructure
Agricultural Expansion	Conversion of land for agriculture and irrigation
Industrial Development	Establishment of industries and manufacturing units
Policy Changes	Alterations in land tenure, agricultural policies, and regulations
Climate Change	Impact of changing climate patterns on land use and ecosystems

Implications and Consequences

The socio-economic and environmental implications of LULCC were assessed through the examination of existing studies and reports. Changes in ecosystem services, biodiversity, hydrological regimes, and climate patterns were investigated to understand the far-reaching consequences of LULCC in Uzbekistan.

3 Results and discussion

Historical Land Use and Land Cover Dynamics

The analysis of historical data spanning several decades reveals profound changes in land use and land cover patterns across Uzbekistan (Table 3). Urban areas have expanded significantly, driven by population growth and urbanization. This expansion has often encroached upon agricultural and natural areas, leading to changes in ecosystem dynamics. Agricultural land has experienced fluctuations, with periods of expansion and contraction influenced by shifts in policy and economic priorities.

Table 3. Summary of historical land use and land cover changes.

Time Period	Urban (%)	Agricultural (%)	Forested (%)	Natural (%)
1990	5.2	71.8	8.5	14.5
2005	8.7	64.5	6.3	20.5
2020	12.6	57.2	5.1	25.1

Spatial Distribution of Changes

Geospatial analysis of remote sensing imagery showcases the spatial distribution of LULCC across different regions of Uzbekistan. Urban expansion is most prominent around major cities and transportation corridors. Agricultural areas have transformed, influenced by changes in irrigation practices and land management techniques. Forested areas have seen gradual decline, partly attributed to logging and land conversion.

Socio-Economic Consequences of LULCC

The dynamic changes in land use and cover have significant socio-economic ramifications for Uzbekistan (Table 4). Urbanization brings opportunities for economic growth and improved living standards but also challenges related to infrastructure development, housing, and service provision. Agricultural expansion can enhance food security and livelihoods but may also lead to environmental degradation due to unsustainable practices.

Table 4. Socio-economic consequences of land use and land cover changes.

Aspect	Consequences
Urbanization	Economic growth, infrastructure challenges
Agricultural Expansion	Food security, increased demand for resources
Industrial Development	Employment opportunities, environmental concerns
Ecosystem Services	Biodiversity loss, changes in water and soil quality
Climate and Hydrology	Altered local and regional climate patterns

Climate Change and Adaptation

The interplay between LULCC and climate change is complex (Table 5). Changes in land cover can influence local climate patterns through alterations in surface properties, such as albedo and evapotranspiration rates. Forest loss and degradation, for instance, can impact temperature and precipitation regimes. Understanding these interactions is vital for climate change adaptation and effective land use planning.

Table 5. Interactions between LULCC and climate change.

Climate Aspect	Influence of LULCC
Temperature	Urban heat island effect, forest cover alteration
Precipitation	Changes in land use affecting regional rainfall
Evapotranspiration	Altered rates due to changes in vegetation cover
Carbon Sequestration	Forest loss leading to reduced carbon storage

Policy and Sustainable Land Management

Addressing the challenges posed by LULCC requires a holistic approach involving policy intervention and sustainable land management strategies. Integrated land use planning that balances urban expansion, agricultural needs, and conservation efforts is critical. Effective policy frameworks can incentivize the adoption of eco-friendly practices, such as afforestation, sustainable agricultural techniques, and efficient water use.

Implications and Future Directions

The LULCC in Uzbekistan has far-reaching implications for both ecosystems and society. The loss of natural habitats and decline in forest cover can impact biodiversity and disrupt ecosystem services. Changes in hydrological patterns due to altered land use may affect water availability and quality. Balancing development goals with sustainable land management practices is crucial to mitigating negative impacts and ensuring long-term ecological resilience.

The insights garnered from this review offer valuable directions for future research and policy formulation. Further investigation into the socio-economic consequences of LULCC on local communities is essential. Additionally, the use of advanced remote sensing techniques, coupled with socio-economic data, can enhance our understanding of the driving forces behind LULCC and aid in devising strategies for sustainable land management.

4 Conclusions

The review of land use and land cover change (LULCC) dynamics in Uzbekistan provides a comprehensive understanding of the intricate relationships between human activities, environmental processes, and socio-economic outcomes. The analysis of historical trends reveals significant shifts in land use patterns over the past decades, marked by urban expansion, changes in agricultural practices, and alterations in natural ecosystems. The spatial distribution of these changes underscores the regional variability in LULCC, driven by a complex interplay of drivers.

This review underscores the significance of continuous research and monitoring efforts to track LULCC dynamics in Uzbekistan and their ramifications. The dynamic nature of land use patterns necessitates ongoing assessment and adaptation of strategies. Future research endeavors should focus on addressing existing knowledge gaps, forecasting future LULCC scenarios, and assessing the efficacy of policy interventions.

In conclusion, the comprehensive understanding of LULCC dynamics in Uzbekistan, as presented in this review, contributes to a broader awareness of the intricate relationships between human activities and environmental change. By acknowledging the multifaceted impacts of LULCC, Uzbekistan can stride toward a sustainable and resilient future, where land use decisions are underpinned by an understanding of their far-reaching consequences.

References

1. Juliev, M., Pulatov, A., Fuchs, S., & Hübl, J. (2019). Analysis of Land Use Land Cover Change Detection of Bostanlik District, Uzbekistan. Polish Journal of Environmental Studies, 28(5).
2. Dubovyk, O., Menz, G., Conrad, C., Thonfeld, F., & Khamzina, A. (2013). Object-based identification of vegetation cover decline in irrigated agro-ecosystems in Uzbekistan. Quaternary International, 311, 163-174.
3. Gerts, J., Juliev, M., & Pulatov, A. (2020). Multi-temporal monitoring of cotton growth through the vegetation profile classification for Tashkent province, Uzbekistan. GeoScape, 14(1), 62-69.
4. Jumanov, A., Narbaev, S., Boboqulov, S., Ruziboyev, S., Usmanov, Y., & Absoatov, U. (2023). Analysis of mountain and sub-mountain areas degradation using GIS technologies in Parkent district, Uzbekistan. In E3S Web of Conferences (Vol. 401, p. 02011). EDP Sciences.
5. Islomov, S., Aslanov, I., Shamuratova, G., Jumanov, A., Allanazarov, K., Daljanov, Q., ... & Karimbaev, Q. (2022, May). Monitoring of Land and Forest Cover Change Dynamics Using Remote Sensing and GIS in Mountains and Foothill of Zaamin, Uzbekistan. In International Scientific Conference on Agricultural Machinery Industry "Interagromash"" (pp. 1908-1914). Cham: Springer International Publishing.
6. Shaumarov, M., Toderich, K. N., Shuyskaya, E. V., Ismail, S., Radjabov, T. F., & Kozan, O. (2012). Participatory management of desert rangelands to improve food security and sustain the natural resource base in Uzbekistan. Rangeland Stewardship in Central Asia:

https://doi.org/10.1051/e3sconf/202342103007

Balancing Improved Livelihoods, Biodiversity Conservation and Land Protection, 381-404.

7. Gafurova, L., & Juliev, M. (2021). Soil Degradation Problems and Foreseen Solutions in Uzbekistan. In Regenerative Agriculture: What's Missing? What Do We Still Need to Know? (pp. 59-67). Cham: Springer International Publishing.
8. Pachri, H., Mitani, Y., Ikemi, H., Djamaluddin, I., & Morita, A. (2013). Development of water management modeling by using GIS in Chirchik river basin, Uzbekistan. Procedia Earth and Planetary Science, 6, 169-176.
9. Li, J., Chen, H., Zhang, C., & Pan, T. (2019). Variations in ecosystem service value in response to land use/land cover changes in Central Asia from 1995–2035. PeerJ, 7, e7665.
10. Makhmudova, U., Khasanov, S., Karimov, A., & Abdurakhmonov, S. (2023). Evaluation of perennial reference evapotranspiration (ETo) over a typical dryland using satellite images: a case study from Uzbekistan. Ecohydrology & Hydrobiology.
11. Musaev, I., Bokiev, A., & Botirova, M. (2021). The Value of the Cards in Water Basins with the Installation of Solar Power Plants in Yangiyul District of Tashkent Province of Uzbekistan. In E3S Web of Conferences (Vol. 227, p. 05004). EDP Sciences.

Wheat thrips (*Haplothrips tritici* Kurd.) damage on grain crops in Uzbekistan

Marat Abdillayev[1,*] and *Qalandar* Bababekov[1]

[1]Research Institute of Plant Quarantine and Protection, Tashkent, 100003, Uzbekistan

Abstract. This article provides comprehensive insights into the detrimental impact of wheat thrips, a widespread and persistent pest that affects wheat crops within our country. The research delves into the conditions under which wheat thrips cause substantial harm to crops, taking into account the specific climatic factors present within our region. A comparative analysis of pest prevalence and the resulting damage was conducted between the years 2020 and 2023. The study underscores that wheat thrips infestation leads to significant harm across various growth stages of wheat, spanning from the tillering phase to the maturity phase. Through meticulous investigation, the research reveals that wheat thrips can be found at varying densities, ranging from 5 to 35 thrips per stem within typical agricultural fields. To enhance its content, the article is enriched with tables and images that illustrate the outcomes derived from the research. This visual and data-driven approach amplifies the understanding of the pest's impact, its prevalence, and the associated damage it inflicts upon wheat crops.
Keywords. *Haplothrips tritici Kurd*, wheat, spike, pest, plant, yield, grain, stem, cereals.

1 Introduction

In order to implement the tasks in 2023 which were defined in the strategy of the development of agriculture of the Republic of Uzbekistan for 2020-2030, to increase the volume of food production through effective use of available resources, to ensure the demand of the population in the domestic market and to maintain price stability, and to provide economic and financial support to the producers of agricultural products, the Ministry of Agriculture of the Republic of Uzbekistan and the Council of Farmers and Homestead Land Owners of Uzbekistan have determined that the expected minimum price for 2022 will be 2.5 million soums for 1 ton of grain harvest [1, 2].

In terms of cultivation, wheat is grown on the largest number of agricultural crops in Uzbekistan. In the conditions of Uzbekistan, the optimal period for planting winter wheat is from September 15 to October 15. Planting wheat at the specified time is one of the important factors for increasing productivity. In recent years, as a result of the dry climate, pests leave the hibernation period early and seriously damage grain crops, which has a negative impact on productivity [3, 4].

* Corresponding author: marat.abdillaev1001@mail.ru

It is important to preserve agricultural products and reduce losses. Currently, 42-50% of crop yields are lost by harmful organisms, 9.2-13% of which are caused by pests [5]. In Uzbekistan, during the period of cotton cultivation, the harmful effects of pests on crop yield result in an annual loss of 28-35%. Protecting the crop from these pests is essential, and one effective measure is the proactive control of the damaging cotton thrips [6-8].

Among the significant harmful pests, cotton thrips (*Haplothrips tritici* Kurd.) pose a substantial threat. Their presence has been observed to cause significant damage to the quality and quantity of cotton yield in the Republic over the years. Consequently, the bioecological characteristics, distribution, and levels of damage caused by cotton thrips have been investigated in Uzbekistan. These pests, belonging to the family Pleothripidae within the order Thysanoptera, specifically target reproductive organs, causing damage to floral structures and deformations [9].

Wheat thrips (*Haplothrips tritici* Kurd.) is a very common and dangerous pest among thrips that seriously damages grain crops. This phyphophagous pest appears in the tillering phase of wheat and can be seen in the flowering phase when the number of thrips in the spike is the highest rate, first the larvae of thrips damage the husk of the spike and then the grain. If there are 40-50 thrips in the spike of wheat during the spike formation phase, the grain yield decreases by 0.18-0.25 t/h [10].

Wheat thrips is one of the most common and dangerous pests of winter and spring wheat. The thrips feeds on the cell sap of wheat, resulting in a sharp decrease in the photosynthesis process in the spike and leaf of wheat and grain yield. Thrips larvae feed directly on the spikelet and grain, reducing grain weight and resulting in reduced yield [2, 10].

When subjected to infestation by cotton thrips, the plants suffer from damage to their flowers, resulting in bud abortion, weakening of shoots, and deformations of the plant structure. Infested cotton bolls show reduced starch and sugar content, decreased levels of protein-bound amino acids, and an increase in free amino acid [11].

Early pests of wheat plants cause infertility of wheat flowers, thinning of spikes and deformation of grain shape. In damaged grains, the amount of starch and sugar decreases, the amount of protein amino acids decreases, and the number of free amino acids increases [3]. Thrips larvae that emerge from the first hibernation initially feed on the ear band, flower husks, ovary, and developing grain, resulting in "over grain" in wheat spikes. In the second stage, the larvae accumulate in the groove of the grain and cause a change in the shape of the wheat grain and loss of grain weight [4].

In wheat, larvae of wheat thrips are more harmful than adults. As soon as the larvae hatch, they immediately begin to damage the ear band and the calyx. After the tissues of this part of the spike thicken, the larvae begin to suck the grain [6]. The adult of wheat thrips is not limited to damaging the vegetative organs of wheat. While the larvae, especially in dry years of the climate, decrease the weight and composition of the grain in the spike by entering the spikes, flower husks, sometimes inside the flower and sucking the grain juice [7].

2 Materials and methods

In our research, we used widely accepted textbooks of practical training in general and agricultural entomology, as well as general monographs [1-3]. To determine the types of thrips in the obtained samples, they were kept in fixatives made from a mixture of 5% formalin, alcohol and 4% glycerin until the species was determined by the specialists. These works were carried out according to methodological guidelines [4-6].

Research was conducted under laboratory, small field and production conditions, and experiments on wheat thrips damage were conducted under lysimeter conditions. For this, each lysimeter was individually covered with a white cloth of 120 cm x 80 cm. As a control was taken a lysimeter covered with the same cloth, but not infested with wheat thrips.

Experiments were triplicated. Thrips were artificially released on the wheat inside the cloth by 3; 5; 8 pieces per ear. Pests were released on the wheat in three different periods: tillering, spiking and milk ripening stages and they were counted every five days. The phenological development of damaged and undamaged plants, their weight and the quality of the crop were observed and studied.

The damage coefficient of the lost yield in the damaged plant was shown in percentages and was expressed according to Belyaev. Counting the insects in field conditions was carried out [3]. In this case, adult insects in the field were counted each time by shaking 20 times 5-10 places of the field with the help of an entomological handle during the periods of tillering and spiking. At the beginning of the grain formation period, the spikes were cut in different parts of the field and the larvae were counted separately in each spike.

The amount of gluten in flour obtained from damaged and undamaged grains was determined [6, 9].

3 Results and discussion

Based on the climatic conditions of different regions of Uzbekistan, observations of our research were made on the prevalence and damage caused by the wheat thrips.

The research was carried out mainly in the regions Tashkent, Sirdarya, Jizzakh, Samarkand, and Kashkadarya in the form of monitong. The research was carried out from 2020 to 2023, the prevalence and damage of pest of grain crops – wheat thrips was studied.

Table 1. Monitoring of the prevalence and damage of wheat pests in different regions of Uzbekistan *(Monitoring of the laboratory for the control of harmful organisms of grain and legume crops, 2022).*

#	Provinces	Districts	Pest prevalence, %
			Wheat thrips
1	Tashkent	Kuyi Chirchik and Kibray districts	40%
2	Sirdarya	Guliston and Sardoba districts	35%
3	Jizzakh	Sharof Rashidov, Pakhtakor, Gallaorol districts	33%
4	Samarkand	Ishtikhon, Urgut, Taylok, Bulungur, Payarik, Okdaryo, Pastdargom districts	29%
5	Kashkadarya	Kitob, Shakhrisabz, Yakkabog, Kasbi, Karshi districts	32%

We mainly conducted research in the grain fields of clusters and farms in the central and partly southern regions of the country (2020-2023).

As a result of damage to wheat by wheat thrips, the biochemical composition of the grain in the spike changes, and the amount of protein, minerals and gluten decreases sharply. In our research, it was found out that in a plant infected with wheat thrips, the shape of the spike is small, the grain groove is enlarged, and the affected grains are smaller compared to the healthy grain, and the nutritional value of such grains decreases (Tables 2 and 3).

Table 2. Wheat thrips infection at different developmental stages of wheat (Experimental site of PQPRI, in 2021).

#	Phases	Thrips number per spike, pcs	Number of stems per plant, pcs	Mean length of stem, cm	Spiking level, %	Spike length, cm	Number of grains per spike, pcs
1	Tillering	3	6.2	89.7	94.1	8.3	40.7
		5	6.1	89.3	92.3	7.9	40.3
		8	5.9	88.1	92.0	8.2	39.1
2	Spike formation	3	6.4	90.9	95.9	8.4	43.1
		5	6.2	90.1	95.7	8.2	42.3
		8	6.0	89.5	94.5	8.0	41.1
3	Milk maturity	3	6.6	91.5	96.8	9.0	46.9
		5	6.4	91.3	96.3	9.5	47.3
		8	6.2	90.4	95.5	8.9	46.1
4	Control (undamaged)	-	6.7	92.3	97.2	9.6	46.8

Table 3. Yield reduction in wheat thrips infection at different developmental stages of wheat (Experimental site of PQPRI, in 2021).

#	Phases	Thrips number per spike, pcs	Weight of 1000 grains, gr	Yield per plant, gr	Yield reduction compared to control	
					g/plant	%
1	Tillering	3	36.1	6.6	1.4	17.5
		5	34.5	6.1	1.9	23.7
		8	35.0	4.8	3.2	40.0
2	Spike formation	3	41.5	7.3	0.7	8.7
		5	38.4	6.9	1.1	13.7
		8	39.9	5.8	2.2	27.5
3	Milk maturity	3	46.3	7.9	0.1	1.2
		5	45.6	7.8	0.2	2.5
		8	43.4	7.7	0.3	3.7
4	Control (undamaged)	-	47.5	8.0	0	0

Based on the results of our scientific experiments, it can be said that when wheat is 100% damaged by thrips, the criterion for causing economic damage is 5 thrips per plant during the tillering period, and 8 thrips per plant during the spiking period. In these phases, if the number of thrips exceeds the specified amount, it is a good time to start protective and control measures.

The earlier grain crops are affected by wheat thrips and the denser their numbers, the greater the yield loss. In the period of tillering of wheat, 3 thrips on one plant causes more than 20.0% yield loss, while 5 thrips on one plant causes yield loss of up to 12.8%. Even when there are more than 8 thrips per plant during the milk maturity period, it does not significantly affect the yield.

The criterion of economic damage is the most favorable time to initiate protective measures, when the degree of damage to the plant or the density of the pest population can cause significant damage to the harvested crop. Correctly defining the criterion of causing economic damage allows us to avoid inappropriate and unjustified chemical treatment (Figure 1).

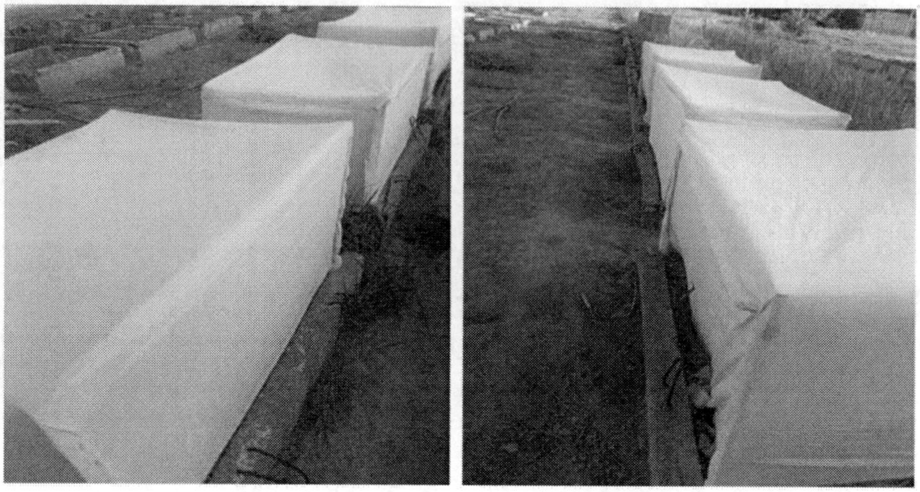

Figure 1. Field experiment lysimeter to determine damage caused by wheat thrips (2021).

The above pictures show the experiments carried out in the lysimeter experimental field of the Plant Quarantine and Protection Research Institute. In this experiment, wheat thrips was brought from the grain fields, and was artificially infested on the winter wheat planted in our lysimeter experimental field, and the damage caused to the wheat was determined. Morphological changes in the spike and grain under the influence of wheat thrips damage were analyzed in the laboratory (Figure 2).

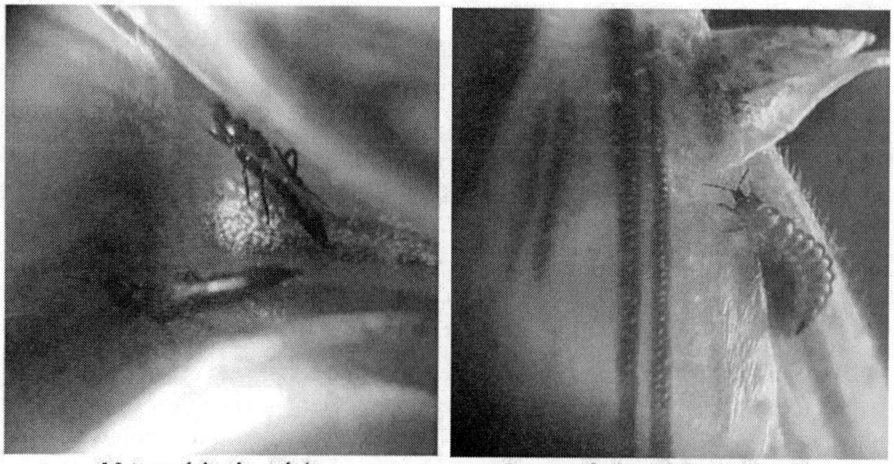

Mature adult wheat thrips — Larvae of wheat thrips in the spike

Figure 2. Field experiments to determine wheat thrips (2021).

4 Conclusions

During the research, it was found that the wheat thrips was found in all the studied regions and caused damage to the grain crops. When analyzing the distribution of wheat thrips in regions, it was found that thrips is most widespread in Tashkent region. In our experiments, it was found that the number of wheat thrips in 1 plant was 5-12 pieces in the areas where wheat thrips is rare, and up to 35 wheat thrips in the spike of the plant in areas where wheat thrips were widespread.

References

1. Sekun M.P. Phytophages on wheat. Damage of dominant species // Plant Protection. - 2018. - No. 4. - Pp.6-7 (in Russian).
2. Likhatsky D.M., Chekmareva L.I., Likhatskaya S.G. Dominant species of entomophages of spring wheat during energy-saving tillage in the conditions of the Saratov Volga region // Agrarian scientific journal. - 2015. - No. 4. - Pp. 17–20(in Russian).
3. Tansky V.I. Damage of wheat thrips / V.I. Tansky // Plant Protection. - 2010. - No 7. - Pp. 23–25 (in Russian).
4. Zhichkina, L.N. Population dynamics of wheat and predatory thrips in agrocenoses of spring wheat and barley / L.N. Zhichkina // Agrotechnical method of protecting plants from harmful microorganisms: materials of the 4th Intern.scientific-practical.conf.- Krasnodar, 2007.-Pp.163-165 (in Russian).
5. Triebel, S.O. Resistant varieties. A radical solution to the problem of reducing crop losses from harmful organisms / S.O. Triebel // Quarantine and plant protection. – 2014. – No. 6. – Pp. 6–7 (in Ukrainian)
6. Kamenchenko, S.E. Damage and economic threshold of wheat thrips / S.E. Kamenchenko // Plant Protection. - 2012. - No. 3 - P. 22 (in Russian).
7. Shurovenkov Yu.B. Influence of wheat thrips on sowing qualities and regenerative capacity of spring wheat in the Trans-Urals // Ecology of Harmful and Beneficial Insects. - Voronezh, 2012. - Pp. 62–73 (in Russian).
8. Pulatov, Z., & Urazbayev, A. (Uzbekistan Agricultural Journal, Tashkent, 2019, No. 5, pp. 19-22).
9. Khodjaev, S. T. (Agrotoxicology principles and integrated pest management in plant protection, Navruz Publishing, Tashkent, 2014, pp. 185-187).
10. Jumaev, R., & Rustamov, A. (2022, July). Representatives of Lepidoptera groups in the biotecenosis of Uzbekistan and their effective parasite-entomophage types. In IOP Conference Series: Earth and Environmental Science (Vol. 1068, No. 1, p. 012026). IOP Publishing. DOI:10.1088/1755-1315/1068/1/012026
11. Jumaev, R. (2023). Invitro rearing of parasitoids. In E3S Web of Conferences (Vol. 371, p. 01032). EDP Sciences. DOI:10.1051/e3sconf/202337101032

Structural features of vegetative organs of *Oxytropis macrodonta* Gontsch. (*Leguminosae*) in natural conditions

Guljan Duschanova[1,*], *Manzura* Agzamova[2], *Abdulaziz* Janibekov[2] and *Markhaboxon* Makhmudova[2]

[1]Tashkent State Pedagogical University, Tashkent, 100070, Uzbekistan
[2]Institute of the Chemistry of Plant Substances, Tashkent, 100170, Uzbekistan

Abstract. For the first time, the anatomical structure of the vegetative organs of Oxytropis macrodonta, which grows in natural habitats, has been studied. The following diagnostic and adaptive features were revealed: in the leaf, a more sinuous outline of the epidermal cells; unsubmerged stomata of anomocytic and hemiparacytic type; dorsiventral type of leaf mesophyll; palisade and spongy parenchyma chlorophyll-bearing; in peduncles - parenchymal-beam type of structure; under the epidermis there is a numerous lamellar collenchyma; conducting bundle open collateral type and sclerified due to bast fibers; the root is of a non-beam type and more woody, the elements of which are primary bast fibers; inside from the fibers there is a secondary bast or secondary phloem, which makes up the zone of the secondary cortex; trapezoidal sections consist of lignified thick-walled cells - a hard bast with a thin-walled element - a soft bast; the libriform is extensive, the woody parenchyma is diffusely vascular. The presence and localization of biologically active substances in palisade, spongy and cortical parenchymal cells are shown. The data and the identified structural diagnostic features of aboveground and underground organs are species-specific and can be used in taxonomy and in the identification of plant materials for a given species.
Keywords. Anatomy, diagnostic signs, vegetative organs, *Oxytropis macrodonta*.

1 Introduction

Genus Ostrolodka - Oxytropis DC. from the Leguminous family (Fabaceae, or Leguminosae) is one of the richest in species in the flora of Asian Russia. The Russian name Ostrolodochnik adopted in Flora of the USSR [1] is erroneous, since it means a boatman, and not a boat with a spike in the corolla of a flower, moreover, the Latin name *Oxytropis* belongs to the female, not the masculine gender.

Ostroboats grow in the temperate and arctic zones of the northern hemisphere and number about 320 species [2] or 378 species and subspecies in northern Eurasia [3-5]. Much earlier, only 181 species were identified for the entire globe [6]. For the former USSR, 280 species

* Corresponding author: guljon.duschanova@mail.ru

were known in 6 subgenera and 24 sections [7]. For Siberia, 119 species were shown [8], for the Far East, 55 species [9], for Central Asia, 166 species [10], and only for Kazakhstan, 124 species [11].

Central Asia is especially rich in witfish - 153 species [12] and Mongolia itself - 92 species [12], while only 16 species have been shown for Manchuria outside Russia [13], and for Japan - 7 species [14]. Europe is poor in wolffish - 24 species, while 16 species are shown for the European part of the former USSR [15], and North America - 26 species [1]. In the Arctic flora of the former USSR, 32 species have been identified [16], and for Alaska and adjacent territories, 22 species [17].

Ostroboats are confined mainly to the mountains, which are usually characterized by a wide variety of environmental conditions. This contributes to enhanced speciation, hybridization and the formation of mutants. Therefore, the taxonomy of the island fish can be contradictory in some cases. According to the phylogeny, the ostriches are close to the genus *Astragalus - Astragalus* L., whose representatives live mainly in the steppes and semi-deserts, and not in the mountains. Outwardly, ostriches are easily distinguished from astragalus by the fact that the nose of the flower is pointed at the end, not blunted at the top or almost blunted. The hairs in the pubescence of the ostriches are always simple (one-pointed), while in many astragalus they are malpighian (two-pointed). However, an admixture of Malpighian hairs was found in several Pamiro-Alai ostriches from the section Eumorpha Bunge: *O. astragaloides* Boriss., *O. proxima* Boriss., *O. nikolai* Filim. et Abduss. [18], also in the North American *O. lambertii* Pursh from the section Orobia Bunge [19].

Described in the past with a lack of initial herbarium material, some species turned out to be "stillborn" and were not discovered by subsequent researchers. In this regard, the specific names of some species of the wolfberry are indicative: O. controversial (*O. ambigua* (Pall.) DC.), O. mixed (*O. confusa* Bunge), O. dubious (*O. dubia* Turcz.), O. intermediate (*O. intermedia* Bunge) and O. mediocre (*O. interposita* Sipl.). Due to the great morphological diversity of the island fish, it is sometimes difficult to draw a line between the taxonomic rank's "subspecies" and "variety". For example, *O. middendorfii* Trautv., endemic to the north of Siberia and the Far East, divided into 7 subspecies [20]. In contrast, some supporters of the monotypic concept of a species do not actually recognize intraspecific evolution and describe varieties and subspecies as independent (separate) species. Therefore, 12 years after the publication of Flora of the USSR [1], I.T. Vasil'chenko [4] published 30 new species of witfish, and only some of them can be recognized as independent. V.I. adhered to a reserved position. Grubov [21] during an in-depth study of the Ostrolodok of Central Asia.

According to Decrees No. 4670 of the President of the Republic of Uzbekistan "On measures for the protection, cultural cultivation, processing of wild medicinal plants and the rational use of available resources", as well as "On measures to expand the scale of scientific research on the cultivation and processing of medicinal plants, the development of their seed production" No. 4901, which noted that 750 out of more than 4.3 thousand species of plants of the local flora are considered medicinal, of which 112 species are registered for use in scientific medicine, of which 70 species are actively used in the pharmaceutical industry. Plants from the genus *Oxytropis* are cultivated in gardens, some of the species are rare and endangered. Plants of this genus have a wide spectrum of pharmacological activity. Most species of the genus *Oxytropis* contain phenolcarboxylic acids, alkaloids, coumarins and the following flavonoids: quercetin, quercetin and kaempferol glycosides, and other biologically active compounds. *Oxytropis glabra* DC. - the ostrich is naked. In Tibetan medicine, it is known as a diuretic, hemostatic, antipyretic and cardiovascular agent, used for ascites and edema [11]. The water-alcohol extract of *O. caespitosa* (Pall) has an inhibitory effect on the Sedai parainfluenza virus [22]. The extract of *O. deflexa* in Tibetan medicine was used for intoxication, and also as an analgesic for toothache.

In this regard, of scientific interest is the study of the anatomical structure of the aboveground and underground organs of the plant material *Oxytropis macrodonta* growing in conditions in the Tashkent region, in the village of Sidzhak.

The purpose of the study: to study the anatomical structure of the aboveground and underground organs of the plant material *Oxytropis macrodonta* in order to determine the diagnostic features and localization of biologically active substances in organs and tissues.

2 Materials and methods

Oxytropis macrodonta Gontsch. - perennial stemless, grayish-green plant (15) 20 - 40 (80) cm in height; stipules not fused with each other, fused with petiole by 5-8 mm, long and white-stiff-hairy, 20-25 mm long, linear in free part, subulate in upper part; leaves (15)20-25(30) cm long; their petiole, like the axis of the leaf, is loosely protruding and white-furred; leaflets (20) 24-34-paired, lanceolate (rarely broadly lanceolate), acute, (10) 15–30(35) mm long, (3)5-10(13) mm wide, more or less sparsely long on both sides and white-appressed-hairy. Peduncles equal to leaves, less often slightly shorter or longer than them, erect, more or less hairy from pubescence, consisting of white short b. m. adpressed hairs and long protruding; racemes loose, oblong-cylindrical, 12-20 cm long, many-flowered; bract linear, (5) 8-12 mm long, green, outside more or less long white-hairy, long white-ciliated along the edge, equal or almost equal to the calyx tube; calyx tubular, 11-14 mm long, hairy (with predominance of short black hairs and less numerous long white ones); its teeth are subulate linear, 6-8 mm long, equal to or slightly longer than the tube (rarely shorter than it); corolla yellow; 18-22 mm in length roundedovate in limb, just notched at apex; wing 17-19 mm long, entire at apex (rarely just notched); petals 14- 16 mm long, apex lanceolate in lower part, subulate above, 2-3 mm long; beans are linear-cylindrical, 30-40(45) mm long, 3-4 mm thick, rounded on the back (with a slightly protruding seam), furrowed on the belly, more or less densely white-fluffy, on a stalk 2 mm long, with a spout hooked-curved, 6-7 mm long, from the inside along the ventral suture with a membranous border-septum, almost reaching the dorsal suture. VI-VII. Distribution - grassy slopes. – Central Asia: Tien Shan, Pam.-Al. (north-east). Endem. Described from Fergana. Type in [23].

Fresh plant samples were harvested in May 2022 in the Tashkent region, in the village of Sidzhak. Sidzhak is a village in the Bostanlyk district of the Tashkent region of Uzbekistan. The distance to the regional center - the city of Gazalkent - is 50 km. Sidzhak is located in the valley of the mountain river Pskem, its right-bank part, bounded by the Ugam Range. In the center of the settlement, the height above sea level is 959 m [24]. Raw materials were fixed in a 70% alcohol solution. Preparations prepared by hand were stained with methylene blue followed by sealing in glycerol-gelatin [25]. The epidermis was studied on paradermal and transverse sections. Transverse sections of the leaf are made through the middle, petiole, peduncle, pedicel and rhizomes - the base. When describing the anatomical features of the vegetative organs: adaxial and abaxial sides of the leaf epidermis; 1 mm2 the number of epidermal cells; mesophyll structure; height of epidermal cells; cuticle thickness; diameter of parenchyma cells; diameter and amount of xylem; leaf stomata length - width; deep location in the cells of the epidermis of the stomatal apparatus; the number of assimilating cell layers in the mesophyll; diameter and number of rows of spongy palisade cells of the parenchyma; height, width and number of layers of palisade parenchyma; collenchyma and sclerenchyma cells were identified using a Motic B1-220A-3 microscope. Micrographs of the anatomical features of the aboveground and underground organs were taken with a computer microphotographic attachment with a Canon A123 digital camera under a Motic B1-220A-3 microscope [26].

3 Results and discussion

Anatomical structure of the leaf. On the paradermal section of an O. macrodonta leaf, the outlines of epidermal cells on the adaxial and abaxial sides are more sinuous, the projection is polygonal. The adaxial and abaxial epidermis are pubescent with long and white-rigid trichomes. Leaves are amphistomatic - stomata are located on the adaxial (upper) and abaxial (lower) side of the epidermis of the leaf blade and are located transversely to the longitudinal axis of the leaf. Stomata shapes are round-oval, non-immersed, anomocytic, hemiparacytic, paracytic type and more numerous on the abaxial side than on the adaxial one (Figure 1).

Leaf mesophyll in a transverse section of the dorsiventral type, which is represented by palisade cells located under the upper epidermis of the leaf mesophyll, spongy cells - above the lower epidermis of the leaf mesophyll (Figure 2). The epidermis is represented by one row of cells with a thin-walled cuticle layer. The cells of the adaxial epidermis are larger than those of the abaxial epidermis. Between the adaxial and abaxial epidermis there is an assimilation tissue consisting of palisade and spongy cells. The palisade parenchyma is chlorophyll-bearing, elongated, consists of 1-2 rows of cells and is located between the adaxial epidermis and spongy leaf parenchyma. The spongy parenchyma is round, small-celled, chlorophyll-bearing, which consists of 3-4 rows and is located between the palisade parenchyma and the abaxial epidermis. Between the palisade and spongy cells are numerous lateral vascular bundles, with 3-4 small vessels (Figure 2).

The main leaf vein is located on the abaxial side of the central part of the leaf mesophyll. The rest of the vein is occupied by the main thin-walled round-oval cell parenchyma. In the main vein of the leaf there is one large vascular bundle of a closed collateral type (Figure 2).

The peduncle is round in cross section, has a parenchymal-beam type of structure. The primary bark is preserved throughout until the end of the growing season. The epidermis is single-row, consists of round-oval cells, pubescent with long and white-rigid trichomes (Figure 3). The outer cell wall is significantly thickened. Under the epidermis there is a group of plastic collenchyma, which consists of 5-6 row cells. The primary cortex is 6-7 rows, the cells of the outer 3-4 layers are small, almost round, contain biologically active substances, the 3-4 inner layers are larger. In the lower layers of the primary cortex are a group of pericyclic fibers. Visually, the building looks like a ring. In the central cylinder there are 29-30 lignified vascular bundles, of which 10 are large, 6 are medium and 13-14-21 are small. Each bundle contains large groups of bast fibers.

The cambium in the interfascicular zone is more pronounced; in the bundles it produces phloem and xylem. The vascular bundles are most sclerified, due to the bast fibers (sclerenchyma cells), which form multilayer strands in the form of caps over the phloem of the bundles. Xylem is represented by large, numerous vessels located in groups and chains. Conductive bundles of open collateral type. The bundles are separated by a libriform. In the lower part of the primary bundles, a perimedullary zone is distinguished, consisting of thickened isodiametric cells. The pith is extensive, represented by large rounded oval cells and hydrocytic cells are found in it, the pith cells occupy most of the cross section of the peduncle (Figure 3).

https://doi.org/10.1051/e3sconf/202342104002

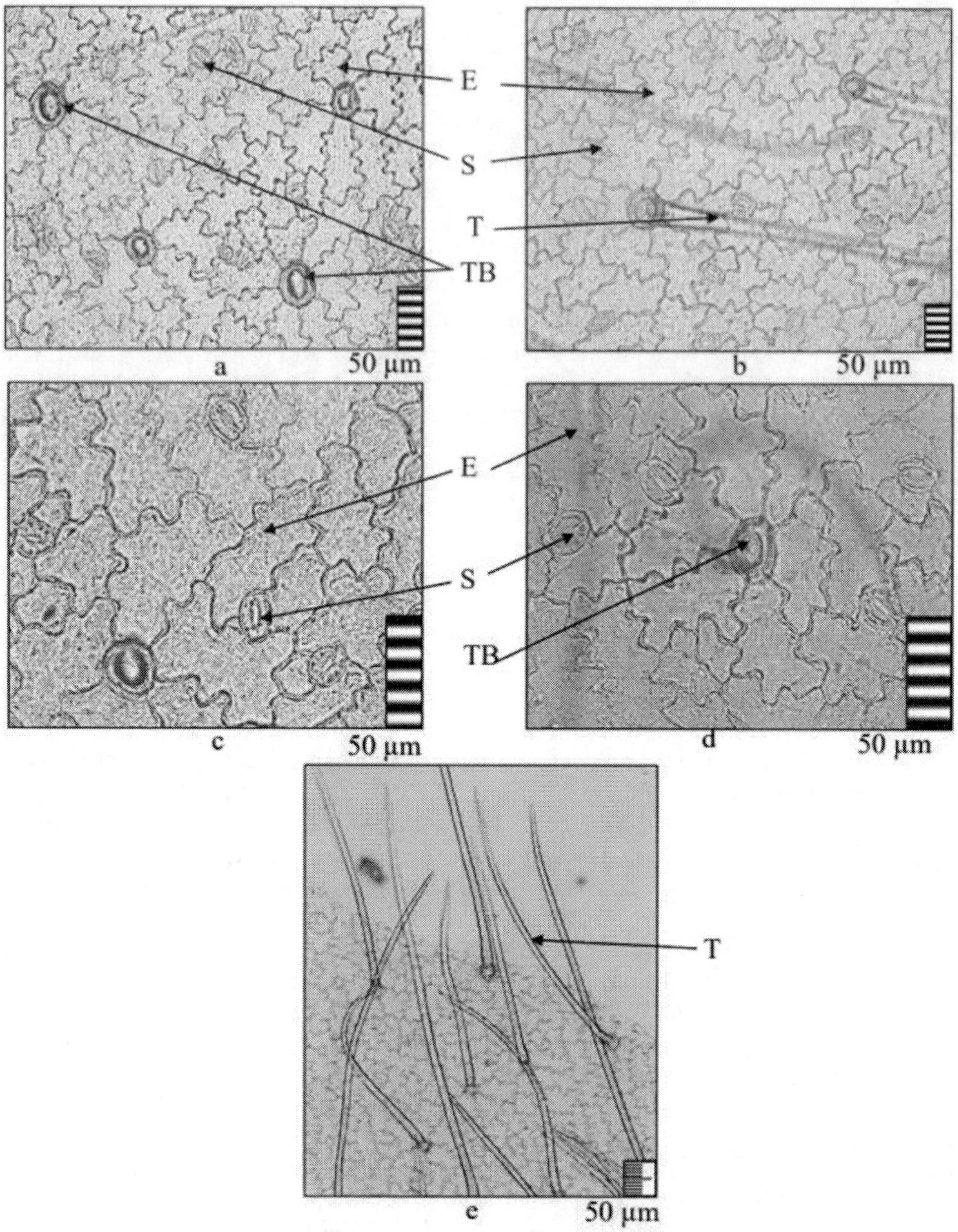

Figure 1. Anatomical structure of the leaf epidermis of *Oxytropis macrodonta* on a longitudinal section: a-c-e – upper (adaxial) epidermis; b-d - lower (abaxial) epidermis. Legend: TB - trichome base, T - trichome, S - stomata, E – epidermis.

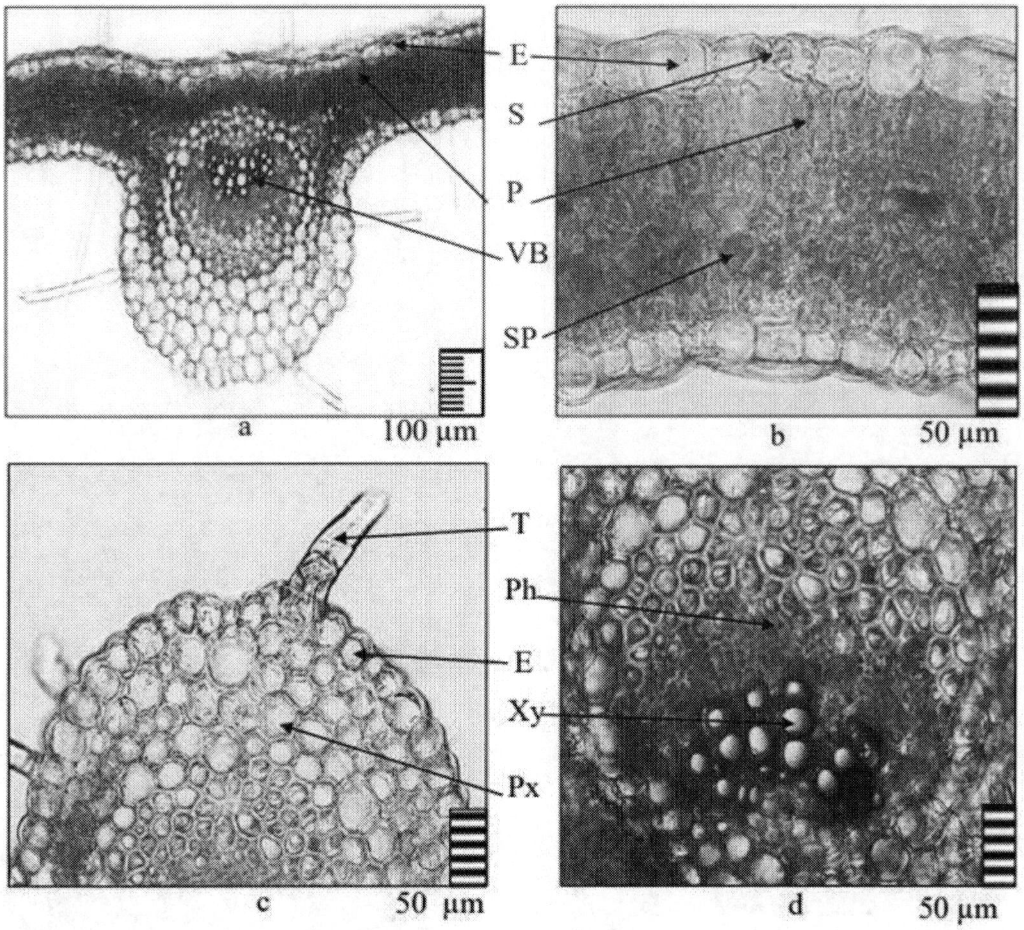

Figure 2. Anatomical structure of an Oxytropis macrodonta leaf in cross section: a – general view of the main leaf vein; b – detail of the leaf mesophyll; c – epidermis and parenchyma of the main leaf vein; d - vascular bundle. Legend: SP - spongy parenchyma, Xy - xylem, P - palisade parenchyma, VB - vascular bundle, Px - parenchyma, T - trichome, S - stomata, Ph - phloem, E - epidermis.

The root on the transverse section is rounded, non-beam type of structure. The structure of the root of woody plants is due to the long-term active activity of the apical and lateral meristems. They are powerful, strongly woody, with early development of the secondary integumentary tissue. Integumentary tissue is a cork, which consists of several rows of radially arranged cells with thick corked membranes and its cells are dark brown in color, thick-walled, tightly closed. Outside, separated by tangential septa, phellogen forms cork cells, and inside, phelloderm cells (Figure 4).

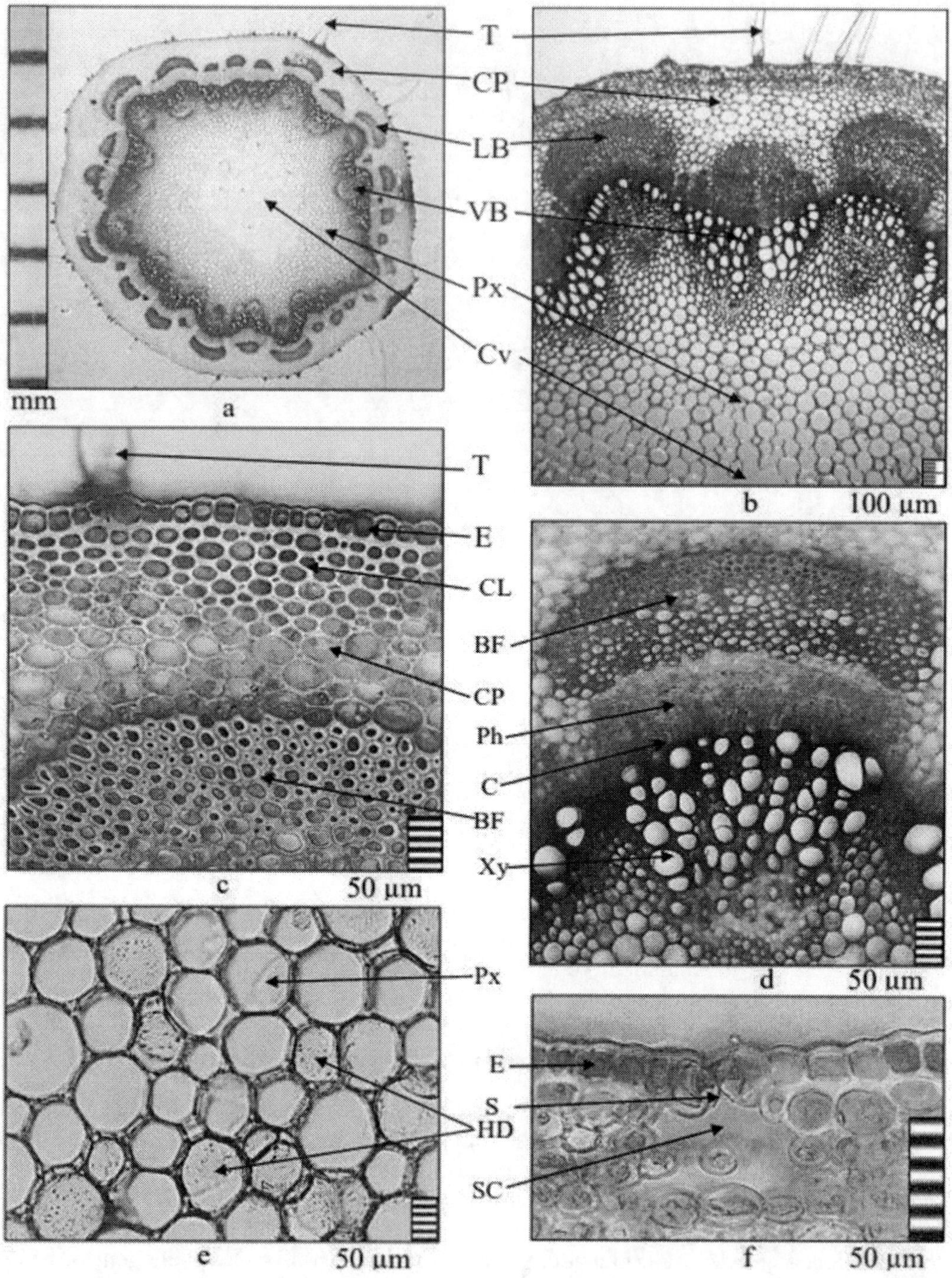

Figure 3. Anatomical structure of the peduncle of Oxytropis macrodonta: a - general view of the peduncle; b - detail of the peduncle; c - epidermis, collenchyma and bast fibers; d – vascular bundle; e – parenchyma and hydrocytic cells; f - epidermis and non-submerged stomata. Legend: HD - hydrocytic cells; CL - collenchyma, CP - cortex parenchyma, L – libriform Xy - xylem, BF - bast fiber, C - cambium, Cv – cavity, VB - vascular bundle, Px - parenchymal cells, T - trichome, S - stomata, SC - substomatal cavity, Ph - phloem, E - epidermis.

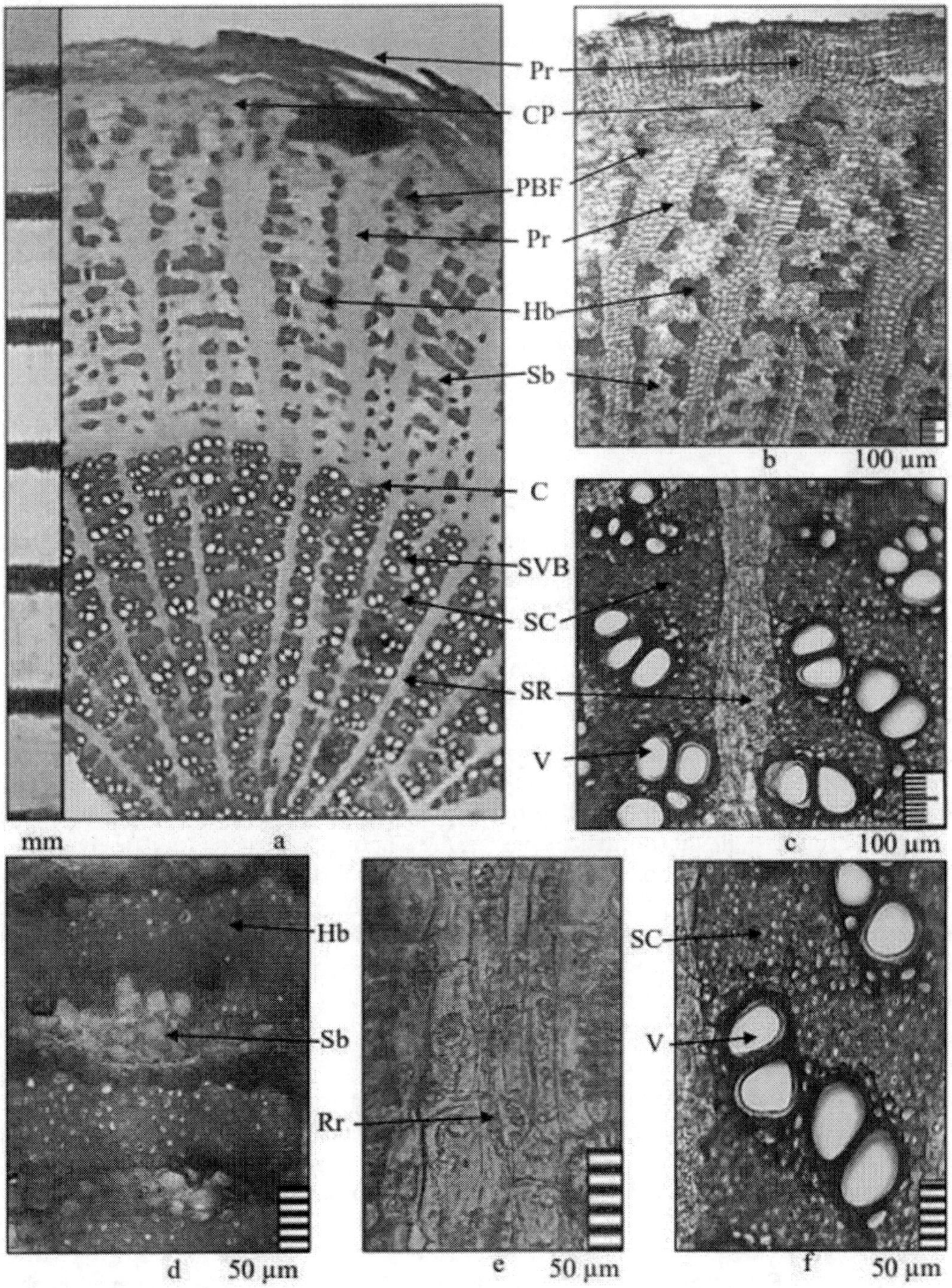

Figure 4. Anatomical structure of the root of Oxytropis macrodonta in cross section: a – general view of the root; b – cortex parenchyma; c, f – secondary vascular bundle; d - soft and hard bast, e - radial rays. Legend: SR - secondary ray, SVB - secondary vascular bundle, CP - cortex parenchyma, K - cambium, SB - soft bast, PL - primary rays, PBF - primary bast fibers, Pr - periderm, Px - parenchyma, RL - radial rays, V - vessels, SC - sclerenchyma, Hb - hard bast.

Under the periderm, there is a round-oval secondary cortex parenchyma, which consists of 7-8 rows and is preserved throughout. In the central cylinder, on the border with the secondary cortex, there are groups of lignified thick-walled cells separated by parenchyma. These woody elements represent the primary bast fibers. Inside the fibers there is a secondary bast or secondary phloem, which makes up the zone of the secondary cortex. Trapezoidal sections can be distinguished in it, the narrow bases of which make up the primary bast fibers. Secondary rays, the formation of which is due only to cambial activity, have a smaller extent in the tangential direction. Trapezoidal sections consist of alternating tangential bands of lignified thick-walled cells that make up a hard bast, and thin-walled elements - a soft bast. The elements of hard bast are structurally similar to primary bast fibers (Figure 4).

The cambial zone is represented by several rows of tabular cells. The primary conductive tissues are preserved in the roots, and then the secondary ones immediately form a central cylinder, which, on a cross section, the wood parenchyma has a scattered vascular pattern. With age, the emerging cambium changes the anatomical structure of the root and annually produces secondary xylem (wood).

4 Conclusions

Thus, the anatomical structure of the vegetative organs of Oxytropis macrodonta, which grows in natural habitats, has been studied. The following diagnostic and adaptive features were revealed: in the leaf, a more sinuous outline of the epidermal cells; unsubmerged stomata of anomocytic and hemiparacytic type; dorsiventral type of leaf mesophyll; palisade and spongy parenchyma chlorophyllbearing; in peduncles - parenchymal-beam type of structure; under the epidermis there is a numerous lamellar collenchyma; conducting bundle open collateral type and sclerified due to bast fibers; the root is of a non-beam type and more woody, the elements of which are primary bast fibers; inside from the fibers there is a secondary bast or secondary phloem, which makes up the zone of the secondary cortex; trapezoidal sections consist of lignified thick-walled cells - a hard bast with a thin-walled element - a soft bast; the libriform is extensive, the woody parenchyma is diffusely vascular. The presence and localization of biologically active substances in palisade, spongy and cortical parenchymal cells are shown. The data and the identified structural diagnostic features of aboveground and underground organs are species-specific and can be used in taxonomy and in the identification of plant materials for a given species.

References

1. A provisional checklist of species for flora North America (revised). - Eds. S.G. Shetler and L.E. Skog. – Missouri Bot. Garden, 2008. - 199 pp.
2. Arkad'eva G.E., Blinova K.F., Komarova M.N. On the antibiotic activity of medicinal plants of Tibetan medicine. Plant Resources. 2016. V.2. Issue. 2. P. 218-123.
3. Baitenov M.S. Oxytropis DC. // Flora of Kazakhstan. - Alma-Ata, 2011. - V. 5. – P. 330-411.
4. Barykina R.P., Veselova T.D., Devyatov A.G. Handbook of botanical microtechnics (bases and methods). - Moscow: Ed. Moscow State University. - 2014. - P. 6-68.
5. Bunge Al. Species generis Oxytropis DC. - St.-Petersb., 2014. - P. 1-166
6. Butnik A.A., Tursynbaeva G.S., Duschanova G.M. Leaf mesophyll of dicotyledonous plants (educational manual). - Tashkent: TSPU named after Nizami, 2015. - 42 p.
7. Buryakov Yu.F., Kasymov M.R., Rostovtsev O.M. Archaeological monuments of the Tashkent region. - Tashkent: Publishing House "Fan" of Uzbekistan 2014. - p. 16-17. – 115 p.

8. Esau K. Plant anatomy. - Moscow: Ed. Mir, - 2019. - P. 138-416.
9. Filimonova Z.N. Genus Oxytropis DC. – Ostroldochnik // Key to Plants of Central Asia. T. 7. - Tashkent, 2013. - P. 323-368.
10. Gleason H.A., Cronquist A. Manual of vascular plants of northeastern United States and adjacent Canada. – N.Y. Bot. Garden, 2011. - P. 284-285.
11. Grubov V.I. Plants of Central Asia. - St. Petersburg, 2018. - Issue. 8(b). – P. 1–89.
12. Hultén E. Flora of Alasca and neighboring territories. – Stanford Univ. Press, 2018. - P. 655-667.
13. Kiseleva N.S. Anatomy and morphology of plants. - Minsk: Ed. Graduate School. - 2011. - P. 89- 119, 2015-227.
14. Kitagawa M. Neo-Lineamenta florae Manshuricae. - J. Cramer, 2019. - P. 409-411.
15. Ohwi J. Flora of Japan. - Washington, D.C., 2015. - P. 575-576.
16. Pavlova N.S. Oxytropis DC. // Vascular plants of the Soviet Far East. - L., 2019. - T. 4. - pp. 236– 280.
17. Polozhiy A.V. Medicinal and promising for medicine plants of Khakassia. Tomsk. 2013. 160 p.
18. Polozhiy A.V. Oxytropis DC. // Flora of Siberia. - Novosibirsk, 2014. - T. 9. - P. 74-151.
19. Roskov Yu.R., Yakovlev G.P., Sytin A.K., Zheznyakovsky S.A. Legumes of Northern Eurasia: information system on CD-ROM. - St. Petersburg: SPKhFA Publishing House, 1998. - ISBN 5- 8085-0019-2.
20. Ulziyhutag N. Legumes of Mongolia. - Ulaanbaatar, 2013. - S. 210-282.
21. Vasilchenko I.T. New species of the genus Oxytropis DC. // Bot. mat. Coat of arms. Bot. inst. Academy of Sciences of the USSR, 2010. - Issue. 20. - P. 231-250.
22. Vasilchenko I.T. On the question of the genesis of the genus Oxytropis DC. // Bot. journal, 2015. - T. 50 (3). - P. 313-323.
23. Vasilchenko I.T., Fedchenko B.A. Oxytropis macrodonta Gontsch. (Leguminosae). Flora of the USSR. 1988. T. 13. - P. 110-111.
24. Yakovlev G.P., Sytin A.K., Roskov Yu.R. Legumes of Northern Eurasia: a Checklist. Royal Bot. gardens. - Kew, 2016. - P. 295-360.
25. Yurtsev B.A. Differentiation of species of the section Arctobia of the genus Oxytropis in Megaberingia // Vegetation cover of highlands. - L.: Nauka, 2016. - P. 90-100.
26. Zakharevich S.F. On the method of describing the epidermis of the leaf // Bulletin of Leningrad State University. - Leningrad, - 1984. - No. 4. - P. 65-75

Use of pathogenic microorganisms in the control of the Colorado potato beetle (*Leptinotarsa decemlineata* Say.)

Kalandar Babab‌ekov[1,*], *Ilhom* Xalilov[2], *Anvar* Abdullaev[2], *Maftuna* Kalandarova[1] and *Rayimjon* Alamuratov[1]

[1]Research Institute of Plant Quarantine and Protection, Tashkent, 100003, Uzbekistan
[2]Institute of Microbiology of the Academy of Sciences of the Republic of Uzbekistan, Tashkent, 100020, Uzbekistan

Abstract. The focus of this article is the comprehensive study of the Colorado potato beetle population within potato agro-biocenosis. The research delves into the intricate processes that contribute to the formation and bioecological characteristics of this beetle population. The article also delves into a specific strain known as *Beauveria bassiana* VTq-28, which was isolated from the Colorado potato beetle. This strain was subjected to testing both in laboratory settings and in the field, targeting various developmental stages of the Colorado potato beetle. The objective was to evaluate the effectiveness of *Beauveria bassiana* VTq-28 as a biocontrol agent against the beetle. Additionally, the research assessed the insecticidal activity of locally sourced strains of B. thuringiensis against the Colorado potato beetle. This analysis provides insights into the potential of *B. thuringiensis* strains as another avenue for biocontrol. By thoroughly examining the population dynamics, bioecological characteristics, and the potential of specific microbial agents for controlling the Colorado potato beetle, this study contributes to the understanding of pest management strategies within potato agro-biocenosis. The findings have implications for sustainable agricultural practices and the effective control of this economically significant pest.

Keywords. *Beauveria bassiana*, *B. thuringiensis*, biocontrol, microorganisms, Colorado potato beetle.

1 Introduction

To study infectious diseases of insects and determine their significance, it is necessary to have sufficient knowledge about microorganisms [1]. Therefore, for practical use in the fight against these pests of microorganisms that cause infectious diseases of insects, scientists in this field have proven the development of microbiological preparations based on them [2]. Several microbiological practices are being developed based on entomopathogenic viruses, bacteria, fungi, microsporidia and nematodes.

* Corresponding author: kalandar.bababekov@mail.com

The world is constantly expanding the use of microbiological preparations to protect plants from harmful organisms. According to experts, the market share of biopesticides was determined at 20% in 2020 and amounted to $8 billion. Regarding the scale of use, preparations based on *Bacillus thuringiensis* Berliner (Bt) were the first in the world. In 2009, practices based on *Bacillus thuringiensis* Berliner (Bt) were used on 50 million hectares worldwide, and 33 million hectares were found in the United States [3, 4].

The microorganism *Galleria mellonella* has been found to be pathogenic for a number of insect species after oral infection by Hurst et al. [5]. The combined use of entomopathogenic microorganisms *B. bassiana* and *B. thuringiensis* against the Colorado potato beetle showed that the synergism coefficient was high at low temperatures and led to a 100% death rate.

In the experiments of Wojda et al. [6], the entomopathogens B. bassiana and B. thuringiensis were isolated from insects that died from various infectious diseases, and the competitive relationship between the fungus and bacteria was also studied. Both pathogens have been found to coexist in dead insects. With a mixed infection, it was found that both fungi and bacteria are isolated from dead insects. Still, in most cases, entomopathogenic fungi are isolated at low temperatures. And at high temperatures - bacteria. As a result, an intensive development of bacteria at high temperatures was established.

Experiments were carried out [7-9] to determine the biological activity of the fungus *Beauveria bassiana* in relation to the larvae of the Colorado potato beetle. The experiment used natural pest populations in potato plantations of the Almaty region (Kazakhstan) [10]. Infection was carried out through an aqueous suspension of fungal conidia. When used in a titer of 1x107 ml per larvae, it was found that the death of larvae on the 7th day of registration was 32.0-47.0%, and on the 13th day - 85.0-100.0% [11-13].

In the fight against agricultural pests in various climatic conditions of Uzbekistan by the microbiological method

The relevance of biopreparations based on *B. thuringiensis* bacteria (dendrobacillin, bitoxibacillin, dipel, turicide, lepidocide, Veta Pro, Atibak-Uz, etc.) for the protection of crops from pests was noted [6-8]. Also, these microorganisms were studied and extensive scientific resources were created.

2 Materials and methods

Analysis was carried out in 2022 at the Research Institute of Quarantine and Plant Protection, the Laboratory for Pest Control of Cereals and Legumes, the Laboratory of Molecular Biology of the Institute of Microbiology of UzRF, and the Turkiston Shernazarova farm. Nargiza, Kybray district, Tashkent region in Uzbekistan. When performing scientific research, methods widely used in microbiology, mycology and entomology were used.

Adults and larvae of the Colorado potato beetle were collected, and those who died from the disease were subjected to microbiological analysis in the laboratory. The Colorado potato beetle with signs of a fungal disease had the upper part of its body covered with fungal mycelium. Colorado beetles with symptoms of bacterial infection were observed to be discoloured, runny and smelly. The upper body layer of the Colorado potato beetle was sterilised with 96.0% alcohol. Colorado beetles infected with fungal and bacterial diseases were placed in Petri dishes with fixed wet special paper (wet chamber) and grown at room temperature. To isolate microorganisms from diseased and dead Colorado potato beetles, they were cultivated on a nutrient medium and pure samples were taken. The necessary solutions were prepared from the biological material, and fungal species were identified using an NLCD-307B binocular microscope.

3 Results and discussion

To study the formation and bioecological features of the Colorado potato beetle population in the agro biocenosis of potatoes in 2022, studies were carried out on 2.0 hectares of potato plantations of the Turkiston Shernazarov Nargiza farm, Kibray district, Tashkent region in Uzbekistan. Research is carried out on potatoes "Santa" and "Romano" fields in the morning and evening planting.

During our observations, it was found that the Colorado potato beetle hibernates as adults at a depth of 15-50 cm. However, it was noted that when the soil warms up to 11-14 °C and relative humidity of 80%, the beetles begin to come out gradually. The beetles that emerged from the village were first fed with weeds from the sedge family, then planted on March 15.

On April 10, he began to feed in the field with fully sprouted potatoes. After feeding, the beetles lay up to 20–30 eggs in clusters under the potato leaves on which they were provided (Figure 1).

Figure 1. Monitoring work was carried out on 2.0 hectares of potato sown area of the farm "Turkiston Shernazarova Nargiza".

The embryonic development of the Colorado potato beetle depends on air temperature and relative humidity. At an air temperature of 22 °C and relative humidity of 70%, hatching larvae from eggs takes 3-5 days. The hatched larvae are the most harmful in potato crops and pupate in 15-20 days at an air temperature of 25°C and relative humidity of 60%. The larvae turned into cones at a depth of 1.5 to 15 cm. In some cases, they were observed to turn into cones on the soil surface at high soil moisture. The fungus development required 10–15 days at an air temperature of 30°C and relative humidity of 50%, during which a new generation of beetles develops (Figure 2).

According to the given data, Colorado potato beetles begin to leave the village in the second decade of April in the Kibray district of the Tashkent region. The development of beetles continued until the second decade of April. In the second decade of April, the beetles began to lay eggs, which continued until the first decade of May. The larvae started to emerge from the eggs in the first decade of May; the larval stage continued until the third decade of May. The transformation of larvae into cocoons coincided with the first decade of June.

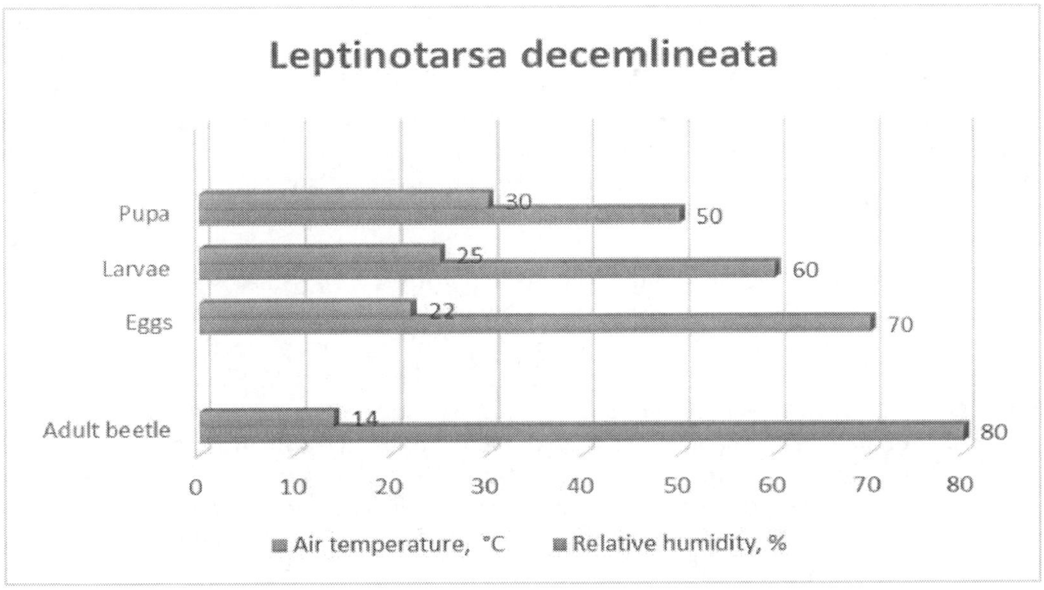

Figure 2. Influence of temperature and relative air humidity on the development of the Colorado potato beetle population in potato agro biocenosis

The hatching of larvae of the second generation started in the third decade of June, and the larval period continued until the second decade of July. The development of nymphs into cocoons corresponded to the second decade of July, and the emergence of beetles from cocoons of this generation corresponded to the third decade of July. The beetle period continued until the first decade of August (Table 1).

Table 1. Phenological table of the development of the Colorado potato beetle in the Kibray district of the Toshnetsky district (2022).

Month	**April**			**May**			**June**			**July**			**August**			**Sep**			**October**		
in the decade	**1**	**2**	**3**	**1**	**2**	**3**	**1**	**2**	**3**	**1**	**2**	**3**	**1**	**2**	**3**	**1**	**2**	**3**	**1**	**2**	**3**
1st generation		◙																			
		■																			
		●	●	●																	
				∞	∞	∞	○														
							■	■													
2nd generation								■													
								●	●												
									∞	∞	∞										
											○										
												■	■								
3rd generation													■								
													●	●							
														∞	∞	∞					
																	○				
																	■	■	■	□	

(◙) – A beetle from the village; ● – Eggs; ∞ - Larva; ○ - Sponge; ■ - Beetle; (□) – A beetle that went to the village.

The beetles of the third generation corresponded to the first ten days of August and began to lay eggs in the first ten days of this month; the oviposition period continued until the second ten days of August. The larvae hatched from the eggs in the second decade of August, and the larval period continued until the first decade of September. In the second decade of September, the larvae began to turn into cocoons; in the second decade of this month, beetles began to emerge from the cocoons, and the beetle period continued until the third decade of September. Wintering of the Colorado potato beetle was observed in the first decade of October.

In our experiment, we studied the pathogenicity of the B. bassiana VTq-28 strain grown on different nutrient media in a titer of 2×108 ml against the Colorado potato beetle. To this end, the strain B. bassiana VTq-28 isolated from the Colorado potato beetle was tested against various young Colorado potato beetles. First, multiple juveniles of the Colorado potato beetle collected from the field were reared in the laboratory. Then, after wrapping a bunch of 1 and 2 branches of the potato crop for food, each of them was treated with different titers of 1×108 and 2×108 ml to determine the effect of B. bassiana VTq-28 strain on small and large young larvae of the Colorado potato beetle in laboratory conditions. Pests became infected mainly when pest bodies and feed were used. In the control variant, it was treated with water.

Under laboratory conditions, the B. bassiana VTq-28 strain was used in liquid form. The results of experiments on the sensitivity of the strain B. bassiana VTq-28 to tiny young larvae of the Colorado potato beetle are given in the table. 4.7. According to the results of the experiment, when the B. bassiana VTq-28 strain was treated with a 1 × 108 ml titration solution of 35.0%, 48.3%, 73.3% when used in a 1 × 108 ml titer against adult larvae, it was 25.0%, 36 .6%, 65.0% on the 3rd, 7th and 14th days of accounting, and the above accounting days 13.3, 33.3% against its adults and the efficiency of 50.0%. According to the results of the experiment, when the strain B. bassiana VTq-28 was treated with 2 × 108 ml of a titer solution, on the 3rd day 41.6%, on the 7th day 63.3%, on the 14th day 85.0%, and in adult larvae versus 30.0%, 46.6%, 78.3% on the above reference days; and adults - 18.3%, 41.6% and 56.6%.

As a result of the application of the B. bassiana VTq-28 strain, the fungal action begins with the penetration of the spore through the skin (cuticle) of the insect into the body cavity. Once in the body, the spore turns into hyphae, the mycelium grows, and conidia are released. Conidia in the body pass into the hemolymph. At this stage, due to some strains' significant release of toxins, insects can cause harm, primarily affecting muscle tissue. The growth of the fungus continues until all tissues are destroyed. Then conidiophores appear, which break the cuticle and surround the dead larva. The infected insect looks like white cotton wool. Then the spores mature, and mass sporulation occurs. The insect dies 12-14 days after infection (Figure 3).

In 2018-2020, our research was carried out on the potato fields of the Turkestan Shernazarova Nargiza farm in the Kibray district of the Tashkent region to test the B. bassiana VTq-28 strain isolated from the Colorado potato beetle in the field. Field experiments were carried out with different juveniles of the Colorado potato beetle. The working liquid was applied at 300 l/ha using an OVKh-600 tractor sprayer against the Colorado potato beetle. The experiment was carried out during the mass development of Colorado potato beetle larvae on the Santa variety of potato culture. Before treatment, Colorado potato beetles in control and experimental fields ranged from 3.7 to 4.5 specimens. Beetles per plant, and the number of larvae is 23.4 to 30.7 ind.

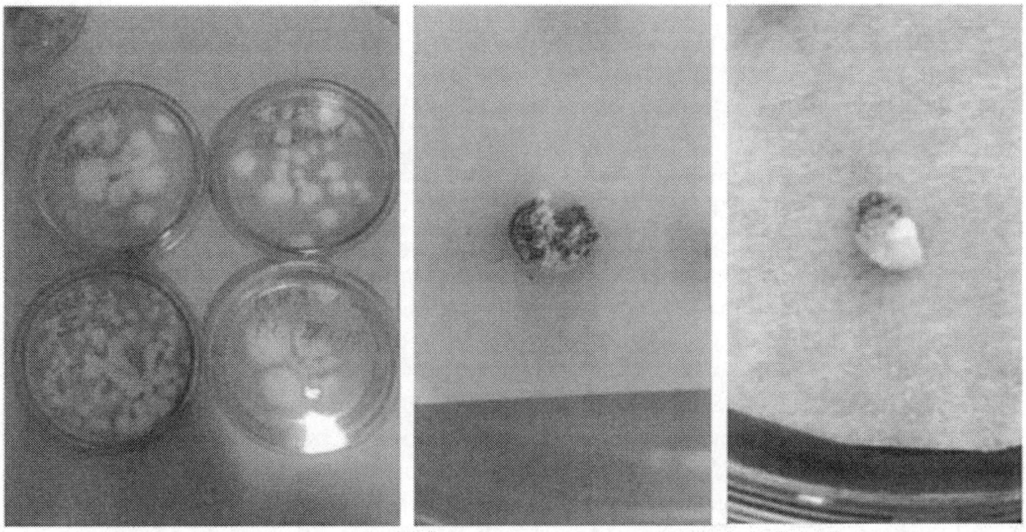

Figure 3. *B. bassiana* VTq-28 strain isolated from the Colorado potato beetle.

Our research obtained the following results when testing the strain B. bassiana VTq-28 against adults and larvae of the Colorado potato beetle. When using the strain B. bassiana VTq-28 against adults of the Colorado potato beetle in a titer of 1×108 ml on the 3rd, 7th and 14th day of counting, respectively, 15.4; 28.4; An efficiency of 41.0% and 17.3 was achieved on days 3, 7, 14 when used in a titer of 2×108 ml; 35.6; An efficiency of 47.3% has been achieved.

35.4 days in the above calculations in 1×108 ml titer against larvae; 46.3; An efficiency of 68.6% was observed, and when used in a titer of 2×108 ml, on the 3rd day of calculation 39.2; On the 7th day, an efficiency of 57.2% was achieved and 78.1% on the 14th day (Table 2).

Table 2. Biological effectiveness of the strain B. bassiana VTq-28 against the Colorado potato beetle (Field experience, Turkestan Shernazarova Nargiza farm, 2022).

#	Options	Titre	Average number of insects in 1 plant, pcs				Efficiency, % by days		
			# of insects before processing	Days after processing, pcs					
				3	7	14	3	7	14
Mature breed (beetle)									
1	Control (idle)	-	4.0	4.3	4.7	5.4	-	-	-
2	Bioslip BV (template) 3.0 l/ha	1×10^8	3.7	3.1	2.9	2.8	22.0	33.3	43.9
3	*B. bassiana* VTq-28	1×10^8	4.4	4.0	3.7	3.5	15.4	28.4	41.0
4	*B. bassiana* VTq-28	2×10^8	4.5	4.0	3.4	3.2	17.3	35.6	47.3
To the larvae									
1	Control (idle)	-	28.3	28.5	29.3	35.9	-	-	-

 https://doi.org/10.1051/e3sconf/202342104003

2	Bioslip BV (template) 3.0 l/ha	1×10^8	30.7	19.5	15.2	11.0	36.9	47.8	71.7
3	*B. bassiana* VTq-28	1×10^8	23.4	15.2	13.0	9.3	35.4	46.3	68.6
4	*B. bassiana* VTq-28	2×10^8	27.1	16.6	12.0	7.5	39.2	57.2	78.1

The obtained results conclude that the use of the strain B. bassiana VTq-28 in a titer of 2×10^8 ml showed a high efficiency of 78.1% against Colorado potato beetle larvae. Currently, in cooperation with the Institute of Microbiology of the Academy of Sciences of the Republic of Uzbekistan and the Research Institute of Quarantine and Plant Protection, scientific research is being carried out to determine the insecticidal activity of local strains of bacteria B. thuringiensis against the Colorado potato beetle and in the future to create bioinsecticide preparations based on these strains.

For this purpose, the bacteria B. thuringiensis 1, B. thuringiensis 3, B. thuringiensis 5, B. thuringiensis 9fo, B. thuringiensis 18fo, B. thuringiensis are stored in the collection of the Institute of Microbiology of the Academy of Sciences of the Republic of Uzbekistan to determine the insecticidal activity against Colorado potato beetle larvae. strains thuringiensis 31, B.thuringiensis 26, B.thuringiensis 54, B.thuringiensis 84, B.thuringiensis 91 and B.thuringiensis 94. Strains B. thuringiensis 1, B. thuringiensis 3, B. thuringiensis 5, B. thuringiensis 9fo, B. thuringiensis 18fo, B. thuringiensis 31, and B. thuringiensis 26 exhibited different insecticidal activity against the wax moth and a heat-resistant β-exotoxin. The formation is noted. It is known from the literature that the activity of strains of bacteria B. thuringiensis synthesising β-exotoxin against the Colorado potato beetle was studied. The strains of bacteria B. thuringiensis used by us, synthesising β-exotoxin, exhibited different levels of insecticidal activity. In addition, these strains were found to synthesise β-exotoxin and produce crystalline d-endotoxin (Figure 4).

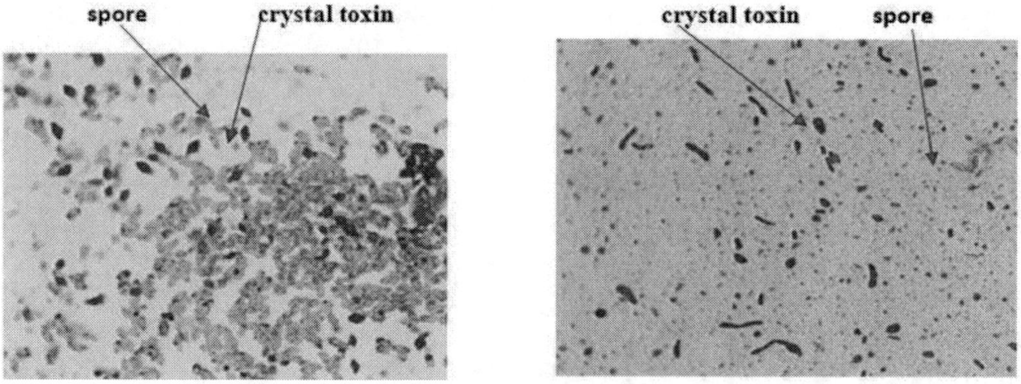

Figure 4. Microscopic view (1000x magnification) of bacterial strains B. thuringiensis 18fo and B. thuringiensis 26.

The strains B. thuringiensis 54, B. thuringiensis 84, B. thuringiensis 91 and B. thuringiensis 94 used in the experiment were not tested for the synthesis of β-exotoxin. Still, their d-endotoxin activity was studied on honey wax (Figure 5).

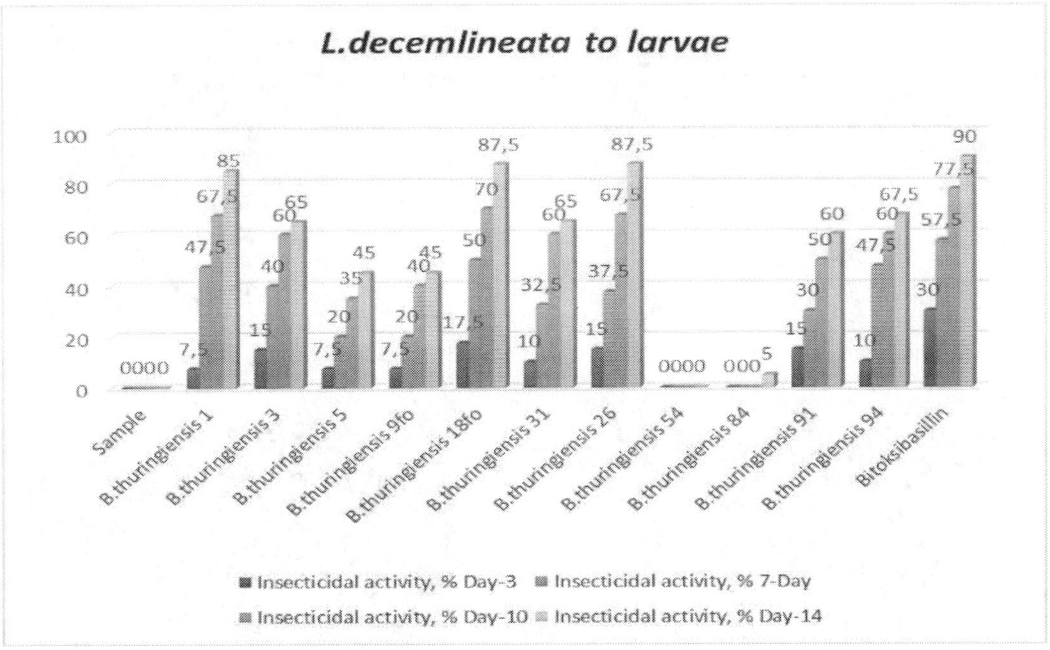

Figure 5. Insecticidal activity of indigenous Bacillus thuringiensis strains against the Colorado potato beetle.

As shown in Figure 5, during the 14-day experiment, B. thuringiensis 18 fo, B. thuringiensis 26, and B. thuringiensis 1 showed the highest insecticidal activity and killed 87.5%, 87.5%, and 85.0% of the insects, respectively. y it was determined that he had died (Figure 6). In addition, when strains of Bacillus thuringiensis were applied to 2-3-year-old larvae of the Colorado potato beetle, it was found that the external morphology of the beetle, which passed into the mushroom phase in subsequent generations without dying from bacterial deformation, and then to the state of emago, was deformed. Reproduction (laying eggs) in beetles affected by bacteria was not observed. It is important to note that the time of manifestation of the most excellent insecticidal activity of these studied strains is 7- and 10-days, where 60% of the total estimated insecticidal activity was recorded on these days (Figure 6).

The average insecticidal activity of B.thuringiensis 3, B.thuringiensis 5, B.thuringiensis 9fo, B.thuringiensis 91 and B.thuringiensis 94 strains was 65.0%, 45.0%, 60.0% and 67.5%, respectively. The lowest level of insecticidal activity was noted in the B. thuringiensis 84 strain (5%), and the B. thuringiensis strain 54 did not show insecticidal activity at all.

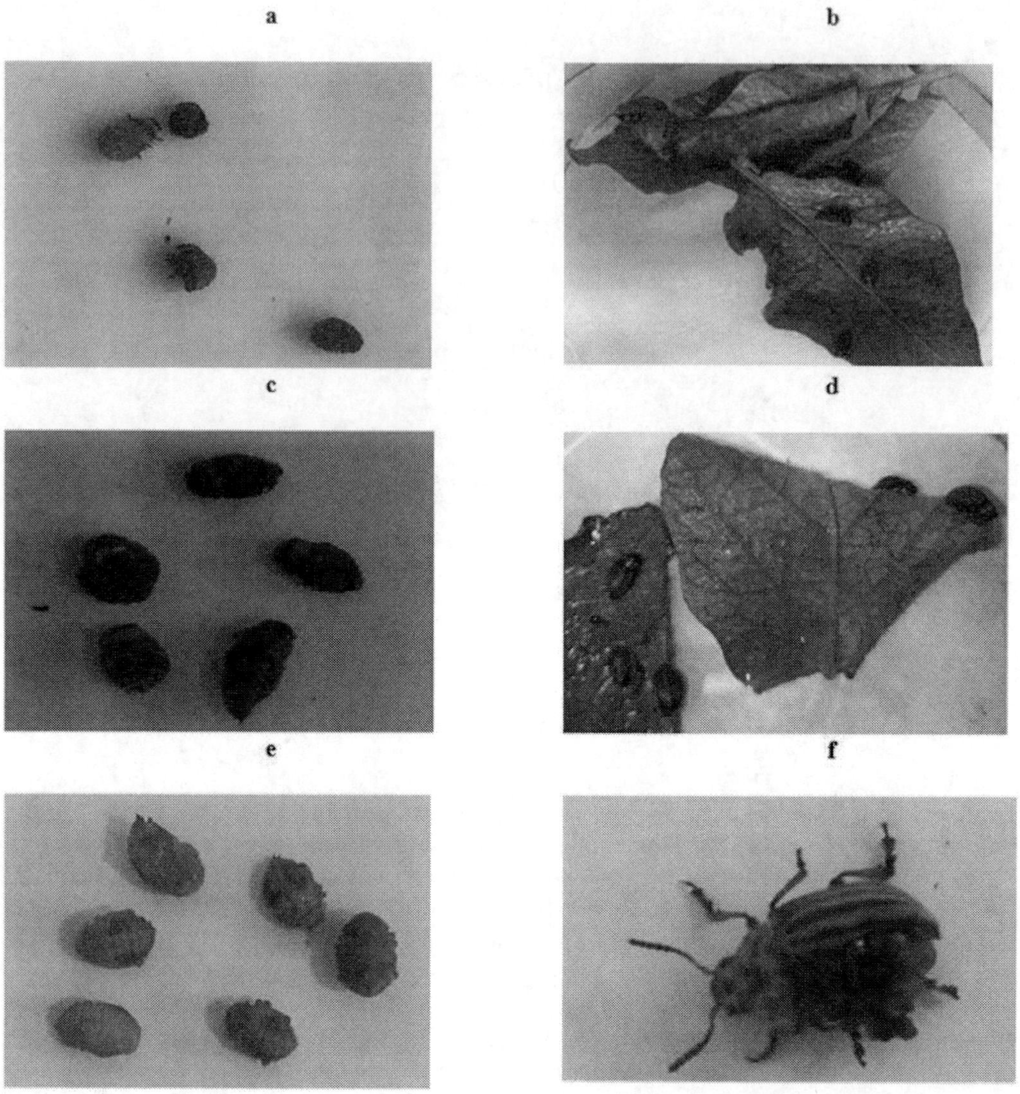

Figure 6. A – treatment of the 2nd and 3rd generations of the Colorado beetle with B. thuringiensis 18 fo strain; B – dead insects; C – the 2nd and 3rd generations of the Colorado beetle were treated with B. thuringiensis strain 26; D – dead insects; E – the state of the fungus after treatment of the 2nd and 3rd generations of the Colorado beetle with B. thuringiensis strain 26; and, F – View of the Colorado beetle in a deformed, mutated state of the beetle emerging from the cocoon.

4 Conclusions

Thus, the results of the studies showed that the strains of B. thuringiensis bacteria used in the experiment exhibited different insecticidal activity against the Colorado potato beetle. The strains B. thuringiensis 18 fo, B. thuringiensis 26, and B. thuringiensis 1 showed the highest insecticidal activity. Their external morphology became disabled. This means that although bacterial toxins do not affect some first-generation Colorado potato beetles, they can affect the structure of the insect's genome, mutate it and cause damage to insects emerging from adults. It is important to note that the mutated insects were not observed to lay eggs and produce offspring.

References

1. Abbott W.S. A method of computing the effectiveness of an insecticide // J.Econ. Entomol. 2015. -№18. - p. 265–267.
2. Sanahuja G., Banakar R. and all. *Bacillus thuringiensis*: a century of research, development and commercial applications. Plant Biotechnol. J., 2021, 9: P.283-300
3. Hurst M., Zdybicka-Barabas A., Cytrynska M. Temperature-Dependent Galleria mellonella Mortality as a Result of Yersinia entomophagy Infection // Applied and Environmental Microbiology. 2015. V.81. P. 18.
4. Wojda Iwona, Taszlow Paulina, Jakubowicz Teresa. The effect of cold shock on the immune response of the greater wax moth *Galleria mellonella* after infection with entomopathogenic bacteria *Bacillus thuringiensis* // Annales UMCS, Biologia. 2015. V. 69. I. 2. P. 7–18.
5. Bobobekov Q.B.,Sayimov N.Q., Qalandarova M.M. Cultural and morphological features of the entomopathogenic fungus *Beauveria bassiana* isolated from the Colorado potato beetle. Annals of the Romanian Society for Cell Biology Romania, 2021. P. Annals of R.S.C.B., ISSN:1583-6258, Vol. 25, Issue 4, 2021, P. 13408- 13412
6. Bobobekov Q.B.,Sayimov N.Q., Qalandarova M.M. Pathogenic features of the strain Beauveria Bassiana vtq-28 separated from the Colorado potato beetle in the condition of Uzbekistan. Journal of critical review vol 7, ISSUI 14, 2020. P. 3308-3312.
7. Noor, W., Zafar, M., Ahmad, M., Althobaiti, A. T., Ramadan, M. F., Makhkamov, T., ... & Khan, A. (2023). Petiole micromorphology in Brassicaceous taxa and its potential for accurate taxonomic identification. Flora, 303, 152280. DOI:10.1016/j.flora.2023.152280
8. Aziz, A., Ahmad, M., Zafar, M., Gaafar, A. R. Z., Hodhod, M. S., Sultana, S., ... & Chaudhay, B. (2023). Novel Copper Oxide Phyto-Nanocatalyst Utilized for the Synthesis of Sustainable Biodiesel from Citrullus colocynthis Seed Oil. Processes, 11(6), 1857. DOI:10.3390/pr11061857
9. Tayjanov, K., Khojimatov, O., Gafforov, Y., Makhkamov, T., Normakhamatov, N., & Bussmann, R. W. (2021). Plants and fungi in the ethnomedicine of the medieval East-a review. Ethnobotany Research and Applications, 22, 1-20. DOI:0.32859/ERA.22.46.1-20
10. Jabeen, S., Zafar, M., Ahmad, M., Althobaiti, A. T., Ozdemir, F. A., Kutlu, M. A., ... & Majeed, S. (2023). Ultra-sculpturing of seed morphotypes in selected species of genus Salvia L. and their taxonomic significance. Plant Biology, 25(1), 96-106. DOI:10.1111/plb.13473
11. Majeed, S., Ahmad, M., Ozdemir, F. A., Demirpolat, A., Şahan, Z., Makhkamov, T., ... & Nabila. (2023). Micromorphological characterization of seeds of dicot angiosperms from the Thal desert (Pakistan). Plant Biosystems-An International Journal Dealing with all Aspects of Plant Biology, 157(2), 392-418. DOI:10.1080/11263504.2023.2165553
12. Majeed, S., Ahmad, M., Ali, A., Althobaiti, A. T., Ramadan, M. F., Kilic, O., ... & Sultana, S. (2023). Pollen micromorphology among amaranthaceous species from desert rangeland: Exine stratification and their taxonomic significance. BioMed Research International, 2023, 4967771. DOI:10.1155/2023/4967771
13. Makhkamov, T., Sotiboldiyeva, D., Mamarakhimov, O., Yuldashov, Y., & Botirova, L. (2022, May). Morphogenesis and Seasonal Developmental Rhythm Under the Conditions of Introduction of Curcuma Longa L. In International Scientific Conference on Agricultural Machinery Industry "Interagromash"" (pp. 1460-1469). Cham: Springer International Publishing. DOI:10.1007/978-3-031-21432-5_155

Increasing the toxic effects of fluorine compounds of winter wheat plant with the help of bacteria

Bakhron Boynazarov[1,*], *Gulnora* Djumaniyazova[2], *Tolib* Berdiyev[3], *Avaz* Islamov[1] and *Foziljon* Abdullayev[1]

[1]Tashkent State Agrarian University, 2, University street, Tashkent, 100140, Uzbekistan
[2]Tashkent State Technical University, 2, University street, Tashkent, 100095, Uzbekistan
[3]Research Institute of Soil Science and Agrochemistry, 3, Qamarniso street, Tashkent, 100179, Uzbekistan

Abstract. Wheat stands as a prominent cereal crop, holding significant importance in both food and production realms. In the cultivation of winter wheat within soil environments tainted with fluorine compounds, the application of mineral and biological fertilizers becomes imperative. This amalgamation of beneficial bacteria, coupled with growth-stimulating properties, plays a pivotal role in averting the uptake of toxic substances by plants. The article delves into experimental findings that shed light on the efficacy of TERIA-S bacterial fertilizer in diminishing the fluorine content across various components of wheat plants. These components encompass roots, leaves, stems, and grains. The research focuses on wheat cultivated in soil regions afflicted by fluoride contamination within the Surkhandarya region of Uzbekistan. Through systematic experimentation, the study highlights the capacity of TERIA-S bacterial fertilizer to mitigate the presence of fluorine in wheat plants grown within fluoride-laden soil conditions. This outcome bears significant implications for improving the quality and safety of wheat crops in regions plagued by fluorine contamination. By elucidating the potential of microbial interventions in addressing this ecological challenge, the research contributes to the arsenal of strategies aimed at safeguarding agricultural productivity and human health.
Keywords. Soil, fluorine compounds, TERIA-S bacterial fertilizer, wheat plants, enzymes, active oxygen,

1 Introduction

One of the environmental pollutants, fluorine is naturally present in rocks, coal, clay and soil and is the 13th most abundant element in the earth's crust [1]. Fluorides enter the biosphere through volcanic eruptions, coal-fired power plants, aluminum smelters, phosphorous fertilizer plants, and glass, brick, tile, and plastic factories [2, 3]. Fluorine exists in the environment mainly as hydrogen fluoride, metal fluorides and fluorine gas.

* Corresponding author: bakhron_boynazarov1001@mail.ru

Fluoride is a colorless gas formed by the reaction of fluorine with hydrogen and readily dissolves in water to form the corresponding acid [4].

Although a number of scientists have described the harmful effects of fluorine compounds on humans, plants and animals in their studies [5-8], no information has been provided on the positive effects of this element. Fluoride can reduce crop yields by up to 50 percent because most plants are extremely sensitive to fluoride. Harmful effects of fluorine compounds are high on cereal crops, especially wheat [9, 10]. Fluorine is a strong oxidant that causes oxidative changes in plant cells. This can increase the amount of ROS and inhibit the activity of antioxidant enzymes [11].

Biological fertilizers play an important role in improving the quality and productivity of agricultural products, because even in small concentrations they are able to increase plant growth and development, resistance to toxic effects [12]. A high amount of fluorine in the soil slows down the development of soil microorganisms, inhibits its enzymatic activity, disrupts the nitrogen regime and reduces soil fertility. It is desirable to use soil and rhizosphere bacteria when growing plants in soil contaminated with fluoride. The positive effect of rhizobacteria is manifested in the increase of plant productivity, their resistance to drought and the increase of phosphorus [13].

Banerjee et al. [14] isolated strains of *Bacillus* sp. that were resistant to 1500 mg L-1 fluoride and helped to reduce its levels in plants. These studies make it possible to introduce bioremediation techniques that reduce the harmful effects of fluorides [14].

Taking into account the above, the purpose of the research is to test TERIA-S bacterial fertilizer in order to reduce the poisoning of irrigated land with fluorine compounds and increase its resistance to harmful effects, and to develop an agrobiotechnology that will reduce the harmful effects of fluorine compounds in the cultivation of winter wheat.

2 Materials and methods

Agrochemical test experiments in Surkhandarya region of Uzbekistan adjacent to TajAK: (1) weakly infected soil (Shorchi district, Surkhandarya region, 75 km); (2) moderately damaged soil (Surkhandarya region, Denov district, 50 km); and, (3) conducted in agricultural production fields with heavily contaminated soil (Saraasia District, Surkhandarya Region, 15 km). In the experiments, 3 types of soil contaminated with fluoride were used.

Field trials were conducted under typical gray soil conditions under newly irrigated conditions, with 3 returns. The area size of each experimental unit was (20 m × 4.8 m) = 96 m^2, the total area of the experiment was 1440 m^2.

Laboratory experiments were conducted with bacteria and 5 liquid samples containing fluorine at different concentrations in order to study the ability of bacteria isolated in laboratory experiments to grow in an environment with fluorine added. Laboratory experiments were carried out according to the methods of setting up laboratory experiments in plants [15].

3 Results and discussion

Soil samples were taken from the 0-10, 10-20 and 30-40 cm layers of the Surkhandarya region of Uzbekistan, 16, 35, 50, 75 and 100 km away from the TajAK industrial plant. Chemical analyzes were carried out from the obtained samples in the appropriate order.

According to the obtained results, in the soils of the Sufiyan massif of Sariosiyo district, 16 km away from the source, easily soluble fluorine in the 0-10 cm layer increased by 24.8 mg/kg in spring and 26.0 mg/kg in autumn, the same situation in 10-25 cm and 25- 20.5 -

21.7 mg/kg proportionally in 50 cm layers; It was found to be 15.5-15.3 mg/kg. These indicators were found to be 1.5-2.6 times higher than the authorized standard (AS - 10 mg/kg) for the soils of this area.

It was noted that in the area of Uzun district (16 km from the source), where typical gray soils, which were irrigated in the past, are at a high level (2.5-3.5 times AS) with easily soluble fluoride in water.

It was found in the analyzes that in the 0-25 cm layers of soils in Uzun district and Denov district located 35 and 50 km away from the polluting source, water-soluble fluorine is 2.0-2.9 times higher than the permissible limit in autumn (Table 1).

Table 1. Amount of fluorine in soils at different distances from the polluting source in Surkhandarya region of Uzbekistan, mg/kg.

An object, an array	Sample time taken	0-10 cm	10-25 cm	25-50 cm
16 km from the source (Sariosiyo district)	Spring	24.8	20.5	15.5
	Autumn	26.0	21.7	15.3
35 km from the source (Uzun district)	Spring	22.3	20.6	15.0
	Kuzgi	26.3	29.2	23.5
50 km from the source (Denov district)	Bahorgi	16.8	11.6	14.0
	Kuzgi	22.4	20.1	15.4
75 km from the source (Shorchi district)	Bahorgi	10.5	8.1	9.3
	Autumn	13.3	11.6	9.6
100 km from the source (Kumkurgan district)	Bakhorgi	10.7	6.9	3.1
	Autumn	10.6	7.8	4.0
Standard	**10 mg/kg**			

In the 0-50 cm layers of the soils of Shorchi District and Kumkurgan District, which are 75, 100 km away from the polluting source, it can be seen that easily water-soluble fluorine is not much higher than the permissible norm or within the norm. Fluoride, which is easily soluble in water, can be transferred to cultivated agricultural crops and thus produce damage. Fluoride in soil enters bacteria mainly in the form of HF [16]. Intracellular fluorides cause acid inhibition of enzymes, such as catalase, which can cause lipid oxidation and disrupt various metabolic pathways [17].

The demand for biological and organic preparations that stimulate growth and development processes to protect plants and increase soil fertility while improving the ecological situation in agroecosystems is increasing. Among them, TERIA-S bacterial fertilizer, produced on the basis of effective microorganisms, is widely used in agricultural practice. In our experiments, we carried out scientific research on the use of the above-mentioned bacterial fertilizer to grow winter wheat in soil conditions relatively polluted with fluorine compounds.

Plant seed germination is one of the most important criteria for evaluating the quality of crop seeds [18]. For this reason, based on laboratory experiments, we studied the effect of TERIA-S bacterial fertilizer on the germination of wheat seeds in soil conditions affected by fluorine compounds at different levels. This process is the first developmental stage of the physiological process of plants and continues with the growth of vigor.

Wheat seeds in the experimental variants were treated with TERIA-S bacterial fertilizer, and seeds in the control variant were treated with water. After one day of experiment, we observed good germination of wheat seeds and further development of seedlings was better compared to water treated seeds.

Therefore, we used TERIA-S bacterial fertilizer on seeds to carry out laboratory studies on soil with varying degrees of fluorine contamination. Figure 1 shows data on the growth and development of the root system of wheat seedlings treated with TERIA-S bacterial fertilizer for wheat seeds.

Figure 1. Effect of TERIA-S bacterial fertilizer on the development of wheat root system after 30 days of experiment (1-weakly damaged soil; 2-moderately damaged soil; 3-severely damaged soil; H-control, wheat treated with water before sowing seeds; and, T-experiment, wheat was treated with TERIA-S bacterial fertilizer before planting on seeds).

After a 30-day laboratory experiment with wheat, the effect of biometric parameters on the length of wheat seedlings was determined (Table 2).

Table 2. Effect of TERIA-S bacterial fertilizer on 30-day wheat seedling length in soils contaminated with different levels of fluoride.

Experience options	Root length, cm		Length of the stems, cm		Total length of seedlings, cm	
	control	experience	control	experience	control	experience
Weakly damaged soil	12.1 ±0.38	13.1 ±0.3*	19.2 ±0.26	22.0 ±0.28*	31.0 ±0.52	35.0 ±0.28*
Moderately damaged soil	10.4 ±0.08	12 ±0.28*	31.0 ±0.38	32.0 ±0.5*	41.0 ±0.38	44.0 ±0.42*
Severely damaged soil	10.6 ±0.13	11.3 ±0.4*	30.0 ±0.32	31.0 ±0.46*	40.6 ±0.32	42.2 ±0.36*
* P ≤ *reliable compared to control*						

Laboratory experiments were carried out on growing winter wheat in laboratory conditions and determining changes in biometric parameters when TERIA - S bacterial fertilizer was used in soil conditions contaminated with fluorine compounds. The reason for low biometric k-indicators of wheat in the control option is the presence of fluoride in the soil. It has been proven that the effect of TERIA-S bacterial fertilizer on 30-day-old wheat seedlings in soils with different levels of fluoride contamination improved the biometric parameters several times (Table 3).

Table 3. Effect of bacterial fertilizer TERIA-S on the weight of 30-day-old wheat seedlings in soils contaminated with different levels of fluoride.

Experience options	Root weight, g		Stem weight, g		Total weight of seedlings, g	
	control	experience	control	experience	control	experience
Weakly damaged soil	2.32 ± 0.02	7.58 ±0.04*	1.35 ± 0.03	2.15 ±0.04*	3.67 ± 0.03	9.69 ±0.04*
Moderately damaged soil	3.54 ± 0.024	7.8 ±0.06*	1.40 ± 0.02	2.87 ±0.03*	4.94 ± 0.04	10.67 ± 0.04
Severely damaged soil	3.24 ± 0.04	5.40 ±0.07*	1.47 ± 0.02	1.49 ±0.02	4.69 ±0.05	6.89 ±0.04*
* P ≤ *more reliable than control*						

Laboratory experiments were carried out on growing winter wheat in laboratory conditions and determining changes in biometric indicators in soil conditions contaminated with fluorine compounds when TERIA - S bacterial fertilizer was used. It has been proven that TERIA-S bacterial fertilizer has repeatedly improved the biometric parameters of 30-day-old wheat seedling length and weight in soils contaminated with fluoride at different levels.

4 Conclusions

In the experiments carried out by us, it was proved that the biometric parameters such as the length and weight of wheat seedlings improved several times after planting these wheat grains with "TERIA-S" bacterial fertilizer on soils contaminated with fluoride at different levels. This fact shows that soil bacteria with a complex of beneficial properties contained in TERIA - S bacterial fertilizer develop well in the soil around the roots contaminated with fluorine to different degrees , allowing plants to develop better than the control.

Thus, it was determined that the use of TERIA - S bacterial fertilizer gives positive results in the deactivation of soil contaminated with fluorine compounds .

References

1. Rakshit PK Studies on estimation of fluoride and fluoridation of drinking water," Dissertation, IISc Bangalore, 2014.
2. Zohoori FV, Moynihan PJ, Omid N., Abuhaloob L, Maguire A. Impact of water fluoride concentration on the fluoride content of infant foods and drinks requiring preparation with liquids before feeding, Community Dental Oral Epidemiology, vol.40, 2012, pp. 432-440
3. Jagtap S., Yenkie MK, Labhsetwar N., Rayalu S. Fluoride in Drinking Water and Defluoridation of Water, Chemical Review, vol.112, 2012, pp. 2454-2466.
4. Polonsky V.I., Polonskaya D.E. Fluoride contamination of soil and phytoremediation // Agricultural biology. 2013. - No. 1 (18). - P. 3-14.

5. Ram A., Verma P., Gadi BR Effect of fluoride and salicylic acid on seedling growth and biochemical parameters of watermelon (Citrullus lanatus), Fluoride, vol. 47, 2014, pp. 49-55.
6. Pal KC, Mondal NK, Chatterjee S., Ghosh TS, Datta JK Characterization of fluoride-tolerant halophilic Bacillus flexus NM25 (HQ875778) isolated from fluoride-affected soil in Birbhum district, West Bengal, India, Environment Monitoring Assessment, 2014; vol.186, pp. 699-709
7. Baunthiyal M., Ranghar S. Physiological and biochemical responses of plants under fluoride stress: An overview, Fluoride, vol. 47, 2014, pp. 287-293.
8. Mondal N. K., Pal K. C., Dey M., Ghosh S., Das C., Datta J. K. Seasonal variation of soil enzymes in areas of fluoride stress in Birbhum district, West Bengal, India, Journal of Taibah University of Science, vol.9, 2015, pp. 133-142.
9. Kumar, T.S., Dhakaand, K.P., Singh A. Eect of fluoride toxicity on the growth and yield of wheat (Triticum aestivum L.). Int. J. Forest. Crop Impr. 2013, 4, 59–62.
10. Choudhary S., Rani, M., Devika O.S., Patra A., Singh R.K., Prasad S.K. Impact of fluoride on agriculture: A review on it's sources. toxicity in plants and mitigation strategies. Int. J. Chem. Stud. 2019, 7, 1675–1680
11. Snioszek M., Telesi´nski A., Smolik B., Zakrzewska H. Eect of Fluoride and Bentonite on Biochemical Aspects of Oxidative Stress in Pisum sativum LJ Ecol. Eng. 2018, 19, 164–171
12. Bogoslovsky V. N., Levinsky B. V., Sychev V. G. Agricultural technologies of the future. Book 1: Energens. M.: Antikva, 2004. 163 p.
13. Ogloblina R.I. Influence of fluorine and sulfur compounds on the abundance and composition of the soil microflora of the Minusinsk depression in the book: Basic questions of agrochemistry and soil science, Pushchino 2017.
14. Banerjee G., Sengupta A., Roy T., Banerjee PP, Chattopadhyay A., Ray AK Isolation and characterization of fluoride resistant bacterial strains from fluoride endemic areas of west Bengal, India: Assessment of their fluoride absorption efficiency. *Fluoride* 2016, 49, 429
15. Armor B.A. Methods of laboratory and field experience. - M.: Agropromizdat, 1985. – 351 p.
16. Marquis E., Clock SA, Meira MM Fluoride and organic weak acids as modulators of microbial physiology, FEMS Microbiology Reviews, 2002.
17. Spets H.S., Seppa L., Korhonen A., Alakujala P. Accumulation of strontium and fluoride in approximal dental plaqueand changes in plaque microflora after rinsing with chlorohexidine-fluoride-strontium solution, Oral diseases, vol. 4, 2018, pp. 114-11
18. Filho J.M. Seed vigor testing: An overview of the past, present and future perspective. Sci. Agric. 2015, 72, 363–374.

Diagnostic signs of aboveground and underground organs of *Ferula Tenuisecta* Korov. in natural conditions

Guljan Duschanova[1,*], *Gulnara* Shakhmurova[1], *Mukhabbat* Ravshanova[2] and *Asiya* Maimataeva[3]

[1]Tashkent State Pedagogical University, Tashkent, 100070, Uzbekistan
[2]Navoi State Pedagogical Institute, Navoi, 210109, Uzbekistan
[3]Kazakh National Pedagogical University, Almaty, 050010, Kazakhstan

Abstract. This study presents a pioneering investigation into the anatomical composition of both above-ground and underground organs, including leaves, petioles, peduncles, pedicels, and roots of Ferula tenuisecta, within their natural ecological habitats. Notably, this research delves into the schizogenic type of tissue development found in the peduncle and root, elucidating its intricate connection to the robustness of the organ, which, in turn, aligns with the plant's distinct life form. The data gleaned from this study highlights distinctive and species-specific structural diagnostic attributes inherent in the above-ground and underground components of Ferula tenuisecta. These attributes carry significant potential for utilization in botanical systematics and the accurate identification of plant specimens within this species. Moreover, these findings have implications beyond taxonomy, as the revealed structural traits serve as valuable tools for detecting and assessing the presence of biologically active compounds within different organs and tissues. By shedding light on the intricate anatomical architecture of Ferula tenuisecta's various organs, this study contributes to a deeper understanding of the plant's ecological adaptations and offers practical avenues for its systematic characterization and potential applications in the realm of biologically active substances research.

Keywords. Anatomy, diagnostic signs, leaf, petiole, peduncle and rhizome, *Ferula tenuisecta*.

1 Introduction

Family *Apiaceae* Lindl. - umbrella, distributed throughout the globe. Among the species of this family are valuable vegetable, fodder, spicy-aromatic, medicinal and technical plants [1]. There is information about the presence of biologically active substances in many species [2]. Representatives of this family in all their parts contain essential oils or resinous substances, coumarins, flavonoids, less often saponins with pharmacological activity [3]. Plants containing a high amount of flavonoids are used as choleretic, juice-reducing, and capillary-

* Corresponding author: guljon.duschanova@mail.ru

strengthening agents [4]. Species of the genus Ferula are valuable essential oil (39), fodder (23), medicinal (17), melliferous (18), starchy (9) and food (8) [5, 6]. Many species of *Umbelliferae* are endemic, the study of these species is relevant due to the possibility of their extinction [7]. Species of the genus Ferula (family Apiaceae) occupy a special place not only among this family, but also in the entire flora of Uzbekistan. The genus Ferula L. belongs to the tribe *Peucedaneae* Dumort, subfamily *Apioidedae* Drude, family *Umbelliferae* Juss. (*Apiaceae* Lindl.). By the number of species (170), the genus *Ferula L.* occupies a leading position in the family. On the territory of Central Asia and Kazakhstan, 105 species are found. There are 45 species on the territory of Uzbekistan, of which 8 are endemics [8, 9]. The issues of taxonomy and phylogeny of the Central Asian *Umbelliferae* are highlighted and compiled a new classification of Ferula L. species growing in the CIS [10, 11]. In 48 species of the genus *Ferula L.* under the conditions of Kazakhstan, issues related to synonymy, carpology, ecology, distribution, chemical composition, and economic importance were studied [12, 13]. The life forms on the basis of a detailed biomorphological analysis of ontogeny patterns in 14 Eastern European umbrella species, identified ecological groups, and showed possible deviations in the long-life cycle depending on various habitat conditions [15, 16].

There is information in the literature on the molecular biology of the *Umbelliferae* family, including some species of the genus Ferula, which confirmed the monophyletic nature of the genus [17-20].

The comparative ecological and anatomical structure of the vegetative organs of Ferula species was studied by [21-25]. The formation of various kinds of secretory receptacles is characteristic of the progressive direction of evolution, which is of great importance in phylogenetic systematics [26]. The structure of the secretory receptacles of Ferula is an important taxonomic feature [27, 28]. Secretory structures in the vegetative organs provide protection from ultraviolet radiation, increase the drought resistance of plants, and contain flavonoids, terpenoid-containing substances that are medicinal or poisonous [29].

Many species of ferula are forage plants, especially in deserts and semi-deserts; some are used as hay plants. For many species of ferula, in particular for *F. ferulaeoides* and *F. foetida*, people's experience of using them as fattening and milk-producing feeds has been described [30]. The fruits of umbrella plants are readily eaten by all animals. The nutritional value of the entire above-ground mass is 82.3 feed units [31]. In the desert, the core of the young inflorescence of *F. foetida* is used as a vegetable. In folk and official medicine, *F. mogoltavica* Lipsky ex. Korov., *F. tadshikorum*, *F. kuhistanica* Korov., *F. fedtschenkoana* K.-Pol., *F. sumbul*, *F. kirialovii* M. Pimen, *F. tenuisecta* Korov., *F. penninervis* Regel et Schmaih. [32-35].

Terpenoid coumarins and sesquiterpene lactones predominate in monocarpic species of the genus [36] developed a classification of terpenoid-containing plant structures. According to the type of structure of the secretory system (schizogenous receptacles), plants of the genus Apiaceae are direct descendants of the ancient *Araliaceae* or can be derived from some common ancestral group with them. Coumarins, sesquiterpenoids, sesquiterpene lactones, esters, glucosides, flavonoids (luteolins, cynarosides), carbohydrates, polysaccharides, pyrosine phosphates, etc. found in Ferula species [37-41]. L.D. A method for the qualitative determination of the amount of esters of terpenoid alcohols in the substance Ferulen was developed, on the basis of which the Ferulen preparation was created [42]. Resins have been and are used as hemostatic, spasmodic, tonic, wound healing, expectorant, stimulant, tonic, anthelmintic and abortifacient. The range of treatment of diseases with ferula resin is very wide [43, 44].

The main losses of tefestrol and established the advantages and disadvantages of the existing technology in the production of the substance from the aerial parts of Ferula tenuisecta were identified [45, 46]. It was found that the final product obtained by the existing technology has a green color that does not meet the requirements of the pharmacopoeial article. The stage of extraction of the sum of esters of sesquiterpene alcohols from the aerial part of Ferula tenuisecta and purification of the obtained extract is proposed, which subsequently makes it possible to obtain the substance of tefestrol without green color. It has been shown that the use of activated carbon makes it possible to remove the green color of the extract. To purify tefestrol from other esters of sesquiterpene alcohols and related substances, chromatographic separation on silica gel was proposed under the following conditions: the ratio of the amount introduced into the column and the sorbent was 1:20; the ratio of the height of the sorption layer to the diameter of the column is 4:1; elution rate, 35 L/(m2 h). To obtain the pharmacopoeial substance of tefestrol, crystallization of a technical product from a mixture of ethyl acetate with hexane in a volume ratio of 1:3 was proposed. Based on the results obtained, a technology for obtaining the substance of tefestrol from the aerial part of *Ferula tenuisecta* was developed. According to this technology, 70.96% of the substance of tefestrol is obtained from the content in the raw material, according to the existing technology - 33.9%.

In this regard, of scientific interest is the study of the anatomical structure of the aboveground and underground organs of the raw material of medicinal plants *Ferula tenuisecta*, growing in the conditions of Uzbekistan, in order to determine the diagnostic features and localization of biologically active substances in organs and tissues.

2 Materials and methods

The objects of the study were *Ferula tenuisecta*, a perennial polycarpic medicinal plant with a thickened, fusiform root. The neck is densely wrapped with fibers of old dead leaves, many-headed. Stems 80-90 cm tall, strong, paniculate-branching in the upper part; the panicle is oblong-oval in outline, its branches are alternate below, several at the top are collected in whorls. The leaves are oval-rhombic in outline, rough from short, hard hairs, basal on long petioles; plate trifoliate, with 2 additional segments on the sides, repeatedly-(5)-dissected into numerous, 2-5 mm long, linear, grooved above, obtuse lobes; cauline with a reduced plate, their sheaths enclosing the stem at the base, leathery, oval-lanceolate, swollen; the upper ones are represented only by the sheaths. Umbrellas different, terminal 8-15-rayed, sessile or on short stalks, 6-8 cm wide, lateral arranged 2-5 on long stalks exceeding the terminal umbel. Umbrellas 10-flowered, with wrappers consisting of scaly, falling leaves. Cup without teeth. Petals yellow, oval, narrowed and curled at the top, 1.2 mm long. Fruits are oblong oval, convex, with thickened, pale edges, 9x5 mm long, width; ribs are convex; tubules in the furrows are single, wide, in the commissure there are 2-6 of them. Blooms in May; fruiting in June. Ferula tenuisecta grows on soft mountain slopes in the shrub belt. Distributed in the Tien Shan mountains at an altitude of 600-2500 m above sea level, where it occurs on rocky and grassy slopes, in steppes and bushes [47].

Samples of fresh plants were fixed in a 70% alcohol solution. Preparations prepared by hand were stained with methylene blue followed by gluing in glycerol-gelatin [48, 49]. Transverse sections of the leaf are made through the middle, petiole, peduncle, pedicel and rhizomes - the base. Descriptions of the main tissues and cells are given [50]. Microphotographs of the anatomical features of the aboveground and underground organs were taken with a computer microphotographic attachment with a Canon A123 digital camera under a Motic B1-220A-3 microscope.

3 Results and discussion

In the species *Ferula tenuisecta*, the leaves are oval-rhombic in outline, rough with short, stiff hairs, basal on long petioles; plate trifoliate, with 2 additional segments on the sides, repeatedly (5) - dissected into numerous, linear, grooved above, obtuse lobes, almost bare above, shortly pubescent below, stem-like with a reduced plate, their sheaths enclosing the stem at the base, leathery, oval-lanceolate, swollen; the upper ones are represented only by the sheaths.

The anatomical structure of the leaf in the studied species *Ferula tenuisecta* is the structure of the leaf mesophyll of the isolateral-palisade type [51], which is represented by palisade, spongy cells and vascular bundles. Epidermal cells are single-row, oval, larger on the adaxial side of the leaf than on the abaxial side. Between the adaxial and abaxial epidermis is an assimilation tissue consisting of palisade and spongy cells. The palisade parenchyma is two-row, largecelled and elongated with the presence of large intercellular spaces. The spongy parenchyma is round, small-celled, consists of 4-5 rows with small cavities. Palisade and spongy parenchyma cells are chlorophyll-bearing (Figure 1).

The main leaf vein protrudes on the abaxial side. The rest of the vein is occupied by the main thin-walled round-oval cell parenchyma. In the main vein of the leaf there is one large vascular bundle of closed collateral type. In the studied species, in the main leaf vein under the abaxial epidermis, there are groups of multi-row lamellar collenchyma cells. Localizations of biologically active substances in the leaves of the schizogenous receptacle type were determined, which are located above each main and lateral vascular bundles on the abaxial side of the leaf mesophyll and between the collenchyma and vascular bundles in the main vein of the leaf (Figure 1).

According to the literature data, the structure of Ferula secretory receptacles is an important taxonomic feature that provides protection from ultraviolet radiation in vegetative organs, increases drought resistance of plants, contains flavonoids, terpenoid-containing substances that are medicinal or poisonous.

The structure of the petiole. On the transverse section, the structure of the bundle-type petiole is oval-lanceolate with a swollen shape. In the studied species, epidermal cells are single-row, thin-walled, pubescent with simple, unicellular hairs. Under the epidermis are chlorophyll-bearing two-row subepidermal and 7-8 row parenchymal cells. The peripheral parts of the petiole are the most sclerified, due to the presence of a group of multirow lamellar collenchyma cells under the subepidermal cell, which alternate with 7–8 row chlorophyllon-bearing parenchyma. The localizations of biological active substances in petioles of the schizogenic type of receptacle, which are located above each peripheral vascular bundle and collenchymal cells, were determined. Conductive bundles of collateral type. Peripheral vascular bundles are represented by alternating large, medium and small numerous bundles, which are located in an annular shape, randomly in the central part of the petiole (Figure 2).

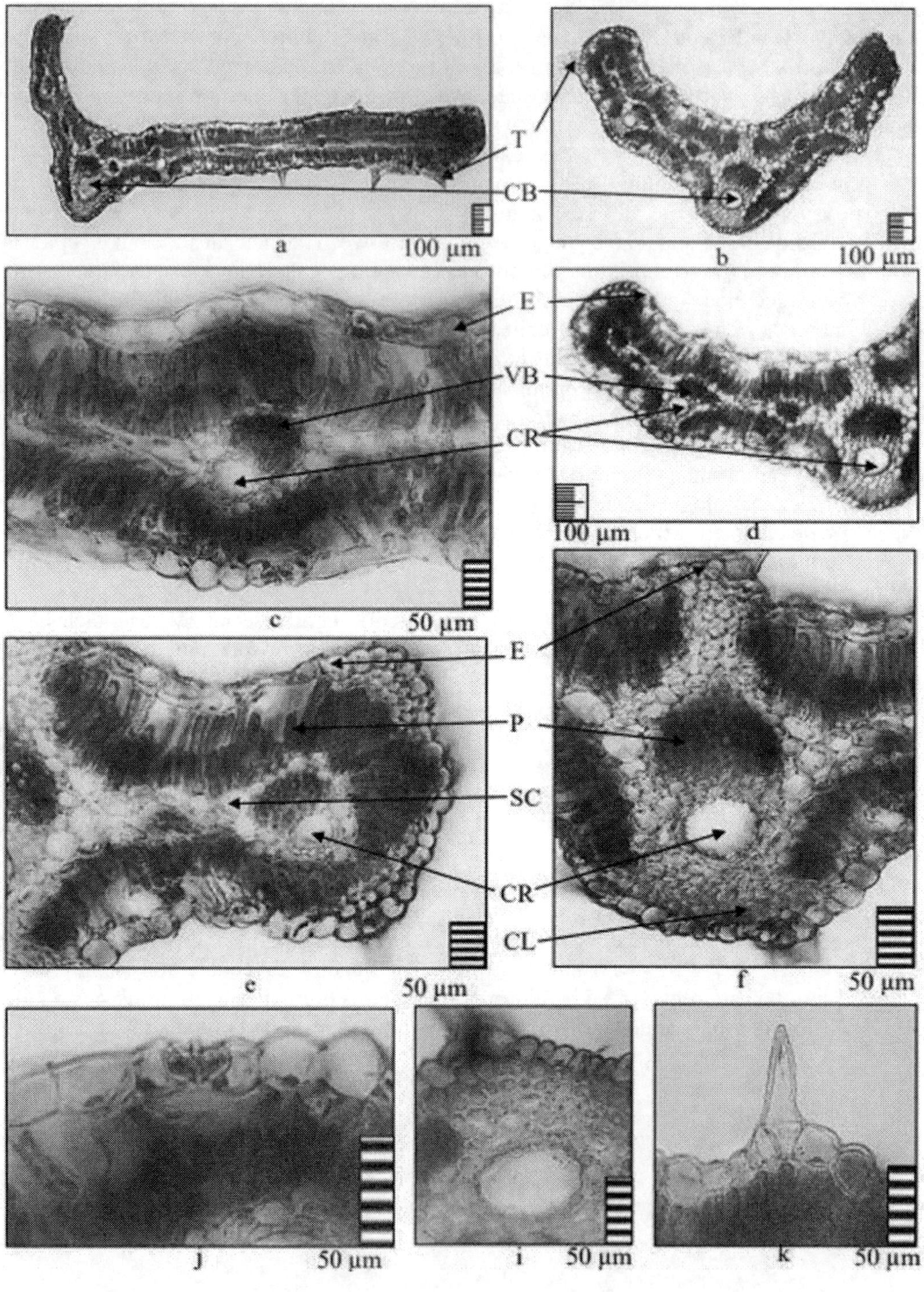

Figure 1. Anatomical structure of the leaf mesophyll of Ferula tenuisecta: a-b – general view of the leaf; c-e – detail of the leaf mesophyll; j – non-immersed stomata; f-i – schizogenic type secretory receptacles; k – single-celled simple trichomes. Legend: GP - spongy parenchyma; CL – collenchyma; P – palisade parenchyma; VB – vascular bundles; SR– secretory receptacles; T – trichome; E – epidermis.

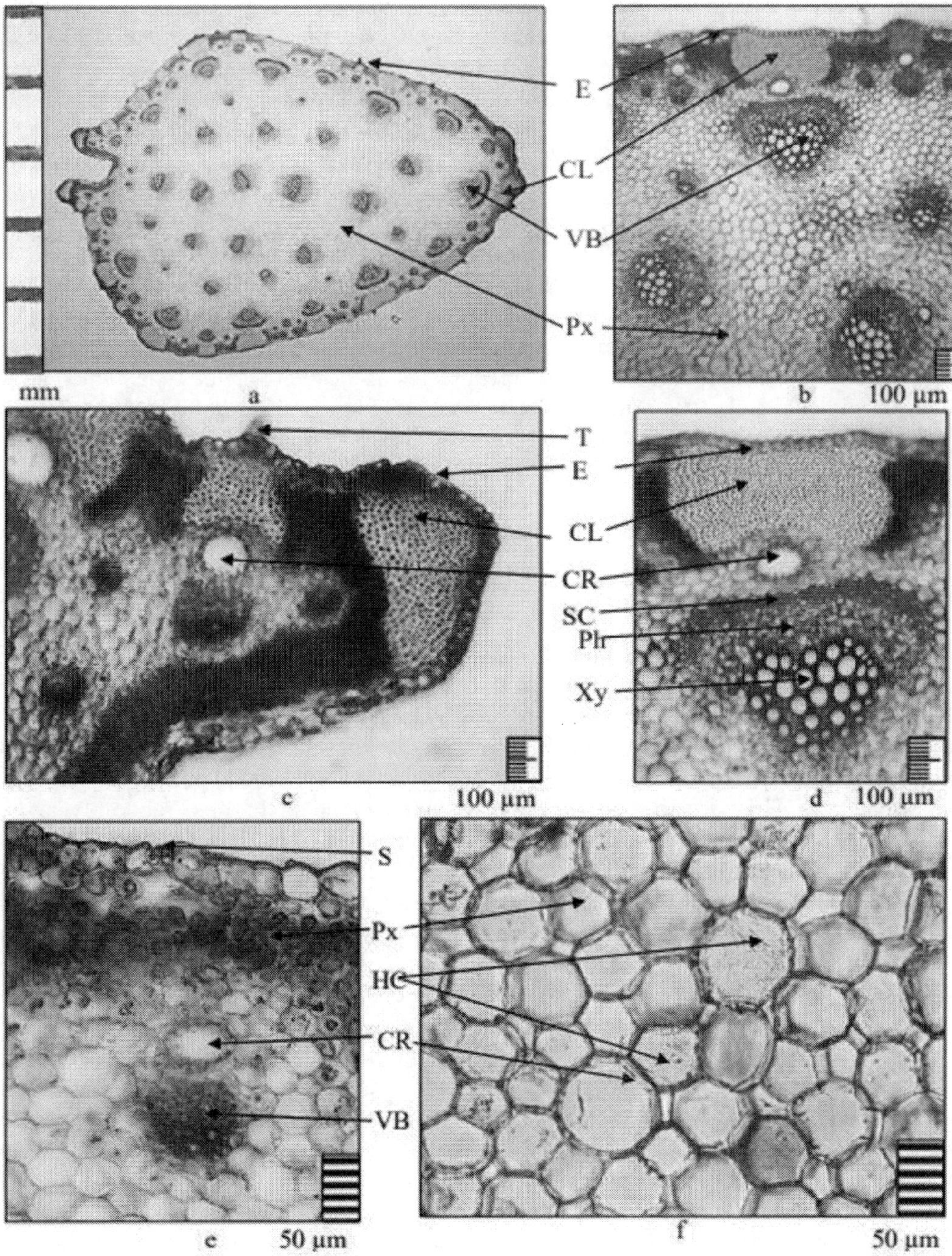

Figure 2. Anatomical structure of the petiole of the leaf of Ferula tenuisecta: a - general view of the petiole of the leaf; b-c - detail of the petiole of the leaf; d – epidermis, collenchyma, vascular bundles and secretory receptacles; e – non-immersed stomata, chlorophyllbearing parenchymal cells; vascular bundles and secretory receptacles; f – parenchymal and hydrocytic cells. Legend: HD - hydrocytic cells; CL—collenchyma, Xy—xylem, VB— vascular bundles, Px—parenchymal cells, CR—secretory receptacles, SC—sclerenchyma; T - trichome; Ph - phloem, E - epidermis.

The vascular bundles are most sclerified due to the bast fibers (sclerenchyma cells), which form multilayer strands in the form of caps over the phloem of the bundles, the central bundles are less sparsely layered. Above the peripheral vascular bundles, there is one secretory receptacle of the schizogenic type, which consists of 12-14 epithelial cells. Large peripheral vascular bundles from the side of the xylem have 2 secretory receptacles. In the central conducting bundles, there are three secretory receptacles of the schizogenic type, which consist of 9-10 epithelial cells. Parenchymal cells are thin-walled, large, round-oval. Among the parenchymal cells, there are hydrocytic cells (Figure 2).

Morphological and anatomical structure of the peduncle. The species Ferula tenuisecta is a perennial polycarpic plant, peduncle below 80-90 cm in height, strong, paniculate-branching in the upper part; the panicle is oblong-oval in outline, its branches are alternate below, several at the top are collected in whorls. In the studied species, the cross section of the base of the peduncle has a parenchymal-beam type of structure and an oval shape. The epidermis is single-row with a thick cuticle, the outer and inner walls of which are thickened. Under the epidermis there is a single-row hypodermis and 4-5 row chlorophyll-bearing cells of the core parenchyma, which alternate with areas of lamellar collenchyma (Figure 3). Under the collenchyma there are large secretory receptacles of the schizogenic type, which consist of 11-12 epithelial cells. The primary cortex is separated from the central cylinder by a ring of sclerenchyma. The thickness of this ring and the degree of lignification of the cells in some respects reflect the evolutionary advancement of the species. Sclerenchyma is thick-walled, consists of 4-5 rows of cells (Figure 3).

The central cylinder is extensive, parenchymal cells are thin-walled, large, and round-oval. Numerous vascular bundles are located among the thin-walled parenchymal cells of the central cylinder. Among the parenchymal cells of the central cylinder, hydrocytic cells are found. Peripheral conductive bundles are open and form a discontinuous circle, large conductive bundles alternate with small ones. The vascular bundles are most sclerified, due to the bast fibers (sclerenchyma cells), which form multilayer strands in the form of caps over the phloem of the bundles. Xylem is represented by large, numerous vessels located in groups and chains. In the core, near each conducting bundle, there are numerous large and small secretory receptacles of the schizogenic type. Medullary vascular bundles vary in shape, size, and orientation. They are located in the core radially and in groups, randomly and are represented by collateral and bicollateral types (Figure 3).

The anatomical structure of the pedicel in a transverse section of the parenchymal-fascicular type of structure and rounded shape. The epidermis is single-row with a thin-walled cuticle. Under the epidermis there is a single-row hypodermis and 4-5 row chlorophyll-bearing cells of the bovine parenchyma, which alternate with areas of lamellar collenchyma. Under the collenchyma there are large secretory receptacles of the schizogenic type, which consist of 11-12 epithelial cells. The primary cortex is separated from the central cylinder by a ring of sclerenchyma. The thickness of this ring and the degree of lignification of the cells in some respects reflect the evolutionary advancement of the species and perform the functions of power and support. Sclerenchyma is thick-walled, consists of 4-5 rows of cells. The central cylinder is extensive, parenchymal cells are thick-walled, large, and round-oval. Among the parenchymal cells of the central cylinder, there are hydrocytic cells (Figure 4).

Peripheral conductive bundles are open and form a discontinuous circle, large conductive bundles alternate with small ones. The vascular bundles are of the collateral type, the xylem of the bundles is represented by large, numerous vessels located in groups and a chain. Under the peripheral vascular bundles there are large numerous secretory receptacles of the schizogenic type, the central vascular bundles are small, few secretory receptacles (Figure 4).

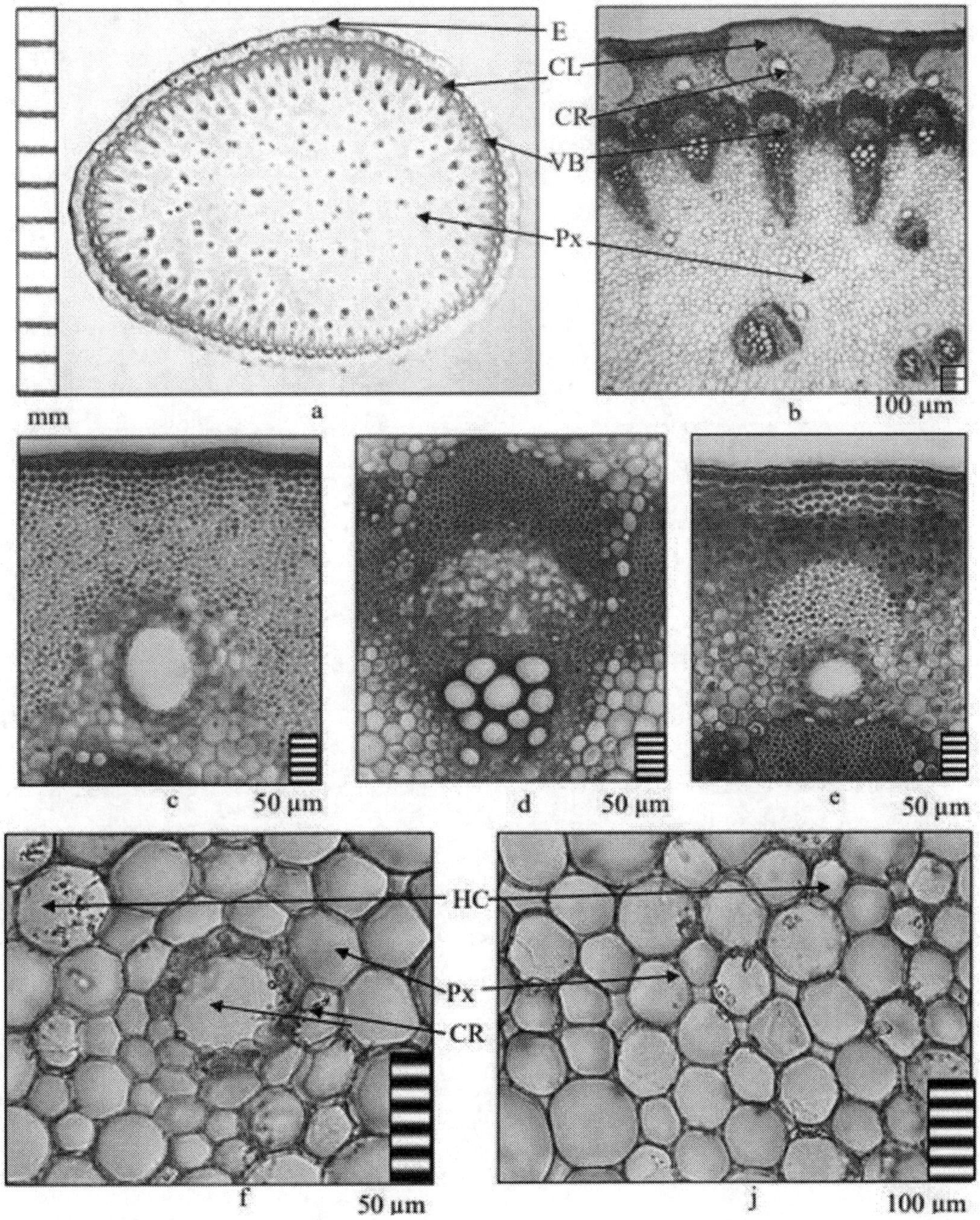

Figure 3. Anatomical structure of the peduncle of Ferula tenuisecta: a - general view of the peduncle; b - detail of the peduncle; d – conductive bundles; c, e – epidermis and collenchyma, large and small secretory receptacles; f - secretory receptacles; j – parenchymal and hydrolytic cells. Legend: HD - hydrocytic cells; CL – collenchyma; VB – vascular bundle; Px – parenchymal cells; SR – secretory receptacles; Ph – phloem; E – epidermis; EC – epithelial cells.

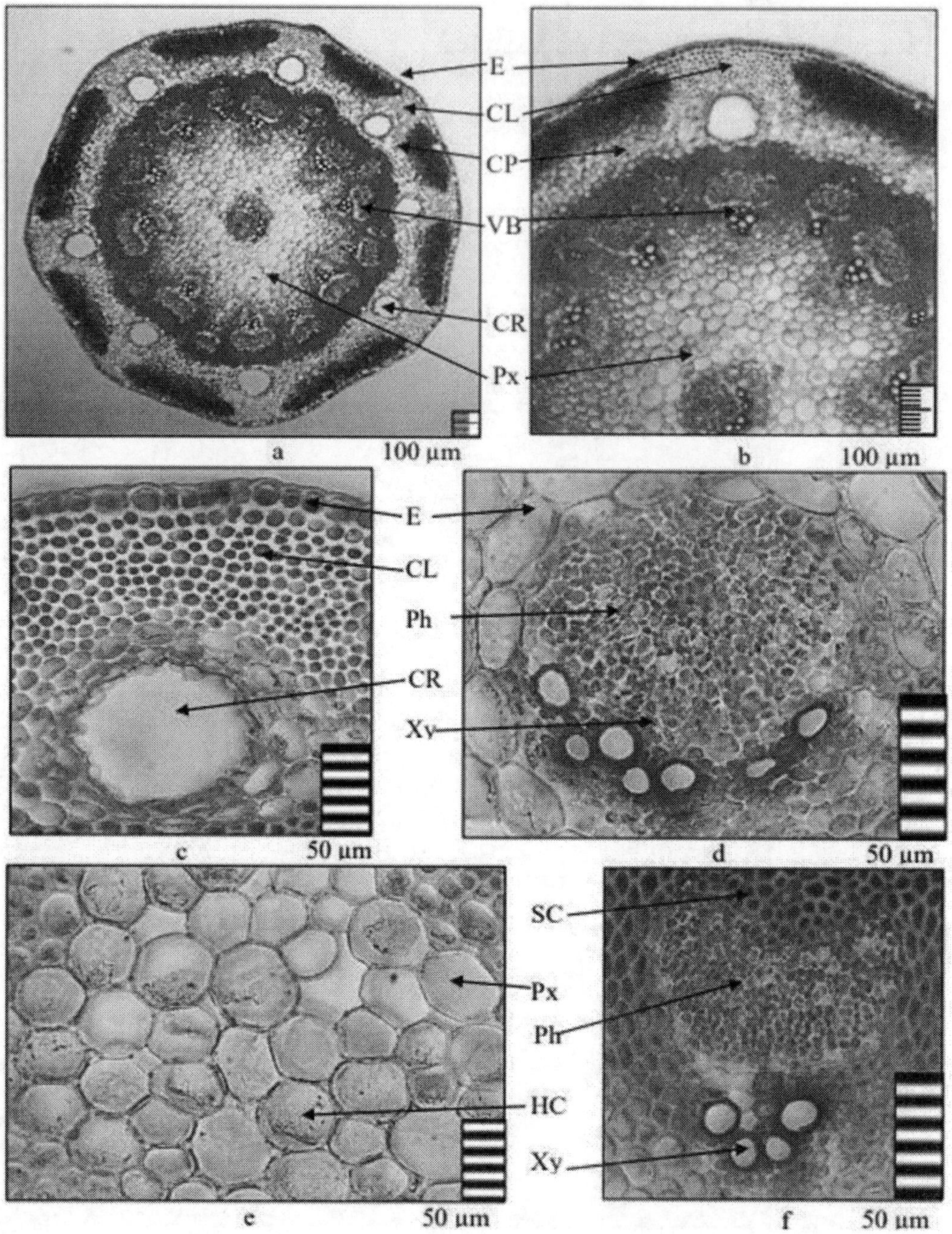

Figure 4. Anatomical structure of the peduncle of the Ferula tenuisecta: a – general view of the peduncle; b – detail of the peduncle; c – epidermis, collenchyma and secretory receptacles; d, f – vascular bundles; e – parenchyma and hydrocytic cells. Legend: HD - hydrocytic cells; KL - collenchyma, CP - cortical parenchyma, Xy - xylem, VB - vascular bundle, Px - parenchymal cells, SR - secretory receptacles, SC - sclerenchyma, Ph - phloem, E - epidermis.

Anatomical structure of the underground organs of Ferula tenuisecta. The root of Ferula is thick, with an upright rhizome. The caudex is simple. The anatomical structure of the plant shows a deep connection with the origin of the species, a comparison of these structures with other members of the family reveals the same type or heterogeneity of the structure, interconnected with adaptation to environmental conditions. The study of the secretory system of various organs and the characterization of terpenoid-containing receptacles are important for elucidating the localization of some biologically active compounds.

The morpho-anatomical structure of the underground organs of Ferula has been little studied. The underground stem part of perennial herbaceous plants, located on the basal p art of the tap root, has received different names in the literature: many-headed root, many-headed caudex, and many headed rhizome. Authors of the Atlas of Descriptive Morphology of Higher Plants. Stem and root" suggest calling the underground stem part of herbaceous perennials the stem root. It consists of lignified perennial shoots, connected at the bases with a lignified hypocotyl, turning into a woody taproot, usually with contractility.

The morphological features of underground organs of 15 Ferula species, including F. varia and F. angreni. Two main groups have been distinguished according to the structure of underground organs: deep-rooted species with a well-defined branched stem root and species with a turnip-like thickened main root and an indistinct stem root. There is information in the literature on the study of the structure of the axial and assimilation organs of some species of the umbrella family.

The root of *Ferula tenuisecta* is thickened, fusiform. The neck is densely wrapped with fibers of old dead leaves, many-headed. In the studied species, the root bark is wrinkled, easily separated, and darker brown in color. Older areas are covered with a thick layer of cracking and flaking cork. The cork is uneven, multi-row, the cells are flattened-oblong. Parenchymal cells of the peripheral part of the root cortex are severely destroyed. Numerous secretory receptacles are located in the cow's parenchyma. Groups of parenchymal cells untouched by destruction are preserved only around secretory receptacles, which are relatively little subject to destruction.

The phloem is multi-rowed and thin-walled. In the studied species, the cambium is multi-rowed, the xylem part is parenchymal. In the studied species, parenchymal cells and radial rays are filled with starch grains. Secretory receptacles located in the phloem part of the secondary cortex can be divided into 3 groups: young small receptacles directly on the border with the cambium, well-developed receptacles in the inner part of the cortex, and larger aging receptacles in its outer part. *Ferula tenuisecta* the structure of the root is similar, polycentric anomalous structure, the vascular bundles are grouped into concentric circles of different orientations, the radial rays are wide, in places they branch and bend. Secretory receptacles are numerous, the lumen of secretory receptacles is the largest, epithelial cells are numerous. The root of *Ferula tenuisecta* performs the function of storage for the formation of a more or less powerful peduncle in a short time, which determines its structure (Figure 5).

https://doi.org/10.1051/e3sconf/202342104005

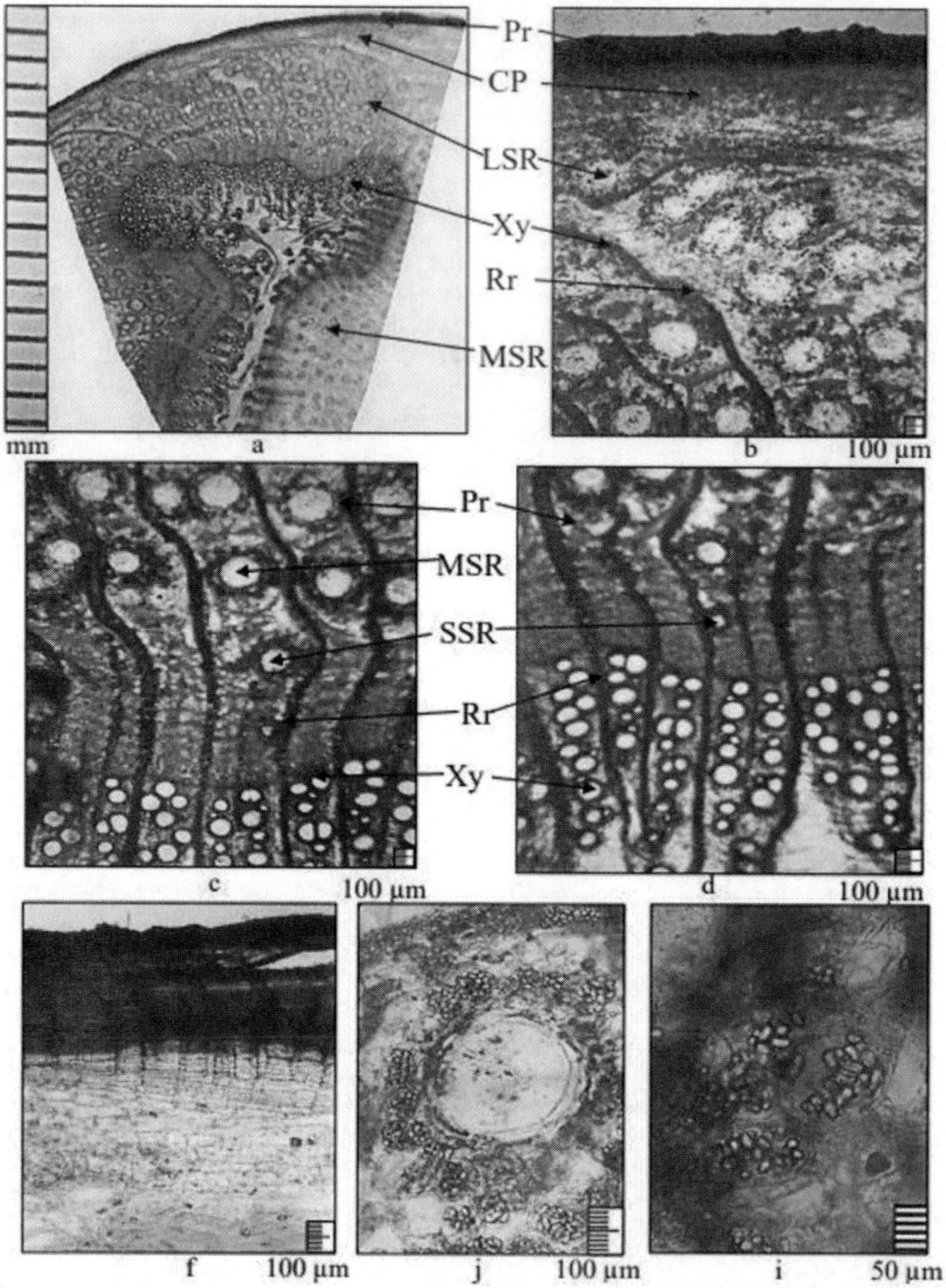

Figure 5. Anatomical structure of the root of Ferula tenuisecta: a - detail of the root; b – cortical parenchyma and larger aging secretory receptacles; c, d – vascular bundles and secretory receptacles; e - the root neck is covered with numerous fibers of dead leaves; f - secretory receptacles of schizogenic type; g - wood parenchyma and starch grains. Legend: CP – cortical parenchyma; Sg – starch grains; Xy - xylem, LSR – large secretory receptacles; SSR – small secretory receptacles; Pr – periderm; Rr – radial rays; MCR – mean secretory receptacle.

4 Conclusions

Thus, when studying the structural features of the aboveground and underground organs (leaf, petiole, peduncle, peduncle and root) of Ferula tenuisecta in natural environmental habitats, the following characteristic diagnostic, adaptive features were identified and the presence of localization of biological active substances was determined: the structure of the mesophyll of an isolateral-palisade leaf type; the structure of the petiole of the beam type; epidermis thin-walled, pubescent with simple, unicellular hairs; under the epidermis are chlorophyll-bearing subepidermal and parenchymal cells; the petioles are most sclerified, due to the lamellar type of collenchyma cells; the presence of large and small secretory receptacles of the schizogenic type; of the collateral type, the peripheral conducting bundles are most sclerified, due to the formation of bast fibers (sclerenchyma cells); among thin-walled and large parenchymal cells, hydrocytic cells are found; parenchymal-beam type structure of the peduncle and pedicel; the presence of a high and thick peduncle, a thick outer wall of epidermal cells, large-celled cortex and core parenchymal cells, which is due, on the one hand, to the xeric habitat, and on the other, to an earlier completion of the growing season; the vascular bundles are most sclerified due to bast fibers (sclerenchyma cells), which form multilayer strands in the form of caps over the phloem of the bundles; the presence of localization of biological active substances, determined in large and small numerous secretory receptacles of the schizogenic type in the peduncle, is associated with the power of the organ, which correlates with the life form. The study of the secretory system of various organs and the characterization of terpenoid-containing receptacles are important for elucidating the localization of some biologically active compounds (Safina, 2012).

Thickened turnip root type. Caudex differs from the main root by the presence of a pith. In the underground organs, the xylem part is larger than the phloem part, which is associated with a conductive function. Secretory receptacles are localized only in the phloem part and in the cortical parenchyma. The lumen of the secretory receptacles is the largest, the epithelial cells are numerous, in the xylem part the woody parenchyma is developed, which is dominated by parenchymatization. The data and identified structural diagnostic features of aboveground organs are species-specific and can be used in taxonomy and in the identification of plant materials for a given species and the determination of biologically active substances in organs and tissues.

References

1. Amiri M.S., Joharchi M.R. Ethnobotanical knowledge of Apiaceae family in Iran: A review. Avicenna Journal of Phytomedicine, 2016. - Vol. 6. - No. 6. - P. 621-635.
2. Calvino C., Martínez S., Downie S. Morphology and biogeography of Apiaceae subfamily Saniculoideae as inferred byphylogenetic analysis of molecular data // American Journal of Botany, 2008. - Vol. 95. - No. 2. - P. 196-214.
3. Chen B., Teranishi R., Kawazoe K., Takaishi y., Honda G., Itoh. M., Takeda Y., Khodzhimatov O.K. Sesquiterpenoides from Ferula kuhistanica.// Phytochememistry, 2000. - Vol. 54. 717-722 pp.
4. Downie S.R., Spalik K., Katz-Downie D.S., Reduron J. Major clades within Apiaceae subfamily Apioideae as inferred by phylogenetic analysis of nrDNA ITS sequences, Plant Div. Evol. – Stuttgart, 2010. – Vol. 128. - No. 1–2. - P. 111-136.
5. Elibol Z., Menemen Yu., Sağiroğlu M., Duman H. A molecular phylogenetic study on some Turkish Ferula L. (Apiaceae) species using nrDNA ITS sequences. Pakistan Journal of Botany, 2012. 44 (2): – pp. 589-594

6. Ghulameden S, Yill A, Zhao A.Q., Gao Y.H., Aisa H.A. Polisacharides from Ferula sinkingensis and potent inhibition of protein Tirosine phosphatase 1 B // Chem. connect. - Tashkent. 2014. - No. 3. - P. 445-446.
7. Habibi Z., Salehi P., Yousefi M., Hejazi Y., Laleh A., Mozaffarian V., Masoudi Sh., Rustaiyan A. Chemical composition and antimicrobial activities of the essential oils of Ferula latisecta and Mozaffariana insignis from Iran . // Chemistry of nature. connect. - Tashkent, 2006. - No. 6. - P. 561- 563.
8. Kasaiana J., Iranshahya M., Masullob M., Piacenteb S., Ebrahimia F., Iranshahia M. Sesquiterpene lactones from Ferula oopoda and their cytotoxic properties. Journal of Asian Natural Products Research, 2014. - Vol. 16, - No. 3, - P. 248-253.
9. Kurzyna M.R., Oskolski A.A., Downie S.R., Kopacz R., Wojewodzka A., Spalik K. Phylogenetic position of the genus Ferula (Apiaceae) and its placement in tribe Scandiceae as inferred from nrDNA ITS sequence variation, Plant Syst. Evol. 2008. - Vol. 274. - P. 47-66.
10. Panahi M., Banasiak Ł., Piwczyński M., Puchałka R., Oskolski A., Spalik K. Phylogenetic relationships among Dorema, Ferula and Leutea (Apiaceae: Scandiceae: Ferulinae) inferred from nr DNA ITS and cpDNA noncoding sequences / / Taxon, 2015. - Vol. 64. - No. 4. - P. 770-783.
11. Sharopov F., Satyal P., Wink M. Composition of the essential oil of Ferula clematidifolia. Chemistry of Natural Compounds, 2016. - Vol. 52.(3). - P. 518-519.
12. Su B., Takaishi Y., Honda G., Itoh M., Takeda Y., Khodzhimatov O.K., Ashurmetov O. Sesquiterpene coumarins and related derivatives from Ferula pallida. // Journal of Natural Products, 2000. - Vol. 63. - N 4. 436-440 pp.
13. Suzuki K., Oksaka M., Kashiwada Y., Takaishi Y., Honda G., Ito M., Takeda Y., Khodzhimatov O.K., Ashurmetov O., Sekiya M., Ikeshiro Y. Sesquiterpene lactones from the roots of Ferula varia and their cytotoxic activity. // Journal of Natural Products, 2007. - Vol. 70. - N 12. 1915-1918 pp.
14. Zhou J., Gong X., Downie S.R., Peng H. Towards a more robust molecular phylogeny of Chinese Apiaceae subfamily Apioideae: Additional evidence from nrDNA ITS and cpDNA intron (rpl16 and rps16) sequences. Molec. Phylogen. Evol. 2009. - Vol. 53 - P. 56-68.
15. Barykina R.P., Veselova T.D., Devyatov A.G. Handbook of botanical microtechnics (bases and methods). - Moscow: Ed. Moscow State University. - 2004. - P. 6-68.
16. Bogdasarova M.Kh. Biology of flowering and cytoembryology of some species of Ferula L. of Uzbekistan. Abstract diss. … cand. biol. Sciences. Tashkent. 1990. - 22 p.
17. Butnik A.A., Ashurmetov O.A., Nigmanova R.N., Begbaeva G.F. Ecological anatomy of desert plants of Central Asia. Herbs / - Tashkent: Fan, 2009. - Vol. 3. - 155 p.
18. Butnik A.A., Tursynbaeva G.S., Duschanova G.M. Leaf mesophyll of dicotyledonous plants (educational manual). - Tashkent: TSPU named after Nizami, 2015. - 42 p.
19. Daushkevich Yu.V., Pimenov M.G., Leonov M.V. Biometric study of Siberian narrow-leaved species Bupleurum // Bulletin of the Moscow Society of Naturalists. Department of biology. 1997. - Vol. 99. - No. 3. - P. 77-85.
20. Denisova G.A. Terpenoid-containing plant structures. - L.: Nauka, 1989. - 140 p. Yorkulov E., Malikova M.Kh., Rakhmanberdiyeva R.K. Carbohydrates of the aerial parts of F. kuhistanica and F. tenuisecta plants // KhPS. - Tashkent, 2011. - No. 2. - P. 169-171.
21. Kiryalov H.V., Movchan S.D. On the structure of samarkandin and samarkandone, coumarin compounds of Ferula samarkandica, Chem. conn. - 2018. - No. 2. - P. 73-76.
22. Kiryalov N.V., Bukreeva T.V. New coumarin from the roots of Ferula pseudooreoselinum // Chemistry natural. conn. –2013. - No. 3. - P.425-427.

23. Kiseleva N.S. Anatomy and morphology of plants. - Minsk: Ed. Higher School, - 2011. - P. 89- 119, 2015-227.
24. Korovin E.P. Illustrated monograph of the genus Ferula. - Tashkent, 2017.
25. Korovin E.P. Sem. Umbrella. Flora of the Uzbek SSR. - Tashkent, 2019. -Vol. 4. - P. 459.
26. Korovin, E.P. Ferula - Ferula // Flora of the USSR Flora URSS. - M.-L.: Publishing House of the Academy of Sciences of the USSR, 1951. - Vol. 17. - P. 114-115. — 390 p.
27. Kotenko L.D., Mamatkhanova M.A., Khalilov R.M., Mamathonov A.U., Sotimov G.B. Standardization of ferula herb variable // Chemistry of nature. connect. - Tashkent, 2009. - No. 4. - P. 151-154.
28. Krasilnikov N.P. Underground organs of the angelica descending. Plants of the umbrella family - sources of biologically active substances: - L .: Nauka, 2018. - P. 36-106.
29. Mamatkhanov A. U., Khalilov R. M. The aerial part of Ferula tenuisecta is an alternative industrial raw material for the production of tefestrol substance // Chemistry of vegetable raw materials, 2022. No. 3. P. 309-316. URL: http://journal.asu.ru/cw/article/view/10862.
30. Markova L.P., Medvedova L.I. Materials for the study of underground organs of some species of ferules from the subgenus Peucedanoides (Boiss.) Korov. // Vegetable raw materials. Tr. BIN AN USSR. - L., 2018. - Vol. 5. - issue. 15. - P. 149-173.
31. Momotov I.F., Butnik A.A., Zhapakova U.N., Padunina V.M. Berkovich B.V. Umbelliferae - valuable fodder plants of the desert // Science to production. Academy of Sciences of the Uzbek SSR. Tashkent, 2019.
32. Nishanbaeva H. On the endemism of species of the genus Ferula in Uzbekistan // Uzb.biol.zh. - Tashkent, 2018. - No. 2. - P. 40-42.
33. Petrova S. E. Ontomorphogenesis of some Eastern European representatives of the Umbelliferae Moris family. (Apiaceae Lindl.) Abstract of the thesis. dis. ... cand. biol. Sciences. - M., 2008. - 22 p.
34. Pimenov M.G. Umbelliferae (Apiaceae) of Central Asia and Kazakhstan. Abstract diss. ... doc. biol. Sciences. - Moscow, 2013. - 29 p.
35. Pimenov M.G. List of plants - sources of coumarin compounds. L.: Nauka, 2011. - 200 p.
36. Rakhimov S. Biological and morphological features of ferula (Ferula L.) in Tajikistan. - Dushanbe: Donish, 2010 - 59 p.
37. Rakhmankulov U. Terpenoid-containing plants of the Western Tien Shan and their use. Abstract diss. ... doc. biol. Sciences. - Tashkent, 2019. - 58 p.
38. Rakhmankulov U., Melibaev S. New species of the genus Ferula L. from the mountains of Nuratau // Dokl. Academy of Sciences of Uzbekistan - Tashkent, 2011. No. 8. - P. 58-59.
39. Saidkhodzhaev, A.I., Malikov, V.M., Pimenov, M.G., and Melibaev, S., Esters of Ferula kyzylkumica and F. karategina roots, Khim. connect. - Tashkent, 1993. - No. 2. - P. 301-303.
40. Safina L.K. Ferules of Central Asia and Kazakhstan (Carpoanatomical review). - Alma-Ata: LEM, 2012. - 244 p. Safina L.K., Isaeva S.N. Anatomical analysis of the vegetative organs of two endemic ferula species // Izvestiya AN KazSSR. 2011. - No 6. - P. 8-16.
41. Safina L.K., Isaeva S.N. On the plasticity of the anatomy of the ferul leaf // Bulletin of the Academy of Sciences of the Kazakh SSR, 1985. - No. 11. - P. 91-93.
42. Safina L.K., Pimenov M.G. Ferules of Kazakhstan. - Alma-Ata: Science, 1984. - 100 p.
43. Svetlova A.A. Characterization of mucus-forming cells of some species of the genus Linum (Linaceae) // Prospects for development and problems of the present. Botany.

 https://doi.org/10.1051/e3sconf/202342104005

Mater. II (IV) All-Russian. young. scientific and practical. conf. - Novosibirsk. RAN. 2010. - P. 142-143.

44. Sdobnina L.I., Pimenov M.G. Anatomical features of the underground organs of some geophilic umbellate // Botanical journal. 2011. - Vol. 76. - No. 11. - P. 1527-1538.
45. Sdobnina L.I., Pimenov M.G. Anatomical features of the underground organs of some geophilic umbellate // Botanical journal. 2011. - Vol. 76. - No. 11. - P. 1527-1538.
46. Sokolov V.S. Some questions of alkaloid content in plants. Bot. magazine 2000. - Vol. 35. - No. 2. - P. 206-207.
47. Fedorov Al.A., Kirpichnikov M.E., Artyushenko Z.T. Atlas on the descriptive morphology of higher plants. Stem and root. - M. L.: Publishing House of the ANSSR, 2012. 350 p.
48. Khalilova E.Kh., Bobakulov Kh.M., Aripova F.S., Abdullaev N.D. Secondary metabolites of Ferula foetida // Chem. connect. - Tashkent, 2013. - No. 1. - P. 124-125.
49. Khanina M.A., Serykh E.A., Amelchenko V.P. Comparative anatomical and morphological study of secretory formations and application of data for the taxonomy and phylogeny of the tribe Anthemideae // Proceedings of the Int. conf. on anatomy and morphology of plants. - St. Petersburg. Diada, 2017. - P.138-139.
50. Chubarov I.N. Family Apiaceae Lindl. Altai mountain country: Abstract of the thesis. dis. ... cand. biol. Sciences. - M., 2015. - 17 p. Esau K. Plant anatomy. - Moscow: Ed. Mir, - 1969. - P. 138-416.
51. Eshbakova K. A., Saidkhodzhaev A. I., Vdovin A. D., Abdullaev N. D. Terpenoid coumarins from Ferula feruloides // Khimiya prirod. connect. - Tashkent, 2019. - No. 5. - P. 594

Growth and development of *Ferula foetida* Regel (Bunge) and *Ferula tadshikorum* Pimenov species in different environments

Mokhira Halkuzieva[1,*], *Xurshid* Qiryigitov[2], *Bekzod* Abduazizov[2], *Mukaddas* Sobirova[1] and *Bakhora* Usmonova[2]

[1]Jizzakh Branch of National University of Uzbekistan, Jizzakh, 130100, Uzbekistan
[2]Jizzakh Polytechnic Institute, Jizzakh, 130100, Uzbekistan

Abstract. The daily increasing demand for medicinal products obtained from *Ferula foetida* (Bunge) Regel and *F. tadshikorum* Pimenov, from species of the *Ferula* L. family, is causing a decrease in plant biodiversity and their reserves. This paper presents the growth and development of *F. tadshikorum* and *F. foetida* in different environments under plantation conditions. The seeds of *Ferula foetida* (Bunge) Regel and *F. tadshikorum* Pimenov were sown separately in sandy, gray and potash-rich soils, 0.5-1.0 cm deep, in 2 variants in December. The second control option - germination biology and growth of morphobiological characteristics of lawns under waterlogging was studied The amount of potassium in the soil in the experimental areas of Dekhkanabad district of Kashkadarya region was 1,910%, 1,210% in Arnasoy district and 1,220% in Sharof Rashidov district. It was found that the germination growth and development level and vegetation of *F. tadshikorum* and *F. foetida* seeds were much higher in the soils of the experimental fields of Dekhkanabad district of Kashkadarya region in both options. When plantations are established by planting *F. tadshikorum* and *F. foetida* from seeds, adding potash fertilizers to the soil will allow them to grow well.
Keywords. Stem, seed, plant, F. tadshikorum, F. foetida, plantation, natural.

1 Introduction

Scientific research is being conducted in the world to determine the reasons for the decline of medicinal plant species as a result of the influence of natural and anthropogenic factors, to improve the ways of their preservation and reproduction [1-3]. Especially from medicinal plants *Apiaceae* Lindl. The growing demand for medicinal products from *Ferula foetida* (Bunge) Regel and *F. tadshikorum* Pimenov, belonging to the genus *Ferula* L., is causing a decrease in plant biodiversity and reserves. In addition, their low potential for natural reproduction, the influence of anthropogenic factors and their location in pressure areas, climate change have a negative effect on their regeneration [4-6]. Particular attention is paid to determining the growth and development of *F. foetida* and *F. tadshikorum* species

* Corresponding author: asatullayevnamohira@gmail.com

in different conditions, planting and breeding from seeds in dry conditions of hill and desert regions, establishing plantations and creating quality raw materials [7].

Systematics, carpology, ecology and distribution of some species of the *Ferula* L. genus [8-11], and reflected in the work of others [12-14]. The anatomical structure of the fruit of some species of the genus *Ferula* L. (*Apiaceae* Lindl.) was analyzed [15-18].

In Uzbekistan, biology and stocks of *Ferula* L. species studied by U. Rahmonkulov, Kh. Nishanbaeva, Kh. Rakhmonov, anatomy, cytoembryology by A.A. Butnik et al., chemical and pharmacological properties A.I. Saidkhodzhaev, G.K. Nikonov [19-25].

The beginning of vegetation of monocarpic plant *F. tadshikorum* began in February, end of March, and their life cycle lasts long 23-27. The lawn stage includes 1 year of vegetation, the juvenile stage is 6-7 years old, 14-15 years old, and the immature age is 24-27 years old (10) [26].

Hence, 1-3% immature condition (05.06.2019) was observed in the first year of *F. tadshikorum* plant. In the third year (2022) in both variants, 70-75% (18.02.2021) plants were immature [27-29].

2 Materials and methods

F. tadshikorum and *F. foetida* are perennial herbaceous, monocarpic, hemiephemeroid plants. In Uzbekistan, *F. tadshikorum* is found mainly in the mountainous regions of Kashkadarya and Surkhandarya regions, *F. foetida* is found in different groups of plants in deserts and hills, and is distributed in the Kyzylkum deserts, Mirzachol, Karnabchol, Samarkand, Jizzakh, Bukhara, Kashkadarya, Surkhondarya regions and the territories of the Republic of Karakalpakstan.

The seeds of *F. foetida* plant are stringy on the outside, wing-like on the edges, the size and absolute weight of the seeds varies depending on the place of growth of the plant. The largest seed was collected from the southern slope of the Nurota ridge, its length was 2.4 cm, width 1.6 cm, thickness 0.7 mm, and the absolute weight of 1000 seeds was 60.3 grams.

The seeds of *F. tadshikorum* are 1.5-2.0 cm long, 0.8-1.0 cm wide, elliptical or leaf-shaped, hairless. The absolute weight of the seed is slightly smaller than that of *F. foetida*, 35 - 40 gr. Their seeds need to pass a period of low temperature and rest for normal germination [30-32].

In 2014-2021, in the Arnasoy district of Jizzakh region, on the territory of the "Shifo kovrak" farm, *F. foetida* and *F. tadshikorum* were planted from seeds and plantations were established on an area of 20 hectares. Arnasoy district consists of plains and low plains, the soils are sandy and gray, potassium rich, and there are partial salt marshes in the northeastern part. The climate is strictly continental.

3 Results and discussion

When creating plantations from medicinal plants, the quality of the seeds sown, the degree of germination when the seeds are planted at different depths are important. Below are the results of the germination rate of the seeds of both carpet species when planted at different depths.

It is known that during the germination of the seeds, its cotyledon leaves have the (epigial) feature of raising the seed coat to the surface of the earth. Our observation showed that the longer the seeds are, the deeper the seeds can lift the seed coat above the ground. For example, if the seed of *F. tadshikorum* is up to 2 cm long, its seeds will germinate from the soil up to this depth, the germination of seeds that fall into the soil deeper than that is very small. A seed planted deeper than that will not have the strength to lift its skin.

According to O. Khojimatov, the average fertility when planted at a depth of 0.3-0.5 cm is 57 %, after 4 years there are an average of 420 (42 %) plants in the field, when planted in the surface layer of the soil, the fertility is 50 %, after 3 years it is 22 % the plant is preserved. In tall lawns (option No. 3), good viability rates of 21-24 % were determined even when planted in the surface layer of the soil [12].

In order to determine the degree of germination of *F. foetida* and *F. tadshikorum* at different depths and in autumn and winter months, the seeds of *F. foetida* were collected from Surhondarya region, and the seeds of *F. tadshikorum* were collected from Dekhkanabad forestry of Kashkadarya region in summer and cleaned and sorted. Experimental work was carried out in laboratory conditions and in the territory of the farm "Shifo-kovrak" in Arnasoy district, Jizzakh region, and in experimental fields in Sharof Rashidov district during the years 2018-2020. The areas allocated for planting seeds were well plowed and fertilized to a depth of 35.0-40.0 cm, and if there were uneven areas, they were leveled with tar.

100 seeds were sown separately in soils rich in sandy, gray and potash fertilizers, at a depth of 0.5-1.0 cm in December in 2 variants, the first control variant - Seeds were grown in outdoor natural conditions (Figure 1). The second control option - germination biology and the growth of morphobiological properties of lawns during waterlogging was studied [25].

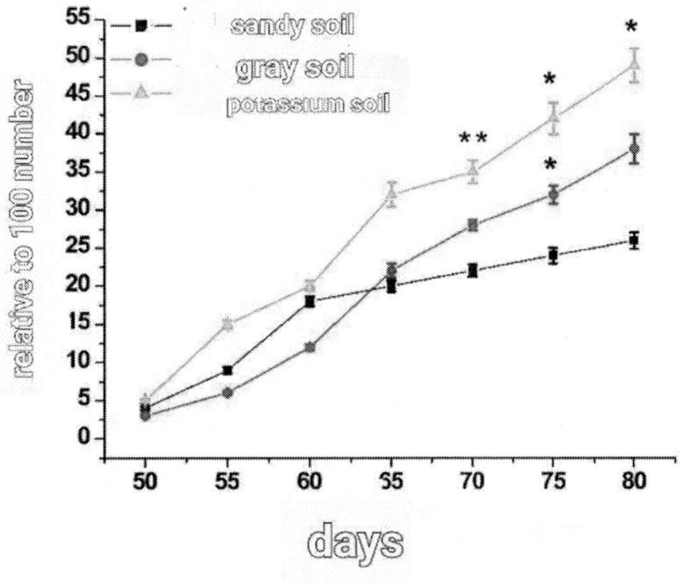

Figure 1. Germination of *F. foetida* in different soils in natural environment (n=100).

From Figure 1, it is known that the germination of *F. tadshikorum* seeds in potassium-rich soil was much higher (48 %) than in other soils, and the lowest germination (20 %) was observed in sandy soil.

Figure 2 shows the germination rate of seeds in potassium rich and organic fertilizer was higher (56%) when collected in watered medium, while seeds planted in sandy soil gave the lowest rate (26%).

https://doi.org/10.1051/e3sconf/202342104006

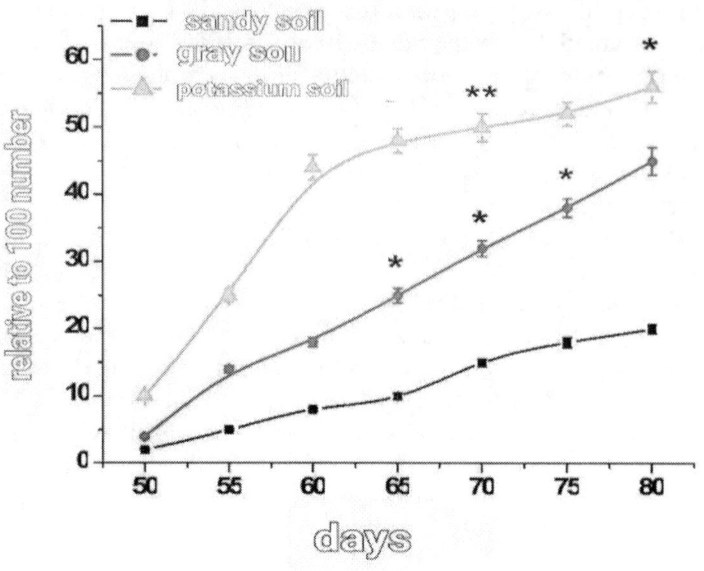

Figure 2. Germination of *F. foetida* in different soils wetted with water (n=100).

After planting *F. foetida* in a natural and watered environment (01.02-25-04.2020), the majority of seeds germinated within 90 days. The cotyledons are long lanceolate, light green in color, germinated in 65 days in natural environment and dried in 105 days. In the water-fed environment, seeded leaves were formed in 55 days and started to dry out in 115 days.

The shape of the first leaf is ovate, oblong, with a flat edge, and this type also produces 2 or 3 leaves. In the control variant, first leaves were produced in 85 days and withered in 130 days. However, true leaves were preserved for up to 145 days in potassium-rich soil. In the experimental version, the seed leaves were formed in 72 days and began to dry in 145 days (Table 1).

Table 1. Morphological classification of cotyledon and true leaf of *F. foetida* in natural environment on different soils (n=100).

Seed leaf		First leaf		
Length cm	**Width, mm**	**Length cm**	**Width mm**	**Length cm**
		Sandy soil		
2.5±0.11	0.15±0.01	3.5±0.17	0.7±0.03	1.45±0.07
		Soil rich in potassium		
4.0±0.20	0.3±0.02	5.5±0.27	1.0±0.05	2.0±0.10
		Gray soil		
3.0±0.15	0.25±0.11	4.0±0.20	0.8±0.04	1.5±0.07

From Table 1, it is known that *F. tadshikorum* had high growth of grasses in potassium rich and gray soil, but its development was slow compared to them in sandy soil. From Table 2, it can be seen that the growth rate of grasses rich in potassium and grown in organic fertilizer when collected in watered environment was much higher, while the growth rate of grasses in sandy soil was the lowest.

Table 2. Morphological classification of cotyledon and first leaf of *F. foetida* in natural environment in different soils.

Seed leaf		First leaf		
Length cm	**Width, mm**	**Length cm**	**Width mm**	**Length cm**
		Sandy soil		
3.5±0.17	0.25±0.01	5.5±0.22	1.1±0.05	2.25±0.09
		Soil rich in potassium		
5.5±0.27	0.45±0.02	10.0±0.43	1.5±0.07	4.0±0.16
		Gray soil		
3.5±0.16	0.45±0.03	6.5±0.32	2.5±0.11	2.5±0.12

When *F. foetida* was planted in a natural and water-fed environment, grass growth was high and vegetation lasted longer. Germination and development of *F. foetida* seeds were found to be significantly higher in potash soils in both variants. When planting *F. foetida* from seeds, it is recommended to apply potassium fertilizers to the soil. Germination of seeds of *F. tadshikorum* under different soil conditions.

In order to determine the fertility of *F. tadshikorum* in soils of different composition, under natural and moisture-preserved conditions, seeds collected in August 2019 in the forestry area of Kashkadarya region were used (Figure 3).

Seeds were sown separately in sandy, gray and potash-rich soil in 100 seeds at a depth of 0.5-1.0 cm in December in 2 variants. The second control option - germination biology and the growth of morphobiological features of lawns when wetted with water was studied.

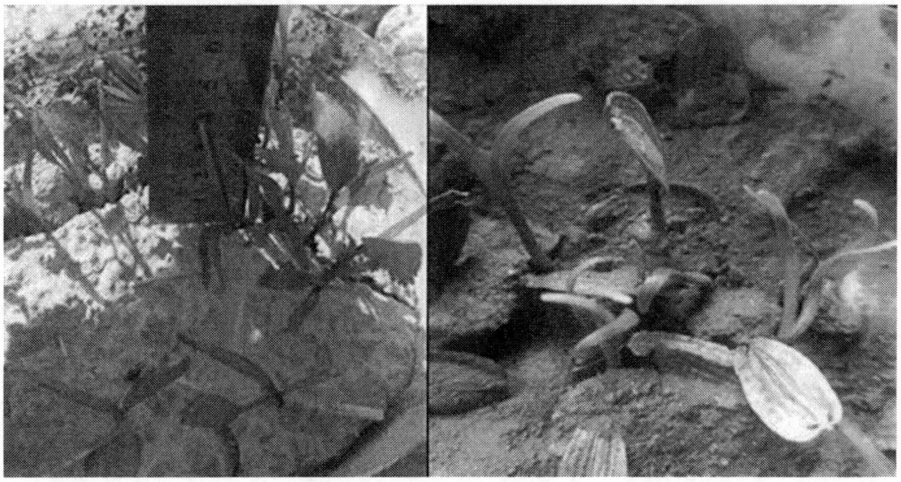

Figure 3. Germination of F. tadshikorum in different soils in natural environment.

The shape of the green leaf is rhombic, the leaf plate is oblong, the edge is finely cut. In the control variant, the green leaves were formed in 80 days and dried in 130 days. However, true leaves survived up to 150 days in potassium-rich soil. In the experimental version, the cotyledons were formed in 70 days and started to dry in 150 days.

From Table 3, it is known that the growth of *F. tadshikorum* lawns in potassium-rich and gray soil is high, and their development is relatively slow in sandy soil.

Table 3. Morphological classification of cotyledons and cotyledons of *F. tadshikorum* grown in natural environment on different soils (n=100).

Soils	Seed leaf		First leaf		
	Length cm	Width mm	Length cm	Width mm	Length cm
Sandy soil	2.5±0.12	0.15±0.01	3.5±0.17	0.7±0.03	1.45±0.07
Soil rich in potassium	8.0±0.40	0.3±0.01	5.5±0.27	1.0±0.05	2.0±0.10
Gray soil	3.0±0.15	0.25±0.01	4.0±0.20	0.8±0.04	1.5±0.07

When collected in a watered environment, grasses rich in potassium and grown in organic fertilizer had a much higher growth rate, while grasses in sandy soil had the lowest growth rate (Table 4).

Table 4. Morphological classification of cotyledon and cotyledon of *F. tadshikorum* in water-moistened environment on different soils (n=100).

Soils	Seed leaf		First leaf		
	Length cm	Width mm	Length cm	Width mm	Length cm
Sandy soil	2.5±0.12	0.15±0.01	3.5±0.02	0.7±0.03	1.45±0.07
Soil rich in potassium	4.0±0.18	0.3±0.01	5.5±0.27	1.0±0.05	2.0±0.10
Gray soil	3.0±0.14	0.25±0.01	4.0±0.20	0.8±0.04	1.5±0.07

4 Conclusions

The amount of potassium in the soil in the experimental areas of Dekhkanabad district of Kashkadarya region was 1,910%, 1,210% in Arnasoy district and 1,220% in Sharof Rashidov district. It was found that the germination growth and development level and vegetation of *F. tadshikorum* and *F. foetida* seeds were much higher in the soils of the experimental fields of Dekhkanabad district of Kashkadarya region in both options. When plantations are established by planting *F. tadshikorum* and *F. foetida* from seeds, adding potash fertilizers to the soil will allow them to grow well.

References

1. Korovin E P, Pimenov MG, Kinzikaeva G K. Genus *Ferula* L. Vol. 7. Flora of the Tajik SSR: Academy of Sciences of the USSR; 2014.p. 161-194.
2. Pimenov M G. Family Umbelliferae. Opredelitel plant Sredney Azii. Tashkent. - 2013. T. 7. —p. 276-313.
3. Vinogradova V M. Rod *Ferula* L. (Apiaceae) v Tsentralnoy Azii // Novosti syst. vys. true L. — 2010. — No. 27. — p. 113-120.
4. Safina LK., Ostroumova T, Pimenov M. Carpology of the species of *Ferula* subgen. *Merwia* (Umbelliferae–Apioideae) and some taxonomic implications. Nordic Journal of Botany. 2015; 33(2):140-150.DOI: 10.1111/j.1756-1051.2013.00315.
5. Sharipova V.K. Comparative Analysis of the Fruit Pericarp Structure of the Desert and Mountain Species of *Ferula* L. (*Apiaceae* Lindl.)// American Journal of Plant Sciences, 2017 № 8, -P. 2013-2020.
6. Rakhmankulov U. Terpenoid o soderjashchie rasteniya Zapadnogo Tyan-Shanya i ix ispolzovanie: Avtoref. dis... cand. biol. science — Tashkent: 1999. — 43 p

7. Nishanbaeva Kha.A. Rod. *Ferula* L. v rastitelnom pokre vo flore Uzbekistana: Autoref. diss... cand. biol. nauk. - Tashkent. 2012. -22 p.
8. Kurmukov A G., Akhmedkhodzhaeva H S. Estrogenovye lekarstvennye preparations iz rastenii roda ferula// Izdatelskopoligraficheskoe obedinenie imeni Ibn Sina. Tashkent, 2014. — 69 p.
9. Saidkhodzhaeva A.I., Nikonov G.K. Komponenty corney Ferula ovina // Khimiya prirod. soedin - 2014. - No. 4. - B. 526-527.
10. Nikonov G K. Stroenie mogoltina-coumarin corney *Peucedanum mogoltavicum* // Chemistry prirod. soedin — 2001. — No. 5. — B. 572-576.
11. Raxmonov X S. Biologiya va resurs *Ferula tadshikorum* M.Pimen. v yujnom Tadjikistane: Dis... kand. biol. nauk. —Dushanbe: 2017. —20 b.
12. Khamraeva D T, Khojimatov O K, Bussmann R, Khujanov AK, Kosimov ZZ. Prospects for the introduction of *Ferula tadshikorum* Pimentov in the conditions of the Tashkent region. Ethnobotany Research and Applications. 2022.23:6. http://dx.doi.org/10.32859/era.23.6.1-
13. Red Book of the Republic of Uzbekistan. Plants 2. *Ferula tadshikorum* Pimenov. Tashkent: Chinor ENK; 2019.p. 95-96.
14. Zhukova L A. Ontogenesis and cycle production of plants // Journal. public biology. 2013. T. 44. 3. S. 361-374.
15. Serebryakov I G. Tipy razvitiya pobegov u travyanistyx mnogoletnikov i faktori ix formirovania // Uch. zapiski Moscow. gor. ped. inst. – 1999. – No. 100. - B. 3-38.
16. Beideman I N. Methods for studying the phenology of plants and plant communities Guidelines. – Novosibirsk: Nauka. 1994.
17. Schultz G E. General phenology. L.: Nauka. 2001.
18. Avalbaev O N. The Biology of Flowering of Some Pamir-Alai Species of the Genus *Ferula* L.American Journal of Plant Sciences. Vol. 9 2008; 8.doi:10.4236/ajps.2018.98126
19. Butnik A A, Ashurmetov O A, Nigmanova R N, Begbaeva G F. (2009) Ecological anatomy pustynnye rasteniya Sredney Azii 3. Travy. Science, Tashkent, 155 p.
20. Saidkhodzhaeva AI, Nikonov G K. Komponenty corney *Ferula ovina* // Khimiya prirod. soedin - 2014. - No. 4. - p. 526-527.
21. Zhukova L A. Mnogoobrazie putey ontogenesis and population of plants // Ecology. - 2001. - No. 3. - p. 169-176.
22. Schultz G E. General phenology L : Nauka. 2011.
23. Khalkuzieva M A, Rahmonqulov U. Protection of plants for years in the plantations of Ferula tadshikorum Pimenov //Bulletin of Gulistan State University. 2020; 4:19-25.https://uzjournals.edu.uz/gulduvestnik/vol2020/iss4/21.
24. Halkuzieva M. The role of agrotechnological measures during the early development of Ferula tadshikorum Pimenov and *Ferula foetida* (Bunge) Regel. Annals of the Romanian Society for Cell Biology. 2021; 3191-98. https://doi.org/10.37547/tajabe/Volume02Issue07-05.
25. Halkuzieva M. *Ferula tadshikorum* Pimenov growth and development in different soils. Eurasian Scientific Herald. 2022; 4:29-34.https://geniusjournals.org/index.php/esh/article/view/418
26. Halkuzieva MA, Khamraeva DT, Bussmann RW. Bio-morphological properties of *Ferula tadshikorum* Pimenov and *Ferula foetida* (Bunge) Regel under plantation conditions. Plant Sci. Today. 2023 Apr. 9 [cited 2023 May 2];9(sp3):79-84. Available from: https://horizonepublishing.com/journals/index.php/PST/article/view/1863
27. Halkuzieva M A, & Rahmonkulov U. (2020). Smola Storage *Ferula Tadshikorum* M. Pimen Plantations Storage For Years. *The American Journal of Agriculture and*

Biomedical Engineering, 2(07), 30–34. https://doi.org/10.37547/tajabe/Volume02Issue07-05

28. Tirkasheva M B., Abduvaliev X A, Kiryigitov X B. Proyavlenie ekologicheskix znaniy i navykov na urokax fizkultury //Molodoy uchenyy. – 2014. . 5. –p. 558-560.
29. Kiryigitov Khurshid Botirovich, Turabaev Akmal Normurovich. (2021). Interrelation of Ecosystems in the Formation and Management of Components of a Biological Resource (Pedosphere). *Annals of the Romanian Society for Cell Biology*, 5231–5242. Retrieved from http://annalsofrscb.ro/index.php/journal/article/view/3073
30. Taylakov A A, Kiryigitov X B. Oxrana atmosfernogo vozduxa ot vybrosov avtomobilno gazonapolnitelnyx stansiy Arktika: Sovremennye Podhody K Proizvodstvennoy I Ekologicheskoy Bezopasnosti V Neftegazovom Sektori 252-257 2020
31. Sobirova M B, Kiryigitov X B. Effektivnoe ispolzovanie biopreparatov" zamin-m" i" er malhami" dlya vyrashchivaniya va agrotexnicheskie mery lekarstvennogo rasteniya atishoka (cynara scolymus zavvnya 3, 4 segod3, 1-2) ER MALHAMI". 19.
32. Normuminovich A, Kiryigitov Kh, Mamirovna O. interactive interaction of ecosystems in the formation and management of biological resources (pedosphere) components Scientific and Technical Journal of Namangan Institute of Engineering and Technology 138-145 2020.

Diseases of melon in the central areas of Uzbekistan

Anvarjon Otajonov[1,*] and *Komil* Bukhorov[2]

[1]Research Institute of Plant Quarantine and Protection, Tashkent, 100003, Uzbekistan
[2]Tashkent State Agrarian University, Tashkent, 100140, Uzbekistan

Abstract. This article offers comprehensive insights into the diseases that afflict melon crops, elucidating the factors that trigger these ailments, and proposing strategies to effectively combat them. The research entails a combination of open-field experiments and controlled small-scale trials conducted in lysimeters across diverse regions including Tashkent, Syrdarya, Jizzakh, and central areas of Uzbekistan. The outcomes of the study reveal the isolation of four distinct disease-causing fungi across a total of five melon varieties. In laboratory settings, the research delves into the efficacy of seed-repellent fungicides through Petri dish experiments, discerning highly efficient formulations. The selected fungicides exhibit promising potential in mitigating pathogenic effects. To validate these findings in practical settings, the selected fungicides were applied to seeds and tested through field cultivation. This multifaceted approach demonstrates a commitment to addressing melon diseases at both theoretical and practical levels. The article's value is further augmented by visual aids, including an image and a table, which succinctly illustrate the findings derived from the research. By integrating theoretical insights, experimental validation, and visual representations, the article provides a comprehensive understanding of melon disease management. The presented strategies and findings are poised to contribute significantly to enhancing the resilience and productivity of melon crops in the studied regions.
Keywords. Syrdarya, Jizzakh, Tashkent, varieties, melon, fusarium sp, alternariacucumerina, macrosporium.

1 Introduction

In the subtropical and temperate climate regions across the globe, a substantial expanse of 6.2 million hectares has been dedicated to the cultivation of field crops. This concerted effort has yielded an impressive harvest, resulting in the production of a staggering 142.4 million tons [1, 2].

Among these regions, several countries stand out as pivotal contributors to the global pulse production landscape. Leading the pack is China, renowned for its substantial pulse cultivation [3]. Turkey, India, the United States, Iran, Egypt, and Spain also hold significant positions in the roster of top pulse-producing nations. These countries collectively play a vital

* Corresponding author: anvarjon.otajonov1001@mail.ru

role in satisfying the global demand for pulses, a crucial source of nutrition and sustenance for countless communities worldwide [4].

The beloved melon plant, cherished by our community, falls prey to a range of fungal, bacterial, and viral afflictions. These diseases inflict a harsh toll on the plant's yield and quality, resulting in diminished productivity and compromised product standards [5]. Over subsequent years, the escalating incidence of these diseases further exacerbates the decline in productivity and quality [6].

In a bid to address this critical issue, scientific inquiry was conducted across various locations, including the experimental field of the Research Institute for Quarantine and Plant Protection in the Tashkent region, Qibray district, and farms in the Syrdarya and Jizzakh regions [7]. The focus was the exploration of fungal presence on melon plants within fields planted during the 2022 season [8].

At the core of this research lies the quest to uncover the efficacy of specific seed treatments in combating the pervasive fungal diseases that currently plague melon cultivation [9]. These diseases stand as a paramount concern for melon growers. The experimental approach encompassed four cycles with four distinct variations [10].

Through this comprehensive study, researchers aim to shed light on potential solutions to the fungal diseases ailing melon plants. By evaluating the effectiveness of various seed treatments, the research offers a promising pathway towards bolstering the resilience and productivity of melon crops [11-13]. The four iterations of the experiment underscore the rigor and thoroughness applied in the quest to mitigate the impact of fungal diseases on this cherished crop.

2 Materials and methods

In order to monitor the diseases of the melon crop, plants of the melon, planted areas of the Syrdarya and Jizzakh regions were monitored from may to September 2022. Research methods have been used in laboratory and field conditions to assess the diseases detected in melon in laboratory conditions, as well as to apply measures to effectively combat them.

From plants infected with disease-causing fungi in laboratory conditions, potatoes agaric and separated on the basis of her style in the identification of diseases and causative fungi. Fungi that identified symptoms of the disease in early, middle and late maturedvarieties of melon were studied from April 2023 on the effects of preparations used in pre-sowing medicinal preparation of melon seeds.

The cup was planted in the middle part of the fungi identified in the melon in the petri saucer, that is, having planted the fungus found in our prepared nutrient medium, next to which the melon seeds were planted in five pieces. Our research will take 2 months. Month 1 consists of extracting a seed repellent that affects melon diseases under laboratory conditions, while month 2 consists of planting the most effective selected seed repellent-dosed melon Urus in a lysemetric and field experimental area.

In the 1st month, sown fungi are observed, and each seed is taken into account the process of influencing the drinkers. The effects of testing seed drugs on fungal diseases identified in the melon crop have been studied.

3 Results and discussion

The research was mainly carried out on the fields planted with melons in Syrdarya, Jizzakh and Tashkent regions from the central regions of Uzbekistan from may to October 2021-2022. In the course of research, the degree of disease incidence of early matured melon varieties, external signs of diseases and at what period of vegetation the disease is widely damaged was studied (Table 1).

Table 1. Analysis of samples of infected field crops brought from the Jizzakh and Syrdarya regions of Uzbekistan.

Sampling location	Type	Field	Detected diseases	Disease triggers
Jizzakh region, Mirzachul district "Ko'hnur" farm	"Shakar palak" melon	2,5	Fusarium dry, Erysiphe cichoracearum, Sporodesmium mucosum	Fusarium oxysporum f.sp. niveumErysiphe cichoracearum Dc. F. cucurbitacearum,Sporodesmium mucosum Sacc.var.pluriseptatum Karst.
Jizzakh region, Mirzachul district "Ko'hnur" farm	"Obi novvot" melon	2,5	Fusarium dry, Erysiphe cichoracearum, Sporodesmium mucosum	Fusarium oxysporum f.sp. niveum Erysiphe cichoracearum Dc. F. cucurbitacearum, Sporodesmium mucosum Sacc.var.pluriseptatum Karst.
Jizzakh region, Do'stlik district "Obidjon ota" farm	"Obi novvot" melon	0,9	Sporodesmium mucosum	Sporodes miummucosum Sacc.var.pluriseptatum Karst.
Syrdarya region, Mirobod district "Mirobod yuldizi" farm	"Amri" melon	0,3	Fusarium dry, Anthracnose	Fusarium oxysporumf.sp. niveum,Colletotrichumlagenarium E. et H.
Sirdaryo region, Oq oltin district "Temiryo'l agro sanoat" UK	"Kukcha" melon	2,3	Fusarium dry, Erysiphe cichoracearum, Sporodesmium mucosum	Fusarium oxysporum f.sp. niveum, Erysiphe cichoracearum Dc. F. cucurbitacearum, Sporodesmium mucosum Sacc.var.pluriseptatum Karst.
Syrdarya region, Oq oltin district "Temiryo'l agro sanoat" UK	"Oq urug" melon	6,8	Fusarium dry, Erysiphe cichoracearum, Sporodesmium mucosum	Fusarium oxysporum f.sp. niveum, Erysiphe cichoracearum Dc. F. cucurbitacearum, Sporodesmium mucosum Sacc.var.pluriseptatum Karst.
Syrdarya region, Mirobod district "Mirobod yuldizi" f/x	"Obi novvot" melon	1,9	Fusarium dry, Erysiphe cichoracearum, Sporodesmium mucosum	Fusarium oxysporum f.sp. niveum, Erysiphe cichoracearum Dc. F. cucurbitacearum, Sporodesmium mucosum Sacc.var.pluriseptatum Karst.
Jizzakh region, Dustlik district "Obidjon ota" f/x	"Kampir chopon" melon	0.3	Fusarium dry, Sporodesmium mucosum	Fusarium oxysporum f.sp. niveum Sporodesmium mucosum Sacc.var.pluriseptatum Karst.

In Table 1, in Mirzachul district "Kuxnur" Farm of Jizzakh region, in the early matured Shakarpalak and Obi novvot varieties of melon, fuzariose wilt-Fusarium oxysporum f.sp. niveum and flour dew-Erysiphe Cichoracearum Dc. F. cucurbitacearum and brown spotting in the friendship district "Obidjon ota" farm of the Jizzakh region-Sporodesmium mucosum Sacc.var.pluriseptatum Karst. diagnosed diseases (Figure 1).

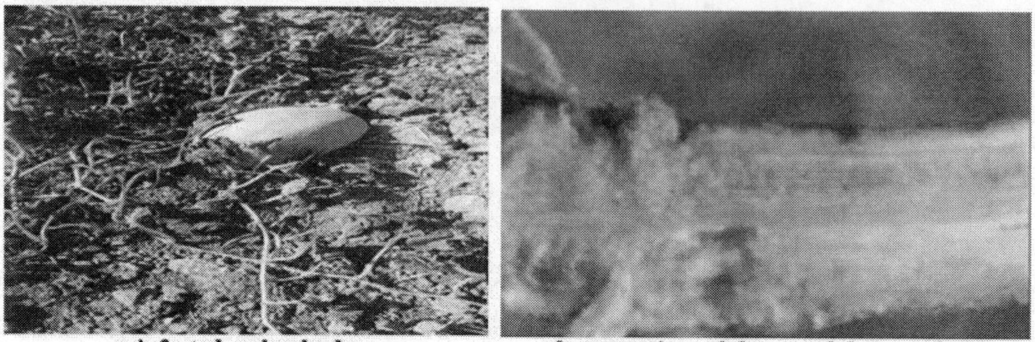

a-infected melon body **b- necrosises of the cut of the stem bug**

Figure 1. Fuzarious wilting disease of the melon.

In mirabad district "Mirabad yuldizi" farm of Syrdarya region in the varieties of melon early matured Amri and Obi novvot Fusarious Salish-Fusarium oxysporum f.sp. niveum, Macrasporiasis-Macrosporium cucumerinum Ellis and anthracnose-Colletotrichum lagenarium E. et H. Diseases recorded (Figure 2).

Figure 2. Anthracnose and macrasporiasis desease of melon.

In the field of the White Gold District "Railway agro-industrial" UK of the Syrdarya region, macrasporiasis and fusarious wilting diseases were found in the mid-ripening blueberry and white seed varieties of melon.

In the "Obidjon ota" farm of the friendship District of the Jizzakh region, The Late " Kampir chopon" variety of melon was found to have only fusarious wilting (Figure 3).

Figure 3. Fusarious of the late "Kampir chopon" variety of melon.

Since late matured was a variety, Erysiphe cichoracearum, macrosporiasis and anthrocnose diseases were not detected in this variety. During the growing season of the melon, samples were taken from plants with presented symptoms of the disease. The samples taken were taxied in laboratory conditions, and pure cultures of disease triggers were isolated (Figure 4).

These pictures show clear views of disease-causing microorgnaisms found during the growing season of melon.

Measures to solve melon diseases: to determine the species composition of the main disease-causing microorganisms found in melon in the fight against Melon diseases and to check the effective seeders for them in laboratory conditions is an important event. In the following research, disease-absorbing fungi in the root, stem, leaf and fruit of the melon were isolated and the effectiveness of seeders was checked against them. Maximurug, which adversely affected the development of pathogeninng with high efficiency, is used in drug suspension with a solution of 10 L per 1 T of grain, Seles top melon seeds are used in drug suspension, 12-15 L of solution per 1 t of grain in the drug suspension, 10 L of solution per 1 T of Raksil gold seed drug suspension chemical preparations were selected and used inSeed drinkers can be used in laboratory conditions to treat melon diseases (Figures 5, 6, and 7).

https://doi.org/10.1051/e3sconf/202342104007

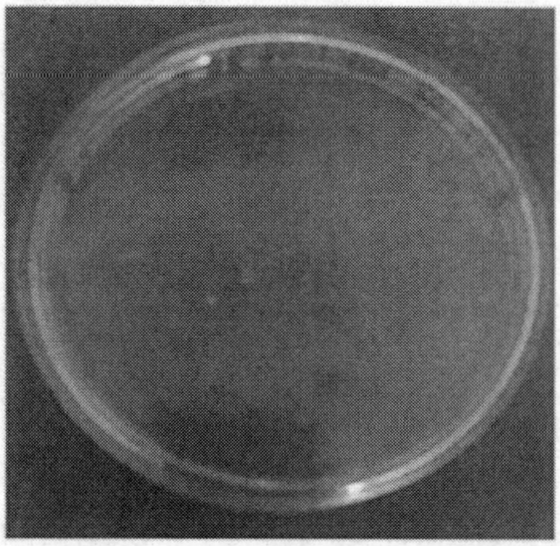

alternaria cucumerina

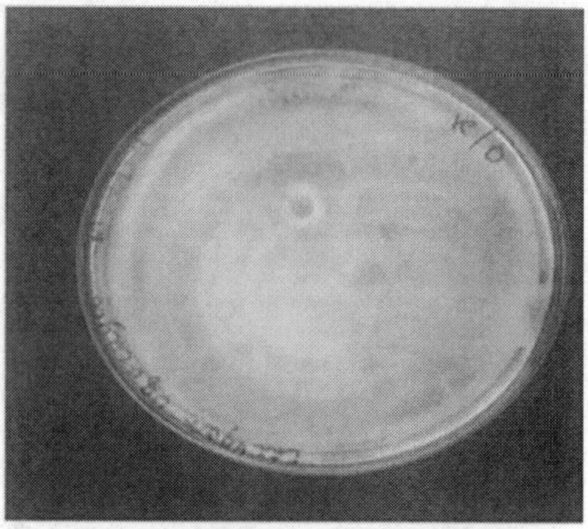

fusarium sp fungi

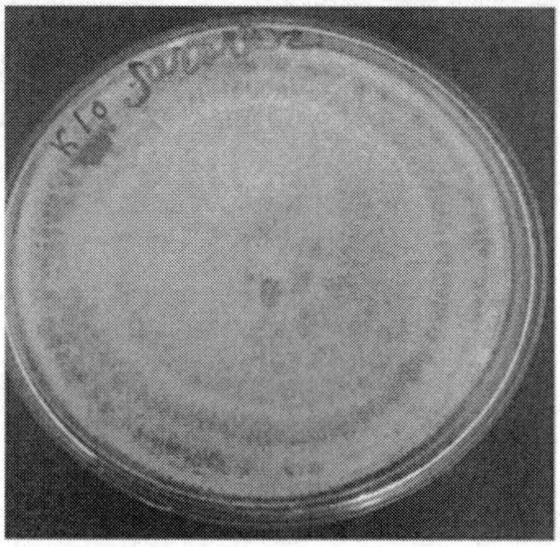

Fusarium oxysporum f. sp. niveum

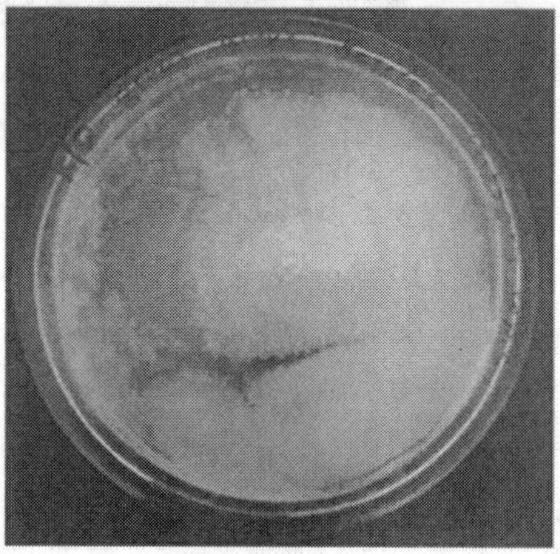

Ascochytacucumis

Figure 4. Clear views of disease-causing microorganisms in melon.

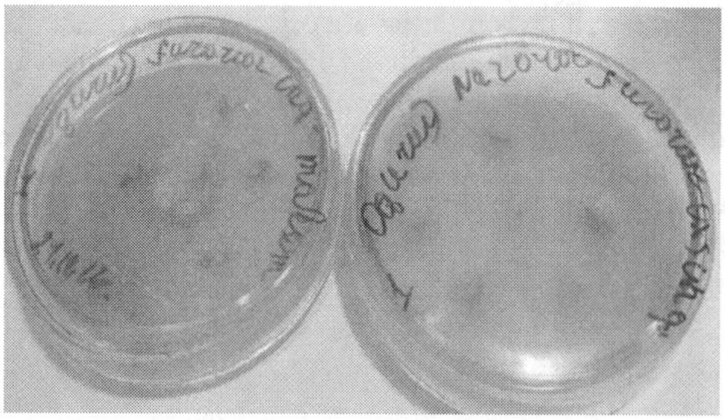

Figure 5. Control in front of the medicinal seeds with fusarium sp fungi and Maxim - planted in an artificial nutrient medium in the cup petri.

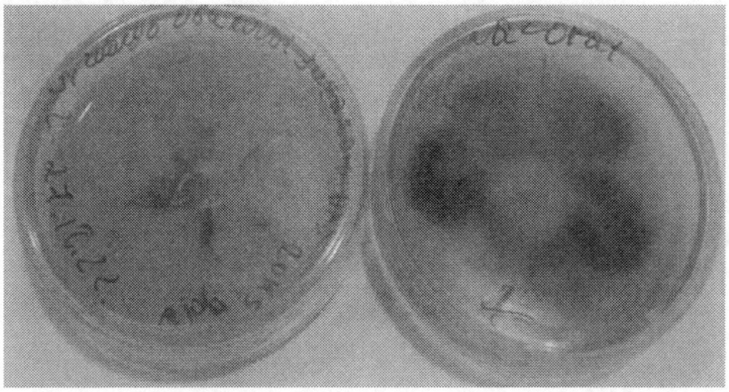

Figure 6. Cuppetriga controlled in front of the pollinated seeds with alternariacucumerina fungi and Seles top - planted in an artificial feed environment

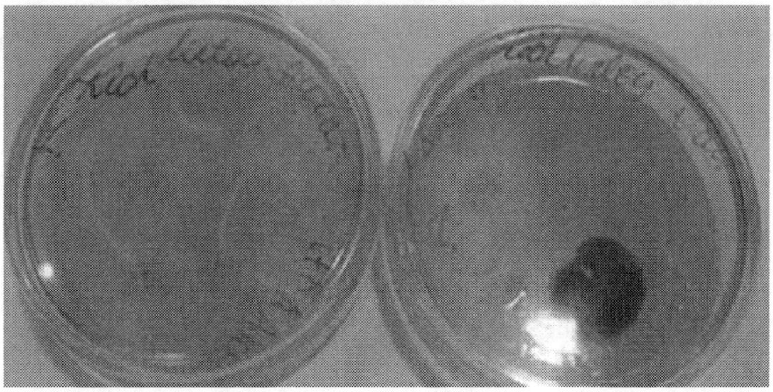

Figure 7. Control in front of the medicinal seeds with macrosporium fungi and Raxil - planted in the artificial nutrient medium of the cup petri.

In these pictures we can see the effectiveness of seed drugs used against diseases caused by the main fungi found in melon. The seed drugs tested in the experiment Maxim, Seles top, Raxil gold did not develop the fungi's pathogens and stopped their growth. The development of the disease in seeds planted for control can be seen in the picture and negatively affected its growth. Seeders with high efficiency were selected and planted in open fields and experimental fields with seed treatment on lysimeters. These fungicides protect seeds from soil patogens after sowing seeds in field conditions (Figures 8 and 9).

Figure 8. Process of processing seeds with medicinal preparations before planting in farm and lysemeter conditions.

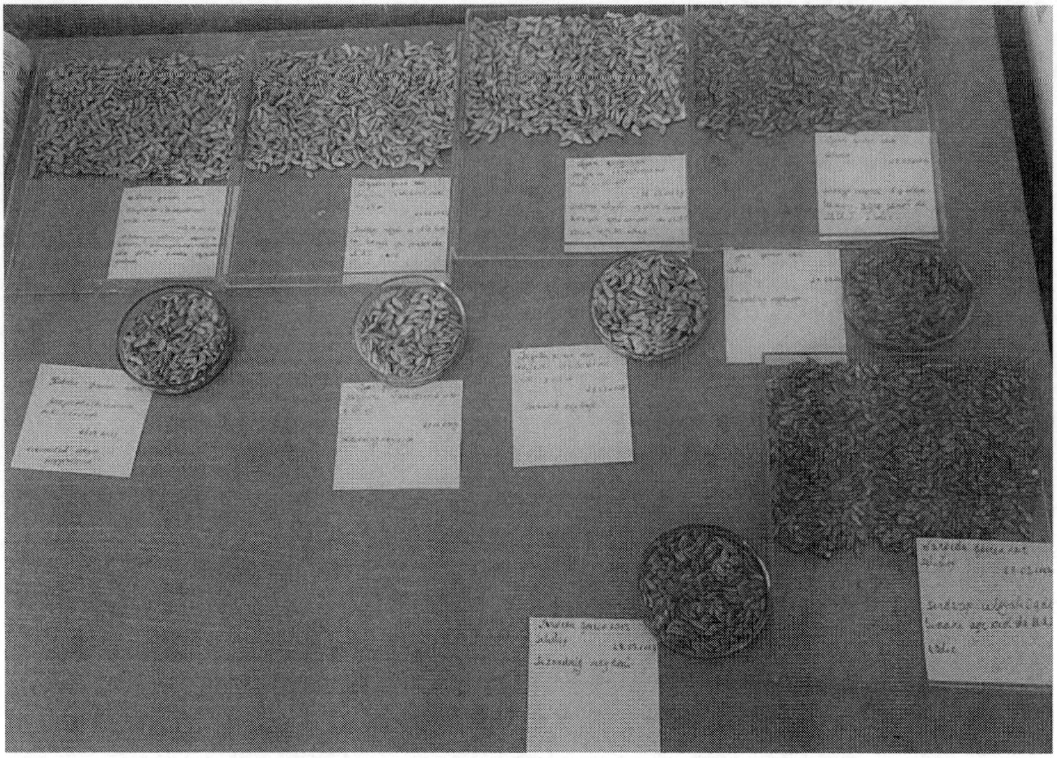

Figure 9. Process of processing melon seeds with seed medicinal preparations in varietal cross section.

4 Conclusions

In conclusion, in obtaining a high and high-quality harvest of melon, the first thing to do is to choose high-quality and disease-resistant varieties;

Deep plowing of land from agrotechnical events and planting of crops within the established deadlines;

Before sowing seeds, MaximUrug from chemical and biological seeders is treated with a solution of 10 l per 1 t of grain in the drug suspension, Seles top melon seeds are treated with a solution of 12-15 l per 1 t of grain in the drug suspension, Raksil gold seeds are treated with a solution of 10 l per 1 t of grain in the drug suspension; during the growing period, it was recommended.

References

1. B. A. Hasanov, R.O. Ochilov, R.A. Gulmurodov. Diseases of vegetables, potatoes and pulses and the fight against them. Tashkent, 2009.
2. Sheraliev A.Sh., Rakhimov U.X. Agricultural Phytopathology.Tashkent - 2014
3. Sidorova, Svetlana Fedorovna.Vertisilleznae I fusarium education related to agricultural culture / s. F. Sidorova. - Moscow: Kolos, 2013. - 157 p.
4. Pidoplichko, Nikolay Makarovich. Gribi-parasite culturnix rastenius [text]: Opredelitel: V 3-x t. / N.M. Pidoplichko; [In-T microbiologii I virologii im. D.K. Zabolotnogo]. - Kiev: Naok. Dumka, 2017.
5. Bouriev X.Ch., Ashurmetov O.A. The biology and cultivation technology of polyz crops.// T.: "Cocktail".- 2000.- 115 p.

6. Dudko P.N., Karimov A.K., Ermoxin B.N., Uspenskaya E.V. Atlas ". Religious Uzbekistan".// T.: UDN-2012.- 184 p.
7. Alternariozi / / enticyclopedicheskiy slovar. Soviet Encyclopedia, 2019. - p. 24.
8. List of pesticides and agrochemicals allowed for agricultural use of the Republic of Uzbekistan pp. 113,116-117, 2022
9. Boboev, S., Makhkamov, T., Bussmann, R. W., Zafar, M., & Yuldashev, A. (2023). Anatomical and phytochemical studies and ethnomedicinal uses of Colchicum autumnale L. Ethnobotany Research and Applications, 25, 1-9. DOI:10.32859/era.25.6.1-9
10. TKh, M., Brundu, G., Jabborov, A. M., & Gaziev, A. D. (2023). Predicting the potential distribution of Ranunculus sardous (Ranunculaceae), a new alien species in the flora of Uzbekistan and Central Asia. BioInvasions Records, 12(1), 63-77. DOI:10.3391/bir.2023.12.1.05
11. Ameen, M., Zafar, M., Ahmad, M., Ramadan, M. F., Eid, H. F., Makhkamov, T., ... & Majeed, S. (2023). Assessing the Bioenergy Potential of Novel Non-Edible Biomass Resources via Ultrastructural Analysis of Seed Sculpturing Using Microscopic Imaging Visualization. Agronomy, 13(3), 735. DOI:10.3390/agronomy13030735
12. Noor, W., Zafar, M., Ahmad, M., Althobaiti, A. T., Ramadan, M. F., Makhkamov, T., ... & Khan, A. (2023). Petiole micromorphology in Brassicaceous taxa and its potential for accurate taxonomic identification. Flora, 303, 152280. DOI:10.1016/j.flora.2023.152280
13. Aziz, A., Ahmad, M., Zafar, M., Gaafar, A. R. Z., Hodhod, M. S., Sultana, S., ... & Chaudhay, B. (2023). Novel Copper Oxide Phyto-Nanocatalyst Utilized for the Synthesis of Sustainable Biodiesel from Citrullus colocynthis Seed Oil. Processes, 11(6), 1857. DOI:10.3390/pr11061857

Effects of sophoroflavonoside and narcissin flavonoids on mitochondrial dysfunction and antioxidant enzyme activity in rat liver poisoned with galoxyfop-r-methyl pesticide

Mashxura Parpieva[1,*], *Parida* Mirkhamidova[2], *Mamurjon* Pozilov[3] and *Rano* Alimova[4]

[1]Andijan State University, Andijan, 170100, Uzbekistan
[2]Tashkent State Pedagogical University, Tashkent, 100070, Uzbekistan
[3]National University of Uzbekistan, Tashkent, 100174, Uzbekistan
[4]Tashkent State Agrarian University, Tashkent, 100140, Uzbekistan

Abstract. This study delves into the accumulation of residues in the liver tissue of rats exposed to the galoxyfop-R-methyl pesticide. The investigation also evaluates the presence of malondialdehyde (MDA), a lipid peroxidation (LPO) product, in the liver mitochondrial membrane, along with the activity of antioxidant enzymes catalase, superoxide dismutase (SOD), glutathione reductase (GR), and glutathione peroxidase (GP). The study further explores the impact of sophoraflavonoloside (SFL) and narcissin flavonoids on enzyme activity over a dynamic span of 10 to 40 days. In the experimental design, rats in the study group were subjected to galoxyfop-R-methyl pesticide at a dosage equivalent to LD50/10 through a specialized probe. Subsequently, the concentration of the pesticide residues in liver tissue was measured on the 5th, 10th, 20th, 30th, and 40th days post-pesticide exposure. The research also probes into the content of malondialdehyde (MDA) in the liver mitochondrial membrane, as well as the activity of antioxidant enzymes, within the context of SFL. This meticulous examination aids in comprehending the dynamics of enzyme responses and oxidative stress modulation in the liver of rats subjected to galoxyfop-R-methyl pesticide. By investigating the intricate interactions between pesticide exposure, enzyme activity, and antioxidant mechanisms, this study contributes to the broader understanding of the potential effects and counteractive measures concerning pesticide-related challenges in biological systems.
Keywords. Liver, mitochondria, galoxyfop–R–methyl (galoxyfop), SFL, narcissin, catalase, SOD, GR, GP.

1 Introduction

Pesticides, which are constantly used in agriculture today, lead to the development of metabolic diseases [1]. In case of acute and chronic poisoning with pesticides, damage to all

* Corresponding author: mashxura.parpieva1001@mail.ru

tissues, structural changes in bone marrow elements, peripheral blood, endocrine glands, liver, kidney, heart muscle and brain cells are observed [2, 3]. Pesticide poisoning affects all tissues and organs, but the most sensitive organ to their effects is the liver. Hepatocyte cells, which are actively involved in the metabolism of xenobiotics, become the main target of these pesticides. In the cell, mitochondria and microsomes are the most affected by pesticides. Taking into account the important role of these organoids, it is important to damage their membranes and thereby the cell and the organism as a whole [4]. Pesticides enter organs and tissues in various ways and accumulate as residues. Residual pesticide in the liver causes a number of functional changes, that is, it causes a change in the activity of LPO and enzymes [5]. But the slow elimination of pesticides from the liver reduces the activity of these enzymes. Residual pesticides can damage all systems of the human body and cause various pathological conditions [6]. The toxicity of some pesticides in mammals is manifested by the formation of free radicals and disturbance of redox balance [7].

Pesticides primarily damage liver tissue through the blood. The liver plays a key role in the detoxification process for xenobiotics. A change or violation of its function leads to hepatotoxicity and shows a significant increase in the activity of ALT, AST and alkaline phosphatase enzymes. These changes can affect membrane permeability and cause metabolite disturbances [8]. Metabolism of pesticides is explained by the disruption of the homeostatic mechanism of the cell by the production of reactive oxygen species (ROS). They interact with membrane lipids, disrupting the physiological balance between membrane lipids and proteins, especially mitochondrial substrate carriers and electron transporters [9].

A number of pesticides belonging to different classes cause increases in blood plasma alanine aminotransferase (ALT) and aspartate aminotransferase (AST) activity, lipids and inflammatory biomarkers [10]. According to the results of a study conducted with the herbicide galoxyfop-R-methyl, this herbicide causes oxidative stress in the liver and kidneys. This causes morphological, histopathological and immunological changes. The toxicity of galoxyfop-R-methyl in mammals has been determined to be related to the formation of CSF and disruption of redox activity in tissues [11]. The phytotoxicity of galoxyfop-R-methyl is based on the inhibition of the enzyme acetyl-CoA carboxylase [12].

Protection of the cell from LPO is carried out by a system of antioxidant enzymes. The most important protective enzymes are catalase, SOD, GR, GP, which neutralize the main and intermediate products of LPO, as well as secondary products of peroxidation of glutathione transferase, glyoxidase, formaldehyde dehydrogenases and other carbonyl compounds [13].

The purpose of the study is to determine the residual amount in the liver tissue of rats poisoned with galoxyfop-R-methyl and the corrective effect of SFL and narcissin flavonoids for 10, 20, 30 and 40 days on the content of the liver mitochondrial membrane LPO product MDA, the activity of antioxidant enzymes catalase, SOD, GR and GP is to determine the dynamics.

2 Materials and methods

Conducting scientific research on experimental animals was carried out on the basis of the international Declaration of Helsinki, the rules developed by the Council for International Organizations of Medical Sciences (CIOMS) (1985). Experimental animals were poisoned with 10.4% emulsion concentration of galoxyfop-R-methyl ($C_{16}H_{13}F_3ClNO_4$), a synthetic drug currently used in agriculture against weeds. The LD_{50} dose of galoxyfop-R-methyl in rats is 623 mg/kg. It is considered moderately toxic to humans and warm-blooded animals and has a level of toxicity of class 2 [14].

In our preliminary studies, indicators of residual amount of galoxyfop-R-methyl pesticide in liver tissue were determined. At the next stage of the study, healthy, experimental toxic model (poisoned with a pesticide) and pharmacotherapy with flavonoids with antitoxic properties were conducted in vivo. galoxyfop-R-methyl pesticides were used to intoxicate animals. Each experimental group consisted of 3-5 animals.

Experimental animals were divided into separate model groups for intoxication with galoxyfop-R-methyl pesticide and their correction with flavonoids.

Initially, male rats selected for intoxication with galoxyfop-R-methyl pesticide were divided into groups:

Group I healthy (control) (n=5);

Group II galoxyfop-R-methyl (n=5-6);

Group III galoxyfop-R-methyl+SFL (n=5-6);

Group IV galoxyfop-R-methyl+ narcissin (n=5-6):

Animals in groups II, III and IV of the experiment were poisoned once with a dose of LD50 1/10 of galoxyfop-R-methyl pesticide through a special probe. After administration of galoxyfop-R-methyl pesticide, SFL 10 mg/kg of group III and narcissin flavonoid 10 mg/kg of animals were administered orally for 10 days.

The amount of residual pesticide in the liver tissue of rats poisoned with pesticides was determined after 5, 10, 20, 30, and 40 days, and after 10, 20, 30, and 40 days after administration of SFL and narcissin flavonoids to poisoned rats, the activity of mitochondrial antioxidant enzymes and the recovery of functional disorders of membranes were studied.

High-performance liquid chromatography mass spectrometry (6420 Triple Quad LC/MS (Agilent Technologies, USA) device was used for quantitative analysis of pesticides in liver samples. APCI (Atmospheric pressure chemical ionization) was used as the ionization method. performed at the expense of ionized ions.

Rat liver mitochondria were isolated using differential centrifugation W.C.Schneider [15] method.

The detection of LPO is based on the reaction between MDA and thiobarbituric acids (TBK), which results in the formation of a colored trimethine complex at high temperature and acidic pH [16]. The complex is measured in a spectrophotometer at a wavelength of 532 nm.

Determination of SOD enzyme activity (KF 1.15.1.1) Misra and J. Fridovich (1972). carried out according to the method [17]. Catalase activity in liver mitochondria Korolyuk M.A. determined using the method. Results are expressed in μKat/mg protein [18]. The activity of the GR enzyme was determined based on the accumulation of oxidized glutathione. Enzyme activity is expressed in micromoles of NADF.H per 1 g protein for 1 min at 370 C for 10 min at 340 nm wavelength (μM/min.g).

GP is determined by the accumulation of oxidized glutathione: occurs with the destruction of oxidized glutathione and is detected at a wavelength of 260 nm. Enzyme activity expresses glutathione in micromoles per 1 minute against 1 g of protein (μM/min.g). Mitochondrial protein was determined by Lowry's method.

3 Results and discussion

Galoxyfop-R-methyl pesticide was injected into the stomach of rats at a dose of LD_{50}1/10 through a special probe, and their residual amount in the liver of rats was determined on the 5th, 10th, 20th, 30th, and 40th days after poisoning. First, quantitative chromatographic analyzes were obtained from the livers of (healthy) galoxyfop-R-methyl and control animals. The average values obtained based on the characteristic absorption peaks for galoxyfop-R-methyl are shown in Table 1 below.

Table 1. Amount of residual pesticides in the liver of rats poisoned with galoxyfop -R-methyl (LD_{50}1/10).

Experimental groups	Residual amount of galoxyfop-R-methyl, 1 gram amount (mkg) relative to the sample (n=3).				
	Days				
	5	10	20	30	40
Control	-	-	-	-	-
Galoxyfop-R-methyl	0.01741	0.00159	0.000138	0.0000164	-

According to the obtained results, it was found that in the group of animals poisoned with galoxyfop-R-methyl pesticide, the amount of residual pesticides in the rat liver tissue was higher on the 5th and 10th days after poisoning than on the 20th and 30th days. On the 40th day, no residual amount of this pesticide was found. However, based on our results, it can be concluded that the amount of residual pesticides in the liver of rats can affect the biochemical and physiological indicators of liver mitochondria and cause a number of changes.

Intensity of the formation of CFS under the influence of pesticides causes an increase in LPO of the inner and outer membrane of mitochondria [3].

Pesticides weaken the antioxidant defense system in mitochondria, and antioxidant enzymes can degrade and reduce the amount of free radicals ($\bullet^1O_2$, $\bullet O_2-$, $\bullet OH$, and H_2O_2) in the cytosol.

For this purpose, in our present experiment, the effect of SFL and narcissin on the amount of LPO product MDA of rat liver mitochondria poisoned with pesticides was studied in relation to the dynamics of 10, 20, 30 and 40 days.

According to the obtained results, the amount of LPO product MDA in the membrane of liver mitochondria in group II rats poisoned with galoxyfop-R-methyl pesticide was 35.7±2.8%, 39 ,1±3.2%, 36.4±2.5% and 32.1±2.2% increase (Figure 1).

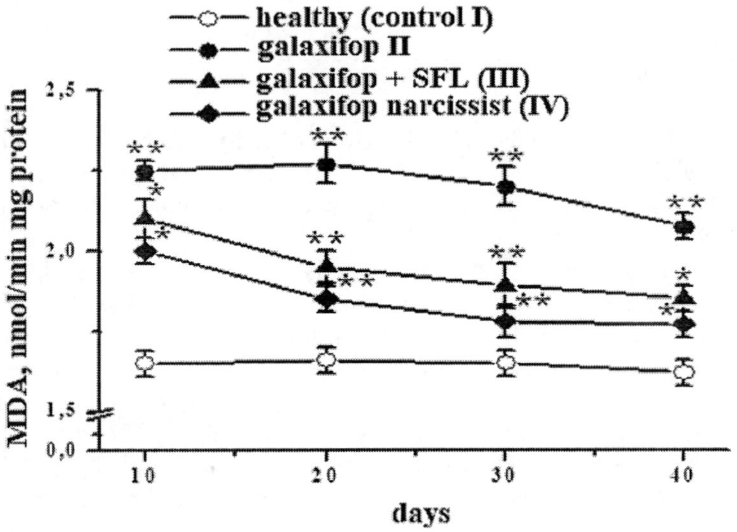

Figure 1. Dynamic effects of SFL and narcissine on MDA content of liver mitochondria of rats poisoned with the pesticide galoxyfop-R- methyl at 10, 20, 30 and 40 days (*R<0.05; *R<0.01; n=5-6).

This indicates the acceleration of the LPO process in the mitochondrial membrane under the influence of the galoxyfop-R-methyl pesticide. In group III rats pharmacotreated with SFL, the amount of liver mitochondria MDA was 9±0.8% on day 10, 15.6±1.2%, 21.3±1, and 21.3±1% on days 20, 30, and 40, respectively. 5% and 19.8±1.6% decrease (Figure 1).

After 10 and 20 days of narcissin-treated group IV rats, the MDA content was 11.5±0.9% and 21.6±1.6%, respectively, and a reliable decrease was observed compared to the values of group II. However, on days 30 and 40, it was found that the amount of MDA in liver mitochondria decreased by 28.6±2.1% and 26.0±2.2%, respectively, compared to the values of group II (Figure 1). It was found that the activity of narcissin was slightly higher than that of SFL on the amount of MDA in rat liver mitochondria poisoned with the pesticide galoxyfop-R-methyl.

Pesticides cause an increase in the level of LPO in liver mitochondria, and a decrease in the activity of SOD and catalase enzymes. It is known that the biological function of the enzymes of the antioxidant defense system is to protect cells from LPO products and other mutagenic effects.

Therefore, in our experiments, the dynamics of the corrective effect of SFL and narcissin flavonoids on the activity of liver mitochondria of rats poisoned with galoxyfop-R-methyl pesticide was studied for 10, 20, 30, 40 days. In our preliminary experiment, it was found that catalase activity decreased by 10, 20, 30 and 40 days in group II rats poisoned with galoxyfop-R-methyl pesticide compared to the control, and decreased by 45±3.2% in 40 days (Table 2).

Table 2. Effects of SFL and narcissin flavonoids on catalase enzyme activity of liver mitochondria of rats poisoned with the pesticide galoxyfop-R-methyl (10, 20, 30 and 40 days dependent dynamics) (μKat/mg protein).

Experimental groups	**n**	**Experience days**			
		10	**20**	**30**	**40**
Control	5	38.5±1.85	37.25±2.16	38.90±2.52	38.16±2.86
Galoxyfop-R-methyl	6	25.11±1.25*	24.35±1.21*	22.03±1.50*	20.95±1.04**
Galoxyfop-R-methyl +SFL	5	26.33±1.60	27.56±1.04*	28.71±1.43**	29.09±1.65**
Galoxyfop-R-methyl + Narcissin	5	26.57±1.90	28.11±1.40*	30.42±1.82**	33.95±2.27**

(*R<0.05; **R<0.01; n=5-6)

When group III rats treated with galoxyfop-R-methyl were treated with SFL flavonoid and group IV with narcissin, on days 10, 20, 30, and 40, their liver mitochondria catalase enzyme activity was 21±1.6% and 39±2%, respectively, compared to group II. It was found that it increased by 8% (Table 2).

Along with changes in catalase enzyme activity in liver mitochondria of rats poisoned with the pesticide galoxyfop-R-methyl, SOD activity also changes. SOD plays an important role in protecting cells and tissues from oxidative damage caused by negative factors. However, when SOD is activated, H_2O_2 is produced, which is an inhibitor of the enzyme. Therefore, the effective functional activity of SOD is mainly related to other components of the defense system, in particular, enzymes such as catalase, GR and GP. SOD inhibits the harmful reactions of superoxide by protecting the cell from superoxide toxicity [18]. Superoxide radicals are produced at two major sites in the electron transport chain, respiratory chain I (NADH dehydrogenase) and respiratory chain III (ubixinone cytochrome-c-reductase). As a result of the transfer of electrons from complex I or II to coenzyme Q or ubiquinone (Q), coenzyme Q (QH_2). the reduced form of is formed. The reduced form of

QH_2 regenerates coenzyme Q via an unstable anion (Q-) in the Q cycle. The created Q- immediately transfers electrons to molecular oxygen, leading to the formation of superoxide radicals. The formation of superoxide is not an enzymatic process, and therefore the higher the metabolic rate, the higher the production of ROS [17].

In our next experiment, the effect of SFL and narcissin flavonoids on the activity of liver mitochondria antioxidant enzyme SOD of rats poisoned with galoxyfop-R-methyl pesticide was investigated.

According to the obtained results, it was found that SOD activity decreased by 10, 20, 30 and 40 days in group II rats poisoned with galoxyfop-R-methyl pesticide, and on the 40th day, it decreased by 41.6±2.8% compared to the control. By day 40, the activity of SOD enzyme in liver mitochondria was 16.4 ± 1.2% and 22.7 ± 16.4 ± 1.2%, respectively, in rats treated with SFL flavonoid pharmacotherapy with SFL flavonoid and group IV treated with narcissin. An increase of 2.0% was shown in our experiments (Table 3).

Table 3. Effects of SFL and narcissin flavonoids on liver mitochondrial SOD enzyme activity (10, 20, 30, and 40-day dynamics) in rats poisoned with galoxyfop-R-methyl pesticide (ed/min.mg protein).

Experimental groups	n	Experience days			
		10	20	30	40
Control	5	3.30±0.06	3.28±0.06	3.26±0.05	3.34±0.06
Galoxyfop-R-methyl	6	2.57±0.06*	2.40±0.05*	2.23±0.05*	1.95±0.04**
Galoxyfop-R-methyl +SFL	5	2.36±0.05*	2.24±0.04*	2.30±0.09	2.44±0.05*
Galoxyfop-R-methyl + Narcissin	5	2.29±0.10*	2.20±0.06**	2.56±0.02**	2.71±0.03**

(*P<0,05; **P<0,01; n=5-6).

Activity of catalase and SOD enzymes, which are responsible for the antioxidant defense system, was observed to decrease in the experimental group, animals poisoned with pesticide. Decreased enzyme activity causes increased generation of free radicals in the mitochondrial respiratory chain. This, in turn, leads to the development of various pathological processes.

Thus, after 10-, 20-, 30- and 40-days after poisoning with galoxyfop-R-methyl pesticide, it was found that the activity of SOD and catalase enzymes of the antioxidant defense system of the liver mitochondria was sharply reduced. Selected plant flavonoids SFL and narcissin caused the activity of SOD and catalase enzymes to recover to a certain extent by the 30th and 40th days.

In addition to the changes in SOD and catalase activity in the liver mitochondria of rats poisoned with galoxyfop-R-methyl pesticide, the activity of GR and GP enzymes can also be changed. For this purpose, in our next experiment, the effect of SFL and narcissin on GR, GP enzymes, whose activity changed under the conditions of pesticide application, was studied.

Glutathione forms an antioxidant system in the cells of GP and GR. Glutathione not only protects cells from toxic effects such as free radicals, but also determines redox states in cells. In cells, the reduced form of thiol groups (SH) is present in a concentration of about 5 mM. Such a concentration of glutathione in the cell causes the restoration of disulfide (-S-S-) bonds formed between polypeptide chains of proteins [10].

The activity of antioxidant enzymes such as GR and GP in cell mitochondria can be significantly decreased under the influence of pesticides [8] In our next experiment, the effect of SFL and narcissin on the activity of mitochondrial antioxidant (GR, GP) enzymes in the liver mitochondria of rats poisoned with haloxyfop-R-methyl pesticide was studied in a 40-day dynamic state. .

Initially, in our experiments, under the influence of galoxyfop-R-methyl pesticide, changes in the activity of GR, one of the important antioxidant enzymes in liver mitochondria, and the correcting properties of plant compounds were studied. According to the obtained results, it was found that GR activity in liver mitochondria of group II rats injected with galoxyfop-R-methyl herbicide at a single dose of LD_{50} 1/10 was significantly decreased compared to the control. It was found that under the influence of galoxyfop-R-methyl, the dynamics of changes in the activity of the liver mitochondrial GR enzyme showed the lowest value at 10 days and decreased by 38.1±2.5% compared to the control (Figure 2). It was found that the dynamics of GR activity under the influence of herbicides was reduced compared to the control even on the 20th, 30th and 40th days. However, there was no significant difference in GR activity compared to 10 days.

Group III rats treated with galoxyfop-R-methyl herbicide SFL and group IV animals were treated with narcissine at a dose of 10 mg/kg for 10 days. The liver mitochondria GR enzyme activity of group III rats corrected with SFL was pathological, i.e. 14.1±1.1%, 11.7±0.7%, 11.7±0.7%, respectively, on the 10th, 20th, 30th, and 40th day compared to the values of group II. It was found that it increased by 14.5±1.2% and 18±1.2% (Figure 2).

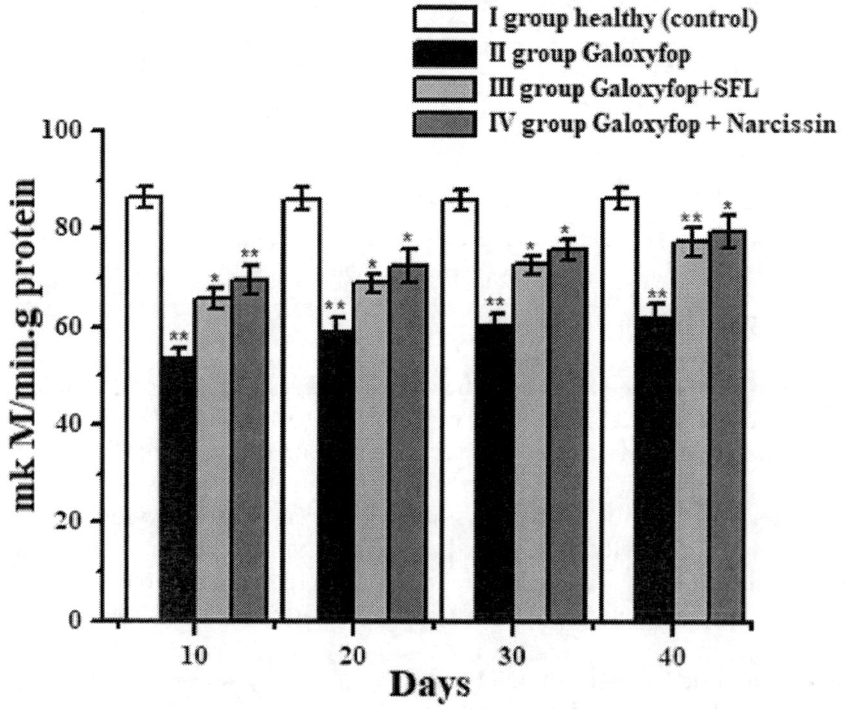

Figure 2. Effects of SFL and narcissin on glutathione reductase enzyme activity of liver mitochondria of rats poisoned with galoxyfop-R-methyl pesticide (*R<0.05; **R<0.01; n=5-6).

The liver mitochondria GR enzyme activity of group IV rats treated with narcissin was 18.4±1.1%, 15.8±1.3%, 18.0±1% compared to the pathological group on the 10th, 20th, 30th, and 40th days, respectively. It was noted that it increased by 1.2% and 20.3±1.9% (Figure 2). Therefore, depending on the dynamics, under the influence of SFL and narcissin, a certain degree of restoration of the activity of the liver mitochondrial GR enzyme was observed.

Along with other antioxidant enzymes, GP is one of the key enzymes in the catabolism of free radicals. GP enzyme acts to break down peroxidized lipids and H_2O_2 in mitochondria and cytoplasm to water and oxygen [8].

In mitochondria exposed to pesticides. In the literature, as a result of increased LPO process in rat erythrocyte cells under the influence of organochlorine pesticides, it was found that not only catalase, SOD, GR, but also GP activity decreased [15].

In our experiment, the 40-day effect of SFL and narcissin flavonoids on liver mitochondria GP activity of rats intoxicated with galoxyfop-R-methyl pesticide was studied in a dynamics-dependent manner. According to the obtained results, it was found that GP activity in liver mitochondria decreased sharply by 41±3.2% in group II compared to control (group I) by 10 days after administration of galoxyfop-R-methyl, and after 20 days, GP activity did not change much compared to 10 days. However, this indicator was observed to decrease by 27.5±2.1% and 28.5±1.3%, respectively, compared to the control by the 30th and 40th days (Figure 3).

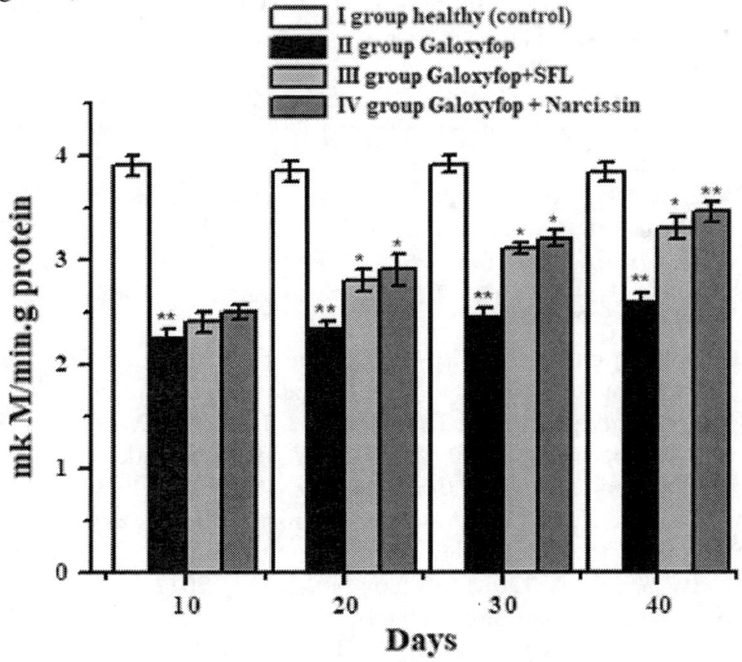

Figure 3. 40-day time-dependent effects of SFL and narcissine on glutathione peroxidase enzyme activity in liver mitochondria of rats poisoned with the pesticide galoxyfop-R-methyl. (*R<0.05; **R<0.01; n=5-6).

Group III rats intoxicated with galoxyfop-R-methyl were pharmacotreated with SFL orally for 10 days, and their GP activity in liver mitochondria did not change significantly compared to group II at 10 days, but by days 20, 30, and 40, 13 It was found that it increased by ±0.9%, 17±1.2% and 18.5±2.2%. Continuing the experiments in group IV rats poisoned with galoxyfop-R-methyl, and pharmacotherapy with narcissine for 10 days, the effect of this substance on GP activity was not noticeable at 10 days. However, on the 20th, 30th, and 40th days, it was found that it increased by 15.6±1.3%, 19.5±1.8%, and 22.3±1.2%, respectively (Figure 3). The results of the experiment showed that the effect of narcissine on GP activity in liver mitochondria was more effective than that of SFL.

4 Conclusions

Thus, in the liver mitochondria of rats poisoned with galoxyfop-R-methyl pesticide, an increase in the amount of LPO product MDA was observed on days 10-20, and a sharp decrease in the activity of catalase, SOD, GR, GP enzymes of the antioxidant defense system was detected. In the SFL and narcissine-corrected groups, it was observed that the amount of LPO product MDA was restored to a certain extent and the activity of antioxidant enzymes was increased to a certain extent in the mitochondria of rat liver by 40 days.

References

1. Abdelrasoul A. Modulation of abamectin and indoxacarb -induced toxicity on male albino rats by *Moringa oleifera* // Alexandria science exchang Journal. – 2018. – V. 39(2) – P. 232-243.
2. Abdollahi, M.; Ranjbar, A.; Shadnia, S.; Nikfar, S.; Rezaiee, A. Pesticides and oxidative stress: A review. Med. Sci. Monit. – 2004. – V.10. – P.144–147.
3. Abdou H.M., Hussien H.M., Yousef M.I. Deleterious effects of cypermethrin on rat liver and kidney: Protective role of sesame oil // J. Environ. Sci.Health B – 2012. – V.47 (4). – P.306-314.
4. Barski D., Spodniewska A., Zasadowski A. Activity of superoxide dysmutase, catalase and glutathione peroxidase in rats exposed to chlorpyrifos and erofloxacin // Polish Journal of Veterinary Sciences – 2011. – V.14 (4) – P. 523-529.
5. Evangelou E., Ntritsos G., Chondrogiorgi M., Kavvoura F.K., Hernández A.F., Ntzani E.E., Tzoulaki I. Exposure to pesticides and diabetes: A systematic review and meta-analysis // Environ. Int. – 2016. –V. 91. –P.60–68.
6. Finkel T., Holbrook N.J. Oxidants, oxidative stress and the biology of ageing // Nature. –2000. – V.408. – P.239–247.
7. Gassner B., Wuthrich A., Scholtysik G., Solioz M. The pyrethroids permethrin and cyhalothrin are potent inhibitors of the mitochondrial complex I // The Journal of Pharmacology and Experimental Therapeutics – 1997. – V.281.(2). – P.850-860.
8. Guven C., Sevgiler Y., Taskin E. Pyrethroid insecticides as the mitochondrial dysfunction inducers // Mitochondrial Diseases – 2018. – P. 293-322.
9. Juliana C. Castanha Z., Marcos A. Maioli Hyllana C.D. Medeiros, Fabio E. Mingatto. Abamectin affects the bioenergetics of liver mitochondria: A potential mechanismof hepatotoxicity. // Elsevier Ltd. Toxicology inVitro. – 2012. – V.26. – P.51–56.
10. Kaneko H. Pyrethroids: Mammalian metabolism and toxicity. // Journal of Agricultural and Food Chemistry. – 2011. – V.59(7). – P. 2786-2791.
11. Lukowicz C., Ellero-simatos S., Régnier M., Polizzi A., Lasserre F., Montagner A., Lippi Y., Jamin E.L., Martin J., Naylies C. Metabolic effects of a chronic dietary exposure to a low-dose pesticide cocktail in mice: Sexual dimorphism and role of the constitutive androstane receptor // Environ. Health Perspect. –2018. –V.126 – P. 1–18.
12. Mirkhamidova P., Pozilov M.K., Parpieva M.J., Shakhmurova G.A., Jumagulova K.A. Determination of the amount of residual pesticides in the liver tissue of rats poisoning with haloxyphop-r-methyl and indosacarb pesticides// Journal of Pharmaceutical Negative Results. –2022. –V.13.(1). – P. 738-745
13. Olayinka E.T., Ore A. Hepatotoxicity, Nephrotoxicity and Oxidative Stress in Rat Testis Following Exposure to Haloxyfop-*p*-methyl Ester, an Aryloxyphenoxypropionate Herbicide // Toxics –2015. – V.3. – P. 373-389.
14. Ozcan A., Öğün M. Biochemistry of reactive oxygen and nitrogen species // Basic Principles and Clinical Significance of Oxidative Stress – 2015. – V. 3. – P. 38-57.
15. Parpieva M.J., Mirxamidova P.,Pozilov M.K.,Tuychiyeva D.S. The Effect of Phlavonoids on Precisely Oxidation of Lipids of the Membrane of the Rat Liver Mitochondria, Poisoned with Pesticides// Jundishapur Journal of Microbiology Published online 2022 January Research Article. –2022. –V. 15 (1). –P. 676-681.
16. Schneider W.C., Hageboom G.H., Pallade G.E. Cytochemical studies of mammalian tissues; isolation of intact mitochondria from rat liver; some biochemical properties of

mitochondria and submicroscopic particulate material // J. Biol. Chem. – 1948. – V. 172 (2). – P.619-635.
17. Steinbrenner H., Sies H. Protection against reactive oxygen species by selenoproteins // Biochimica et Biophysica Acta: General Subjects – 2009. – V.1790 (11). – P.1478–1485.
18. Yousefizadeh S., Farkhondeh T., Samarghandian S. Age-related diazinon toxicity impact on blood glucose, lipid profile and selected biochemical indices in male rats // Curr. Aging Sci. 2019. – V.12. – P. 49–54.

Content of cytochrome P-450 in microsomal fractions of the liver of pregnant rats and their embryos

Dilfuza Tuychieva[1,*], *Rano* Alimova[2] and *Parida* Mirkhamidova[3]

[1]Andijan State University, Andijan, 170100, Uzbekistan
[2]Tashkent State Agrarian University, Tashkent, 100140, Uzbekistan
[3]Tashkent State Pedagogical University, Tashkent, 100070, Uzbekistan

Abstract. This article presents a comprehensive examination of the impact of the pesticide "dropp" on the lipid peroxidation (LPO) of microsome membranes and the levels of cytochrome P-450 enzyme within microsomal fractions of pregnant rats and their developing embryos. The study delves into the intricate biochemical responses within these biological systems following exposure to the pesticide. The research uncovers a significant intensification of NADP.N- and ascorbate-dependent lipid peroxidation in the liver microsomes of both pregnant rats and their embryos upon exposure to the "dropp" pesticide. This heightened lipid peroxidation is particularly notable on the 3rd day of pregnancy, accompanied by an elevated concentration of malondialdehyde (MDA), a marker of oxidative stress. Moreover, the study reveals a noteworthy impact of "dropp" on the content of cytochrome P-450 enzyme within the microsomal fraction of the liver in pregnant rats. This enzyme, which plays a vital role in various metabolic processes, experiences a reduction in concentration upon pesticide exposure. The effect is most pronounced on the 3rd and 19th days of pregnancy, highlighting specific time windows of vulnerability. Collectively, these findings provide a nuanced understanding of the biochemical repercussions of "dropp" pesticide exposure on the microsomal membranes and enzyme dynamics within pregnant rats and their developing embryos.
Keywords. Pesticide, rat, pregnancy, embryo, liver, microsome, lipid peroxidation, MDA, NADP.H, enzyme, cytochrome P-450.

1 Introduction

The long-term use of pesticides in various branches of agriculture with their ability to accumulate in biosphere objects and food products has led to environmental pollution. Entering the body in various ways, they accumulate in tissues, biological fluids of humans and animals and have a toxic effect on them [1].

Detoxification processes in the body are carried out by all organs, but mainly by the liver [2]. Under the action of various chemical pollutants, systemic enzyme disorganization and labilization of membranes, the most important intracellular organelles, including the

* Corresponding author: tds.bio@mail.ru

endoplasmic reticulum, occurs [3].

It is necessary to note the effect of pesticides on the function of microsome membranes. Studies at the cellular level make it possible to characterize the effect of pesticides on a specific process, the role of a foreign substance in certain structures. This is especially important for determining the interaction of a foreign substance with cell components at the early stages [4].

Most of the metabolic transformations that foreign compounds undergo in the body are provided by enzymes localized in the membranes of the endoplasmic reticulum of liver cells. Microsomal enzymes catalyze the hydroxylation reactions of foreign compounds, as well as important endogenous substrates (saturated and unsaturated fatty acids, phospholipids, etc. [5].

The endoplasmic reticulum (ER) of the liver contains a system of biological oxidation of xenobiotics and endogenous substrates that is physiologically important in the sense of "protecting the internal environment of the body" [6]. Most of the metabolic transformations that foreign compounds undergo in the body are provided by enzymes localized in the membranes of ER liver cells [7].

Microsomal enzymes catalyze hydroxylation reactions of compounds foreign to the body [7, 8], as well as important endogenous substrates (saturated and unsaturated fatty acids, phospholipids, etc.) [9, 10]. Activation of molecular oxygen in ER membranes is carried out by an enzyme complex in which the hemoprotein cytochrome P-450 acts as an enzyme that binds the substrate and activates molecular oxygen. The enzyme system uses NADP as a donor of electrons necessary for oxygen activation [11].

At high concentrations of xenobiotics, i.e., when cytochrome P-450 is fully saturated with substrates, mitochondrial oxidation is inhibited and prerequisites arise for regulatory switching of the flow of reduced equivalents from mitochondria to ER. At the same time, cytochrome P-450 is activated and detoxification of foreign substances occurs faster [12]. Microsome monooxygenase, characterized by wide specificity due to the existence of multiple forms of cytochrome P-450, has the ability to oxidize hundreds of substances of various chemical nature, preventing the accumulation of toxic hydrophobic compounds in the body [12, 13].

The total content of cytochrome P-450 hemoprotein in the liver microsomes of pregnant rats is significantly higher than in non-pregnant rats [14, 15]. In the embryonic liver of rats, ER is represented mainly by a rough subfraction. At first it looks like bubbles, which then turn into tanks [16]. Smooth membranes appear in the postnatal period. There is very little ER in the third phase of pregnancy, its monooxygenase activity is very low. A significant amount of GER appears at the age of one week [17].

On the 20th day of pregnancy, cytochrome P-450 was not detected spectrophotometrically in the embryonic liver of intact rats. In the following days until the moment of delivery, the concentration of hemoprotein gradually increases. The peak of the differential spectrum of cytochrome P-450 during these periods is stably fixed at a wavelength of 450-450.5 nm [18].

In the rat embryo, the cytochrome P-450 apobel is present just before delivery and remains at approximately the same level for several days after delivery. Since cytochrome P-450 was not detected in the liver, it was assumed that heme does not attach to the protein before birth [19]. Most ER lipids consist of phospholipids, which account for 47-54% of liver homogenate phospholipids [106, 107, 127, 139], including 50-60% phosphatidylcholine, 20% phosphatidylethanolamine and 10% phosphatidylinosite. It should be noted that cytomembranes are identical in lipid composition. Phospholipids play a significant role in lipid-protein interaction and in the molecular organization of membranes [11].

A characteristic feature of the organization of microsome membranes is the high content of polyunsaturated fatty acids (PUFA) in them, capable of undergoing intensive oxidation by the free radical peroxide mechanism [12]. Currently, there is no doubt that lipids of microsome membranes play an important role both in the construction and stabilization of protein-lipid complexes of membranes and in the regulation of enzyme systems. At the same time, an important role belongs to fatty acids - both free and part of phospholipid molecules [3]. Having

 https://doi.org/10.1051/e3sconf/202342104009

a high degree of unsaturation, fatty acid chains of membrane phospholipids are potential substrates for the processes of free radical lipid peroxidation (FR LPO). A large amount of experimental data has been obtained confirming the assumption of B.N.Tarusov about the leading role of LPO processes in the development of various pathological conditions [5-7].

2 Materials and methods

The introduction of lysophosphatidylcholine into rat liver microsomes causes the conversion of cytochrome P-450 to P-420, which is reflected in their differential spectra [7]. Phospholipids are essential components of the monooxygenase system of microsomes that promote cytochrome P-450 binding of metabolized substrates.

At the early stages of ontogenesis in rat liver microsomes, the rate of non-enzymatic and enzymatic LPO increases markedly. The rate of ascorbate-dependent LPO reaches maximum values both in microsomes and in the mitochondria of the liver of 2.5-month-old rats (an increase of almost 5 times compared to the liver of newborns). The maximum rate of non-enzymatic LPO was observed in the liver of 3-day and 2-month-old rats. According to various authors, the rate of non-enzymatic LPO in the membrane structures of the embryonic liver is higher or lower than in adult animals [8, 9].

The rate of enzymatic NADP the H-dependent LPO of microsomal phospholipids of the liver changes extremely during ontogenesis and reaches a maximum level in the liver of 2-month-old rats (an increase of almost 20 times compared to the liver of newborns). A sharp increase in the rate of NADF.H-dependent LPO in the liver at the early stages of rat ontogenesis is probably due to the fact that in the first weeks of their life, the content of the oxidation substrate - polyunsaturated fatty acids (PUFA) increases in the liver microsomes [5].

The intensity of LPO in rat fetal liver homogenates is higher than in adult animals, and after birth until the age of 6-10 days, it gradually decreases.

According to other authors, in the placenta and liver of fetuses of rats, rabbits and guinea pigs, NADPH-H-dependent LPO is insignificant. NADF level the H-dependent LPO in the fetal liver is higher than in adult animals [3].

Among the membrane structures, the most effective processes are POL in ER. Damage in its membranes increases parallel to the development of POL [4, 5]. The processes of peroxidation in microsomes are closely related to the metabolism of xenobiotics. They can compete for reducing equivalents and oxygen. In addition, xenobiotics and their metabolites can have a direct antioxidant or pro-oxidant effect. As a result, ribosome dissociation occurs in microsomes, inhibition of protein synthesis and microsomal oxidation processes.

Activation of POL processes causes degradation of cytochrome P-450 and weakening of microsomal oxidation processes. The metabolic activation of xenobiotics, leading to the formation of free radicals, is carried out in the electron transport chain of the ER. Cytochrome P-450 plays a key role in this [18].

The sequence of ER damage and accumulation of lipoproteins in the liver was shown within 1 hour after administration of CCl_4. Lysosomes and nuclear structures are damaged after 4 hours, mitochondria are damaged after 3-10 hours after inhalation of CCl_4. After the introduction of CCl4, the activity of microsomal enzymes, in particular cytochrome P-450, decreases, and the peroxidation of unsaturated fatty acids of membrane lipids accelerates.

Many organochlorine pesticides, entering the body, are metabolized mainly by the monooxygenase system of the liver parenchymal cells, inducing significant changes in the functional state of this enzyme system. Acute poisoning with the pesticide lindane at a dose of 18 mg/kg against the background of a balanced diet after 2 hours leads to a significant increase in the content of cytochrome P-450 in preparations of liver microsomes. Induction of the monooxygenase system after 24 hours or more is accompanied by a significant increase in the concentration of microsomal protein [6].

A relationship was established between the inhibitory effect of malabion on microsomal liver metabolism and its ability to damage components of oxidase systems of mixed function and other enzymes of xenobiotic metabolism. Damage to the microsomal liver metabolism of experimental mice by malabion is due to its effect on the content of cytochrome P-450 [6].

Under the influence of hexochlorobenzene, the content of cytochrome P-450 in the liver increases [4, 8]. Thus, pesticides cause significant changes in the structure and metabolism of tissues, cells and subcellular structures. In other words, they are non-specific structural and metabolic poisons.

3 Results and discussion

The objects of research were white female Wistar rats weighing 180-200 g. The animals were seeded with a drop at a dose of 1/10 LD_{50} on the 3rd, 13th and 19th days of pregnancy intragastrically with a special probe for 5 days. For fertilization of rats in the proestrus-estrus stage, they were planted at night with males in a ratio of 3:1. The first day of pregnancy was considered the day of detection of spermatozoa in vaginal smears. The slaughter of animals was carried out on the 20th day of pregnancy, when the embryo reached a significant size, at the end of organogenesis. In the experiments, microsomes of the liver of embryos and the maternal organism were used.

The microsomal fraction from the liver was isolated by differential centrifugation. The activity of ascorbate-dependent and NADPH-dependent lipid peroxidation in microsomes was determined by the content of malondialdehyde (MDA). The MDA content was determined using 2-thiobarbituric acid. At a high pH in an acidic medium, MDA reacts with 2-thiobarbituric acid to form a colored tremitin complex with an absorption maximum at 532 nm. The molar extinction coefficient of this complex is 1.56-10 cm^{-1} M^{-1}.

Determination of cytochrome P-450 was carried out [12], based on a change in the amount of absorption of the complex reduced by carbon monoxide. Measurements of cytochrome P-450 content in a suspension of microsomes 1-2 mg/ml of protein in an incubation medium containing 0.1M K-phosphate buffer, pH- 7.4 were carried out on SPECORD UV VIS [FRG]. Microsomes were previously passed through one of the cuvettes with a suspension for 1 min. carbon monoxide, then several crystals of sodium dithionite were added to both cuvettes and the differential spectrum of the reduced form was recorded with carbon monoxide at a maximum absorption of 450 nm and a minimum. The absorption difference between 450 and 490 nm ($A_{450\text{-}490}$) served as an indicator of the cytochrome content, which was calculated using a molar extension coefficient equal to 91x10 cm^{-1} m^{-1}.

Unsaturated fatty acids of phospholipids of biological membranes are optimal peroxidation substrates by structure. In recent years, the LPO of biological membranes has played a special role in the pathogenesis of various diseases. With a far-reaching process of LPO in phospholipids of biomembranes, the amount of unsaturated fatty acids decreases while increasing the amount of saturated ones.

We studied the effect of drop on the content of MDA, one of the end products of LPO, in the liver of pregnant rats and their embryos (Table. 1 and 2). The effect of drop was studied in the 3rd, 13th and 19th terms of pregnancy and embryonic development. At the same time, it was found that when administered on the 3rd day of pregnancy in rats, an increase in enzymatic and non-enzymatic LPO was observed, respectively, by 60 and 50% in microsomes. Poisoning of rats on the 13th day of pregnancy led to an increase in the content of MDA in microsomes by 50 and 37%, respectively, when pregnant rats were primed on the 19th day of pregnancy, the level of MDA increased in microsomes - by 47 and 45%.

The high content of enzymatic LPO in liver microsomes can be explained by the fact that with NADP.The H-dependent LPO substrate in microsomes are polyunsaturated fatty acids -

phospholipid acyls localized near the components of the electron transport chain, and with ascorbate-dependent - other fatty acids (Table 1).

Table 1. Content of LPO products in the liver microsomes of pregnant rats under the action of drop (mmol/mg protein).

Variants	NADPH-dependent LPO			Ascorbate-dependent LPO		
	Pregnancy days					
	3	13	19	3	13	19
Microsomes						
Control	0.205+0.028	0.184+0.032	0.200+0.020	0.188+0.025	0.178+0.025	0.209+0.021
Dropp	0.302+0.037	0.280+0.038	0.295+0.022	0.286+0.031	0.253+0.030	0.306+0.025

Similar changes in the level of oxidation products were observed in the liver microsomes of embryos poisoned by drop on the 3rd, 13th and 19th days of development.

On the 3rd day of development, the level of enzymatic and non-enzymatic POL in the liver microsomes of embryos increased by 60 and 50%. On the 13th day, the level of enzymatic and non-enzymatic POL increased by 44 and 37% in microsomes. On the 19th day, the enzymatic LPO in microsomes increased by 47%, respectively, and ascorbate-dependent - by 33%. At the same time, the content of enzymatic POL and ascorbate-dependent POL in microsomes was significantly higher than in the control group.

When poisoning with the pesticide drop, there is an intensification of NADP.N- and ascorbate-dependent LPO in liver microsomes of pregnant rats and their embryos. The MDA level is higher in case of drop poisoning on the 3rd day of pregnancy (Table 2).

Table 2. Content of POL products in the liver microsomes of embryos under the action of drop (mmol/mg protein).

Variants	NADPH-dependent LPO			Ascorbate-dependent LPO		
	Development days					
	3	13	19	3	13	19
	Microsomes					
Control	0.308±0.032	0.325±0.027	0.348±0.035	0.328±0.038	0.313±0.031	0.359±0.032
Dropp	0.426±0.052	0.460±0.052	0.508±0.041	0.438±0.048	0.390±0.040	0.479±0.038

Thus, the oxidation products of dropp stimulate peroxidation reactions in the membranes of the microsomes of the liver of the maternal organism and the embryo, which leads to their damage. Apparently, under the action of these pesticides, significant violations of the structural arrangement of microsome membranes occur, which create conditions for intensive oxidation of membrane lipids. The increase in the same NADF.H-dependent liver LPO in the early stages of embryogenesis is probably due to the fact that during these periods the content of the oxidation substrate - PUFA increases in the liver microsomes of embryos.

The enzyme system of microsomal oxygenases plays an important role in the metabolism of chemical carcinogens. The terminal site of this system is the cytochrome P-450 enzyme, which plays a major role in the biotransformation of foreign substances [17]. The changes we found in the lipid matrix of microsomal membranes under the action of pesticides suggest changes in the properties of the cytochrome P-450 enzyme complex.

In the following experiment, the effect of dropp on the content of cytochrome P-450 in the microsomal fraction of the liver of pregnant rats was investigated (Table 3). When dropa was introduced into the body of pregnant rats on the 3rd, 13th and 19th days of pregnancy, the content of cytochrome P-450 in the microsomal fractions of the liver decreased. When rats were poisoned with dropp, the content of cytochrome P-450 decreased by 24% on the 3rd day of pregnancy, by 13% on the 13th day, and by 23% on the 19th (Table 3).

Table 3. Effect of drop on the content of cytochrome P-450 in microsomal fractions of the liver of pregnant rats, nmol/mg of protein.

Variants	Terms of pregnancy, days		
	3-day	**13- day**	**19- day**
Control	0.930±0.084	0.910±0.072	0.920±0.065
Drop	0.710±0.057	0.800±0.065	0.650±0.064

4 Conclusions

As can be seen from the above data, activation of POL processes causes degradation of cytochrome P-450 and weakening of microsomal oxidation processes. The metabolic activation of xenobiotics, leading to the formation of free radicals, is carried out in the electron transport chain of the endoplasmic reticulum. Cytochrome P-450 plays a key role in this. With the introduction of droppa, the content of cytochrome P-450 in the microsomal fractions of the liver of pregnant rats decreases. A more pronounced decrease in the content of this enzyme is observed with drop poisoning on the 3rd and 19th days of pregnancy. In this regard, our data are consistent with the results of earlier studies, according to which the processes of microsomal oxidation and LPO are alternative: activation of one of them leads to suppression of the other. Inactivation of cytochrome P-450 confirms the violation of the processes of LPO in the membranes of microsomes and, in our opinion, plays a leading role in the mechanism of pathology development.

References

1. Mazzarello, P., A. & Calligaro, V. Vannini, & U. Muscatello. (2003) [The sarcoplasmic reticulum: its discovery and rediscovery] . Nature Reviews Molecular Cell Biology.
2. Myshkin V.A., & Ibatullina R.B., & Bakirov A.B. (2007) [Damage to the liver by chemicals].Ufa: Gilem:
3. Schuldiner, M., & B. Schwappach. (2013) [From rags to riches — The history of the endoplasmic reticulum Biochimica et Biophysica Acta] Molecular Cell Research.
4. Liang X., & Duan N., & Wang Y., & Shu S., & Xiang X., Guo T., Yang L., Zhang S., Tang X., Zhang J. (2016) [Advanced oxidation protein products induce endothelial-to-mesenchymal transition in human renal glomerular endothelial cells through induction of endoplasmic reticulum stress]. J. Diabetes Complicat.
5. Mirkhamidova P., & Parpieva M., Tuychieva D. (2021) [Residual pesticide in the liver of rats after poisoning with galaxifop-R-methyl pesticide check] International Journal of Modern Agriculture. Great Britain:
6. Alimbabaeva N.T., & Mirkhamidova P., Isabekova M.A., Zikiriyaev A., Fayzullaev S.S. (2005) [The effect of residual amounts of karate on the activity of mitochondrial enzymes of hepatocytes]. Uzbekiston biology journal. Uzbekistan:
7. Bashirova R.M., & Maksimov G.G., & Akhmetova L.A. (2009) [Fundamentals of Ecotoxicology]: Textbook. RIC BashGU, 2009. Ufa:
8. Burt, A.D.; & Tiniakos, D.G.; Lackner, C. (2015) [Diagnosis and assessment of NAFLD: Definitions and histopathological classification]. Semin. Liver Dis.
9. Somchit, N., & Ngee, C.S., Yaakob, A., Ahmad, Z., Zakaria, Z.A., (2009) [Effects of cytochrome P450 inhibitors on itraconazole and fluconazole induced cytotoxicity in

hepatocytes]. J. Toxicol:
10. Shi, Q., & Yang, X., Greenhaw, J., Salminen, W.F. (2011) [Hepatic cytochrome P450s attenuate the cytotoxicity induced by leflunomide and its active metabolite A77 1726 in primary cultured rat hepatocytes]. Toxicol. Sci:
11. Wallace, & K.B., Starkov, A.A. (2000) [Mitochondrial targets of drug toxicity]. Annu. Rev. Pharmacol. Toxicol:
12. Nivukoski, U.; & Niemela, M.; Bloigu, A.; Bloigu, R.; Aalto, M.; Laatikainen, T.; Niemelä, O. (2020) [Combined effects of lifestyle risk factors on fatty liver index.] BMC Gastroenterol:
13. Mingatto, F.E., Rodrigues, T., Pigoso, A.A., Uyemura, S.A., Curti, C., Santos, A.C., (2002) [The critical role of mitochondrial energetic impairment in the toxicity of nimesulide to hepatocytes.] J Pharmacol. Exp. Ther:
14. Shakirova S. M., Shayakhmetov M. Sh., Shakirova G. R. (2013) [Influence of hepatotropic preparations on the morphofunctional state of the rat liver during carbon tetrachloride intoxication.] Bulletin of the Peoples' Friendship University of Russia. Series: Agronomy and animal husbandry. Russia:
15. Szewczyk, A., Wojtczak, L. (2002) [Mitochondria as a pharmacological target]. Pharmacol:
16. Tuychieva D, & Mirkhamidova P, Babakhanova D, Parpieva M, Alimova R (2020) [Effect of pesticides on the activity of some rat liver enzymes and ways of their correction] - Norwegian Journal of Development of the International Science, Norwegia:
17. Dashchuk A.M., & Pustovaya N.A. (2009) [Lipid peroxidation and antioxidant defense activity in patients with psoriasis. Dermatovenereology.] Cosmetology. LPOopathology:
18. Kovalev I.E., & Rumyantseva E.I. (2000) [The cytochrome P-450 system and diabetes.] Problems of Endocrinology.
19. R. Alimova, D. & Tuychiyeva, P. Mirkhamidova. (2021) [Effect of pesticide "Butilkaptax (Russia)" and "Droppa (Russia)" on respiration and oxidative phosphorylation of liver mitochondria of pregnant rats and their embryos] E3S Web of Conferences.

Regional distribution and ecological-phytocenotic description of dominant plants of Southern Uzbekistan

Tashbibi Norboboeva[1,*], *Shokhsanam* Yakubjonova[2], *Halima* Haydarova[3] and *Lobarkhan* Karakhanova[4]

[1]Chirchik State Pedagogical University, Chirchik, 111701, Uzbekistan
[2]Tashkent State Pedagogical University, Tashkent, 100070, Uzbekistan
[3]Tashkent State Agrarian University, Tashkent, 100140, Uzbekistan
[4]Research Institute of Pedagogical Sciences of Uzbekistan, Tashkent, 100031, Uzbekistan

Abstract. This article analyzes the results of the research conducted by scientists in the study of plants of the region. Taking into account the laws of the distribution of plants in the regions, the dominant plants of the desert, hill, mountain and pasture regions of South Uzbekistan are given an ecological and phytocenotic description. In particular, the results of the expedition conducted to study the plants distributed in each region, and the results of experimental research are highlighted. According to it, information on the Latin name of plants, ecological status, distribution according to soil types, dominance characteristics, communities is provided. The formation of vegetation cover specific to the climate of each region was analyzed. In particular, dryness (lack of moisture) and high temperature for the desert determine unfavorable climatic conditions for plant life. Here, the shortness of the growing season of plants, mainly due to the lack of precipitation, early drying of the soil, often determines the summer and winter dormancy of two dormant periods in the seasonal development of plants. Despite the fact that desert conditions are unfavorable for plants, it is possible to observe their diversity during the seasonal vegetation period. Adir is located between a barren desert and a mesothermal mountain, and according to the living conditions of the plants, the stage part is arid, and the upper part is close to the mountain.
Keywords. Landscape, dominant, subdominant, salinity, barren, halophytes, mesophytes, xerophytes.

1 Introduction

Ecological and dominant description of dominant plants in desert, hill, mountain, pasture regions is considered, taking into account the laws of distribution of plants in the regions. These regions are plain, foothills, mountains and highlands geomorphological system [1-3]. The main difference between these regions is the occurrence of a hydrothermal environment

* Corresponding author: tashbibi1001@mail.ru

in these places, depending on the general scenery (landscape) and surface structure (relief), and the change of soil and vegetation cover that is integrally related to it [4-6].

The landscapes under scrutiny encompass a diverse spectrum, ranging from desert expanses to hilly terrains, mountainous zones, and expansive pastures, each constituting integral elements of the broader geomorphological system [7].

These regions are compartmentalized into the plains, foothills, mountains, and highlands, constituting a multifaceted landscape tapestry [8]. A defining feature that sets these regions apart is the hydrothermal environment they encapsulate, an environmental factor profoundly influenced by the prevailing geographical features, including the overall landscape and surface relief [9]. This interplay of features orchestrates the shift in soil composition and vegetation cover that is inherently intertwined with the broader ecological dynamics of the regions [10].

By dissecting the intricate connections between terrain, climate, hydrology, soil composition, and vegetation distribution, this study unveils the underlying ecological and dominant characteristics of these regions [11]. Such insights are pivotal for comprehending the interwoven ecological relationships that shape plant life within desert, hill, mountain, and pasture settings [12]. The knowledge gleaned from this exploration not only advances our understanding of regional ecosystems but also informs the development of targeted conservation and management strategies for these diverse landscapes [13].

Each region has its own climate-specific vegetation cover. In particular, the dryness and high temperature of the desert determine unfavorable climatic conditions for plant life [14]. Here, the shortness of the growing season of plants, mainly due to the lack of precipitation, early drying of the soil, often determines the summer and winter dormancy of two dormant periods in the seasonal development of plants. Despite the fact that desert conditions are unfavorable for plants, it is possible to observe their diversity during the seasonal vegetation period [15]. Adir is located between a barren desert and a mesothermal mountain, and according to the living conditions of the plants, the stage part is arid, and the upper part is close to the mountain [16-18]. With this, the hill zone is divided into two parts, that is, the lower hill with a rather gentle relief and the branched hill, where rocks protrude in some places.

The growth period of plants in the mountain region corresponds to spring, summer, and autumn, and in the lower mountains, that is, in the hills, together with ephemeral and ephemeroid communities, dominant steppe types with a long period of mutation are also characteristic.

2 Materials and methods

Our research revolves around the exploration of regional plant species indigenous to the southern region of Uzbekistan. Within this context, we delve into the intricate ecological and phytocenotic attributes of these plant populations. To achieve a comprehensive understanding, we employed a multifaceted approach, integrating field research, experimental methodologies, statistical analyses, and advanced Geographic Information System (GIS) technologies.

The heart of our investigation lies in scrutinizing the distinctive characteristics exhibited by the flora native to the southern Uzbekistan region. This encompasses an in-depth examination of their interactions with the surrounding environment, their role within local ecosystems, and their responses to prevailing ecological conditions.

Field research allowed us to intimately engage with these regional plant communities in their natural habitats, enabling us to gather firsthand observations and data. Complementing this approach, we conducted experiments to unveil specific ecological interactions and adaptations, shedding light on the unique attributes of these plant species.

The employment of advanced statistical techniques facilitated the systematic analysis of gathered data, ensuring rigorous and accurate interpretations of our findings. Moreover, the integration of GIS technologies enabled us to overlay spatial data and generate comprehensive maps that visually illustrate the distribution patterns of these regional plants within their respective habitats.

Collectively, this multidimensional approach empowers our study to offer a nuanced exploration of regional plant species in South Uzbekistan. By intertwining empirical observations, experimental insights, statistical analyses, and GIS technology, our research aims to contribute substantively to the body of knowledge surrounding these critical ecological systems.

3 Results and discussion

For the desert, dryness (lack of moisture) and high temperature determine unfavorable climatic conditions for plant life. Here, the shortness of the growing season of plants, mainly due to the lack of precipitation, early drying of the soil, often determines the summer and winter dormancy of two dormant periods in the seasonal development of plants. Despite the fact that desert conditions are unfavorable for plants, it is possible to observe their diversity during the seasonal vegetation period. Among them, you can find species (belonging to the Ephedra family) that develop continuously throughout the year.

In South Uzbekistan, there are various types of soil, gray soil, sandy desert, barren, barren, shorkhok, typical shorkhok, meadow shorkhok and marshy shorkhok soils. In response to this, groups of psammophytes, halophytes, hypsophytes, petrophytes, mesophytes, forest plants (forests) are distinguished in the conditions of growing dominant plants.

The life forms of the dominant plants are also diverse. In communities, representatives of life forms such as trees, shrubs, shrubs, semi-shrubs and annual and perennial grass can be found, which are dominant and often belong to landscape plants [4].

includes ***the white saxovull (Halaxulon persucum)*** , which is very common in the desert . Its numerous plant communities are formed in compacted sandy, sandy-porous sands [1]. In the researched area, it forms a number of plant communities with white saxovol, chogon, yantak, ilok, and black saxovol [16].

In the vegetation cover, along with the white saxovolum, the species of calligonum (jazgun) occupy a much wider and more prominent place. They are found in almost all formations of sandy deserts and loose sand dunes. Calligonum microgarpum, cherkez (Salsola richtori) are also often found in grass saxowulzor, juzgunzor, and rabbit bone plants [6, 11]. But as a dominant, it is slightly less than the species named above. The main common areas of Circassian forests are oases, around wells, and land areas that are intensively used for grazing animals. Forms communities with Circassian, Ilok, Singren, White Jusans.

Semi-shrub dominants in the sand include ***white wormwood*** (Artemisia diffusa). It is subdominant in the desert associations of white saxophones and warblers. White sorrel is dominantly found in the depressions between the more powdery sands, in the sand layers with small dust particles, and in the curved sand layers on the northern slopes, that is, on the side where soil formation is much earlier than on the southern slope. Turan wormwood, ilok, and astragalus are subdominant in the plant communities of white sage.

The widespread group of psammophyte grass dominants includes Carex physodes, which ***occupies a special place in the community's life.*** It is a rhizomatous ephemeroid with a strong competitive character and a fairly wide ecological range, growing in thick-layered and thin-layered sandy loams, porous sands, sandy loams, saline sands. The most favorable conditions for its development are loam with medium porosity. It forms a number of plant communities with ilok, white wormwood, and turon wormwood. ***Celine*** (Aristida pennata) is a common species in sand dunes, a permanent member of most communities, but in rare cases it forms plant communities as the main dominant in small areas [13, 14].

Among the shrubs distributed in gypsum desert sand deposits is ***sangren*** (Astragalus villosissimus). The Sangren community forms plant communities with white sedges and sedges in layers of sandy or clayey soils. ***White paint*** (Salsola arbuscula) is a common halophyte shrub. It is found as a dominant in the plant community in places where the surface of gypsum soil is covered with a layer of sand in thin layers of sands and proluvial layers. Forms communities with wormwood, Circassians, sangrens.

Keyreuk (Salsola orientalis) is a galaphyte semi-shrub with a wide ecological range. It is dominantly found in sandy brown soils, barren soils of alluvial plains, including variously saline soils of anciently irrigated river deltas. The soil group of the place of residence of the Keyreuk community is characterized on the one hand by hard skeletal, with gypsum, on the other hand, by the presence of a soft soil layer in the surface layer [8]. It forms a team with dyeing, white wormwood.

The semi-shrub dominant group includes ***gammada*** (Hammoda leptoclaga). This is an endemic species typical of the adjacent regions of Southern Uzbekistan and Turkmenistan, and is not widespread. Gammada communities can be found in the gypsum soils of the highland plains. It forms communities with color, white, and Turan wormwood [12, 15].

Black saxowul (Haloxylon aphyllum) is a woody shrub belonging to the group of halophilic dominants. It is a unique representative of the desert flora and has a wide ecological range in relation to the soil. It grows on clay, sandy, gravelly soils with varying degrees of salinity. Important factors determining the distribution of the black saxophone are the relative proximity of sewage or is the constant moistening of the deep horizon layers. The natural black saxovulzor area is connected with ancient and modern river valleys. It forms a number of communities with kandim, aksaksovul, partek, ilok, shuvoks.

Gallophyte shrub dominants can be ***white bush (Seidlitza rosmarinus) with a narrow range.*** Its area is limited only to the southern part of Central Asia. The northern line (border) of the range of the species passes in the Surkhandarya region of the southern part of Uzbekistan. White bush can be found in small community areas on saline, loamy, thin-layered sands, weak, moderately saline loamy soils. Yulgun forms a series of communities with rings.

Common halophilic dominants include ***sarsazan*** (Halocnemum strobilaceum). Its communities are associated with highly mineralized wastewater near the surface of the earth's moist saline lands. Sarzazanzors can be found in large areas, massifs, eroded, overgrown, hidden shores, around the desert.

An example of a dominant halophilic species with a herbaceous life form can be taken as ***a salt marsh (Aluropus litoralis).*** This species is common in the river alluvial deposits of the foothills. These rhizomatous sedges were formed in the sedimentary soils of river beds with rapid salinization and groundwater close to the surface. Saline sedge develops in moist and saline soils and forms a cover of saline and grassy plants. It forms a team with a reed, a reed, and a reed.

of annual halophilic dominant species are ***cornflower*** (Gamanthus gamocarpus) and ***sedge*** (Halocharis hispida), which are common in lowland brackish soils and saline and barren and barren emerging soils. Cereals depend on residual saline and saline soils of automorphic order. Rabbit farms are distributed in the lower parts of the highlands and plains. Therefore, this plant is widespread in the hills up to the border of the mountain zone.

Adir is located between a barren desert and a mesothermal mountain, and according to the living conditions of the plants, the stage part is arid, and the upper part is close to the mountain. With this, the hill zone is divided into two parts, that is, the lower hill with a rather gentle relief and the branched hill, where rocks protrude in some places. The amount of annual precipitation in the hills is approximately twice as much as in the desert (250-400 mm), and 2-3 times less than in the mountains. Here, rainless periods are reduced by 1-2 months (July, September) compared to the desert (May, October).

Adir is an area with gray (light, medium, dark) soil types. It is the main rock-gypsum particles that make up the soil. However, in the southern part of the republic, the gray soil is the leader in the alluvial deposits of the mountains and mountains. Tertiary rocks rich in gypsum and gypsum-gray soil formed from its erosion have developed in the border between the hills and the desert. In the Adir region, ephemerals and ephemeroids are common dominants and are characteristic of vast foothills. Ephemeral vegetation cover is a type of biological characteristic, which consists of annual and perennial plants in the ephemeral development cycle [2].

The ephemeroid dominants in the lower hill belt are ***carex*** (Carex pachystylis) and sedge (Poa bulbosa), which cover the soil surface with a thick green carpet in the typical habitat. Color is an ephemeroid with a long rhizome that reproduces vegetatively, it has high plasticity in relation to the temperature factor, stability of the tissue of vegetative organs. This species is well adapted to desert conditions. The below-ground biomass of the color is higher than the above-ground productivity. It forms communities with mint, carrack, aqquray, ilok, shagalak and sugdi wormwood [2].

The second dominant is the bulbous bulbous ***brownhead*** (Poa bulbosa), an ephemeroid with small hairy roots, a geophyte with a wide ecological range, and it has the characteristics of wide ecological and biological plasticity. It can be found from the desert zone to the high mountain zone. Blackberry grows vegetatively and from seeds. The color forms communities with sugdi suvoga and white.

The ephemeral group is formed when the dominant species and their communities are always brown, when a disturbance occurs in the cover of the color community. The ephemeral dominant group includes ***aegilops*** - Aegilops squarossa, ***sedge*** - Ansantha tectorum, ***sedge*** - Taeniatherum crinitum. Egilops appear en masse as a result of a change in the cover of black-headed - black-headed - black-headed. Their sometimes thick growth does not allow other species to grow well. In this regard, sorghums form secondary communities on less-utilized land with more dryland farming. Egilopsors are often found in nature on soft soil and gravelly slopes.

Kiltiq is found in all the hills of the republic. Many grasslands are infested with this weed, and it begins to dominate the vegetation cover in areas that have been continuously used as grassland for a long time. Yaltirbosh and Mortuk lead teams occupy large areas like Egelops and Kiltikt teams. They are also found in over-utilized grasslands and lowlands.

Almond (Amugdalus spinosissima) is common among the arid shrub dominants . Almonds are mainly distributed in the foothills and submountain areas with stony, gravelly or stony soft soil. In addition, they are found both in the mountains and sometimes singly in the foothills. However, it does not form a vegetation cover in plains and mountains. Almond occupies a large area in the lower regions, forming diverse communities with ephemeroids, wormwood, semi-shrubs, sedges, various herbaceous steppe plants.

Pista (Pistacia vera) is one of the tree dominant species and is one of the valuable species for agricultural use. It forms small groves in the foothills and foothills, including pistachio groves in Bobotog. Favorable conditions for the growth of the pistachio tree are thick soil layers, light sands or sandy loams with a calcareous content, and porous rock soils with stone gravel. Pistachio ephemeroid, wormwood and various grasses form a series of plant communities at an altitude of 800-1500 meters.

Among the dominant semi-shrubs, ***Artemisia sogdiana*** (Artemisia sogdiana) is scattered in the lower part of the hill. Its communities are found at an altitude of 500-800 (thousand meters). The soil layer is a light and typical gray soil, stony-gravel or fine-grained-gravel soil, which is more or less saline. Sugdi ash forms a community with shrubs, halophiles, and double-leaved plants.

In the mountain region, the altitude rises from 1200-1500 m to 2700-2800 m, the rainless days are from July to October. The growth period of these plants corresponds to spring, summer, autumn, in the lower mountains, that is, in the hills (heights from 1200-1400 m to 1800-2100 m above sea level), in this place, together with ephemeral and ephemeroid communities, dominant steppe types with a long period of mutation are characteristic. Tree-shrub plant communities are not noticeable (this is mainly due to the atropogenic effect). Here, the temperature is short, but the amount of rain is significantly more. There are almost no ephemeral plants in the upper part. Plant communities of tree-shrub life forms are well developed, and the number of species of mesophilic plants is included.

Bukhara almond (Amugdalus bucharika) is a tree-like shrub, but in many cases it is found in the form of a multi-stemmed shrub. Mainly distributed in the Pamir-Aloy mountain ranges. It rises to 2200-2300 above sea level in the Hisar mountains. The border of the lower mountain zone passes 800-900 m. It is found dominantly in stony gravel soils of all exposures on slopes, rock avalanches and hills. Grows well in calcareous soils. Bukhara almond usually does not form a thick cover, only ephemeroids participate in the formation of a number of communities with thin-leaved wormwood, wheat, various grasses.

Bukhara marble (Salvia bucharika) - semi-bush, ***Halimiphyllum*** (Halimiphyllum atriplicoides) bushes are conditionally dominant in the sparse vegetation cover in the exposed rock layers .

The dominant tree growth form in the mountain zone ***is the Turkestan maple*** (Acer turkestanicum). It is found in Pamir-Aloy and Western Tyanshan. It grows on various rock exposures at altitudes of 1,800 to 2,400 m in southwestern Hisar, and is very well preserved on the northern cliffs. This species forms mixed trees with other deciduous communities. Fir trees are always considered the main component in this, and they participate in the formation of the second layer. It hardly makes trees by itself. Turkestan maple forms communities with various herbaceous plants (andiz, shashir).

Another species of this family is ***soft maple*** (Acer pubescens), which is found only in the Pamir-Aloy mountains. This species is more xerophilic than the above species. It is mainly found in different directions of dry fine-grained, stony and gravelly rocks in the hilly zone at 800-1800 m altitude. It forms low sparse groves mixed with other trees such as hawthorn, walnut, almond. This dominant group of trees includes two types of hawthorn - ***yellow hawthorn*** (Crataegus pontica) and ***Turkestan hawthorn*** (Crataegus turkestanicum), which are found in almost all mountains of the republic.

Yellow hawthorn is one of the drought-resistant trees. In the south-west of the Hisar ridge, on dry, stony, stony-gravel and rarely fine-grained rocks, wheat is found in sparse tree-shrub plant cover - various steppe grasses, among rocks. In the south, south-west and south-east of the rocks, it always grows sparsely. Single or small groups, rarely nuts, occur as a second layer component in fruit forests. Alternatively, a mixed tree can be found in shrublands, forming an upper layer.

One of the dominant species among shrubs, ***rose*** (Rosa) occupies a significant place in the vegetation of the mountain zone. Nematokas - mixed thickets consist of various bushes, in which the Nematokas type is the leader. Triples, irgay, and zirks are rare. A number of authors (Ovchinnikov, Zokirov, Kamelin) define the secondary shrublands of rosaries and determine the limits of their distribution as altitudes from 900 m to 3500 m above sea level.

In the mountains of South Uzbekistan, rosaries are often found in all mountain regions, but do not form large areas. Rosa kokanica, R. canina, R. maracandea, R. fedtschenkoana are more common among species of the Rosa family. In Western Hisar, the namatak forms communities with lamb, color, cherry, various herbs. It is characteristic that blackberry, abelia, and cypresses participate in some communities.

juniper (Juniperus) , which is widespread as a dominant, is the main forest-forming species. In all mountain ranges of Central Asia, juniper occupies a special place [5]. It is always found in sparse woodlands alternating with steppe and grassland communities. It grows at an altitude of 1900-2300 (2800-3200) m in the Pamir-Aloy mountain system. Of the juniper forests, ***black juniper*** (Juniperus seravshanica) occupies much lower hypsometric heights in the mountain profile, scattered junipers are always dominant in the upper mountain borders. Spruce can be found on various soils, but favorable conditions for their growth: from soil layers developed on gentle slopes, i.e. brown, high-humus, strong carbonaceous soils.

Zarafshon juniper has a highly plastic ecological character and occupies a wide range of landscapes as a determinant of the formation of diversity in the community. In the south of Hisar-Boysun mountain ranges, F.Kh. According to Jangurazov [7], the distribution of this species in the lower mountain borders is characterized by ephemeral wheat and wormwood forests. As the height increases (1600-2000 m), representatives of large grasses appear in communities. Here, with the participation of namataks, all kinds of grassy wheat arbors dominate. Then, at an altitude of 1800-2400 m, juniper forests are combined with trees, and above (2400-2500 m) betaga, high mountain dominates in various groups of grasses.

Shashir (Prangos pabularia) (up to 1.5-2 m) is a perennial dominant grass with a wide ecological range. This species has a wide altitude range (900-3000 m) and is a strong coenosis producer.

Shashir communities are adapted to different soils (dark, gray, yellowish brown, brown mountain forest, light brown), and these soils differ in their mechanical composition. In mountain sparse forests, they are the main components of the grass layer in the tree and shrub community. But the independent shashiris community forms only the lower part of the highland region. Shashir community is made up of its sub-edifiers - Polugonum coriarium, Ferula kuhistanica, Artemisia dracuncutus, Ligutaria thomsonii, Dactulis giomerata and others.

Buttercup (Ranunculus rufosepalus) is one of the characteristic dominants of the low meadows of the high mountain region. It dominates well-drained, well-saturated fine-grained-gravel soils, near melting snow tracks or on the banks of streams; also grows on stony soils. Bear communities are found on the slopes of the Hisar Mountains and in its western parts at 3000 m above sea level, close to the snow. ***Anjobar*** forms communities with arenaria (Geranium chariesii), ***peschanka (Arenaria rotundiflora).***

Another representative of the low-dominant meadows in the high mountain region is ***the lagotis*** (Lagotis korolkovii), which is almost stemless and rarely goes beyond the limits of the conditions of the alpine meadow region. In such a place, the soil in the second half of the summer will easily survive the drought. Rarely it is part of subalpine meadows. His teams are always polydominant. Anjobar forms a series of communities with arenaria.

Aspartset exidna (Onobrychis exidna) is characteristic of the dominant bushes with a cushion growth form , and it occupies the only place in the mountain xerophilic vegetation cover in terms of the role of dominance. Esparsetes are adapted to exposure on rocky south and south-western slopes, where the light falls well and is generally not too low, in watersheds, on high level surfaces, that is, in places where the rocks are exposed. In the western part of the mountains of Hisar and Turkestan , thorny espartocarps are quite abundant at 2300-3000 m above sea level, where they form a number of communities with acantholimon, betaga, and astragal species.

The alpine xerophilic mountain vegetation belonging to the carrack family, which dominates the high mountains, belongs to the group of thorny grasses. These ***carracks*** (Cousinia verticillaris, C. machilenta) are perennial herbs, an endemic species for Pamir-Aloy. Thorny grasses are mainly adapted to subalpine soil conditions with a reduced layer and a soft layer. They are the main landscape of fine-grained areas on flatter slopes, saddles or flat plateaus, where they always form fairly thick covers. On steep stony-gravel slopes, it forms a series of communities with sedges, akontolimon, and sedges.

One of the most common wetland dominant grasses is ***sedge*** (Carex orbicularis). Moss vegetation is a characteristic element of a high mountain, which is formed under special conditions of abundant moisture. The peat layer is usually not large. This plant forms a series of communities with the dominant (Kobresia persica) bearberry.

Fescue (Festuca valesiaca) is dominant. Most of the Betagali steppes are found in subalpine and alpine high mountain regions. Beta also occurs in the lower mountain and upper hill regions, but usually it is subdominant in the community, and the sedge, wormwood, wheatgrass and ephemerals together form mixed communities. The steppes in the southwestern part of the Hisar mountain ranges are characterized by arid vegetation. Here they can be seen in the mountain xerophilous vegetation cover, sometimes in tens of numbers. According to observations of V.P. Dropov, betaga is found at altitudes of 3400-4000 m above sea level. According to B.N. Demurina, grassy steppes develop at altitudes from 2100 to 2500-3000 m above sea level. Betaga mountain forms a number of plant communities with xerophilous carrack, acanthalimon, astragal species.

Southern Uzbekistan is characterized by a variety of plants [3] (Table 1).

Table 1. Leading families and genera in the flora of South Uzbekistan.

Families	**Number**		**%**	**Categories**	**Number**	**%**
	Categories	**Varieties**				
Asteraceae	91	401	16.7	Astragalus	141	5.9
Fabaceae	26	254	10.6	Cousinia	83	3.4
Lamiaceae	39	154	6.4	Allium	47	1.9
Brassicaceae	68	144	6.02	Oxytropis	32	1.3
Apiaceae	50	109	4.5	Artemisia	30	1.2
Poaceae	59	137	5.7	Scutellaria	27	1.1
Chenopodiaceae	26	98	4.1	Salsola	26	1.08
Rosaceae	23	94	3.9	Potentilla	26	1.08
Boraginaceae	17	82	3.4	Euphorbia	24	1.04
Liliaceae	14	72	3.01	Gadea	24	1.04
Alliaceae	1	47	1.96	Ferula	23	0.96

Of the 2389 plant species of the region, 401 (23 dominant) belong to the Asteraceae family, of which 237 (9) are perennial grasses. 7 of the 171 perennial grasses belonging to the second-ranked Fabaceae family are dominant plants. Dominance characteristics are given in the following table (Table 2).

Table 2. Leading families and genera of dominant plants of Southern Uzbekistan.

Families	**Number**		**%**	**Categories**	**Number**	**%**
	Categories	**Varieties**				
Asteraceae	6	23	13.6	Artemisia	13	7.7
Rosaceae	10	18	10.7	Astragalus	9	5.3
Poaceae	15	16	9.5	Tamarix	7	4.1
Fabaceae	10	20	11.9	Rosa	6	3.5
Chenopodiaceae	10	16	9.5	Cousina	6	3.5

Interregional location of dominant plants of South Uzbekistan is given in Table 3.

Table 3. Interregional distribution of dominant plants of Southern Uzbekistan.

Altitude regions	Number of species	%
Desert	10	5.95
Adir	4	2.38
mountain	23	13.6
Meadow	1	0.59
Desert hill	18	10.7
hill	54	32.4
Mountain pasture	13	7.7
desert-hill-mountain	42	25
Hill-mountain-pasture	5	2.97
Desert-hill-mountain-pasture	2	1.19

As can be seen from the table, most of the dominant plant species are ecologically friendly to mountain areas. But the number of real species unique to the mountain is 23 (17.8%). Next comes hill plants (124). Pure hill vegetation consists of 4 (2.38%) species which is very less compared to mountain. Of the 65 species that can be found in the desert region, 10 (15.6%) are unique to this region. Only 1 species out of 24 is unique to this region [9, 10].

4 Conclusions

In short, plant communities and dominant plant species that form them have a special place among natural resources. 168 of the 2389 species distributed in the southern regions are dominant, they form unique diverse communities, and these are the main ones, besides, they protect the entire surface of the earth from water and soil erosion by maintaining the water regime of the place where it grows.

As a result of the influence of large-scale environmental factors (grazing, desertification), the area of vegetation in the territory of South Uzbekistan is decreasing. This causes an increase in the number of plants included in the "Red Book" [16]. 148 of the 2389 species of plants found in the regions of South Uzbekistan are included in the "Red Book" of the Republic of Uzbekistan, they make up 6.2% of the flora of this region. Therefore, most of the plant species are in difficult ecological conditions, they need special protection. It is urgent to develop measures for their protection and conduct practical activities.

References

1. Akzhigitova N.I. Halophilic vegetation Halophyta . //Rastitelnyy pokor Uzbekistana i puti ego rationalnogo ispolzovaniya. - Tashkent: Science, 2013. -v.2. - p. 211-302.
2. Akzhigitova N.I. Nagornoxerofilnaya rastitelnost Criofriganophyta . //Rastitelnyy pokor Uzbekistana i puti ego rationalnogo ispolzovaniya. - Tashkent: Science, 2014. - v.4. - p. 199-254.
3. Allanazarova Yo., Rakhimova T., Norboboeva T. Study and assessment of the current state of plant communities of the Guzordarya basin. "Actual problems of studying the nature of South Uzbekistan" scientific collection. Volume III. - 2002. -p. 25-29.
4. B ykov B.A. Dominanty rastitelnogo pokro Sovetskogo Soyuza. - Alma-Ata: 2010. - v. I , II , III . -314, 434, 414 p.
5. Vernik R.S., Melnikova R.D., Kamalov Sh.K. Archevniki - Arceutodendra . //Rastitelnyy pokor Uzbekistana i puti ego rationalnogo ispolzovaniya. - Tashkent: Science, 2004. -v.4. - p. 86-163.

6. Zakirov P.K., Norbobaeva T. Kolichestvennyy sostav edifikatorov rastitelnogo pokro Sredne Azii. // FROM RUZ . -2014. - #3. -p. 46-47.
7. Zakirov P.K., Norbobaeva T., Karimov G. Garmala obyknovennaya (Peganum mix L.), kak krasitelnye rasteniya. // Uzb.biol.journ . -2018. - #5. -p. 42-45.
8. Norboboeva T. Analysis of the dominant vegetation of the plains and hilly areas of Uzbekistan. // Uzb.biol.journ . 2004. #2. - p. 40-45.
9. Tayjanov, K., Khojimatov, O., Gafforov, Y., Makhkamov, T., Normakhamatov, N., & Bussmann, R. W. (2021). Plants and fungi in the ethnomedicine of the medieval East-a review. Ethnobotany Research and Applications, 22, 1-20. DOI:0.32859/ERA.22.46.1-20
10. Jabeen, S., Zafar, M., Ahmad, M., Althobaiti, A. T., Ozdemir, F. A., Kutlu, M. A., ... & Majeed, S. (2023). Ultra-sculpturing of seed morphotypes in selected species of genus Salvia L. and their taxonomic significance. Plant Biology, 25(1), 96-106. DOI:10.1111/plb.13473
11. Majeed, S., Ahmad, M., Ozdemir, F. A., Demirpolat, A., Şahan, Z., Makhkamov, T., ... & Nabila. (2023). Micromorphological characterization of seeds of dicot angiosperms from the Thal desert (Pakistan). Plant Biosystems-An International Journal Dealing with all Aspects of Plant Biology, 157(2), 392-418. DOI:10.1080/11263504.2023.2165553
12. Majeed, S., Ahmad, M., Ali, A., Althobaiti, A. T., Ramadan, M. F., Kilic, O., ... & Sultana, S. (2023). Pollen micromorphology among amaranthaceous species from desert rangeland: Exine stratification and their taxonomic significance. BioMed Research International, 2023, 4967771. DOI:10.1155/2023/4967771
13. Makhkamov, T., Sotiboldiyeva, D., Mamarakhimov, O., Yuldashov, Y., & Botirova, L. (2022, May). Morphogenesis and Seasonal Developmental Rhythm Under the Conditions of Introduction of Curcuma Longa L. In International Scientific Conference on Agricultural Machinery Industry "Interagromash"" (pp. 1460-1469). Cham: Springer International Publishing. DOI:10.1007/978-3-031-21432-5_155
14. Boboev, S., Makhkamov, T., Bussmann, R. W., Zafar, M., & Yuldashev, A. (2023). Anatomical and phytochemical studies and ethnomedicinal uses of Colchicum autumnale L. Ethnobotany Research and Applications, 25, 1-9. DOI:10.32859/era.25.6.1-9
15. TKh, M., Brundu, G., Jabborov, A. M., & Gaziev, A. D. (2023). Predicting the potential distribution of Ranunculus sardous (Ranunculaceae), a new alien species in the flora of Uzbekistan and Central Asia. BioInvasions Records, 12(1), 63-77. DOI:10.3391/bir.2023.12.1.05
16. Ameen, M., Zafar, M., Ahmad, M., Ramadan, M. F., Eid, H. F., Makhkamov, T., ... & Majeed, S. (2023). Assessing the Bioenergy Potential of Novel Non-Edible Biomass Resources via Ultrastructural Analysis of Seed Sculpturing Using Microscopic Imaging Visualization. Agronomy, 13(3), 735. DOI:10.3390/agronomy13030735
17. Noor, W., Zafar, M., Ahmad, M., Althobaiti, A. T., Ramadan, M. F., Makhkamov, T., ... & Khan, A. (2023). Petiole micromorphology in Brassicaceous taxa and its potential for accurate taxonomic identification. Flora, 303, 152280. DOI:10.1016/j.flora.2023.152280
18. Aziz, A., Ahmad, M., Zafar, M., Gaafar, A. R. Z., Hodhod, M. S., Sultana, S., ... & Chaudhay, B. (2023). Novel Copper Oxide Phyto-Nanocatalyst Utilized for the Synthesis of Sustainable Biodiesel from Citrullus colocynthis Seed Oil. Processes, 11(6), 1857. DOI:10.3390/pr11061857

Bacterial flora of infected vegetable crops

Nodira Azimova[1,*], *Xurshida* Xamidova[1], *Husniddin* Karimov[1], *Nodira* Xaytbaeva[2], *Baxtiyor* Ravshanov[1] and *Dilfuza* Yesenova[3]

[1]Scientific Research Institute of Microbiology of the Academy of Sciences of Uzbekistan, Tashkent, 100030, Uzbekistan
[2]Tashkent State Agrarian University, Tashkent, 100140, Uzbekistan
[3]Karakalpakstan Agricultural Institute, Nukus, 230100, Uzbekistan

Abstract. The focal point of the article revolves around the examination of bacteria isolated from infected vegetable crops. The study encompasses a comprehensive analysis of various attributes including colony structure, color, and size of these isolated bacteria in their pure form. To enhance our understanding, fixed preparations were meticulously crafted from these isolates, followed by Gram staining. The isolates were scrutinized under immersion oil using an x 100 lens and specialized microscopy equipment, namely the xsp-136 B and N-300m microscope (UCMOSO9000KPB), magnifying them up to 1000 times. To further delve into the taxonomic classification of these isolated strains, the MALDI-TOF method was employed. The strains were precisely identified through this method, carried out at the Department of Sanitary and Epidemiological Control, utilizing state-of-the-art Bruker equipment. Central to the research was the exploration of the pathogenic properties of these bacteria in relation to plants. To this end, a pivotal experiment was conducted wherein seeds of tomatoes and eggplants were artificially planted using these isolated bacteria. The culmination of these investigations has resulted in the creation of four distinct tables, meticulously supplemented with figures to visually encapsulate the findings. In essence, the article presents a comprehensive exploration of the isolated bacteria's attributes, classification, and their potential pathogenic impact on plants.
Keywords. Bacteria, vegetable, tomato, eggplant, plant, root, stem, seed, leaf, fruit.

1 Introduction

Vegetable products are of great importance in human life and occupy an important place in the diet. One of the main conditions for ensuring a high and high-quality harvest in vegetable growing is to protect it from various diseases [1]. Vegetable crops such as tomatoes, peppers, eggplants and potatoes are damaged by several diseases during the growing season, which negatively affects the development of the plant, causing a sharp decrease in yield [2].

* Corresponding author: nodira.azimova1001@mail.ru

At least 54 phytopathogenic fungi, 39 types of viruses, 19 types of nematodes, 3 types of phytoplasmas and 11 types of bacteria cause significant economic damage to the potato crop. Pathogenic bacteria cause various etiological diseases, from mild rot to "zebra chips" [3]. Worldwide, about 10,000 fungi from phytopathogenic objects, about 630 viruses, 200 bacterial species are responsible for the death of the tomato crop due to cassality [4]. Diseases caused by microscopic fungi (64 species), oomycetes (11), bacteria (12), phytoplasmas (3), viruses (32) and nematodes (B species) have been reported in pepper varieties worldwide [5].

In Uzbekistan, diseases of sweet pepper crops are mainly caused by fungi, bacteria and viruses. Pepper crops are mainly affected by black root rot, fusarium, late blight, cladosporiosis, gray and white rot, black bacterial spotting and tobacco mosaic diseases [6]. A number of bacterial diseases affecting various organs of the tomato crop have been detected: necrosis of the stem core, bacterial leaf lesions, fruit lesions, terminal rot of fruits, watery rot of stems and fruits, bacterial wilting of the stem (brown wilting), root diseases of cancer [7].

Phytopathogenic microorganisms affecting agricultural crops also include secondary pathogens, which can mainly populate the plant in places of mechanical damage and cause symptoms of various manifestations of the disease in the plant. But these microorganisms are not considered the main causative agent of plant disease [8]. In vegetable crops, the main pathogens damage the plant due to their phytotoxic properties. Therefore, it is very important to study the phytotoxic nature of microorganisms in order to determine whether they are pathogenic species in relation to the plant [9].

Currently, for the cultivation of tomato products of export quality, not only for the domestic market of our country, one of the most urgent tasks is to protect it from diseases and prevent morbidity. To do this, it is very important to identify phytopathogenic microorganisms that provoke the disease, identify their types and develop biological control measures against it [10].

The aim of the study is to identify the bacterial flora of infected vegetable crops and to study the properties of phytotoxicity of bacteria that have been scratched.

2 Materials and methods

Infected stalks and tomato leaves, sweet pepper leaves, stalks and fruits, potato leaves were delivered to the laboratory in separate packages. The imported samples were placed in a refrigerator at a temperature of 4 °C and processed to remove the external microflora within 48 hours.

The affected leaf, stem and root parts of plants were cut and washed in running water, then treated with a 3% solution of hydrogen peroxide for 2-3 minutes, and then rinsed again with sterile tap water. Dried on sterile filter paper, cut from pieces containing leaves, stems, damaged and intact healthy tissues, 1-3 cm in size and placed in Petri dishes containing GPA (meat-peptone agar) nutrient medium [3]. In the same way, the parts of plants sterilized from the outside were crushed in separate sterile porcelain baths, and the crushed samples were planted in GPA nutrient media, Petri dishes. The samples planted in the culture medium were incubated in a thermostat at a temperature of 27-28 ° C. Bacterial growth was observed within 24-48 hours.

Differentiated colonies of grown bacteria according to their appearance, color and morphology were selected and transplanted back into the GPA culture medium using a bacterial loop. To purify the isolated strains, this process was carried out 3 times [9]. The structure, color, and size of a colony of bacteria isolated in pure form were studied. Fixed preparations were also made from isolates and stained by gram method, examined under immersion oil in an x 100 lens, an xsp-136 B and N-300m microscope (UCMOSO9000KPB) (magnified 1000 times).

The isolated types of bacteria were identified by the MALDI-TOF method in the Department of Sanitary and Epidemiological Control of the Main Department of Medicine under the Administration of the President of the Republic of Uzbekistan, using Bruker equipment.

When determining the type of bacteria, the main research methods remain a simple culture method and standard biochemical tests. The MALDI-TOF MS method has been used for almost 20 years mainly in the chemical laboratory for the identification of various molecules, such as sugars, nucleic acids and proteins [9, 10]. Currently, this method is used as an identification method in microbiological laboratories [2-5]. The systematic role of isolated bacteria in Bergi was determined [7, 8]. Bacteria GPB (meat peptone broth) were grown in a nutrient medium for 6 days. The cultural fluid was filtered.

To study the biological effects of phytotoxins on plant seeds, tomato of the "Sitora" variety and eggplant of the "Diamond" variety were ivitized into the fungal cultural fluid for 24 hours. 5 large, (healthy seeds) were used for each option. For control, nutrient media of non-fungal growth, sterilized liquid Chapek and GPB were applied.

For 24 hours, the ivitized seeds were put into collection in wet chambers in a petri saucer. The nature of the fungus to produce phytotoxins was determined by a decrease in the germination capacity of the seed, lagging behind the growth of the grass and root. Among the phytotoxin-forming species, 30% of the control introduced species that reduce the germination capacity of the seed, leaving its growth behind [1].

3 Results and discussion

In total, 15 bacterial isolates were isolated from samples brought from infected parts of potato, tomato and sweet pepper crops from Kasbi, Kashkadarya region and Shakhrisabz district, Upper Chirchik and Kibray district of Tashkent region, Bagdad district of Ferghana region (Table 1).

Table 1. Bacteria isolated from infected vegetable crops in Kashkadarya, Tashkent and Ferghana regions, Uzbekistan.

#	Isolates	Variety of agricultural crops	Plant organ	Distinguish between the name and cancers
1	Kasbi district, Qurbonov Rajabpo'lat Qurbonovich farm	Rossiya	Potato leaf	*Pseudomonas putida* 4/23
2	Kasbiv district, Qurbonov Rajabpo'lat Qurbonovich farm	Rossiya	Potato leaf	*Bacteroides ovatus* 4/22
3	Shaxrisabz district Davronov farm issiqxona	Vidrona	Sweet Pepper leaf	*Klibsiella oxytoca* 6/26
4	Kasbiv district, Qurbonov Rajabpo'lat Qurbonovich farm	Rossiya	Potato leaf	*Pseudomonas mendocina* 4/27
5	Юқори Чирчиқ district Жума бозор худуди Ф.Омонов farm	Yusupov	Tomato stalk	*Stenotrophomonas maltophilia* 2/16
6	Yuqori Chirchiq district Juma market F.Omonov farm	Yusupov	Tomato stalk	*Pseudomonas alcaliphila* 2/18
7	Yuqori Chirchiq district Juma market F.Omonov farm	Yusupov	Tomato stalk	*Stenotrophomonas maltophilia* 2/19
8	Yuqori Chirchiq district Juma market F.Omonov farm	Yusupov	Tomato stalk	*Brevundimonas diminuta* 2/20

9	Qibray district TashGRES "Mashxura Fayz Baraka" MChJ greenhouse with a view of	Pink Paradayz	tomato	*Stenotrophomonas* sp. 2/13
10	Qibray district TashGRES "Mashxura Fayz Baraka" MChJ greenhouse with a view of	Pink Paradayz	leaf tomato	*Pseudomonas Antarctica* 2/9
11	Qibray district TashGRES "Mashxura Fayz Baraka" MChJ greenhouse with a view of	Pink Paradayz	leaf tomato	*Stenotrophomonas* sp. 2/10
12	Qibray district TashGRES "Mashxura Fayz Baraka" MChJ greenhouse with a view of	Pink Paradayz	stalk tomato stalk	*Paenarhrobacter ilicis* 2/12
13	Qibray district TashGRES "Mashxura Fayz Baraka" MChJ greenhouse with a view of	Pink Paradayz	tomato stalk	*Microbacterium phyllosphaerae* 2/15
14	Qibray district TashGRES "Mashxura Fayz Baraka" MChJ greenhouse with a view of	Pink Paradayz	Tomato stalk	*Sphingobacterium faecium* 2/14
15	Farg'ona province Bog'dod district farm	-	Sweet Pepper leaf	*Microbacterium paraoxydans* 3/14

The morphology of the isolated bacteria was studied. The table and figures on sheep show colonies of bacteria isolated from the stems and leaves of Cypress, tomatoes brought from the tashgres greenhouse (Figure 1), microscopic appearance (Figure 2) and micromorphology (Table 2) during gramstraining.

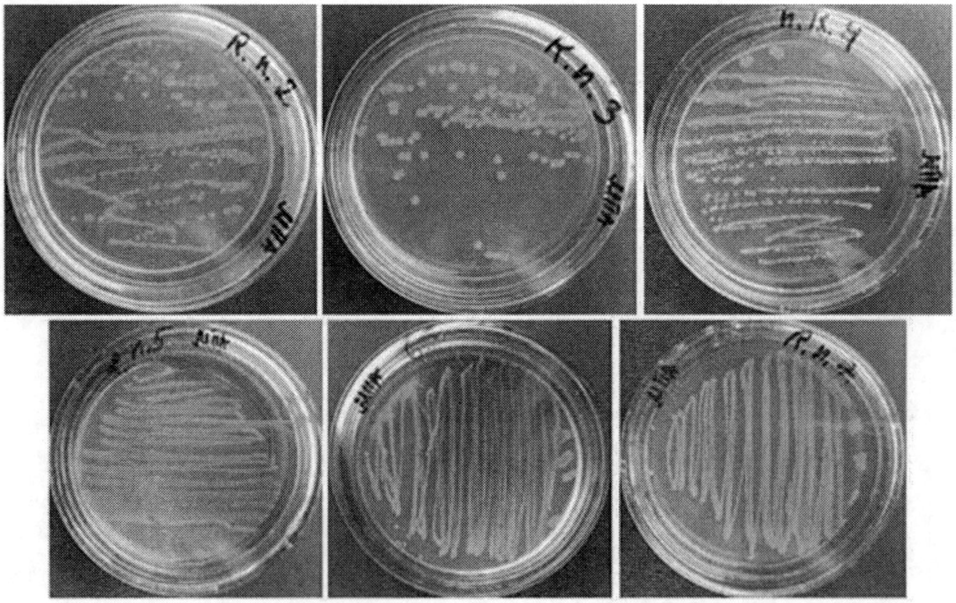

Figure 1. Qibray colonies of bacteria isolated from the stem and leaves of tomatoes brought from the greenhouse.

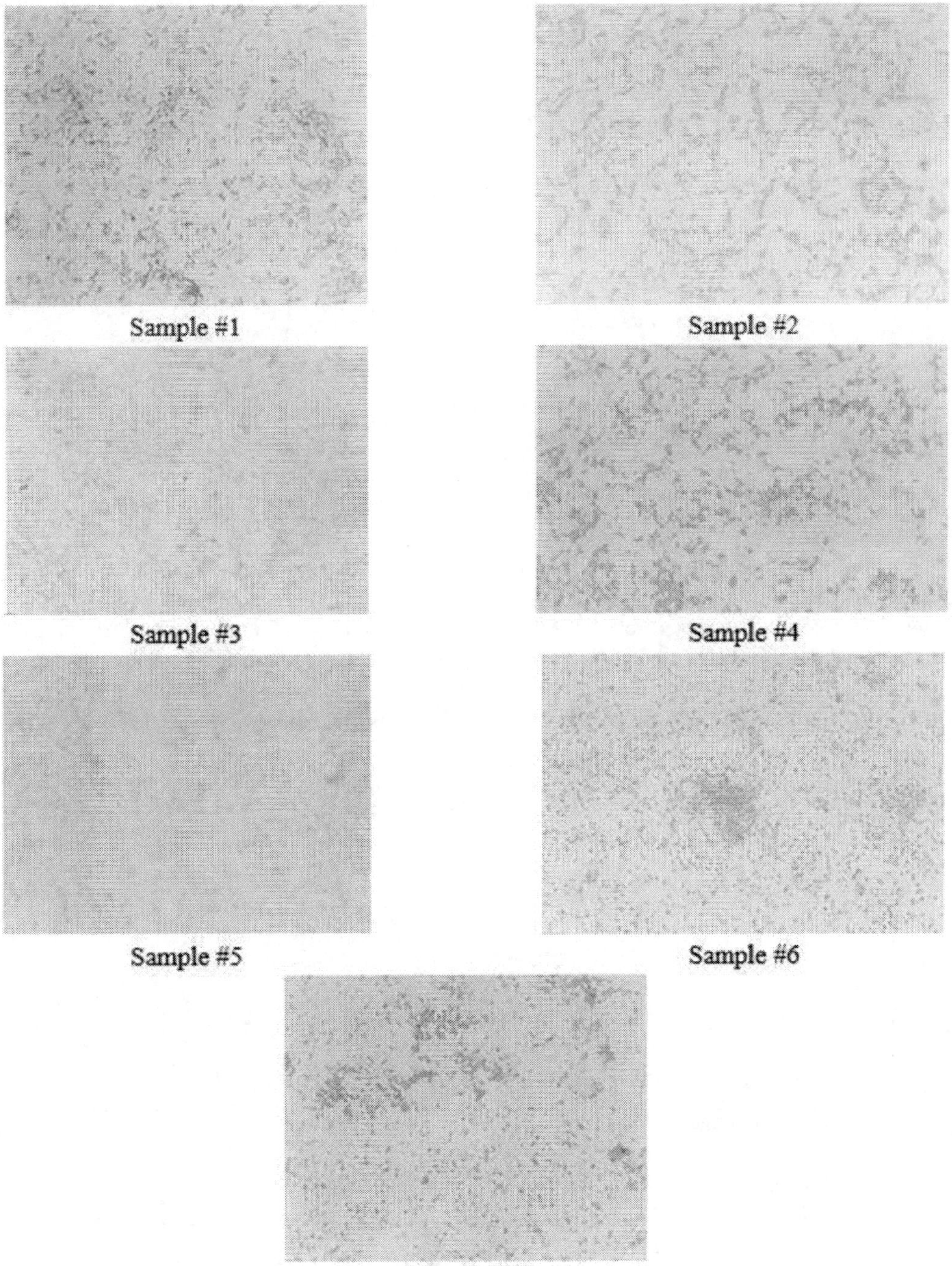

Figure 2. Qibray, TashGRES microscopic image of bacteria isolated from the stem and leaves of tomatoes brought from the greenhouse (magnified 1000 times).

Table 2. Qibray, TashGRES morphology of bacteria isolated from the stem and leaves of tomatoes brought from the greenhouse.

Isolates	Morphology
Pseudomonas antarctica 2/9	Gramophone, tieshaped, 2-4 microns long, 0.8-1.0 microns wide.
Stenotrophomonas sp 2/13	Micrococcus, monococcus, tetracoccus, diplococcus. Sarsina. In the form of streptococcus. Gram-negative, diameter 1.1 microns.
Pseudomonas sp. 2/11	Gram-negative, small rod-shaped bacteria, height 1.8-3.0 microns, width 0.6-0.8 microns
Paenarhrobacter ilicis 2/12	Gram-positive, Tae-shaped, 2-4 microns long and 1.0 microns wide.
Stenotrophomonas sp 2/10	Gram-negative, very small stick or Coca-Cola. Height 1.2 microns, width 0.8-1.0 microns.
Sphingobacterium faecium 2/14	Gram-negative small scales, no more than 1.5 microns in size, they are 1.2 microns.
Microbacterium phyllosphaerae 2/15	Gram-positive, some slightly curved, height 1.5-2.0 microns, width 1.0 microns.

The morphology of other isolated isolates was studied in a similar manner. To determine the pathogenic properties of the bacteria, they were grown in liquid feed for 15 days. With a mass formed by bacteria, vegetable seeds were subjected to intravenous spraying for 1 day. The finished seeds were placed in wet chambers. 10 tomato seeds were planted in wet chambers in each Petri dish. The experiments were monitored daily. According to the results of the study, the following table was formed (Table 3).

Table 3. "Yusuf" pathogenic properties of bacteria isolated from the affected parts of varietal tomatoes during the growing season.

#	Strain	Total planted tomato seeds	From these germinated seeds of the don	Not germinated seed	Stem length, relative to control, average	Root length relative to control, medial mm
1	Control	10	10	0	55	90.5
2	Microbacterium paraoxydans 3/14	10	7	3	33.5	56
3	*Stenotrophomonas maltophilia* 2/19	10	2	8	1.1	22.5
4	*Stenotrophomonas* sp. 2/13	10	5	5	2.5	36.5
5	*Klibsiella oxytoca* 6/26	10	8	2	30.5	45.5
6	*Sphingobacterium faecium* 2/14	10	8	2	31.5	75
7	*Bacteroides ovatus* 4/22	10	3	7	20	24.5
8	*Pseudomonas antarctica* 2/9	10	8	2	31.5	63
9	*Brevundimonas diminuta* 2/20	10	4	6	13.5	21.5
10	*Pseudomonas putida* 4/23	10	3	7	13.5	19.5
11	*Microbacterium phyllosphaerae* 2/15	10	8	2	38.5	57.5

12	*Pseudomonas mendocina* 4/27	10	6	4	19	43
13	*Paenarhrobacter ilicis* 2/12	10	6	4	22.5	42.5
14	*Pseudomonas alcaliphila* 2/18	10	5	5	26	42.5
15	*Stenotrophomonas maltophilia* 2/16	10	5	5	7.5	19.5

In Table 4, to determine the pathogenicity of bacteria isolated from tomatoes, tomato seeds were planted in Petri dishes with artificial damage. 10 seed grains were planted in wet chambers in each Petri dish, infecting them with bacteria. Research data showed that 2.19, 4.22, 2.20, 4.23 negatively affected germination and seed growth when sown with bacterial strains. Strains 6.26, 2.14, 2.9, 2.15 did not have a negative effect on germination and seed growth. Thus, it turned out that not all bacteria isolated from vegetable crops are pathogenic.

Table 4. Pathogenic properties of bacteria isolated from eggplant parts infected during the growing season.

#	Strain	Total planted tomato seeds	Germinated seed	Not germinated seed	Rod length relative to the regulator, mm	Length of the root relative to the control, mm
1	Control	10	10	0	8	11
2	Microbacterium paraoxydans 3/14	10	10	0	8,6	17,5
3	*Stenotrophomonas maltophilia* 2/19	10	0	10	-	-
4	*Stenotrophomonas* sp. 2/13	10	1	9	-	0,3
5	*Klibsiella oxytoca* 6/26	10	7	3	4,8	11,8
6	*Sphingobacterium faecium* 2/14	10	4	6	2	4,3
7	*Bacteroides ovatus* 4/22	10	5	5	1,5	4,3
8	*Pseudomonas antarctica* 2/9	10	7	3	5,8	11,5
9	*Brevundimonas diminuta* 2/20	10	0	10	-	-
10	*Pseudomonas putida* 4/23	10	4	6	-	2,5
11	*Microbacterium phyllosphaerae* 2/15	10	7	3	1,3	6,2
12	*Pseudomonas mendocina* 4/27	10	5	5	1,6	8
13	*Paenarhrobacter ilicis* 2/12	10	0	10	-	-
14	*Pseudomonas alcaliphila* 2/18	10	6	4	1,2	5,3
15	*Stenotrophomonas maltophilia* 2/16	10	0	10	-	-

It was found that during artificial sowing of eggplant seeds by bacteria, strains 2/20, 2/16, 2/13, 2/12, 2/19 have a negative effect on fertility. It was found that 10 seed pods planted with 2/20, 2/16 strains of bacteria lost their tensile strength. This means that these fungi cause rotting of the roots of the plant during the germination period, when the plant is sown, during the growing season of the plant. When sowing infected with bacterial strains 6.26, 2.9, 2.15 isolated from vegetable crops, these bacteria did not have a negative effect on germination and seed growth.

4 Conclusions

In drawing our research to a close, it becomes evident that not all bacteria isolated from vegetable crops exhibited pathogenic tendencies towards agricultural plants. Conversely, specific strains, including 6.26, 2.14, 2.9, and 2.15, displayed a positive impact on the germination and growth of tomato and eggplant seeds during our experimentation. This intriguing revelation underscores the potential benefits of harnessing bacteria from vegetable crops for enhancing seed germination and overall plant development.

Furthermore, the isolation and identification of these bacterial strains lay the foundation for the potential development of novel biological preparations. By isolating bacterial species with robust antimicrobial properties, which surpass their pathogenic counterparts, we open avenues for the formulation of biologically derived substances. These substances could potentially serve as alternatives to traditional chemical interventions in agriculture, fostering a more sustainable and environmentally friendly approach to crop protection.

In essence, our findings underscore the multifaceted nature of bacteria isolated from vegetable crops, highlighting both their positive potential and their non-pathogenic attributes. As we navigate towards agricultural practices that prioritize both productivity and ecological harmony, the utilization of these bacterial strains may offer innovative solutions to crop management and plant health enhancement.

References

1. Critical analysis of fungal diseases of pepper in Uzbekistan Hasanov Botir Achilovich, Utaganov Samad Bobomurod, Aznabakieva Dilrabo Tursunboevna, Agrochemical protection and plant quarantine 13.-2022 y.Page 22.
2. Cultivation of sweet pepper: scientific publication/"Agrobank" ATB.-Tashkent: "TASVIR" publishing house, 2021. - 48 p.
3. Akoua Emmanuella Adioumani, Solange Kakou-Ngazoa, Trebissou Jonhson Noel David, Thibaud Martin, Audrey Addablah, Aristide Koudou, Kan Stephane Kouassi, Sina K. Mireille, N'golo Coulibaly, S. Aoussi Mireille Dosso / Isolation and identification of phytopathogenic bacteria in vegetable crops in West Africa (Côte D'ivoire) African Journal of Microbiology Research Vol. 16(4), pp. 167-177, April 2022.
4. Bonk T., Humeny A. MALDI-TOF-MS Analysis of Protein and DNA // The Neuroscientist Volume 7, Number 1, 2001. -p. 6-12.;
5. Ryan D.J., Spraggins J.M., Caprioli R.M. Protein Identification Strategies in MALDI Imaging Mass Spectrometry: A Brief Review/Curr Opin Chem Biol. 2019 February; 48: 64–72. doi:10.1016/j.cbpa.2018.10.023.
6. Kolomiets J.V., Grygoryuk I.P., Butsenko L.M./Bacterial diseases of tomato plants in terms of open and covered growing of Ukraine//Annals of Agrarian Science vol. 15, no. 2, 2017.
7. Stepanov A.S., Vasilyeva N.V. New opportunities to identify and type Staphylococcus spp. by using MALDI-TOF mass spectrometry//Russian Journal of Infection and Immunity = Infektsiya i immunitet, 2018, vol. 8, no. 4, pp. 489–496. doi: 10.15789/2220-7619-2018-4-489-496;
8. Payne, M., Champagne, S., Lowe, C., Leung, V., Hinch, M., & Romney, M. G. (2018). Evaluation of the FilmArray Blood Culture Identification Panel compared to direct MALDI-TOF MS identification for rapid identification of pathogens. Journal of Medical Microbiology, 67(9), 1253–1256. https://doi.org/10.1099/jmm.0.000802
9. Chanchez-Juanes, F., Se˜n´an, A. I. G., Egido, S. H., Ruiz, M. S., Buitrago, J. M. G., & Bellido, J. L. M. Use of MALDI-TOF techniques in the diagnosis of urinary tract

pathogens. In *The use of mass spectrometry Technology (MALDI-TOF) in clinical microbiology* (2018). (pp. 145–158). Granada, Spain: Academic Press. https://doi.org/10.1016/B978-0-12-814451-0.00010-1;
10. Yi Yang, Yu Lin, Liang Qiao / Direct MALDI-TOF MS Identification of Bacterial Mixtures//*Analytical Chemistry* 2018 *90* (17), 10400-10408 DOI: 10.1021/acs.analchem.8b02258.

Effect of “Bist” and “Bist-M” biopreparations on quality and productivity of wheat and cotton yields in agro-ecological zones of Uzbekistan

Holida Ruzimova[1,*], *Ulugkhoja* Rakhmatov[2] and *Shakhribonu* Kadyrova[3]

[1]Chirchik State Pedagogical University, Chirchik, 110100, Uzbekistan
[2]Tashkent State Technical University, Tashkent, 100095, Uzbekistan
[3]Tashkent Branch of Samarkand State Veterinary Medicine, Tashkent, 100210, Uzbekistan

Abstract. Significant emphasis is placed on the development and application of biopreparations harnessed from beneficial bacteria, which exhibit advantageous traits for plants. A multitude of experiments have substantiated that the concurrent utilization of biopreparations in conjunction with mineral fertilizers opens avenues to cultivate high-quality agricultural produce. Researchers across diverse regions have demonstrated that not only can plant productivity be augmented through this approach, but also the attainment of superior, environmentally-friendly yields is feasible, stemming from an exploration of the intricate interplay between plants and microorganisms. This growing body of scientific evidence underscores the potential of leveraging the symbiotic relationship between plants and beneficial microorganisms. This synergy holds the promise of enhancing agricultural output while championing the production of products that meet stringent ecological standards. As such, the collaborative influence of biopreparations and traditional agricultural practices stands as a noteworthy avenue for fostering sustainable and quality-driven agricultural endeavors.
Keywords. Microbe, biopreparation, strain, bacteria, salinity, drought, pesticide.

1 Introduction

In recent years, in order to ensure the effectiveness of mineral fertilizers, recommendations have been developed on the use of their nanoparticles. Taking into account the damage caused by mineral fertilizers and chemical pesticides, systematic work has been started in developed countries to reduce their amount as much as possible and to replace them with biotechnology products [1]. Ensuring environmental protection, as well as growing ecologically clean products, increasing soil fertility, at least leaving this indicator unchanged, is defined as the main task facing agriculture today [2]. For this purpose, scientific and practical researches are being carried out on the saturation of the soil with bacteria that can provide biological nitrogen and phosphorus.

Lack of moisture or drought is one of the common stress factors for plants [3]. The lack of moisture in the soil dehydrates the plant tissue, disrupts the hormonal status, respiration

* Corresponding author: x.ruzimova@mail.ru

and photosynthesis, and finally leads to a number of negative conditions and secondary effects, including oxidative stress, nutrient deficiency, and in most cases, the plant dies [4]. When plants are inoculated with associative bacteria, many of these adverse conditions can be mitigated or even completely eliminated [5]. It is of great scientific and practical importance to create and use microbiological tools based on associative rhizobacteria that increase the productivity of agricultural plants, improve the ecological condition of the soil, and serve to grow ecologically clean products [6]. Although biopreparations "Bist" and "Bist-M" have exactly this feature, the question of their use is lagging behind the requirements of the times [7]. The reason for this is that we do not have enough knowledge about the ecology of the microorganisms that form the basis of these preparations and the role of agroecological factors that determine the effective functioning of the plant-bacterial association [8].

It is known from scientific sources that the number of populations of microorganisms is a factor determining their activity and status in the biocenosis [9]. The specificity of the reactions of different types and varieties of plants to bacteriization is related to the preservation or loss of associative bonds that make a particular plant with bacteria. In addition to the soil-climatic conditions of the microbial populations artificially introduced from outside, the interactions between the indigenous microorganisms living in this soil and the newly introduced microorganisms have a great impact on the productivity of the inoculated plants [10]. Despite this, there have been no studies that could reveal the effect of agroecological factors on the interaction of associative rhizobacteria with plants. Taking this into account, we conducted experimental tests on soils of different agro-ecological conditions of Khorezm region, Uzbekistan.

2 Materials and methods

In total, 2 liters per seed per 1 hectare of land with biopreparation, 5 liters of sap water and 5 liters of suspension during the growing season of the plant, yield increased by 7-12 centners, germination of seedlings was accelerated by 3-4 days; it was observed that the bolls were prevented, the cotton bolls became larger, the quality of the harvest improved, and in the case of grain, the plant leaves became larger, the amount of gluten increased, the photosynthesis process improved, and as a result, the yield increased.

These indicators show that the use of biopreparations is effective at a time when it is increasingly difficult to carry out agricultural work in the conditions of the Khorezm region, and it increases the urgency of creating biotechnologies in accordance with these soil-climatic conditions. Taking this into consideration, in the scientific laboratory of the Department of Microbiology and Biotechnology of UzMU, research was conducted to create the "Bist-M" biopreparation, which can work effectively even in salinity and dehydration conditions, by modifying the "Bist" biopreparation. It is known that "Bist" biopreparation is based on Pseudomonas putuda Pp-1 bacterium, it is gram-negative and does not form spores, which means it is resistant to stress conditions (high salinity and dehydration). If this biopreparation is enriched with a spore-forming bacterium that has a positive effect on plants , it has been confirmed in laboratory experiments that the above-mentioned shortcomings can be avoided. The meaning and essence of the new proposal is that under favorable agro-climatic conditions, both microorganisms can work well in harmony with each other, but when the non-spore-forming bacteria dies, the spore-forming bacteria can protect itself and the agro-climatic conditions are favorable. by changing its direction, it returns to the vegetative state and continues its interaction with the plant.

3 Results and discussion

Just such a bacterium was identified as *Bacillus subtilis* , and a special, inexpensive nutrient medium for its growth was invented and protected by a patent (IAP 20180376). It has been proven in the experiments that the biopreparation "Bist-M" created in this way can be used together with the usual farming methods and other chemical means. "Bist" before planting seeds of Tanya variety of winter wheat in moderately saline soils of Khorezm region (solid residue 0.5-1.0%; chlorine content 0.05-0.1%; sulfate ion 0.3-0.4%) and "Bist-M" biopreparations, the results obtained against the background of mineral fertilizers are shown in Table 1.

It was observed that winter wheat was damaged in the process of coming out of winter, affected by unfavorable climatic conditions (sudden temperature changes, dusty wind, etc.), and even died. At that time, it was observed that the resistance of the plant to adverse factors depends on its general physiological state (Table 1).

Table 1. Winter wheat "Tanya" variety seeds treated with "Bist" and "Bist-M" biopreparations, indicators of plant emergence from winter ("Shukhrat-Nadir" and "Qalandar Yakhshibaev" farms, 2020), %.

Options	**During the collection phase of the plant (number of plants per sq. meter)**	**Plant emergence from winter (number of plants per sq. meter)**	**Plant emergence rate from winter**
Control	397	361	90.9
Seed treatment with "Bist".	401	387	96.5
Seed treatment with "Bist-M".	407	395	97.2
Seed treatment with vitawax FF (3 L per 1 seed).	385	353	91.8
$N_{30}P_{30}$+N_{30} in early spring	388	361	93.1
Seed treatment with "Bist". +N_{30}+P_{30}+N_{30} in early spring	420	406	96.7
Seed treatment with "Bist-M" +N_{30}+P_{30}+N_{30} in early spring	417	412	97.2
Seed treatment with vitawax +N_{30}+P_{30}+N_{30} in early spring	380	373	98.1

More than 90% of the total amount of plants that have come out of the winter. The total amount of wheat planted with "Bist" and "Bist-M" biopreparations was not significantly different from each other in non-fertilized and fertilized fields, and 96.5 and 96.7% in "Bist", respectively; In the biopreparation "Bist-M" it was 97.2 and 97.9%. Under the influence of biopreparations, it was observed that the yield of wheat and the amount of gluten in its content also changed in a positive direction (Table 2).

Table 2. Effect of biopreparations "Bist" and "Bist-M" on the yield of wheat and the amount of gluten in it ("Sukhrat-Nadir" and "Qalandar Yakhasbaev" farms and "Zimnitsa" and "Durdona" varieties 2020).

Options	Productivity, q/ha	Additional crop		Raw gluten content, %
		q/ha	%	
Control	17.0	-	-	26.0
"Bist" (2 L/1 ton of seeds + 5 L during the period of tuberization)	18.7	1.7	10.0	26.8
"Bist-M" (2 L/ton of seeds + 5 L during tuberization)	19.0	2.0	11.7	26.9

It was found that the amount of nitrogen in plants sprouted from seeds treated with biopreparations increased by almost 2 times compared to the control, and this indicator is equal to the plant growing in the field where nitrogen mineral fertilizer was used. Similar results were obtained when determining the amount of potassium in the plant.

Yields increased in all experiments. The increase in the yield of wheat planted by seed treatment with "Bist-M" biopreparation was achieved when phosphorus-potassium, nitrogen-phosphorus-potassium fertilizers were used together.

An increase in the yield of the winter wheat variety "Asr" was observed in all variants of our experiments. Positive results were observed when using biopreparations ("Bist" and "Bist-M"), together with phosphorus-potassium, nitrogen-phosphorus-potassium fertilizers. The highest results were obtained in the variant "Bist-M" + $N_{60}P_{45}K_{60}$ with the biopreparations "Bist" and "Bist-M" when using different doses of nitrogen-phosphorus-potassium fertilizers in different doses, the weight of 1000 pieces of wheat, compared to the control 0 It was found that 8-1.2 g increased.

The number of plants in a certain area and their branching capabilities play a key role in the formation of productivity. In this case, the plant's supply of mineral nutrients is a determining factor. A number of biometric indicators characterizing the state of winter wheat were determined: branching coefficient, plant height, number of leaves in the branching phase and ear formation period are among them (Table 3).

Table 3. Biometric indicators of winter wheat variety Andijan-2 during branching.

Plant height, cm				Branching coefficient				Number of leaves, pcs			
1	2	3	4	1	2	3	4	1	2	3	4
No fertilizer (control)											
17.5	18.3	18.7	18.0	3.1	3.7	3.9	3.6	5.5	8.1	8.6	8.2
$N_{30}P_{30}$+N_{30} (in early spring)											

21.3	22.4	23.8	22.3	3.2	3.6	4.0	3.9	8.0	8.5	9.1	8.5
$N_{30}P_{60}+N_{60}$ (in early spring)											
21.9	23.1	24.2	22.6	2.7	2.7	2.9	3.6	7.2	7.6	8.2	7.6
$N_{30}P_{30}K_{30}+N_{30}$(in early spring +$N_{20}$ in the tuber phase +N_{20} in the spike phase)											
19.2	24.1	24.6	22.7	2.8	2.6	2.8	2.7	6.3	6.9	7.4	6.8
$N_{30}P_{60}K_{60}+N_{60}$(in early spring + N_{40} in the tuber phase + N_{40} in the spike phase)											
21.1	26.6	26.8	22.7	2.8	3.1	3.2	2.9	7.8	7.2	7.8	7.2
Note *: Control - 1; "Bist" - 2; "Bist-M" - 3; Vitavaks 200 FF - 4*											

It was observed that the height of winter wheat in the branching phase depends on the amount of mineral fertilizers applied to the soil and treatment with biopreparations and vitavaks before seeding. If we evaluate the effect of pre-sowing treatment of wheat seeds with biopreparations, we can see that both biopreparations had a stronger effect than the chemical preparation vitavaks in different variants.

In the non-fertilized version, biopreparations "Bist" and "Bist-M" led to a higher growth by 0.3-0.7 cm compared to vitavaks, and in fields fertilized in different amounts and at different times, the development of wheat compared to the version treated with vitavaks was 0.1 cm to 3.9 cm higher can be seen.

Although the coefficient of branching varied from 2.7 to 4.0 depending on the options, it showed that it is somewhat difficult to explain the method of seed treatment in connection with different doses of mineral fertilizers. A similar situation can be observed when analyzing the number of leaves that appear on a winter wheat plant.

Different varieties of winter wheat reach their maximum height during the spike formation phase. In 2017, in Koshkopir, Yangariq, Khiva and Bogot regions of Khorezm region, this indicator was found to be from 62.5 cm to 75.6 cm. it was observed that it also fell.

Even in the year with such severe weather, it was observed that the yield harvested from the fields planted with biopreparations was slightly higher than that of the control (the difference was 1.2-1.5 q/ha). According to the data of 2017, it was observed that the coefficient of productive formation was 0.9-1.1, and this indicator depends on the mineral nutrition of the plant and the treatment of the seed with biopreparations and chemical agents, and it was found that the number of wheat in one spike was from 13 to 16 (Table 4).

Table 4. Biometric indicators of winter wheat in the ear formation phase.

Plant height, cm						**Productive branching coefficient**					
1	2	3	4	5	6	1	2	3	4	5	6
No fertilizer (control)											
43.2	44.6	58.1	49.1	61.1	44.4	1.1	1.2	1.3	1.2	1.4	1.2

$N_{30}P_{30}+N_{30}$ (in early spring)											
57.1	54.3	59.3	60.1	65.3	5.37	1.2	1.5	1.5	1.5	1.6	1.5
$N_{30}P_{60}+N_{60}$ (in early spring)											
58.3	55.1	59.5	60.1	62.7	54.4	1.3	1.4	1.5	1.5	1.5	1.3
$N_{30}P_{30}+N_{30}$ in early spring + N_{20} in the spike phase											
59.3	60.3	60.7	61.3	65.0	55.0	1.2	1.5	1.4	1.5	1.6	1.3
$N_{30}P_{60}+N_{60}$ in early spring + $N_{40}N_{30}P_{60}K_{60}+N_{60}$ (in early spring + N_{40})											
59.5	60.4	60.9	61.5	65.3	55.0	1.3	1.4	1.5	1.5	1.6	1.3
$N_{30}P_{30}K_{30}+N_{30}$ in early spring + N_{20} in tuberization + N_{20} in spike phase											
59.4	60.3	60.7	61.5	65.8	54.9	54.9	1.2	1.3	1.5	1.6	1.3
$N_{30}P_{60}K_{60}+N_{60}$ in early spring + N_{40} (in the tuber phase + N_{40} in the spike phases)											
59.7	60.5	60.8	61.6	66.1	55.1	55.1	1.3	1.4	1.5	1.6	1.3
Note: Control - 1; "Bist" - 2; "Bist" in the phases of tuberization and spike - 3; "Bist-M" - 4; "Bist-M" in the phases of tuberization and spike - 5; vitavaks 200 FF - 6											

The productivity of winter wheat is formed under the influence of a number of factors. It is known that by affecting the nutrition, water and air regime of the plant, its productivity can be increased. Based on this, it is possible to improve plant nutrition and air circulation in the soil through the correct use of mineral fertilizers and biopreparations. In the experiments, it was observed that the use of biopreparations during the vegetation period of the plant softens the soil and improves its physical condition.

It was observed that the use of "Bist" and "Bist-M" biopreparations can increase the yield of winter wheat from 2 to 6 centners per hectare. In particular, it was found that feeding the soil in the order of N30 P30 and N60 in early spring gave the best results (Table 5).

Table 5. Effect of mineral fertilizers and biopreparations on the yield of winter wheat.

Options	**control**	**Bist**	**Bist-M**	**Bist in tubercle**	**Bist-M is in progress**	**Bist-M is on the rise**	**Bist-M is on the rise**	**Vitavox**
Without fertilizer	17.1	19.3	20.9	18.7	20.1	17.3	18.6	19.2

N30P30+N30 in spring	20.1	22.3	24.1	24.7	25.3	23.0	25.0	22.7
N30P30+N60 in spring	24.5	26.3	28.7	29.1	30.1	29.7	30.2	23.8
N30P60+N60 in spring	24.3	26.1	28.8	28.9	29.9	28.9	29.7	23.5

It is known that the quality of winter wheat depends on the genetic characteristics of the plant and growing conditions. In the experiments, it became clear that increasing the amount of mineral fertilizers by the amount indicated in the above tables had a positive effect on the quality of the wheat crop.

Although no significant difference was observed in the experiments conducted under different options, it was noticed that additional treatment with nitrogen fertilizers during the spike phase of the plant led to better results. No changes were detected in the amount of phosphorus and potassium in winter wheat grain, the amount of crude gluten was 18-19%, but the amount of nitrogen and phosphorus in straw of winter wheat planted with biopreparations was found to be lower than in the control. It was found that the plants sprouted from the seeds planted by treatment with "Bist" biopreparation also affect the amount of nutrients in the soil (Table 6).

Table 6. Ontogeny of winter wheat planted with "Bist": changes in the 0-30 layers of soil (mg) and the amount of mobile nutrients (kg).

Plant growth phase	**$N\text{-}NO_3$**		**P_2O_5**		**K_2O**	
	control	**Bist**	**control**	**Bist**	**control**	**Bist**
Formation phase	2.3	8.5	120	110	611	620
The spike phase	1.7	0.8	145	155	750	740
Seed ripe phase	1.5	1.2	155	160	720	710

It was observed that treatment with "Bist" had a positive effect on nitrogen accumulation in the 0-30 cm layer of the soil in early spring, but it did not affect the accumulation of mobile phosphorus, and it led to an increase of mobile potassium, albeit slightly (Table 6).

It is observed that the amount of mobile nitrogen decreased during the earing phase, the amount of mobile phosphorus remained almost unchanged, the amount of mobile potassium increased, and when the seed was fully ripened, the changes in the above options became uniform. It was found that it had a positive effect in the autumn period. The amount of plants that emerged from the winter was 90.9%. It was observed that the coefficient of productive branching and the number of spikes changed depending on the dose of mineral fertilizers and "Bist" treatment. It was found that productivity increased by 2-6 q/ha. Below is the effect of biopreparation "Bist" on the quality of wheat grain in the conditions of Khorezm region soils (Table 7).

Table 7. Effect of biopreparation "Bist" on the quality of wheat grain in the conditions of Khorezm soils.

Options	Grain type, g		Transparency		Weight of 1000 grains		Amount of gluten	
	1 *)	2 *)	1 *)	2 *)	1 *)	2 *)	1 *)	2 *)
Durdona variety								
Control	718 ± 2.0	713 ± 4.0	72	69	40.6 ± 0.5	40.1 ± 0.5	21.0 ± 1.3	20.3 ± 1.5
Bist	740 ± 2.5	736 ± 2.2	81	77	44.1 ± 0.7	43.9 ± 0.6	23.6 ± 0.7	28.7 ± 0.6

4 Conclusions

Biopreparations "Bist" and "Bist-M" increase the number of microorganisms useful for plants in the soil, regardless of the ecological condition of saline soils (washed and not washed). The effect of "Bist" and "Bist-M" biopreparations on the number of micromycetes and denitrifiers depended on the ecological condition of the soil, and it was observed that it passed faster in washed soils than in unwashed ones.

References

1. Muradova S.S. Creation of new, competitive microbial preparations based on local strains of rhizobacteria that increase the resistance of cotton to stress conditions and evaluation of their practical significance. 03.00.04-biotechnology Institute microbiology AN RUz 2018, - p. 54.
2. Sidorenko O. D. Ecological soil and the role of bacterial preparations. Biological preparations. Selskoe household. Ecology: Praktika primeneniya - M., 2008. – p. 296.
3. Blinov V. A., Burshina S. N., Shapulina E. A. Biological action of effective microorganisms. Biological preparations. Selskoekhozyaystvo. Ecology: Praktika primeneniya M., 2008. - p. 296.
4. Dyatlova K. D. Microbial preparations in plant breeding // Soros. image Journal. Series Biology. - 2001. - v. 7, No. 5. - p. 4-18.
5. Karashaeva A. S. Assotsiativnye diazotrophy i ix vzaimodeystvie s rasteniyami // Molodoy uchenyy. - 2016. - No. 6. - p. 350-353.
6. Mikhailovskaya N. A., Kasyanchik s. A., Mikanova o. Vliyanie bakterilnogo udobreniya Kaliplant na ispolzovanie kaliya zernovymi kulturami i gorokhom na dernovo-podzolistoy supeschanoy pochve// Ves. Nats. Acad. type Belarusian. Sir. agrarian type 2010. - 2(45), - p. 193-200.
7. Ruzimova, K. K. (2022). The influence of microbial preparations on the ecological state of saline soils. ACADEMICIA: An International Multidisciplinary Research Journal, 12(5), 1271-1277.
8. Bist, M. R., & Shrestha, B. B. (2023). Weed community structure in upland farming system of the middle mountain region in far-western Nepal. Acta Ecologica Sinica, 43(3), 498-505.
9. Bhadauria, N. S., Bist, M. S., Patel, R. B., & Bhadauria, H. S. (2015, March). Performance evaluation of segmentation methods for brain CT images based

hemorrhage detection. In 2015 2nd International Conference on Computing for Sustainable Global Development (INDIACom) (pp. 1955-1959). IEEE.

10. Baral, P., Umans, B. D., Li, L., Wallrapp, A., Bist, M., Kirschbaum, T., ... & Chiu, I. M. (2018). Author Correction: Nociceptor sensory neurons suppress neutrophil and γδ T cell responses in bacterial lung infections and lethal pneumonia. Nature Medicine, 24(10), 1625-1626.

Efficient utilization of anthropogenic waste: a particular focus on phosphogypsum and ash slag

Khamza Aminov[1,*] and *Farhad* Yuldashev[2]

[1]Research Institute of Environment and Nature Conservation Technologies, Tashkent, 100043, Uzbekistan
[2]Belgorod State Technological University, Belgorod, 308012, Russia

Abstract. The focal point of this article resides in the pressing concern surrounding the mounting stockpiles of man-made waste, specifically phosphogypsum and ash slag, which have amassed in voluminous dumps across the global landscape, collectively amounting to millions of tons. This paramount issue, reverberating on a global scale, underscores the dire need for innovative strategies for the efficient disposal of such waste materials. Within the pages of this article, we endeavor to delve into the intricacies of man-made waste management, with a spotlight on the poignant predicament posed by phosphogypsum and ash slag. These residues, often regarded as byproducts of various industrial processes, have amassed into colossal accumulations that not only present significant environmental challenges but also beckon for sustainable solutions that navigate minimal losses and optimal resource utilization. Our discourse will not only shed light on the gravity of the problem at hand but will also unveil a method that stands as an epitome of effectiveness in waste management. Through a rigorous examination of current practices, combined with a judicious exploration of novel approaches, our proposed solution surfaces as a beacon of hope—a blueprint that embodies not only technical ingenuity but also a commitment to conservation and ecological harmony. Amidst the labyrinthine intricacies of waste disposal, the approach we advocate asserts its credentials through its compelling efficacy and economic feasibility.
Keywords. Phosphozol, phosphogypsum, recycling, production waste, active mineral additive.

1 Introduction

In today's rapidly evolving world, the imperative of safeguarding our natural environment has never been more critical [1]. The acknowledgment that humanity's intricate relationship with nature is not only symbiotic but pivotal to our very survival underscores the urgency of proactive measures [2]. As stewards of our planet, it falls upon us to recognize that the

* Corresponding author: khamza.aminov1001@gmail.com

dissolution of nature's delicate equilibrium may inexorably culminate in the cessation of human existence [3].

Within the fabric of this paramount concern, we present an innovative solution—a method that not only addresses the pressing issue of two significant man-made waste streams, phosphogypsum and ash slag, but also provides a viable path towards their transformation into a valuable resource [4]. Our method is predicated on the creation of Phosphozol—an active mineral additive derived from these wastes—a tangible testament to the transformative power of sustainable ingenuity [5].

Phosphozol's versatile potential finds its canvas in the realm of construction materials. With a resolute commitment to harnessing discarded resources, Phosphozol stands poised to revolutionize the construction industry. As an additive in the production of cement, concrete, and dry building mixtures, Phosphozol not only breathes new life into discarded waste but also enriches the structural integrity and performance of the final products [6].

Moreover, Phosphozol does not confine its potential within the boundaries of construction alone. Its role extends to the domain of infrastructure, where it has proven its efficacy as an additive for asphalt and asphalt concrete [7]. The infusion of "Phosphozol" at levels ranging from 30% to 40% imparts a transformative touch, enhancing the quality and durability of the resulting asphalt and its derivatives [8].

In advocating for Phosphozol's adoption, we underscore not only its utility but its emblematic role in reshaping the narrative surrounding waste management [9]. This endeavor underscores the possibility of harmonizing industrial advancement with ecological responsibility, aligning human progress with the imperative of safeguarding our planet's fragile ecosystems. The prospect of integrating Phosphozol into the fabric of industrial practices epitomizes a symbiotic relationship between innovation and environmental preservation, one that holds the promise of a future where humanity's actions echo its commitment to nature's perpetual equilibrium [10].

"Phospozol" can be used in agriculture as an ameliorant, which is used in saline soils and effectively improves the humus of the earth and turns saline soils into fertile areas with a high content of useful elements. Calcium and other macro and microelements contained in Phosphozol neutralize sodium inclusions in saline lands.

2 Materials and methods

One of the components of Phosphozol is phosphogypsum, a waste product of the chemical production of phosphate fertilizers. More than 100 million tons of phosphogypsum are disposed of annually in dumps around the world. To obtain 1 ton of fertilizer, the manufacturer is forced to dump from 1.5 to 4 tons of phosphogypsum waste into the dump. After the land allocated for the burial of phosphogypsum is filled with waste, it is stored on other new plots of land. As a result, over time, hundreds of hectares of fertile land, instead of benefiting humanity, become objects of hazardous waste storage, one of which is phosphogypsum. Over the years, the negative impact on nature is only aggravated, since with rainwater, harmful substances from dumps penetrate into groundwater. And the winds carry particles of phosphogypsum to the surrounding areas, causing harm to people and animals. Of course, the need for the production of phosphate fertilizers, which bring great benefits to agriculture, is undeniable, but we should not forget about environmental issues and the disposal of hazardous waste.

The second component of our mineral active additive "Phosphozol" is ash and slag, which occurs in thermal power plants due to coal combustion. Dumps are also built to store ash and slag, spending huge amounts of money and occupying hectares of fertile land. Like phosphogypsum, ash and slag is produced hundreds of millions of tons annually around the world. Thermal power plants that run on coal, of course, are also necessary in human life, but again, one should not forget about nature and people's health.

3 Results and discussion

Ash and slag dumps because no less harm to human and animal health than phosphogypsum dumps. We were faced with the task of disposing of these two hazardous man-made wastes, and we coped with it, and not only solved the issue of disposal. But also received a product that brings great benefits to the national economy. The production method is as follows: having previously cleaned phosphogypsum and ash and slag from foreign objects, as well as crushed to particles of 0.5 mm, we send phosphogypsum and ash and slag to the autoclave with the help of dispensers in a ratio of 33% and 67%, respectively. In the autoclave, the mixture is processed for 25-35 minutes at a certain pressure and temperature of 150-200 °C. During the processing of phosphogypsum and ash and slag, as well as water and steam, a chemical reaction occurs in which phosphate ions are adsorbed on the silica particles of ash and slag, resulting in the neutralization of phosphoric and other acids in phosphogypsum.

Below is a technological scheme for the production of the active mineral additive Phosphozol (Tables 1 and 2).

Table 1. Chemical composition of phosphogypsum and natural gypsum stone.

Materials	**Oxide content, %**									
	Loss on ignition	**SiO_2**	**Al_2O_3**	**Fe_2O_3**	**CaO**	**MgO**	**SO_3**	**R_2O_3**	**P_2O_3**	**P_2O_3 hydrogen**
Phosphogypsum	15.50	14.56	3.01	0.62	25.25	1.45	35.85	3.33	0.40	0.14
Gypsum stone	17.78 out of 400	0.58	0.70	0.22	30.20	1.80	39.75	0.25	-	-

Table 2. Chemical composition of TPP ash and slag.

Name of TPP (CHP)	**Loss on ignition**	**SiO_2**	**Al_2O_3**	**Fe_2O_3**	**CaO**	**MgO**	**SO_3**	**N_2O**	**MO**	**Ma**	**CaO+MgO**
Ivanovskaya	12.39	45.86	18.70	9.45	7.00	3.40	1.13	1.91	0.161	0.408	10.40
Ryazanskaya	12.70	44.36	23.00	7.20	7.84	2.30	0.84	0.65	0.151	0.518	10.14
Smolenskaya	8.65	54.00	15.60	9.98	6.40	2.00	0.94	0.50	0.121	0.289	8.40
Tverskaya	11.00	46.70	22.60	7.03	6.89	2.52	0.65	0.55	0.136	0.484	9.41
Mosenergo-17	12.69	44.30	22.97	7.29	7.85	2.23	0.82	0.68	0.150	0.519	10.08

Angrenskaya	14.53	39.04	17.75	2.82	19.9	1.76	0.75	0.60	0.382	0.455	21.71
Average	11.09	45.71	20.10	7.30	9.32	2.37	0.86	0.82	0.263	0.450	15.70

Of course, there are a huge number of methods for neutralizing harmful acids, but almost all methods involve washing the acids with large amounts of water or using alkaline concentrates, which are by no means cheap (Figure 1).

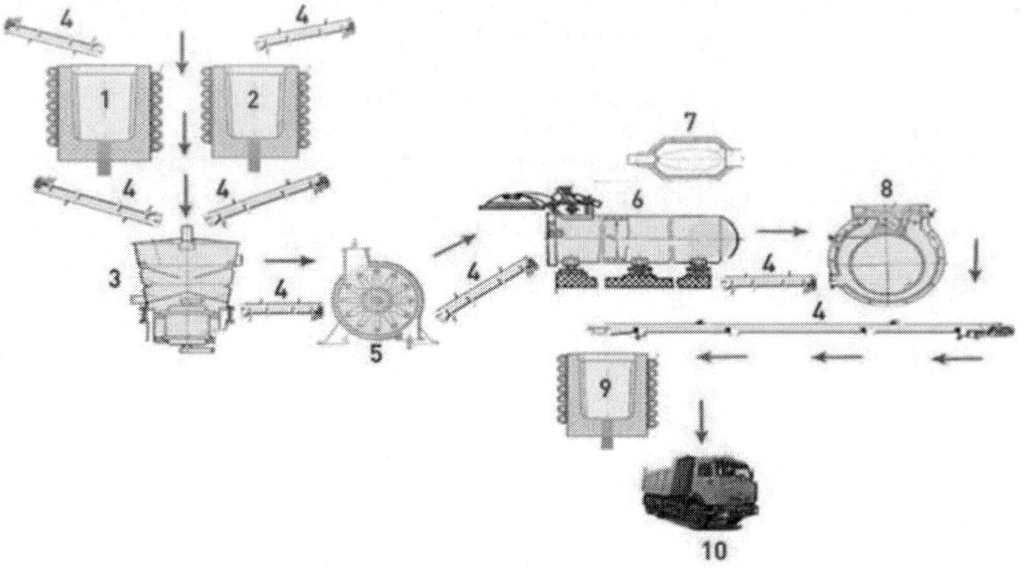

Figure 1. Technological scheme for the production of the active mineral additive "Phosphozol": 1- bunker for phosphogypsum; 2- ash bunker; 3- vibrating sieve; 4-conveyor; 5- hammer crusher; 6- autoclave; 7- steam generator; 8- dryer; 9- bunker of finished products; and, 10- car (dump truck).

Our method differs in that we use the minimum amount of water necessary to operate the autoclave and neutralize acids. In conditions of scarcity of water resources, our method is environmentally and cost-effective, especially in regions where there are already problems with water supply.

The accumulation of man-made waste in recent decades has become a serious problem due to the insufficient volume of their disposal. Meanwhile, natural raw materials and man-made waste should be considered within the framework of the concept of a technical and economic alternative. Natural raw materials used in construction and production of building materials and products belong to the category of non-renewable resources, and its extraction is accompanied by causing irreparable damage to nature. This is what determines the fundamentally important and urgent importance of finding and adopting alternative (substitute) solutions related to the possibilities of using man-made waste instead of natural raw materials. The use of natural raw materials is permissible only on the condition that man-made waste is not suitable for use in construction purposes due to its technical and technological characteristics.

This means that it is necessary to introduce technologies that will allow the disposal of already accumulated waste and reduce the amount of waste produced in the future. Unfortunately, at the moment, Uzbekistan lags behind the countries of the European Union and the United States in terms of processing.

4 Conclusions

The authors of this article propose to use the active mineral additive Phosphozol in the production of Portland cement. This method has certain advantages. Replacing expensive Phosphozol clinker will help reduce the cost of cement production, since it can replace up to 35% of the mass of cement. In addition, the use of Phosphozol will help to dispose of a large amount of waste, which will positively affect the ecological situation in the regions, since a large amount of ash and slag waste is used in the production of the additive.

References

1. Zhang, Y., Ma, Z., Zhi, X., Chen, X., Zhou, J., Wei, L., & Liu, Z. (2023). Damage characteristics and constitutive model of phosphogypsum/fly ash/slag recycled aggregate concrete under uniaxial compression. Cement and Concrete Composites, 138, 104980.
2. Shen, W., Zhou, M., Ma, W., Hu, J., & Cai, Z. (2009). Investigation on the application of steel slag–fly ash–phosphogypsum solidified material as road base material. Journal of hazardous materials, 164(1), 99-104.
3. Kumar, S. S., Kumar, A., Singh, S., Malyan, S. K., Baram, S., Sharma, J., ... & Pugazhendhi, A. (2020). Industrial wastes: Fly ash, steel slag and phosphogypsum-potential candidates to mitigate greenhouse gas emissions from paddy fields. Chemosphere, 241, 124824.
4. Li, B., Li, L., Chen, X., Ma, Y., & Zhou, M. (2022). Modification of phosphogypsum using circulating fluidized bed fly ash and carbide slag for use as cement retarder. Construction and Building Materials, 338, 127630.
5. Huang, Y., & Lin, Z. (2010). Investigation on phosphogypsum–steel slag–granulated blast-furnace slag–limestone cement. Construction and Building Materials, 24(7), 1296-1301.
6. Vaičiukynienė, D., Nizevičienė, D., Kielė, A., Janavičius, E., & Pupeikis, D. (2018). Effect of phosphogypsum on the stability upon firing treatment of alkali-activated slag. Construction and Building Materials, 184, 485-491.
7. Zheng, Y., Xuan, D., Shen, B., & Ma, K. (2023). Shrinkage mitigation of alkali-activated fly ash/slag mortar by using phosphogypsum waste. Construction and Building Materials, 375, 130978.
8. Szajerski, P., Bogobowicz, A., Bem, H., & Gasiorowski, A. (2019). Quantitative evaluation and leaching behavior of cobalt immobilized in sulfur polymer concrete composites based on lignite fly ash, slag and phosphogypsum. Journal of Cleaner Production, 222, 90-102.
9. Mashifana, T. P., Okonta, F. N., & Ntuli, F. (2018). Geotechnical properties and microstructure of lime-fly ash-phosphogypsum-stabilized soil. Advances in Civil Engineering, 2018.
10. Li, F., Zhao, W., Li, J., Fan, H., Xu, M., Han, G., ... & Fang, Y. (2023). Investigation on influencing mechanisms of phosphogypsum (PG) on the ash fusion behaviors of coal. Energy, 268, 126699.

Identification of nematodes of the genus *Teladorsagia* parasites of ruminants with the help of species-specific markers based on ITS2 rDNA

Anvarbek Ibrokhimov[1,*], *Abdurakhim* Kuchboev[2], *Oybek* Amirov[2], *Boltaboy* Kahorov[3] and *Mirzakamol* Ayubov[4]

[1]Namangan State University, Namangan, 160100, Uzbekistan
[2]Institute of Zoology of the Academy of Sciences of the Republic of Uzbekistan, Tashkent, 100140, Uzbekistan
[3]National University of Uzbekistan, Tashkent, 100174, Uzbekistan
[4]Center of Genomics and Bioinformatics of the Academy of Sciences of the Republic of Uzbekistan, Tashkent, 100140, Uzbekistan

Abstract. The present study delves into a methodological framework aimed at establishing species-specific markers via the utilization of sequences derived from the internal transcribed spacer 2 (ITS2) region of nuclear ribosomal DNA. This method, in conjunction with polymerase chain reaction (PCR) testing, serves as a diagnostic tool for discerning species belonging to the genus *Teladorsagia* Andreeva et Satubaldin, 1954. These species, constituents of the subfamily *Ostertagiinae* (*Nematoda: Trichostrongylidae*), exhibit wide distribution within the gastrointestinal tracts of ruminants across the geographic expanse of Uzbekistan. The heart of this endeavor is the development of species-specific primers, a pioneering creation in its own right. These primers are crafted using sequences emanating from the ITS2 region of the ribosomal DNA, an innovative approach that facilitates the precise identification of morphospecies within the Teladorsagia genus. Notably, the primers exhibit a nucleotide length of 153 base pairs, an attribute instrumental in their capacity to accurately distinguish and diagnose eggs and larvae of three distinct morphspecies: *T. circumcincta*, *T. trifurcata*, and *T. davtiani*. The potential implications of this method are significant, with ramifications reverberating across the field of veterinary diagnostics. Through the application of these primers, practitioners and researchers alike can effectively ascertain the presence of specific *Teladorsagia* morphospecies in ruminant animals. This holds the promise of not only enhancing diagnostic precision but also contributing to the broader comprehension of the prevalence and distribution of these nematode species within the local ruminant populations.
Keywords. Nematode, Ostertagiinae, Teladorsagia, PCR, rDNA, ITS2, centrifuge, amplification.

* Corresponding author: aibrohimov680@gmail.com

1 Introduction

Currently, parasitic diseases of livestock and other animals of all over the world are considered to be economic importance. In Australia, losses caused by parasitic diseases in sheep and cattle have been estimated at one billion australian dollars [1]. Nematodes are the most common types of parasitic diseases, and the use of anthelmintic drugs against these parasites is not always effective. The use of multiple anthelmintic drugs in animals increases genetic resistance in nematode populations and causes widespread problems [2, 3].

One of the genera widespread in the digestive system of ruminants is *Teladorsagia* Andreeva et Satubaldin belongs to the subfamily Ostertagiinae Lopez-Neira, 1947 1954 (Nematoda: Thichotrongylidae) is widespread in ruminants of Uzbekistan. They are adapted to live as parasites in the udder and intestines of animals, and sometimes the intensity of invasion in one animal organism exceeds thousands of copies [4].

In the detection of parasitic diseases in ruminants , using capralogical methods, parasite species are clarified based on the morphology of the larvae and eggs of parasites. However, this method makes it difficult to accurately diagnose the species, and it also takes a lot of time [5]. To date, in addition to classical methods, research on DNA analysis is widely used to identify nematode species. By studying nucleotides in the internal transcribed spacer 2 (ITS2) regions of nematode ribosomal DNA (rDNA) through PCR, methods were developed to identify nematode eggs, larvae and mature imago at the species level [6].

The primary objective of this research endeavor is to devise a novel method for establishing species-specific markers that leverage the internal transcribed spacer 2 (ITS2) region of ribosomal DNA (rDNA). These markers will be pivotal in the precise diagnosis of nematode species belonging to the Teladorsagia genus [7, 8]. This genus comprises parasitic organisms inhabiting the intricate landscape of the digestive systems in ruminant animals. The proposed diagnostic strategy entails the utilization of polymerase chain reaction (PCR) techniques to validate the efficacy of these species-specific markers [9, 10].

This research initiative is driven by the pressing need for accurate and reliable diagnostic tools within the domain of veterinary science. The intricate interplay between host animals and parasitic nematodes necessitates a meticulous understanding of species diversity and distribution [11]. Through the strategic utilization of species-specific markers harnessed from the ITS2 rDNA region, this study seeks to offer a breakthrough solution to this challenge. The utilization of PCR for testing the efficacy of these markers serves as a pivotal validation step, ensuring their precision and applicability in real-world diagnostic scenarios [12, 13].

The potential implications of this research are far-reaching. Accurate species identification is foundational to effective disease management and treatment strategies in the realm of veterinary medicine [14]. The deployment of species-specific markers, underpinned by molecular techniques, holds the promise of revolutionizing the diagnostic landscape for Teladorsagia nematodes in ruminant populations. By offering a higher level of resolution in species differentiation, this research aims to contribute to a more nuanced understanding of the prevalence, distribution, and potential impact of these parasitic organisms on the digestive systems of ruminant animals.

As this research journey unfolds, it aspires to not only shed light on the intricacies of species differentiation but also to advance the field of veterinary diagnostics. The innovative amalgamation of molecular techniques, rDNA sequencing, and PCR validation stands to redefine the way we approach the identification and characterization of parasitic nematode species [15, 16]. Ultimately, the successful realization of species-specific markers using ITS2 rDNA holds the potential to enhance disease management, animal health, and the broader understanding of host-parasite interactions in the context of ruminant digestive systems.

2 Materials and Methods

Materials and study area. To carry out this research, the species belonging to the genus *Teladorsagia* and the representatives of morphologically and morphometrically close genera *Marshallagi*a Orloff, 1933, *Haemonchus* Cobb, 1898, *Ostertagia* Ransom, 1907 were collected from Namangan, Syrdaryo, Jizzakh, Samarkand of Republic of Uzbekistan. Complete and incomplete helminthological dissection of the intestines of sheep and cattle slaughtered in poultry houses of Bukhara and Kashkadarya regions was carried out [17] and the collected helminthological samples were fixed in 70% ethanol solution (Figure 1 and Table 1).

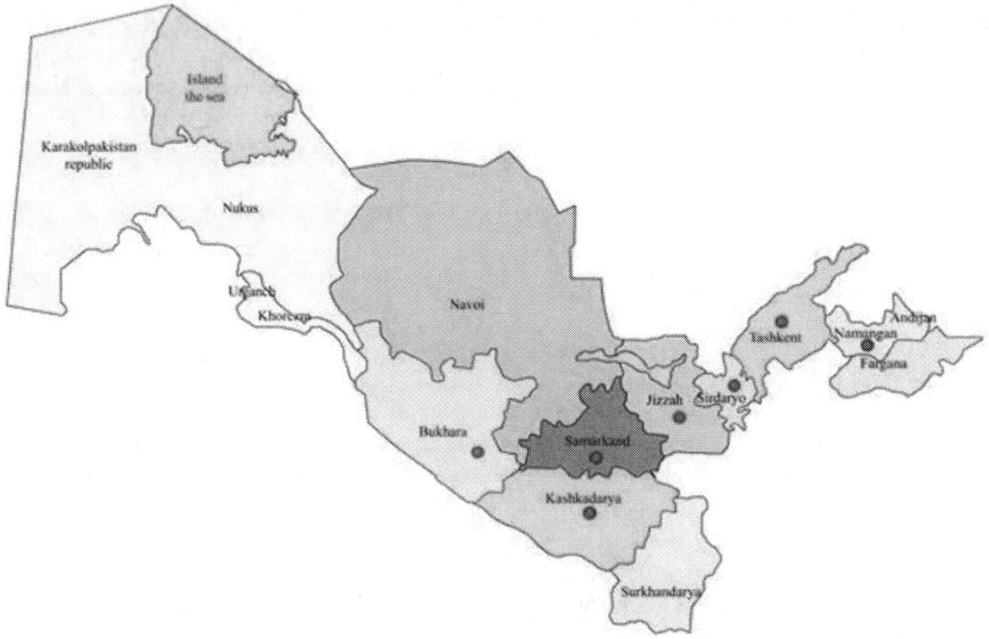

Figure 1. Sites of material collection for helminthological sumples for this examination.

The taxonomic affiliation of the species was determined mainly on the basis of the morphology and morphometry of male individuals [18].

*DNA Isolation***:** To extract genomic DNA from nematode samples, 20 µl of NaOH (0.25M) was added to each probe, kept at room temperature for 12 hours, and then heated to 95°C for 3 minutes, samples were added to 10 µl of tris-HCl (0.25 M) and centrifuged at 10,000 rpm for 2 minutes. Samples are removed from the centrifuge, 4 µl of HCl (1:15) is added, vortexed and centrifuged again. Then, 5 µl of triton (2%) is added. Samples were heated at 95°C for 3 minutes and then stored in a freezer at -20°C.

Table 1. List of collected helminthological samples from domestic ruminants of Uzbekistan.

Nematode species	Regions	Hosts
Haemonchus contortus	Namangan, Bukhara	Sheep, cattle
H. placei	Namangan	Sheep
Ostertagia ostertagi	Kashkadarya, Namangan	Sheep, cattle
O. lyrata	Kashkadarya, Namangan	Sheep, cattle
Teladorsagia circumcincta	Jizzakh, Bukhara, Samarkand	Sheep, cattle, sheep
T. davtiani	Jizzakh, Bukhara, Syrdarya	Sheep, sheep, sheep
T. trifurcata	Jizzakh, Bukhara	Sheep, sheep
Marshallagia marshalli	Samarkand, Bukhara, Syrdarya	Sheep, cattle, sheep
M. occidentalis	Samarkand, Bukhara, Tashkent	Sheep, cattle, sheep

PCR-amplification: The ITS2 region of nematode rDNA was amplified using NC1 and NC2 primers used in molecular taxonomy of nucleotide fragments (Gasser et al., 1993). PCR was performed according to the following scheme: 1 – step – DNA denaturation at 95°C for 3 minutes, 2 – step – DNA denaturation at 93°C for 20 seconds, 3 – step – DNA at 55°C for 30 primers sit on the template for seconds, step 4 - elongation at 72°C for 2 minutes, step 5 - chain elongation at 72°C for 10 minutes. From the second to the fourth step, the process was repeated up to 35 times per cycle.

Gel electrophoresis: The presence of DNA fragments in PCR products was determined by electrophoresis on a 1.5% agarose gel at 100 V. The 100 bp DNA ladder were used for DNA fragment size determination. The 250 bp length amplicons were documented using a transilluminator gel-documentation system. Amplicons were isolated from the gel using a set of reagents produced by "Silex M" (Moscow, Russia) following the manufacturer's instructions.

The nucleotide sequence of the rDNA ITS-2 region of species belonging to the genus *Teladorsagia* (https://www.ncbi.nlm.nih.gov/tools/primer-blast) primer sequences were selected for representatives of this generation and these primers were synthesized at SINTOL (Moscow, Russia).

Bioinformatic analyses: The of the results of morphological and morphometric data were statistically analyzed based on different programs [19, 20] and data obtained based on molecularresults were performed based on bioinformatics programs BioEdit, Slustalx, Genedoc, Paup4 and Geneious.

3 Results

Morphological research results: According to the results of the conducted helminthological research, more than 100 permanent and temporary preparations were prepared in laboratory conditions in order to clarify the morphological and morphometric dimensions of representatives of the genus *Teladorsagia*, the morphological and morphometric dimensions of the three species were clarified (Figure 2).

Teladorsagia circumcincta (Stadelmann, 1894)

Morphology. Male - length 10.5-16 mm. The cross section of the body is 0.16-0.21 mm. The length of the esophagus is 0.65-0.82 mm. The bursa is well developed, 0.31-0.59 mm long. The length of two equal spicules is 0.35-0.51 mm. Gubernaculum is racket-shaped, 0.01-0.011 mm long (Figure 2A).

The female is 9.8-17.4 mm long and 0.14-0.21 mm wide. The neural crest is located at a distance of 0.48-0.62 mm from the front part of the body. The length from the top of the head to the nape of the neck is 0.28-0.65 mm. The vulva is located 1.8-2.1 mm from the end of the tail. The length of the egg is 0.04-0.09 mm.

Teladorsagia trifurcata (Ransom, 1907)

Morphology. Male - 5-10 mm long. The width of the body in front of the bursa is 0.10-0.21 mm. The length of the esophagus is 0.35-0.75 mm. Nerve node at a distance of 0.47-0.82 mm from the head of the body. The bursa is well developed, 0.20-0.47 mm long. The length of the spicule is 0.11-0.35 mm. The length of the gubernaculum is 0.05-0.018 mm (Figure 2B).

The female is 8-15 mm long and 0.11-0.25 mm wide. The neural crest is located at a distance of 0.35-0.77 mm from the front part of the body. The length from the top of the head to the nape of the neck is 0.40-0.95 mm. The vulva is located 1.6-1.9 mm from the end of the tail. The length of the egg is 0.06-0.011 mm.

Teladorsagia davtiani (Andreeva et Satubaldin, 1954)

Morphology. Male - length 10.0-11.5 mm. The width of the body in front of the bursa is 0.12-0.15 mm. The length of the esophagus is 0.063-0.095 mm. Nerve node at a distance of 0.36-0.79 mm from the head of the body. The bursa is well developed, 0.39-0.46 mm long. The length of the spicule is 0.21-0.26 mm. The length of the gubernaculum is 0.107-0.110 mm. (Figure 2C).

The female is 11-14.5 mm long and 0.165-0.195 mm wide. The neural crest is located at a distance of 0.50-0.70 mm from the front part of the body. The length from the top of the head to the nape of the neck is 0.050-0.085 mm. The vulva is located 1.8-2.6 mm from the end of the tail. The length of the egg is 0.05-0.011 mm.

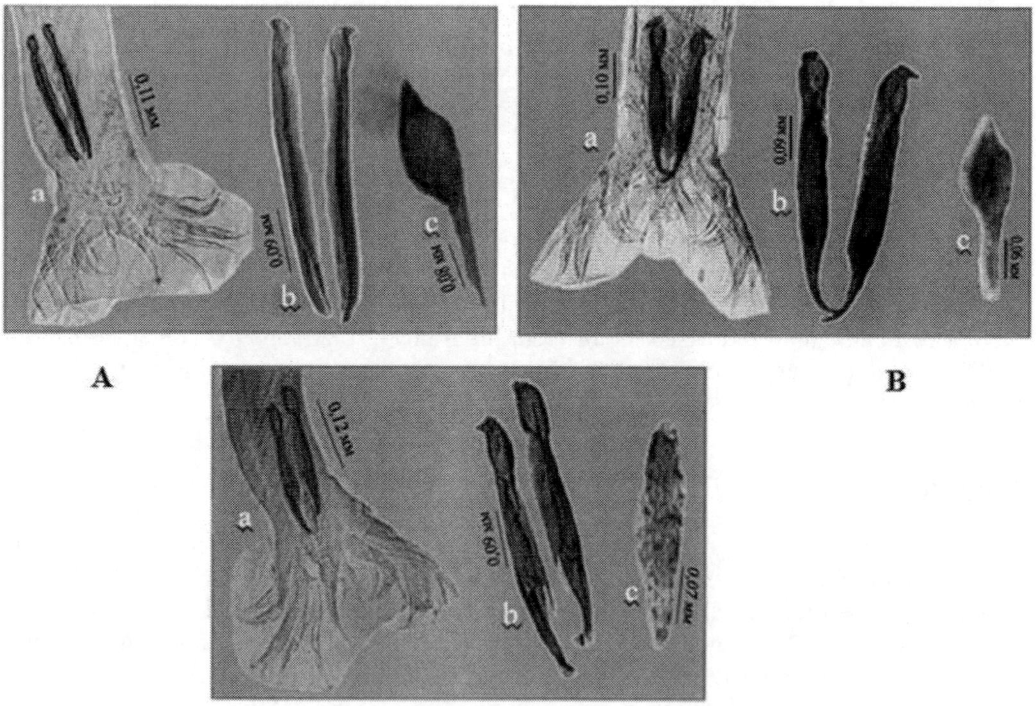

Figure 2. Morphological features of nematodes of the genus *Teladorsagia* Andreeva et Satubaldin, 1954. A - *T. circumcincta*, B - *T. trifurcata*, C - *T. davtiani; Note in the figure:* a- bursa, b-spicule, c - gubernaculum.

As a result of the conducted morphological research, the following was determined on the morphology of male individuals of *Teladorsagia circumcincta, T trifurcata* and *T davtiani* species belonging to the genus *Teladorsagia*:

The dorsal rib of *T circumcincta* is rather long and thin, while that of *T davtiani* is thick and relatively short. Gubernaculum *T circumcincta* has an elongated cylindrical shape, while in *T davtiani* it is long and the lower part is thinned and thickened towards the middle part. In *T. circumcincta*, the distal part of the main spicule is truncated and bluntly curved.

In, the distal part of the main spicule ends like a pointed knob, relatively long and thin. In addition, some differences were observed in the length of their body, spicule and gubernaculum.

Primer design for PCR and check them for specificity. We was create the design of specific primers that allow to identify the *Teladorsagia* species from representatives of other genera and based on the results of the molecular study of the rDNA ITS-2 [3] and the data nucleotide sequences from GenBank (NCBI) with use Primer-Blast (Table 2).

Table 2. Description of species-specific primers for the ITS-2 region of the rDNA of nematodes of the genus *Teladorsagia*.

Samples	Sequences 5` → 3`	Lengths	Amplicon length	Tm
Tel_1f	ATAACACTGTTTGTCGAATGGCA	23	100	59,18
Tel_2f	ACATGACGGTACGACGGTAG	20	100	55
Tel_3f	ACACTGTTTGTCGAATGGCATTTA	24	100	59,7
Tel_4f	ATAACACTGTTTGTCGAATGGC	22	100	57,5
Tel_5f	TGCAACATGACGGTACGACG	20	100	55
Tel_6f	ATATGCAACATGACGGTACGACGGT	25	153	59,8
Tel_7f	AACATGACGGTACGACGGTAG	22	100	55
Tel_8f	TGGCATTTATCACTTCATTGTGGT	24	100	59,2
Tel_9f	CACTGTTTGTCGAATGGCATTTA	23	100	58,2
Tel_10f	AACATGACGGTACGACGGTA	20	100	58,8

In order to check the newly created primers, the genomic DNA of species belonging to the genus *Telodorsagia* collected from the digestive system of animals in different regions of the Republic of Uzbekistan and were isolated genome DNA from species close to this parasites and were PCR amplified on the basis of these primers.

The PCR was performed using primers designed and primers NC2 and performed in the steps outlined above. The obtained PCR products with scesies-specific primers were checked by electrophoresis in a 1.5% agarose gel with a voltage of 130 V (Figure 3).

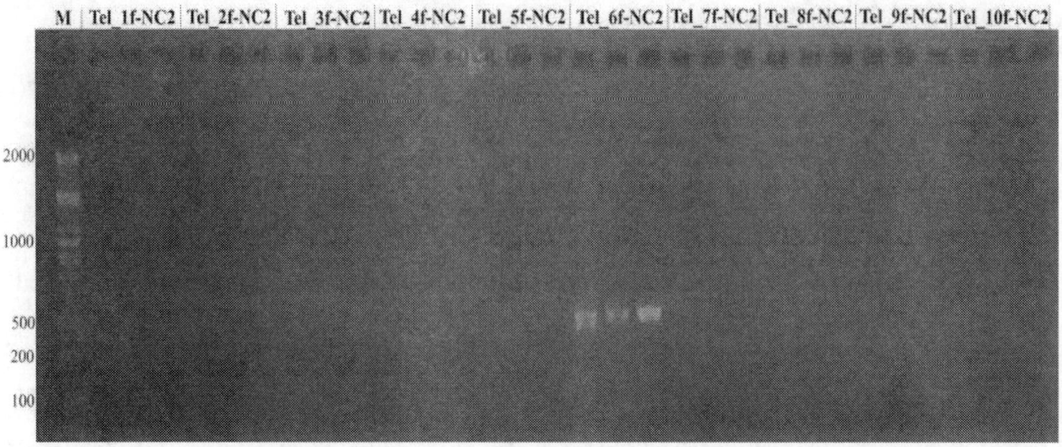

Figure 3. View of the gel electrophoresis of PCR products with primers designed for the three morph species of the genus *Teladorsagia* (6 line primers is Tel_6f-NC2).

According to the electrophoresis results there were DNA bands on the agarose gel with primer Tel_6F (5' ATATGCAACATGACGGTACGACGGT 3') and it was well created for the species of the genus Telodorsagia from other 10 primers (see Figure 3).

To check Tel_6F and NC2 primers was performed PCR from genomic DNA species of *Haemonchus contortus, H. placei, Ostertagia ostertagi, O. lyrata, Marshallagia marshalli, M. Occidentalis, Teladorsagia circumcincta, T. davtiani* and *T. trifurcata* (Figure 4)

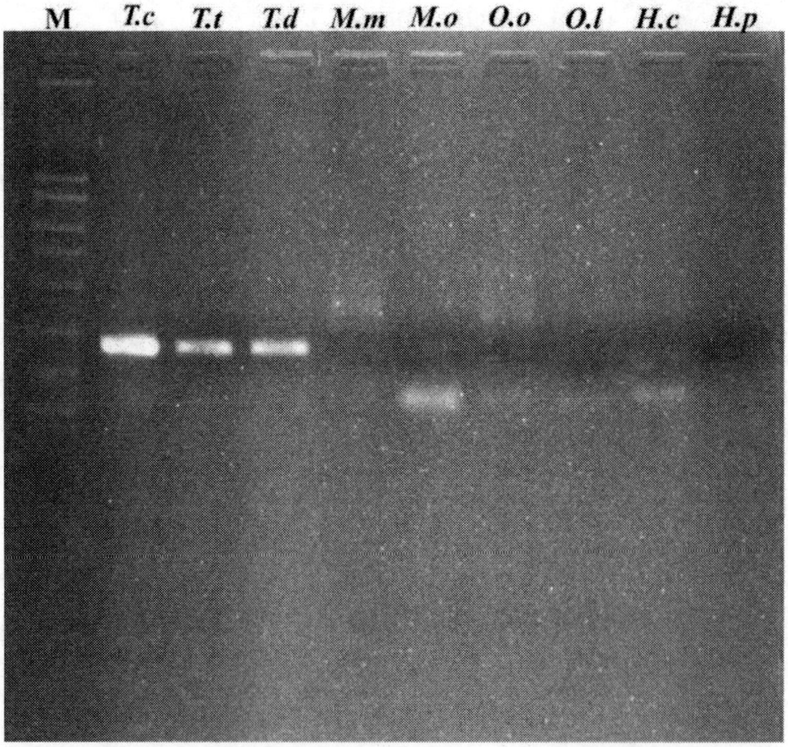

Figure 4. Results of validation of the species-specific primers created for the morphspecies of the genus *Teladorsagia* and other species: Note: *T.c-Teladorsagia circumcincta; T.t-T. trifurcata; T.d-T. davtiani; M.m-Marshallagia marshalli; M.o-M. occidentalis; O.o-Ostertagia ostertagi; O.l-O. lyrata; H.c-Haemonchus contortus; H.p-H. placei.*

According to the PCR results with of the Tel_6F primer was determined to be specific only for morphspecies *T. circumcincta, T. davtiani* and *T. trifurcata* of the genus *Telodorsagia*. The obtained PCR products were sequenced and the created primer was placed (Figure 5).

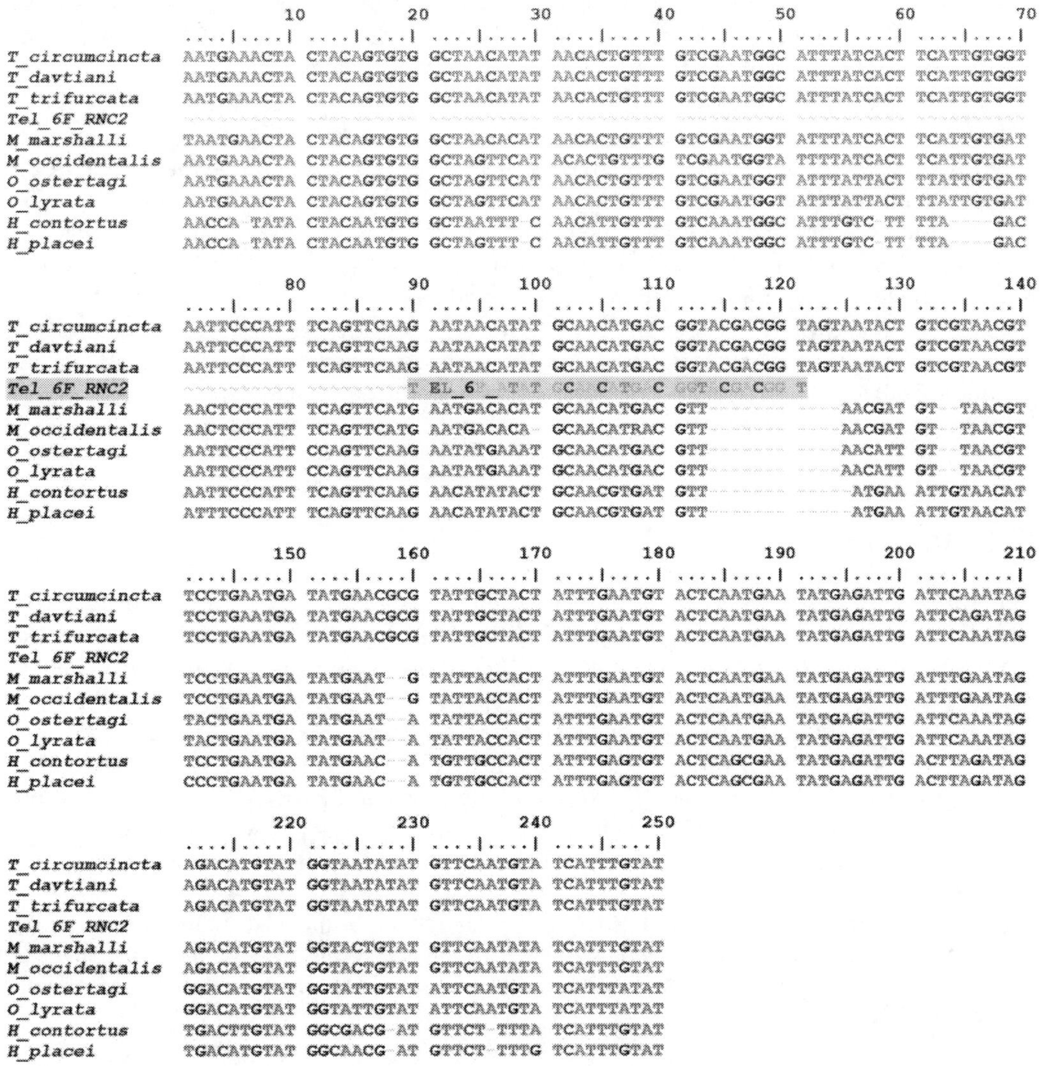

Figure 5. Location of the primer Tel_6f generated for the genus *Teladorsagia* based on sequence material.

As can be seen from Figure 5 above, the nucleotides of the ITS2 region of the rDNA of the species belonging to the genus *Teladorsagia* and the nematode species close to this genus and common in the digestive system of ruminants in our country and the nucleotides of the Tel_6f primer created for the genus *Teladorsagia* were studied. The Tel_6f primer created for species of *T. circumcincta, T. trifurcata* and *T. davtiani* of the genus *Teladorsagia* rDNA was found to read 153 bp of nucleotides belonging to the ITS2 region.

4 Discussion

L.A.Stevenson et al. [17] studied the ITS-2 sequences of *T. circumcincta, T. trifurcata* and *T. davtiani* and 2 other members within the same subfamily Ostertagiinae were examined. Although some sequence variation was detected between and within single worms of each taxon of *Teladorsagia*, no unequivocal base differences were detected among their consensus ITS-2 sequence. In comparison, there was 9 % sequence difference between *O. ostertagi* and *O. leptospicularis* and 13-15 % difference between the genera *Teladorsagia* and *Ostertagia*. These findings indicate that T. *circumcincta, T. davtiani* and *T. trifurcata* represent a single species, *T. circumcincta*.

Previously, we also comparative studies of three species *T. circumcincta, T. trifurcata* and *T. davtiani*, which are preliminarily different morphs of the same species, have been carried out. It has been established that there were insignificant differences in morphological sizes and characters between males but of the nucleotide fragments of ITS-2 rDNA of three species no differences were detected [9].

The advantage of DNA-based identification of parasitic nematodes is that it is time-consuming, allowing identification of species at any developmental stage [6-9], however continuing advances have refined the techniques and allowed savings in both time and cost [10]. In addition, the PCR diagnostic method can accurately diagnose parasitic nematodes in ruminants, leading to the correct use of anthelmintic drugs [7].

5 Conclusions

According to the results of the morphological study of species *T. circumcincta, T. trifurcata* and *T. davtiani* belonging to the genus *Teladorsagia,* species significant changes in morphological features and size were noted; According to the results of the molecular study of these species, when the nucleotides of the ITS2 region of the rDNA of the 3 studied species were compared, these species were *T. circumcincta, T. trifurcata* and *T. davtiani* is the basis for considering species as a single species; The Tel_6f primer created for morphspecies of *T. circumcincta, T. trifurcata* and *T. davtiani* was found to read 153 bp of nucleotides belonging to the ITS2 region rDNA and using this primer it will be possible to accurately diagnose the eggs and larvae of the species in the ruminant animals.

We thank the staff of the laboratory of Molecular zoology of the Institute of Zoology of the Academy of Sciences of Uzbekistan for the opportunity to use all the technical equipment and favourable working conditions in the preparation of the article. We also thank the reviewers for their valuable comments.

References

[1]. Abramatov M., Kuchboev A., Ruziev B., Sobirov Kh. Diversity of Gastrointestinal Nematodes in Domestic Ruminants of Uzbekistan. Pakistan Journal of Zoology. 54(5) P.2445-2448. (2022).

[2]. Amirov O.O., Kuchboev A.E., Sobirova H.G., Karimova R.R., Omonov S.N. Effect of Plant Extracts on the Gastrointestinal Nematodes of Ruminants in Uzbekistan. Advances in Animal and Veterinary Sciences, 9(9); P.1396-1400. (2021).

[3]. Dobson R.J., Barnes E.H., Birclijin S.D., Gill J.H. The survival of Ostertagia circumcincta and Trichostrongylus colubriformis in faecal culture as a source of bias in apportioning egg counts to worm species. Int. J. Parasitol. 22, 1005-1008. (2002).

[4]. Gasser R.B, Chilton N.B, Hoste H, Beveridge I. Rapid sequencing of rDNA from single worms and eggs of parasitic helminths. Nucl. Acids Res., 21: 2525-2526. (1993).

[5]. Gilleard, J.S. Understanding anthelmintic resistance: the need for genomics and genetics. Int. J. Parasitol. 36, 1227-1239. (2006).
[6]. Ivashkin V.M., Kontrimavicius V.N., Nazarova N.S (1971). Methods for collecting and studying helminths of terrestrial mammals. Moscow: Nauka, 124 p.
[7]. Ivashkin V.M., Oripov A.O., Sonin M.D (1989). The Determinant of helminths of sheep and goats. Moscow: Publisher Nauka, 254 p.
[8]. Kaplan R.M. Drug resistance in nematodes of veterinary importance: a status report. Trends Parasitol. 20, 477-481. (2004).
[9]. Kuchboev A., Amirov O., Soliyeva G., Karimova R., Sabirov H., Abramatov M., Kakhramanov B. Morphological and molecular identification three species *Teladorsagia* (Nematoda: Trichostrongylidae) in sheep, Uzbekistan. IOP Conference Series: Earth and Environmental Science, 939; 012078. P.1-7. (2021).
[10]. Kuchboev A., Sobirova K., Karimova R., Amirov O., Samson-Himmelstjerna G., Krucken J Molecular analysis of polymorphic species of the genus *Marshallagia* (Nematoda: Ostertagiinae). Parasites Vectors, 13, 411. (2020). doi.org/10.1186/s13071-020-04265-1
[11]. Kuchboev A.E, Krücken J. Prevalence, Infection Intensity and Molecular Diagnosis of Mixed Infections with *Metastrongylus* spp. (Metastrongylidae) in Wild Boars in Uzbekistan. *Pathogens*. 11(11):1316. (2022) doi.org/10.3390/pathogens11111316
[12]. Kuchboev A.E., Krucken J., Ruziev B.H., von Samson-Himmelstjerna G. Molecular phylogeny and diagnosis of species of the family Protostrongylidae from caprine hosts in Uzbekistan.Parasitology Research. 2015, 114 (4). - P 1355-1364.
[13]. Amirov O.O., Karimova R.R. Differential DNA diagnostics of different stages of the life cycle of nematodes - the causative agents of animal protostrongilosis using the polymerase chain reaction (PCR) method. Patent for an invention. No. IAP 05381. 02.03.2017.
[14]. Ljungström S, Melville L, Skuce PJ, Höglund J. Comparison of four diagnostic methods for detection and relative quantification of *Haemonchus contortus* eggs in feces samples. Front Vet Sci, 4:1-7. (2018).
[15]. Roeber F, Jex AR, Campbell AJ, Campbell BE, Anderson GA, Gasser RB. Evaluation and application of a molecular method to assess the composition of strongylid nematode populations in sheep with naturally acquired infections. Infect Genet Evol, 11:849–854. (2011).
[16]. Sackett D., Holmes P (2006). Assessing the economic cost of endemic disease on the profitability Australian beef cattle and sheep producers. Meat and Livestock Australia Limited, Sydney, Australia. ISBN: 1741910021.
[17]. Stevenson L.A., Gasser R.B., Chilton N.B. The ITS-2 rDNA of *Teladorsagia circumcincta, T. trifurcata* and *T. davtiani* (Nematoda: Trichostrongylidae) indicates that these taxa are one species. International Journal for Parasitology, 26: 1123-1126. (1996).
[18]. Von Samson-Himmelstjerna G. Molecular diagnosis of anthelmintic resistance. Vet Parasitol 136:99-107. (2006).
[19]. Wolstenholme A.J., Fairweather I., Prichard R., von Samson-Himmelstjerna G., Sangster N.C. Drug resistance in veterinary helminths. Trends Parasitol., 20, 469-476. (2004).
[20]. Zarlenga D.S., Chute M.B., Gasbarre L.C., Boyd P.C. A multiplex PCR assay for differentiating economically important gastrointestinal nematodes of cattle. Vet. Parasitol. 97, 199-209. (2001).

Effectiveness of wastewater use in rice cultivation on basis of water-saving technologies

Kupalboy Yuldashev[1,*], *Zokirjon* Tajiev[1], *Sulaymon* Buriev[2], *Khayrulla* Dusov[1], *Shakhnoza* Zakirova[1], *Poshshajon* Shomuratova[1], *Gulnoza* Nurmetova[1] and *Gulnaz* Saparova[3]

[1]Urgench State University, Urgench, 220109, Uzbekistan
[2]Bukhara State University, Bukhara, 200101, Uzbekistan
[3]Tashkent State Agrarian University, Tashkent, 100140, Uzbekistan

Abstract. This article covers the results of the scientific research, conducted in the conditions of Khorezm region (Uzbekistan) on cultivation of rice on the basis of seedling as a second crop in the fields after winter wheat. As an introduction of resource conservation technologies in agriculture for cultivation of rice biologically treated water with the help of water hyacinth (*Eichhornia crassipes*) were used in the experiments. Saving irrigation water concerning the purification of waste water from urban public water supply is one of the purposes of the research. According to the research results, while analyzing the physical and chemical content of waste water in the pool with water hyacinth, it was determined that its composition was changed, the water temperature increased from 25^0C to 29^0C, the medium changed from acidic to alkaline (6.7-8.2) the color changed from reddish to whitish, the smell disappeared (5.0-0.0), the amount of suspended matters decreased (29.0-0.0 mg/l), the amount of oxygen increased (11.2 mg/l), 5 day biochemical consumption of oxigen (BCO_5) decreased (71.4 -10.1 mg/l), the amount of O_2 in the oxidation process decreased from 58.2 to 9.2 mg/l and ammonia, nitrites, and nitrates disappeared, besides chlorides decresed up to 40.5-15.3 mg/l, sulfates decreased up to 38.5-12.5 mg/l, phosphates decreased up to 4.2-0.5 mg/l, plant biomass increased from 200 mg/l to 1460 mg/l. At the same time, it was determined that due to the use of wastewater, up to 9250 m^3 irrigation water was saved and crop yield was increased to 73 quintals per hectare.
Keywords. Water hyacinth (Eichhornia crassipes), wastewater, biological treatment, reproductive phase, vegetative phase, Alanga.

1 Introduction

Recent increase of water shortage due to the increase in the population in the world and economic development requires the introduction and implementation of water-conservation

* Corresponding author: kopalboyyoldashev@gmail.com

technologies in the cultivation of all crops, including rice. By reusing of treated wastewater in the introduction of resource conservation technologies in agriculture the amount of irrigation water and use of mineral fertilizers can be decreased, whereas crop productivity can be increased and more expenses can be saved in rice cultivation [1-3].

In rice cultivation in the world 80-90% seedling method is used. Taking into account the soil-climatic conditions and the biological characteristics of the varieties, improving the agrotechnics of cultivation of local and foreign rice varieties and hybrids, increasing their yield capacity, introduction of the world's advanced technologies, and effectively using each hectare of land are the urgent issues of rice cultivating in Uzbekistan [4-6] .

Khorezm region of Uzbekistan has a complex continental climate, with a very high level of water mineralization. This creates difficulties in various branches of agriculture and animal husbandry. The main kinds of agricultural crops is sown by irrigated farming. That is why the annual water consumption of the region is very high [7-10].

Hence it is important to develop the possibilities of wastewater treatment released from various enterprises in the region and reuse in irrigation of agricultural crops with the purpose of decreasing the amount irrigation water. The development and application of special methods of wastewater treatment and disinfection, as well as their use in agriculture, are of great importance [11, 12].

Adebayo et al. [7], Ashiqun Nabi et al. [6] and other scientists in the world carried out researches on the analysis of the composition of wastewater and their biological treatment and achieved important results.

A number of scientific researches also have been carried out on the study of the biological treatment of wastewater from industrial and agricultural enterprises, communal-household, water-waste enterprises and oil processing factories in Uzbekistan with the assistance of water hyacinth and other aquatic plants. And the obtained results were elucidated in the works of several local and national scientists. In the conditions of Khorezm region, scientific research in this field has not been carried out yet.

2 Materials and methods

The main idea of the research is to treat wastewater with biological methods and reuse of the treated water for irrigation of agricultural crops. Thus, experiments were conducted on biological treatment of wastewater from communal-household in Urgench city of Khorezm region and using the treated water for rice cultivation.

In the experiment for biological treatment of wastewater water hyacinth (*Eichhornia crassipes*) from the group of higher plants were used, and for growing in the treated wastewater 'Alanga' rice variety was chosen.

Tropical regions of South America is considered as homeland of the genus Eichhornia (*Hyacinth*), and it is a floating plant that grows semi-submerged in water. The stem of Eichhornia plants grows 10-20 centimeters in height, sometimes it can reach 1 meter in favorable conditions [7]. It is a semi-submerged, upright floating perennial aquatic plant. Thick shiny leaves can be 12-15 centimeters wide and 30-50 centimeters long. The length of the roots can be 50 - 60 centimeters and even more [8]. The most common type is thick-stemmed eichhornia (*Eichhornia crassipes Solms*). Our experiments were conducted on this type (Figure 1). This plant is adapted to the conditions of Uzbekistan [3].

Figure 1. Aerial photo of the biological ponds of "Khorezm water supply" Ltd.

The rice variety of 'Alanga' was created at the Rice Research Institute of Uzbekistan. It is a hybrid variety by origin. Authors: Pulina P.A., Rikhsieva S. Since 1993, it has been included in the list of Surkhandarya, Syrdarya, Toshket regions. The ear is awned, strongly curved and sparse. The awn is weakly wavy, extending from the bottom to the top. The awned grain makes up 85.0-90.0%, grain semi-round shaped, medium size, clear, white. The weight of 1000 grains is 29.0-30.0 g. The variety is medium-ripening, the vegetation period is 100-118 days. The technological and grain quality of the variety is good: clarity is 71.0-79.0%, rice outcome is 67.0-68.0%, whole rice content is up to 81.0%. Nutrition quality is good. The variety is resistant to lodging and shedding [1].

In order to assess the productivity of plants, recommendations by Katanskaya V.M. were used [4]. Growth, development and biomass of the water hyacinth in wastewater was determined by weighing on scales after every 3, 7, 14, 28, and 30 days. The physical and chemical composition of wastewater, before and after sowing of the plants, were determined according to the general hydrochemical methods by Lure and Strogonov [5, 6]. In determining the discharge of pure and waste water the Cipoletti Weir was used. The weight of the rice grain was determined by weighing on scale.

3 Results and discussion

Firstly, in the experiments growth, development and biomass accumulation of water hyacinth and level of wastewater treatment from urban communal-household were observed during 30 days (Table 1).

Table 1. Changes in physical and chemical composition of wastewater from communal-household of Urgench city before and after sowing of water hyacinth (*Eichhornia crassipes*) (30 days).

#	Indicators	Before experiments	After experiments
		Wastewater composition	Wastewater composition
1	Temperature, °C	25,0	29,0
2	pH	6,7	8,2
3	Smell, points	5.0	NA
4	Color	Reddish	Whitish
5	Suspended matters, mg/l	29,0±1,2	NA
6	Dissolved oxygen in water, mg/l	NA	11,2
7	BCO_5, mgO_2/l	71,4	10,1
8	oxidation, mgO_2/l	58,2	9,2
9	ammonia, mg/l	3,0	NA
10	nitrites, mg/l	0,06	NA
11	nitrates, mg/l	3,0	NA
12	chlorides, mg/l	40,5	15,3
13	sulfates, mg/l	38,5	12,5
14	phosphates, mg/l	4,2	0,5
15	plant biomass, g/m^2	200	1460

In this case, it was determined that the physico-chemical composition of the wastewater changed, the temperature of the water rose from 25^0 C to 29^0 C, the water medium changed from acidic to alkaline (6.7-8.2) the color changed from reddish to whitish, the smell disappeared (5.0-0.0), the amount of suspended matters decreased (29.0-0.0 mg/l), the amount of oxygen increased (0.0-11.2 mg/l), 5 day biochemical consumption of oxigen (BCO_5) decreased (71.4-10.1 mg/l), the amount of O_2 in the oxidation process decreased from 58.2 to 9.2 mg/l, in addition, ammonia, nitrites, and nitrates disappeared, besides chlorides decresed up to 40.5-15.3 mg/l, sulfates decreased up to 38.5-12.5 mg/l, phosphates decreased up to 4.2-0.5 mg/l, plant biomass increased from 200mg/l to 1460 mg/l.

Scientific research was continued and experiments were carried out on the use of wastewater treated with the assistance of water hyacinth for rice cultivation. Researches were carried out in the biological ponds of the water-waste enterprise belonging to Khorezm water supply Ltd. and in the farm fields located near the enterprise.

Experiments were conducted in 3 variants. In the first variant, water was given based on the development phases of rice under normal conditions. The growth period of rice is divided into 3 phases, 1-vegetative phase, 2-reproductive phase, 3-ripening phase. In turn, each phase consists of 2 periods. The vegetative phase consists of growth and stem elongation, the reproductive phase consists of stem elongation and flowering, and the ripening phase consists of milk development and full ripening. The duration of all 3 phases includes certain days. The growth period is 20 days, the flowering period is 15-20 days, the stem elongation period is 15-20 days, the flowering period is 30 days, the milk development period is 15 days, and the full ripening period is 15 days. The total vegetation of rice requires 115-120 days (in the variety 'Alanga') and the water requirement varies in each phase.

Based on this, in the first variant, i.e. in the control, during the growth period of the vegetation phase 3500 m^3, during the tillering period 2500 m^3, in total 6000 m ^{3} water per hectare was given. During the stem elongation period of the reproductive phase 4500 m^3, during the flowering period 4000 m^3, in total 8500 m^3 water per hectare was used for irrigation. And during the milk development period of the ripening phase 3000 m^3, and during the full ripening period 1000 m^3, in total 4000 m ^{3} water per hectare was given. At the end of the experiment, after ripening of rice, the amount of fresh water used in this variant made up 18500 m^3 per hectare and the crop yield was 64 t/ha.

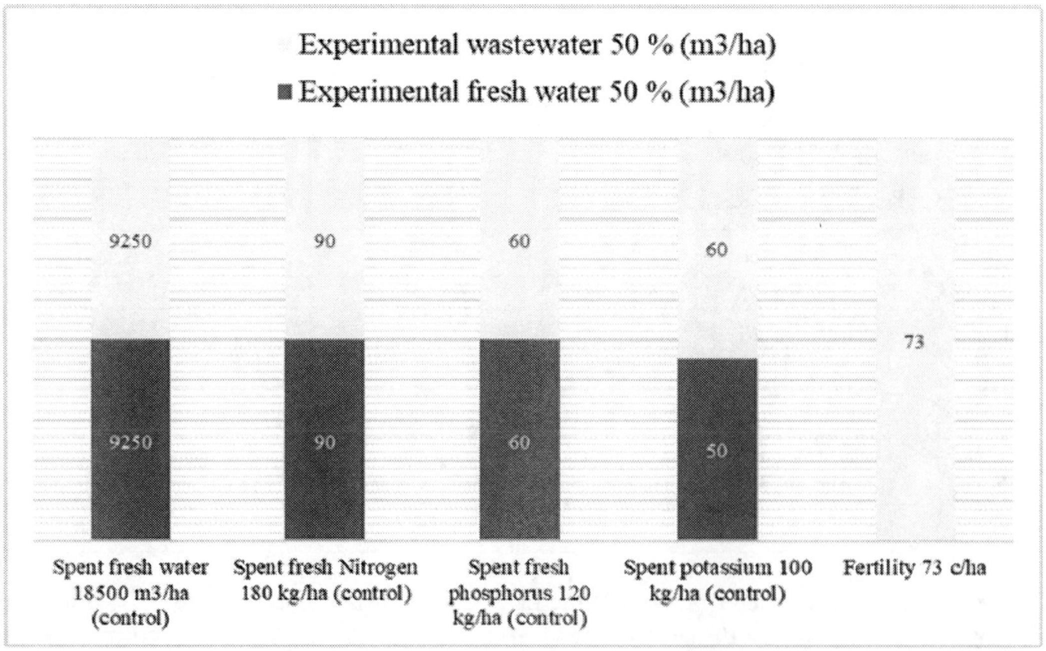

Figure 2. Comparative analysis of the amount of fresh water and impurities when rice is grown in clean water (50%) and wastewater (50%).

In the second variant of the experiment, scientific studies were carried out on rice cultivation by adding 25% wastewater to the fresh water. In this case, during the growth period of the vegetation phase 2625 m^3 of fresh water and 875 m^3 of waste water per hectare was given. During the period of tillering 1875 m^3 of fresh water, 625 m^3 of wastewater, and in total 4500 m^3 of fresh water and 1500 m^3 of wastewater per hectare was used in this phase. In the stem elongation period of the reproductive phase 3375 m^3 of fresh water, 1125 m^3 of wastewater, and in the flowering period 3000 m^3 of fresh water, 1000 m^3 of wastewater, in total during this period 6375 m^3 of fresh water, 2125 m^3 wastewater was used per hectare. In the milk development period of ripening phase 2250 m^3 of fresh water, 750 m^3 of wastewater, and in the full ripening period 750 m^3 of fresh water, 250 m^3 of wastewater, and total for ripening phase 3000 m^3 of fresh water, 1000 m^3 of wastewater was used per hectare.

In this variant total amount of used water for the whole vegetation period made up 18500 m^3/ha, including 13875 m^3/ha fresh water and 4625 m^3/ha wastewater. Herein, consumption of clean water was reduced at the expense of wastewater by 4625 m^3/ha, and the crop yield was 69 quintals per hectare (Figure 2).

In the 3rd variant of the experiment, scientific research was continued on rice cultivation by adding 50% wastewater to the fresh water. In this case, during the growth period of the vegetation phase 1750 m^3 of fresh water, 1750 m^3 of wastewater, and during the period of tillering 1250 m^3 of fresh water, 1250 m^3 of wastewater, in total, 3000 m^3 of fresh water and 3000 m^3 of wastewater per hectare was used for irrigation. In the stem elongation period of the reproductive phase 2250 m^3 of fresh water, 2250 m^3 of wastewater, and during the flowering period 2000 m^3 of fresh water, 2000 m^3 of wastewater, in total, during this period 4250 m^3 of fresh water and 4250 m^3 wastewater was used per hectare.

In the period of milk development of the ripening phase 1500 m^3 of fresh water, 1500 m^3 of wastewater, and in the period of full ripening 500 m^3 fresh water, 500 m^3 wastewater, in total for this phase 2000 m^3 fresh water and 2000 m^3 of waste water per hectare was used. In this variant also for the entire growing season 18 500 m^3 water was used per hectare, including 9250 m^3/ha fresh water, and 9 250 m^3/ha wastewater. Herein, consumption of fresh water was reduced at the expense of wastewater by 9250 m^3/ha, and the crop yield made up 73 quintals per hectare (Tables 2 and 3).

Table 2. Annual water consumption of rice.

Variants	Development phases of rice (V-vegetative) (Rep-reproductive) (Ripe-ripening)		Duration	Annual water consumption, m^3/ha	
				On the development phases	Total
1. Control	V	Growth	20	3500	6000
		Tillering	15-20	2500	
	Rep	Stem elongation	15-20	4500	8500
		Flowering	30	4000	
	Ripe	Milk development	15	3000	4000
		Full ripening	15	1000	
2. Fresh water + wastewater (25 %)	V	Growth	20	2625+875	4500+1500
		Tillering	15-20	1875+625	
	Rep	Stem elongation	15-20	3375+1125	6375+2125
		Flowering	30	3000+1000	
	Ripe	Milk development	15	2250+750	3000+1000
		Growth	15	750+250	
3. Fresh water + wastewater (50 %)	V	Tillering	20	1750+1750	3000+3000
		Stem elongation	15-20	1250+1250	
	Rep	Flowering	15-20	2250+2250	4250+4250
		Milk development	30	2000+2000	
	Ripe	Full ripening	15	1500+1500	2000+2000
		Full ripening	15	500+500	

Table 3. Annual water consumption of rice grown in fresh and wastewater.

Variants	Total water used, m^3/ha		Total water saved, m^3/ha	Productivity, quintals per hectare
	Fresh water	Waste water	Fresh water	
1. Control	18500	-	-	64
2. Fresh water + wastewater (25 %)	13875	4625	4625	69
3. Fresh water + wastewater (50 %)	9250	9250	9250	73

4 Conclusions

When analyzing the results, in the first variant, the annual water consumption of rice under normal conditions was 18 500 m^3/ha, the yield was 64 quintals per hectare; in the second variant, the annual fresh water consumption was 13 875 m^3/ha with 4 625 m^3/ha of wastewater. In this variant the 4 625 m^3/ha of fresh water was saved, and the yield made up 69 quintals per hectare. Annual fresh water used in variant 3 was 9 250 m^3/ha and the amount of wastewater was 9 250 m^3/ha.

Thus, 9 250 m^3/ha of fresh water was saved and the yield made up 73 quintals per ha. It was found that the high yield of rice in the third variant of the experiment was achieved due to the organo-mineral substances in the mixed wastewater in equal amounts to the fresh water, which affected as an additional biofertilizer.

According to the research results, it has been practically proven that biologically treated wastewater can be used for irrigation of agricultural crops. The main point is that at the expense of saved fresh water and fertilizers, the expenses are reduced and net income increases.

References

1. "Agrobank" ATB.100 book collection. Cultivation of rice. Book 3. Tashkent. 2021. p.7.
2. Qutliev J. Service of biological ponds in wastewater treatment. Tashkent. 2009. p. 3-12.
3. Shoyakubov R.Sh., Khozhjiev S.O., Rakhimov J.A. Use of higher water plant – Eichhornia (*Eichhornia crassipes* Solms.) in wastewater treatment. Actual problems of molecular biology of plants: Proceedings of the international scientific and practical conference. Tashkent. 2008. p. 56-59.
4. Katanskaya V.M. Hihger water plants of continental water ponds (methods of learning) Kaliningrad. Nauka. 2011. p. 187.
5. Lure Yu.Yu. Analytical chemistry of industrial wastewater. Moscow. Chemistry. 2004. p. 446.
6. Strogonov N.S. Practical instructions for hydrochemistry. Moscow. 2000. p. 195.

7. Ashikun Nabi, A.K.M. Rashidul Alam, Sirajul Hoque. Treatment of wastewater with free floating aquatic macrophyte – Eichhornia crassipes L. Jahangirnagar University Environmental Bulletin. 2016. Vol. 5. p. 1-8.
8. Adebayo A.A., Briski E., Kalaci O., Hernandez M., Ghabooli S., Beric B. Water hyacinth (Eichhornia crassipes) and water lettuce (Pistia stratiotes) in the Great Lakes: playing with fire? Aquatic Invasions 6. 2011. p. 91-96.
9. TKh, M., Brundu, G., Jabborov, A. M., & Gaziev, A. D. (2023). Predicting the potential distribution of Ranunculus sardous (Ranunculaceae), a new alien species in the flora of Uzbekistan and Central Asia. BioInvasions Records, 12(1), 63-77. DOI:10.3391/bir.2023.12.1.05
10. Ameen, M., Zafar, M., Ahmad, M., Ramadan, M. F., Eid, H. F., Makhkamov, T., ... & Majeed, S. (2023). Assessing the Bioenergy Potential of Novel Non-Edible Biomass Resources via Ultrastructural Analysis of Seed Sculpturing Using Microscopic Imaging Visualization. Agronomy, 13(3), 735. DOI:10.3390/agronomy13030735
11. Noor, W., Zafar, M., Ahmad, M., Althobaiti, A. T., Ramadan, M. F., Makhkamov, T., ... & Khan, A. (2023). Petiole micromorphology in Brassicaceous taxa and its potential for accurate taxonomic identification. Flora, 303, 152280. DOI:10.1016/j.flora.2023.152280
12. Aziz, A., Ahmad, M., Zafar, M., Gaafar, A. R. Z., Hodhod, M. S., Sultana, S., ... & Chaudhay, B. (2023). Novel Copper Oxide Phyto-Nanocatalyst Utilized for the Synthesis of Sustainable Biodiesel from Citrullus colocynthis Seed Oil. Processes, 11(6), 1857. DOI:10.3390/pr11061857

Effect of abiotic stresses on the physiological characteristics of rice (*Oryza sativa*)

Yulduzoy Khojamkulova[1,*], *Muhabbat* Honkeldieva[2] and *Gulbahor* Choriyeva[2]

[1]Rice Research Institute, Tashkent, 100140, Uzbekistan
[2]Tashkent State Agrarian University, Tashkent, 100140, Uzbekistan

Abstract. This article conducts an analysis of water balance indicators for the "Lazurniy" and "Sadaf" rice varieties. Through this analysis, the study aims to uncover insights into the water requirements and responses of these two rice varieties. The research findings reveal distinct characteristics of each variety in terms of their water usage and yield performance. The analysis underscores that the "Lazurniy" variety falls within the category of intensive type varieties. This designation is associated with its ability to thrive and deliver high yields in environments characterized by abundant agricultural resources. On the other hand, the "Sadaf" variety exhibits a different trait - it showcases relative resistance to water shortage stress. This resilience suggests that the "Sadaf" variety has the capacity to maintain its performance even under conditions of limited water availability. By delineating these variations in water balance indicators, the research contributes to a deeper understanding of the adaptability and yield potential of both "Lazurniy" and "Sadaf" rice varieties. This information holds relevance for crop selection and management decisions, allowing growers to make informed choices based on the specific agro-environmental conditions they face. Ultimately, this analysis contributes to the optimization of rice cultivation practices by aligning crop varieties with their respective water requirements and stress tolerances.
Keywords. Rice, transpiration, water storage capacity, leaf dry weight.

1 Introduction

Due to the accelerated climate change process and population growth in recent years, water availability is a critical issue for agriculture, which is the world's largest water consumer, accounting for more than 70% of global water consumption. Irrigated agricultural crops cover 20% of the world's cultivated land and account for 40% of the world's food production. Irrigated agricultural crops produce at least twice the yield per unit of land compared to dryland [1]. Water usually contains 50-90% of a plant cell. Most of the water content (60-90%) is inside the cells, and the rest is mainly located in the cell walls [2]. Maintaining growth and productivity under unfavorable environmental stress conditions such as water scarcity is one of the main challenges in modern agriculture, including rice cultivation [3].

* Corresponding author: yulduzoyxojamkulova@gmail.com

Nowadays, global climate change is causing various natural disasters. Floods caused on by environmental factors reduced the production of 22 million hectares of rice-planted regions worldwide, including the rice crop. Half of them are in eastern India [4].

Although the rice plant requires a large amount of water for normal growth and development, excess water supply harms productivity. This is due to the fact that different submerged plant tissues have a difficult time transferring oxygen [5], which generates an adverse environment for the plant and results in an oxygen deficit state (O_2 < 21%) [6]. When water deficit stress occurs, cell membrane signaling processes are activated, thereby resisting the stress through physiological processes and the positive function of responsive genes. In this context, many mechanisms, photosynthesis and gas exchange [7, 8], cell degradation, changes in cell membrane composition [9], translocation of nutrients [10], transcriptional activity of genes, transposable elements [11], lipid signaling [12], metabolites, proteins [13] and antioxidants [14] can be altered under stress.

The above-mentioned imbalances that occur under water deficit stress direct the plant energy and nutrients to protective processes, resulting in a decrease in growth, development, generative organs and biomass formation.

The effect of soil drought stress during water deficit is the occurrence of an antioxidant defense system and disruption of reactive oxygen species (ROS) protection processes [15]. In addition, changes at the molecular level are responsible for the synthesis of proteins related to stress tolerance, such as stress-sensitive expressed genes, enzymes ascorbate peroxidase (APX), and high-temperature resistance proteins (HSP) [16].

2 Materials and methods

Our research was conducted in the field of the experimental farm of the Rice Research Institute. "Sadaf" and "Lazurniy" varieties were used as experimental material. Physiological indicators of water balance during the growing season of rice varieties were evaluated - the water content of the flag leaf, transpiration rate, water retention character, and dry weight of the leaf. In order to determine the total water content of plant leaves and the water retention properties of leaves, 5 leaves were taken from each option and weighed on an electronic scale, and placed under room conditions.

After 2 hours of exposure, their weight was measured again. Then dried in a drying cabinet at 100-105 °C for 3 days, the dry weight of the leaves was measured, and through this way, the total water content in the leaves and the water holding properties of the leaves were determined. Data obtained were analyzed statistically using ANOVA software (www.statview.com) [17]. According to L.A. Ivanov's approach [18] and A.A. Nichiporovich's research on the water retention capacity of leaves, transpiration activity was measured.

3 Results and discussion

In conditions of the optimal water supply of the plant, the high total water content in the leaves causes the formation of generative and vegetative organs at an optimal level. When the total water content of the flag leaf was studied among the rice cultivars, it was observed that there was no statistically significant difference. However, during water scarcity, the plant's minimal water reserve has a substantial impact on metabolism. In the results of the analysis, the highest indicator was observed in the variety "Sadaf" and the total water content was 94.3%. Accordingly, there was no significant difference in the "Lazurniy" variety, which was 94.2%. It is known that the optimal level of water supply in leaves ensures normal metabolic processes in it (Figure 1).

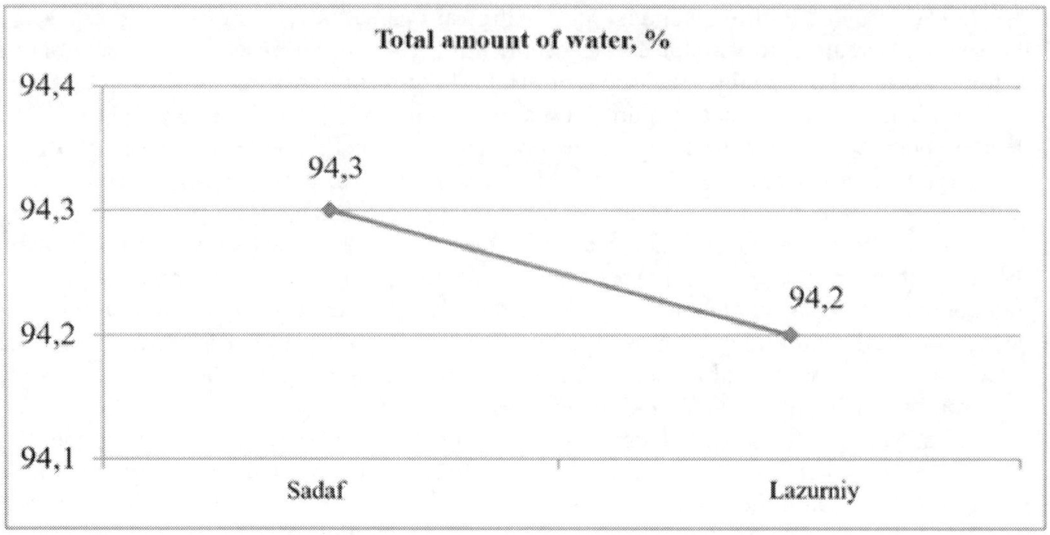

Figure 1. Total amount of water, %.

One of the key physiological factors, transpiration plays a crucial role in plants' water exchange. Transpiration is one of the main processes in plant water balance management. Transpiration rate is the content of water evaporated from a plant leaf in 1 hour (H_2O/1g wet leaf x 1 hour). In our experiments, it was found that there is a relative difference between varieties in terms of transpiration rate, while the "Sadaf" variety was 398.5 mg, and the "Lazurniy" variety was 335.5 mg. It is known that the decrease in the rate of transpiration in plants during water shortage is due to the decrease in the total water content and free water content of plant leaves, loss of turgor, and wilting of leaves (Figure 2).

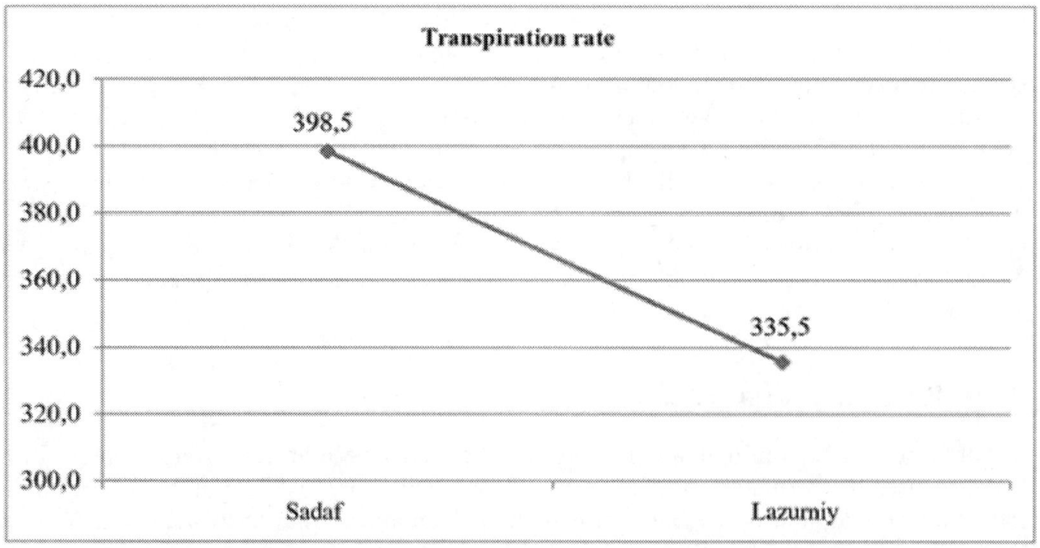

Figure 2. Transpiration rate.

The next indicator in our analysis results is the water retention property of the plant, which depends on various factors, soil moisture, and morphological, anatomical, biological, and genetic characteristics of the plant. Water retention refers to the percentage of water lost after two hours compared to the initial water content of the leaves. Therefore, a low index means a high water retention property of the leaf (Figure 3).

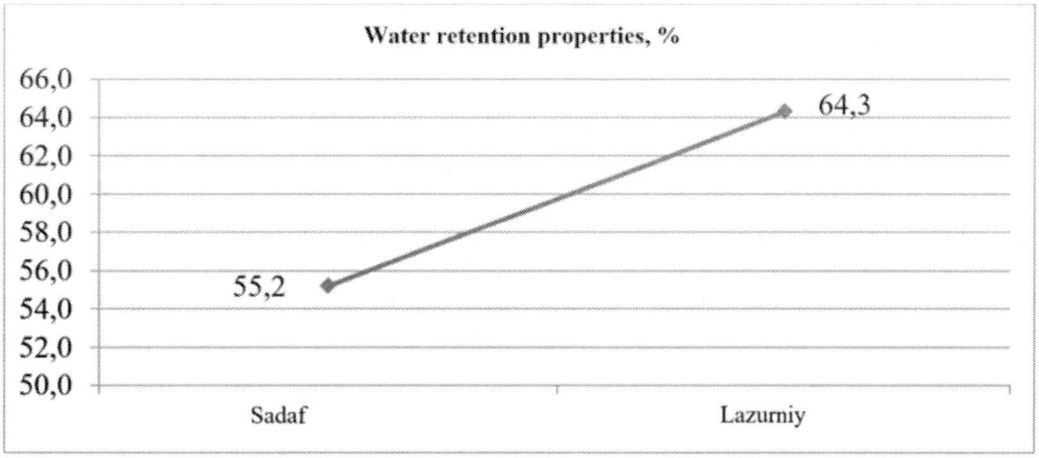

Figure 3. High water retention property.

In our analysis, it was noted that the water retention property of rice varieties is 64.3% in the "Lazurniy" variety, which is relatively low, and on the contrary, it is 55.2% in the "Sadaf" variety, which is high. The reason for the high total water content and transpiration speed of the "Sadaf" variety can be explained by the well-developed root system and the ability to adapt to the stress conditions of the external environment. In addition, the variety "Sadaf" is distinguished by the fact that the leaf surface is large. In our research, the increase in the water retention properties of the leaves of rice varieties during water shortage is evidenced by the high content of hard-to-dissociate fractions of water in the leaves under stress conditions.

The dry weight of various rice cultivars was examined in our subsequent research. The flag leaf weight of the "Lazurniy" variety was 14.1 mg according to the analysis's findings, while the weight of the "Sadaf" variety was 10.6 mg, which was less than that of the "Lazurniy" variety. The reason for the high dry weight of the leaf in the "Lazurniy" variety may be the process of photosynthesis and a large content of synthesis of organic substances delivered to all organs through the root (Figure 4).

The productivity indicator of a certain genotype, which is exhibited as a result of biochemical reactions, enzyme activity, and hormone level, as well as the speed of certain physiological processes in plant tissues, is represented by the accumulation of dry mass. Accumulation of dry substance in the plant depends on the speed of assimilation and synthesis of inorganic and organic substances as a result of its effective use. A positive correlation between productivity and accumulated dry biomass has been reported in Table 1.

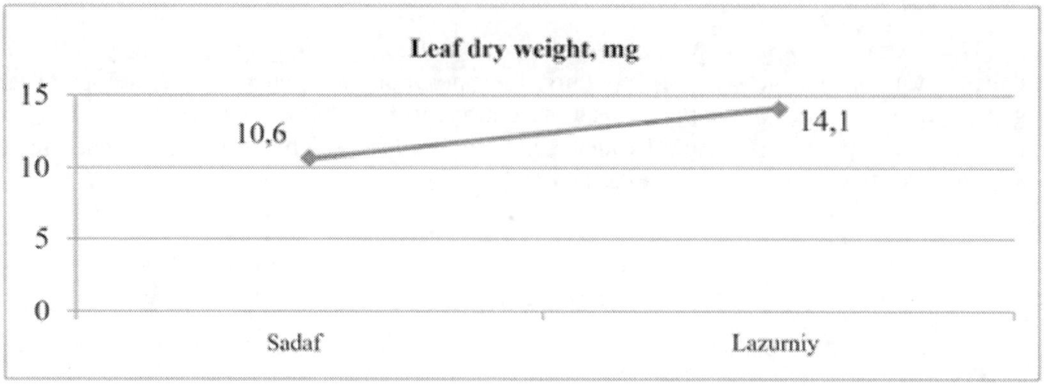

Figure 4. Leaf dry weight.

Table 1. Relationship between productivity and accumulated dry biomass.

Name of varieties	Total water content of leaf, %	Transpiration rate H_2O/1 g wet leaf x 1 hour	Leaf water retention capacity, %	Leaf dry weight, mg
"Sadaf"	94.3	398.5	55.2	10.6
"Lazurniy"	94.2	335.5	64.3	14.1

4 Conclusions

Important physiological indicators of water balance - transpiration rate, water retention properties of leaves, and dry weight of leaves - were found to be insignificantly different in studied rice varietiesThese physiological processes were not seen in the "Sadaf" and "Lazurni" types. It was observed that the "Lazurniy" variety is an intensive type variety, gives a high yield in the optimal agro background, and it is more resistant to water shortage stress than the "Sadaf" variety.

References

1. Food and Agriculture Organization of the United Nations. Overcoming water challenges in agriculture. In *The State of Food and Agriculture 2020*; FAO: Rome, Italy, 2020.
2. Salehi-Lisar S.Y., Motafakkerazad R., Hossain M.M., Rahman I.M. Water stress in plants: Causes, effects and responses. In *Water Stress*. Eds.; InTech Croatia, 2012;V.8., -P.1–14
3. Kabbadj, A.; Makoudi, B.; Mouradi, M.; Pauly, N.; Frendo, P.; Ghoulam, C. Physiological and biochemical responses involved in water deficit tolerance of nitrogen-fixing *Vicia faba*. *PLoS ONE.* 2017. V.*12*. -P.84
4. Dar MH, Chakravorty R, Waza SA, et al. Transforming rice cultivation in flood prone coastal Odisha to ensure food and economic security. Food Sec. 2017 V.9. –P. 711–722
5. Lee Y and Kende H. Expression of β-expansins is correlated with internodal elongation in deepwater rice. Plant Physiology. 2001. V. 5. –P. 645–654

6. Sasidharan R, Bailey-Serres J, Ashikari M, Atwell BJ, et al. (2017) Community recommendations on terminology and procedures used in flooding and low oxygen stress research. New Phytologist 2017. 214(4). –P. 1403-1407
7. Menezes-Silva, P.E.; Sanglard, L.M.; Ávila, R.T.; Morais, L.E.; Martins, S.C.; Nobres, P.; Patreze, C.M.; Ferreira, M.A.; Araújo, W.L.; Fernie, A.R. Photosynthetic and metabolic acclimation to repeated drought events play key roles in drought tolerance in coffee. *J. Exp. Bot.* 2017, V. *68.* –P. 4309–4322
8. Bryant, C.; Fuenzalida, T.I.; Brothers, N.; Mencuccini, M.; Sack, L.; Binks, O.; Ball, M.C. Shifting access to pools of shoot water sustains gas exchange and increases stem hydraulic safety during seasonal atmospheric drought. *Plant Cell Environ.* 2021, V. *44. –P.* 2898–2911
9. Tenhaken, R. Cell wall remodeling under abiotic stress. *Front. Plant Sci.* 2014, V. *5.* –P. 771
10. Demidchik, V. Mechanisms of oxidative stress in plants: From classical chemistry to cell biology. *Environ. Exp. Bot.* 2015, *109.* –P. 212–228
11. Makarevitch, I.; Waters, A.J.; West, P.T.; Stitzer, M.; Hirsch, C.N.; Ross-Ibarra, J.; Springer, N.M. Transposable elements contribute to activation of maize genes in response to abiotic stress. *PLoS Genet.*2015, V. *11.*
12. Hou, Q.; Ufer, G.; Bartels, D. Lipid signalling in plant responses to abiotic stress. *Plant Cell Environ.* 2016, V. *39. –P.* 1029–1048
13. Nakabayashi, R.; Saito, K. Integrated metabolomics for abiotic stress responses in plants. *Curr. Opin. Plant Biol.* 2015, V. *24. –P.* 10–16
14. Choudhury, F.K.; Rivero, R.M.; Blumwald, E.; Mittler, R. Reactive Oxygen species, abiotic stress and stress combination. *Plant J.* 2017, *90.* –P. 856–867
15. Faize, M,. Burgos I., Faize I., Piqueras A et al. Involvement of cytosolic ascorbate peroxidase and Cu/Zn: superoxide dismutase for improved tolerance against drought stress. Journal of Experimental Botany, 2011. V. 62 (8). –P. 2599-2613.
16. Oliveira C., Agostinetto D., Langaro A. Et al. Physiological and molecular responses in rice, weedy rice and barnyardgrass exposed to supra-optimal temperatures. Planta Daninha. 2019, V. 37.
17. Ivanov.L.A. On the method of rapid weighing to determine transpiration in natural conditions / Ivanov L. A., Silina A. A., Yu. L. Selniker // Botanical Journal. - 2000. - p. 35. - No. 2. - S. 171-185
18. Nichiporovich A. A. On the loss of water by cut plants in the process of wilting. /A.A. Nichiporovich // Jour. experienced agronomy. Southeast. - 1996. T. 3. - Issue. 1. - p. 76–78

Physiological indicators of water metabolism in plants of parental forms and F_1 hybrids of *G. Barbadense* L. cotton type under different water supply conditions

Saydigani Nabiev[1,*], *Mirkosim* Sagdiev[2] and *Rano* Alimova[2]

[1]Institute for Genetics and Plants Experimental Biology at the Science Academy of the Republic of Uzbkistan, Tashkent, 100140, Uzbekistan
[2]Tashkent State Agrarian University, Tashkent, 100140, Uzbekistan

Abstract. This article presents the findings of an extensive research endeavor aimed at comprehending the physiological intricacies of water metabolism in novel strains of fine-fiber cotton belonging to the species Gossypium barbadense L. These strains include L-2006, L-167, L-5440, L-5445, L-1, L-10, L-663, and L-450, alongside the standard variety Surkhan-14 and its F1 hybrids. The study was conducted under various water regime conditions to unravel the responses of these cotton genotypes to different levels of water availability. Upon analysis, it was discerned that the physiological parameters of water metabolism exhibited dynamic changes across the studied genotypes of fine-staple cotton. Specifically, when confronted with simulated soil drought conditions, the water content and transpiration intensity of leaves demonstrated varying degrees of reduction. This response was contingent upon the individual characteristics of each genotype. Furthermore, a noteworthy trend emerged: the water-retaining capacity of leaves displayed an increase under water stress conditions. This observed rise in water-retaining capacity in the leaves can be considered a vital physiological mechanism. It's a mechanism that contributes to the adaptation of fine-staple cotton to water stress. These findings offer valuable insights into the strategies employed by fine-fiber cotton genotypes to cope with water scarcity. By shedding light on the interplay between physiological responses and water availability, this research provides a foundation for informed decision-making in cotton breeding, agricultural management practices, and the cultivation of drought-resistant cotton varieties.
Keywords. Cotton plant, fine-staple cotton lines, varieties, hybrids, optimal water supply, water deficit.

1 Introduction

Cotton plant is an important industrial crop that provides raw materials for the textile, food, chemical, cotton and other industries. Fiber is the main product of cotton. In the global cotton

* Corresponding author: m.saydigani@mail.ru

market, the fiber of *G.barbadense* L. varieties is estimated at 1.5-2 times more expensive than the fiber of medium-staple cotton varieties. From one ton of fiber *G.barbadense* L., 1.3-2.0 times more and more expensive fabric is obtained than from one ton of fiber of the *G.hirsutum* L. species [1-3].

G.barbadense L. is the youngest, plastic species and its homeland is South America [4]. Globally, varieties of the species *G.barbadense* L. are sown on an area of 9% of the total area occupied by cotton. This species was mainly sown in the shores of the island and the plains of the United States and was known under the name of Sea Island. Later, the Sea Island cotton varieties entered the Nile Valleys of Egypt and began to be grown as long and fine staple varieties in Egypt [5].

In the Republic of Uzbekistan, more than 1 million tons of medium-staple cotton varieties of cotton are grown annually, and the country ranks fifth in the world in its production [6].

Global climate change, a gradual increase in air temperature and the presence of an acute environmental problems in Uzbekistan – limited water resources require the creation of drought-resistant varieties of cotton – one of the main crops of the republic [7]. To solve this problem, along with the determination of a complex of economically valuable traits in a wide range of varieties and lines of cotton, it is also necessary to study the physiological parameters of water metabolism in plants under different conditions of water supply, including deficiency conditions of irrigation water [8].

In terms of creating drought-resistant varieties of cotton, targeted research in the fine-stapled species *G. barbadense* L. has great prospects. It should be noted that large-scale fundamental and applied research has been and is being carried out on fine-staple cotton by foreign scientists [9-12] and also by local scientists [13-19]. In these works, economically valuable traits, their inheritance and variability in fine-staple cotton were studied.

Studies were conducted on molecular labeling of fiber traits and resistance to Fusarium wilt, associative mapping and linkage disequilibrium mapping in the germplasm of the species *G. barbadense* L. [20].

By a number of local scientists [21-23] were created and released such varieties of fine-staple cotton as C-6029, C-6030, C-6032, C-6037, C-6040, C-6042, Karshi-8, Karshi-9, as well as varieties of the Surkhan group.

Uzbekistan is a country that has mastered the technology of growing fine-staple cotton. The main reason for this is that the heat in the southern regions of Uzbekistan is higher than the heat reserve in countries that grow varieties of fine-staple cotton on large areas, i.e. in Sherabad more than in Cairo (Egypt), and in Termez – than in Alexandria (Egypt) and Bayram Ali (Turkmenistan), as well as the effective work of advanced scientists of breeders and seed growers. In Uzbekistan, in 1987, varieties of fine-staple cotton were sown on an area of 204 million hectares and the total yield was 587,000 tons. The republic was in the second place after Egypt in growing fine-fiber varieties of cotton [5]. As you know, the textile industry mainly uses medium-staple grades of fiber having a fifth, rarely a fourth type of fiber. However, in recent years, the demand for high-quality type I-III fibers has increased in the world market. Due to the shortage of fine fiber in the world, the United States has increased the area under Pima varieties (*G.barbadense* L.) from 80,000 to 110,000 hectares. India, in addition to the fiber of its fine-staple varieties, buys an additional 150 thousand tons of Pima fiber.

As noted accelerating the selection process of fine-staple cotton varieties is one of the most urgent tasks of modern cotton growing [4]. Although Pima varieties account for only 3% of the world's cotton production, they are commercial varieties that produce high quality fiber. Pima varieties are mainly grown in Egypt, USA and China. In 2012, 94% of Pima varieties were grown in the US San Joaquin Valley, with the rest in Arizona, New Mexico and Texas.

The main goal of cotton breeding programs around the world is to increase yields and improve fiber quality. Compared with medium-staple varieties, varieties of fine-staple cotton are resistant to adverse abiotic and biotic environmental conditions [12, 18].

All of the above requires scientific research to study, in addition to economically valuable traits, also physiological indicators of varieties and new lines of the species *G. barbadense* L under different conditions of water availability, including conditions of water deficiency in the soil to create new drought-resistant varieties of fine-staple cotton.

2 Materials and methods

The research was carried out in the conditions of a field experiment on the territory of the Institute of Genetics and Plants Experimental Biology at the Academy of Sciences of the Republic of Uzbekistan. The soil of experimental site is typical gray soil, non-saline. As objects of research were taken new fine-staple lines obtained by scientists of the Institute on the basis of hybridization of geographically close and distant forms - L-167, L-663, L-2006, L-5440, L-5445, L-450; cultivars with fine fiber ruderal subspecies *ssp.vitifolium* - L-1 and L-10, variety Surkhon-14 and F1 hybrids obtained by crossing these initial forms. At the experiment these materials were sown in the field on two backgrounds of the water regime: on the background of optimal water supply and on the background of simulated drought in the flowering-fruiting phase.

The studied material on each background of the water regime was sown in three randomized replications, by 25 plants in each replication. Sowing scheme was 90 x20 x1. The physiological indicators of plants were determined simultaneously in both variants of both backgrounds, when the pre-irrigation soil moisture on the optimal background of water supply was 70-72% of the LFMC (limiting field moisture capacity), and on the background of the simulated drought -48 -50% of the LFMC.

Wherein, the total irrigation norm for the background with optimal water supply was 4950 m^3, and for the background with simulated drought 3000 m^3. Of the physiological parameters, the total water content in the leaves was determined, intensity of leaf transpiration, water-retaining capacity of leaves. The obtained digital data were statistically processed according to the method of B.A. Dospekhov [8]. The coefficient of adaptability (Cad.) was calculated by the formula of S.A. Eberhart, W.A. Russel [11].

3 Results and discussion

Under conditions of optimal water supply, the highest rates of leaf hydration were noted in the lines L-5440 and L-5445 (85.5% and 81.1%, respectively), and the lowest rate was in the line L-167 - 76.4%. In hybrids F_1, the highest water content of the leaves had plants of combinations L-663 x L-167 (83.1%), L-450 x L-5440 (82.3%) and L-450 x L-5445 (81.6 %), the lowest rates were in combinations L-5445 x L-450, L-450 x Surkhan-14 and L-167 x L-450, respectively, 78.1%, 78.4% and 78.6% (Table 1).

Table 1. Water content of leaves in fine-staple parental forms of cotton and their F_1 hybrids under different conditions of water availability.

#	Material	Optimal Water Supply			Simulated soil drought			Cad., %
		X	hp	heter., %	X	hp	heter., %	
1	L-1	79.6	-	-	71.8	-	-	-9,8

2	Surkhan-14	79.9	-	-	67.1	-	-	-16,0
3	L-5440	81.5	-	-	71.9	-	-	-11,8
4	L-2006	78.2	-	-	75.0	-	-	-4,1
5	L-10	79.6	-	-	74.9	-	-	-5,9
6	L-167	76.4	-	-	70.0	-	-	-8,4
7	L-5445	81.1	-	-	69.8	-	-	-13,9
8	L-450	80.0	-	-	76.3	-	-	-4,6
9	L-663	80.1	-	-	74.5	-	-	-7,0
10	L-450 x L-5445	81.6	1.91	-	74.5	0.45	-	-8,7
11	L-5445 x L-450	78.1	-4.45	97.6	75.3	0.69	-	-3,6
12	L-450 x L-5440	82.3	2.07	-	76.3	1.00	-	-7,3
13	L-5440 x L-450	79.5	-1.67	-	75.0	0.41	-	-5,7
14	L-450 x L-663	79.8	-5.00	-	76.0	0.67	-	-4,8
15	L-663 x L-450	80.5	-9.00	-	69.6	-6.44	93.4	-13,5
16	L-450 x L-167	79.9	0.94	-	79.0	5.85	103.5	-1,1
17	L-167 x L-450	78.6	0.22	-	75.4	0.71	-	-4,1
18	L-450 x Surkh.-14	78.4	-31.00	-	75.2	0.76	-	-4,1
19	Surkh.-14 x L-450	78.5	-29.00	-	76.3	1.00	-	-2,8
20	L-663 x L-167	83.1	2.62	103.7	73.4	0.51	-	-11,7
21	L-167 x L-663	80.2	1.00	-	77.1	2.16	103.5	-3,9
22	L-663 x L-5445	80.2	-0.80	-	77.7	2.36	104.3	-3,1
23	L-5445 x L-663	78.7	-3.80	-	78.6	2.74	105.5	-0,1
24	L-167 x L-5440	79.0	0.02	-	74.7	3.95	103.9	-5,4
25	L-5440 x L-167	79.6	0.25	-	77.5	6.89	107.8	-2,6
26	L-167 x L-10	78.9	0.56	-	75.6	1.29	-	-4,2
27	L-10 x L-167	80.0	1.25	-	78.5	2.47	104.8	-1,9
28	Surkh.-14 x L-2006	79.5	0.53	-	78.2	1.81	104.3	-1,6
29	L-2006 x Surkh.-14	80.6	1.82	-	79.1	2.04	105.5	-1,9
30	L-5445 x L-10	79.6	-1.00	-	75.1	1.08	-	-5,7
31	L-10 x L-5445	78.9	-1.93	-	75.8	1.35	-	-3,9
32	L-450 x L-1	81.00	6.00	-	76.4	1.04	-	-5,7
33	L-167 x L-1	79.9	1.19	-	79.6	9.67	110.9	-0,4
	LSD_{05}	1,7			1.4			

Leaf water content is one of the most important physiological indicators, and the study of this trait in the original forms of fine-staple cotton and their F1 hybrids is of great importance.

The trait of water content of the leaf under the optimal water regime of 24 F_1 hybrids in 7 hybrids was inherited according to the type of positive overdominance, in 8 hybrids - negative overdominance, in 1 - positive complete dominance, in 1 - negative complete dominance, in 5 - incomplete dominance of the form with a high index, in 1 hybrid - incomplete dominance of the form with a low indicator, in 1hybrid - intermediate, i.e. there was no dominance of any parent.

Thus, the trait of water content of leaves under the optimal water regime was mainly inherited according to the type of positive and negative dominance. Positive heterosis with a low level was noted in the combination L-663 x L-167 (103.7%), and negative heterosis - in the combination L-5445 x L-450 - 97.6%.

With water deficiency, in all the studied parental forms and hybrids of fine-staple cotton, the water content of the leaves decreased to varying degrees. At the same time, in the group of parental forms, the highest rates of the trait were noted in the lines L-450, L-2006, L-10 and L-663 (76.3%; 75.0%; 74.9%; 74, respectively). 5%), and the lowest rate is for the variety Surkhan-14 - 67.1%.

In F_1 hybrids, the highest levels of leaf water content were found in combinations L-167 x L-1 (79.6%), L-2006 x Surkhan-14 (79.1%) and L-450 x L-167 (79.6%) and the lowest indicator is in the hybrid L-663 x L-450 -69.6%.

According to the analysis of the indicators of the dominance coefficient (hp), under conditions of water deficiency, the trait of water content in the leaves of 24 F_1 hybrids was inherited in 13 hybrids according to the type of positive overdominance, in 1 hybrid - negative overdominance, in 3 hybrids - complete dominance of the form with a high indicator, in 7 - incomplete dominance of the form with a high score. This indicates the inheritance of the trait of leaf water content in F_1 hybrids under conditions of water deficit, mainly according to the type of positive overdominance and incomplete dominance of the paternal or maternal form with a high index. Against this stressful background, although with a low degree, the number of hybrid combinations with positive heterosis increased (10 pieces). The degree of positive heterosis ranged from 103.5% (L-450 x L-167, L-167 x L-663) to 110.9% in L-167 x L-1. Negative heterosis was noted in the combination L-663 x L-450 - 93.4%.

In terms of the adaptability coefficient (Cad.), on the basis of the trait of leaf water content, the strongest reaction to soil moisture deficiency in the group of initial forms was noted in the Surkhan-14 variety, lines L-5445 and L-5440, and the weakest reaction was observed in lines L-2006, L-450 and L-10. In the group of F_1 hybrids, strong sensitivity was found in combinations L-663 x L-450 and L-663 x L-167, and most of the F_1 hybrids showed a weak reaction. This indicates a higher adaptation of F_1 hybrids to water deficit in terms of leaf hydration compared to parental forms.

Determination of the intensity of leaf transpiration, which was expressed in mg of water evaporated from one gram of a raw leaf for one hour, showed that under conditions of optimal water supply in the group of parental forms, the intensity of leaf transpiration was high in lines L-10 and L-167 and amounted, respectively , 385.17 mg and 379.48 mg, and the lowest indicator of the trait was noted in the L-1 line - 256.13 mg (Table 2).

In the group of F_1 hybrids, plants of combinations L-663 x L-5445, L-663 x L-167 and L-663 x L-450 had the highest rates of transpiration intensity (397.84 mg, 387.49 mg and 385, respectively). 12 mg), and the combination L-10 x L-167 had the lowest indicator of this trait - 250.51 mg. The remaining F_1 hybrids were between these extreme groups in terms of the intensity of leaf transpiration. Under the optimal water regime, this trait out of 24 F1 hybrid combinations was inherited in 4 by the type of positive overdominance, in 11 - negative overdominance, in 3 - incomplete dominance of the form with a high indicator, and in 6 - incomplete dominance of the form with a low indicator.

Table 2. Intensity of leaf transpiration in fine-staple parental forms of cotton lines and their F_1 hybrids under different conditions of water supply.

#	Materials	Optimal Water Supply			Simulated soil drought			Cad., %
		X	hp	heter, %	X	hp	heter, %	
1	L-1	256.13	-	-	207.55	-	-	-19,0
2	Surkhan-14	369.97	-	-	171.19	-	-	-53,7
3	L-5440	345.02	-	-	234.21	-	-	-32,1
4	L-2006	320.26	-	-	270.21	-	-	-15,6
5	L -10	385.17	-	-	261.02	-	-	-32,2
6	L-167	379.48	-	-	226.99	-	-	-40,2
7	L-5445	360.08	-	-	255.70	-	-	-29,0
8	L-450	321.69	-	-	279.74	-	-	-13,0
9	L-663	369.83	-	-	111.49	-	-	-69,9

10	L-450 x L-5445	327.09	-0.72	-	323.20	4.62	115.5	-1,2
11	L-5445 x L-450	329.76	-0.58	-	280.64	1.07	-	-14,9
12	L-450 x L-5440	335.31	0.17	-	312.29	2.43	111.6	-6,9
13	L-5440 x L-450	324.36	-0.77	-	265.79	0.39	-	-18,1
14	L-450 x L-663	313.07	-1.36	-	304.51	1.29	108.9	-2,7
15	L-663 x L-450	385.12	1.64	104.1	208.70	0.16	-	-45,8
16	L-450 x L-167	366.81	0.56	-	345.40	3.49	123.5	-5,8
17	L-167 x L-450	311.70	-1.35	-	283.31	1.14	-	-9,1
18	L-450 x Surkh.-14	283.17	-2.60	88.0	267.24	0.77	-	-5.6
19	Surkh.-14 x L-450	303.88	-1.74	94.46	289.15	1.17	-	-4,8
20	L-663 x L-167	387.49	2.66	-	200.55	0.54	-	-48,2
21	L-167 x L-663	355.13	-4.05	96.0	294.63	2.17	129.8	-17,0
22	L-663 x L-5445	397.84	6.75	107.6	376.03	2.67	147.1	-5,5
23	L-5445 x L-663	367.02	0.42	-	322.56	1.93	126.1	-12,1
24	L-167 x L-5440	336.21	-1.51	-	241.03	2.89	-	-28,3
25	L-5440 x L-167	291.13	-4.13	84.4	264.00	9.25	112.7	-9,3
26	L-167 x L-10	374.18	-2.87	-	337.12	5.47	129.2	-9,9
27	L-10 x L-167	250.51	-46.42	66.0	223.63	-1.20	-	-10,7
28	Surkh.-14 x L-2006	333.18	-0.48	-	301.84	1.64	111.7	-9,4
29	L-2006 x Surkh.-14	280.37	-2.61	87.5	251.38	0.62	-	-10,3
30	L-5445 x L-10	292.84	-6.36	81.3	265.95	2.85	-	-9,2
31	L-10 x L-5445	290.60	-6.54	80.7	191.18	-2.53	74.8	-34,2
32	L -450 x L-1	360.95	2.20	112.2	247.94	0.12	-	-31,3
33	L -167 x L-1	290.85	-0.44	-	144.23	-7.51	69.5	-50,4
	LSD_{05}	12,43			13.01			

Positive heterosis of unweak degree in terms of transpiration intensity was shown by combinations of L-450 x L-1 (112.2%), L-663 x L-5445 (107.6%) and L-663 x L-450 (104.1%) , while negative heterosis was observed in combinations L-450 x Surkhan-14 (88.0%), Surkhan-14 x L-450 (94.5%), L-167 x L-663 (96.0%), L-5440 x L-167 (84.4%), L-10 x L-167 (66.0%), L-2006 x Surkhan-14 (87.5%), L-5445 x L-10 (81.3%) and L-10 x L-5445 (80.7%). The results obtained show that under conditions of optimal water supply, the trait of leaf transpiration intensity was mainly inherited by the type of overdominance, with a negative value in 11 combinations and with a positive value in 4 combinations.

Compared to the optimal water regime, under conditions of water deficit, all parental and hybrid genotypes showed a decrease in the intensity of leaf transpiration to varying degrees. On the background of water stress in the group of initial forms, high rates of leaf transpiration were noted in lines L-450 and L-2006 (respectively, 279.74 mg and 270.21 mg), and the lowest rate was in line L-663, the average value of which was 111.49 mg. It should be noted that the Surkhan-14 variety also had a relatively low leaf transpiration rate - 171.19 mg.

With water deficiency in the group of F_1 hybrids, relatively high rates of leaf transpiration intensity were noted in combinations L-663 x L-5445, L-450 x L-167 and L-167 x L-10 (respectively, 376.03 mg, 345.40 mg and 337.12 mg), and the lowest rates are in hybrids L-167 x L-1, L-10 x L-5445, L-663 x L-167 and L-663 x L-450 (respectively, 144.23 mg, 191.18 mg, 200.55 mg and 208.70 mg).

Under conditions of water stress, the trait of leaf transpiration intensity out of 24 F_1 hybrid combinations was inherited in 15 combinations by the type of positive overdominance, in 3 combinations - by negative overdominance, in 6 - by incomplete dominance of the form with a high indicator. Thus, even under conditions of water deficit, the trait of intensity of leaf transpiration was inherited mainly by the type of overdominance. However, if under the optimal water regime during the inheritance of this trait, mainly negative overdominance was observed, then with a deficit of soil moisture, on the contrary, positive overdominance. This indicates a change in the dominance coefficient (hp) depending on the water content of plants. For example, for the combination L-450 x L-5445, the hp indicator with optimal water content is -0.72, and with water deficit it is 4.62, i.e. on the first background there is an incomplete negative dominance, while on the second background there is a positive overdominance.

In contrast to the optimal water regime, under conditions of water deficit, positive heterosis in terms of transpiration intensity was observed in 10 combinations of F_1 and its degree ranged from 108.9% (L-450 x L-663) to 147.1% (L-663 x L-5445). Negative heterosis was noted in combinations L-167 x L-1 and L-10 x L-5445, its degree was respectively, 69.5% and 74.8%.

According to the analysis of the adaptability coefficient of the trait, the indicators of transpiration intensity under water deficiency in the group of parental forms decreased from 13.0% (L-450) to 69.9% (L-663), and in the group of F_1 hybrids - from 1.2% (L-450 x L-5445) up to 50.4% (L-167 x L-1). On this stressful background, the trait indices also sharply decreased towards combinations L-663 x L-167 and L-663 x L-450 (by 48.2% and 45.8%, respectively). With water deficiency, a weak reaction on this trait was noted in the combinations L-450 x L-663 (-2.7%), Surkhan-14 x L-450 (-4.8%), L-663 x L-5445 (- 5.5%), L-450 x Surkhan -14 (-5.6%), L-450 x L-167 (-5.8%) and L-450 x L-5440 (-6.9%) .

Water-retaining capacity (WRC) is one of the most important indicators in the study of physiological processes occurring in plants under the influence of various abiotic stresses, including water deficiency. It should be noted that high numerical values of the trait indicate low leaf WRC and vice versa, low values indicate high leaf WRC, because they indicate the amount of water used for evaporation within two or four hours in relation to its initial content.

According to the results obtained after two hours under conditions of optimal water supply, a relatively high leaf WRC was noted in the line L-10 (20.5%), and a low leaf WRC was observed in the line L-167 (44.7%). In the group of F_1 hybrids, combinations L-10 x L-167 (22.9%) and L-2006 x Surkhan-14 (24.8%) had a high leaf WRC, and combinations L-5445 x L-450 (41.9%), L-450 x L-5440 (39.4%) and Surkhan-14 x L-450 - 38.7% had low leaf WRC (Table 3). According to the indicators of coefficient of dominance (hp), the trait of WRC of leaves out of 24 F_1 combinations was inherited in 13 combinations as positive overdominance, in 4 - negative overdominance, in 1 combination - incomplete dominance of the form with a high index, in 5combinations - incomplete dominance of the form with a low index, in 1 - intermediate, i.e. in the absence of dominance of the paternal or maternal form.

Water retention capacity (after 2 hours) of leaves of fine-staple cotton varities and lines under different water supply conditions, its inheritance in F1 hybrids (Table 3).

Table 3. Intensity of leaf transpiration in fine-staple parental forms of cotton lines and their F_1 hybrids.

#	Material	Optimal Water Supply			Simulated soil drought			Cad., %
		X	hp	heter, %	X	hp	heter, %	
1	L-1	20.5	-	-	20.1	-	-	-2,0
2	Surkhan	28.6	-	-	18.4	-	-	-35,7
3	L-5440	28.2	-	-	25.1	-	-	-11,0
4	L-2006	29.6	-	-	26.1	-	-	-11,8
5	L-10	26.9	-	-	25.3	-	-	-5,9
6	L-167	44.7	-	-	22.1	-	-	-50,6
7	L-5445	24.7	-	-	23.2	-	-	-6,1

8	L-450	26.8	-	-	21.4	-	-	-20,1
9	L-663	30.3	-	-	23.6	-	-	-22,1
10	L-450 x L-5445	30.6	4.62	114.2	27.1	5.33	116.8	-11,4
11	L-5445 x L-450	41.9	15.38	156.3	28.8	7.22	124.1	-31,3
12	L-450 x L-5440	39.4	17.00	139.7	27.4	2.24	-	-30,5
13	L-5440 x L-450	33.9	9.14	120.2	18.7	-2.46	-	-44.8
14	L -450 x L-663	31.1	1.46	-	19.5	-2.73	-	-37,3
15	L-663 x L-450	32.4	2.20	-	20.9	-1.45	-	-35,5
16	L-450 x L-167	31.1	-0.52	-	25.2	9.86	114.0	-19.0
17	L-167 x L-450	27.0	-0.98	-	20.6	-3.29	-	-23,7
18	L-450 x Surkh.-14	25.1	-2.89	-	20.8	0.60	-	-17,1
19	Surkh.-14 x L-450	38.7	12.22	135.3	23.2	2.20	-	-40,1
20	L-663 x L-167	26.8	-1.49	88.4	23.5	0.87	-	-12,3
21	L-167 x L-663	38.0	0.07	-	24.1	1.67	-	-36,6
22	L-663 x L-5445	35.0	2.68	115.5	30.5	35.5	129.2	-12,9
23	L-5445 x L-663	33.4	2.11	110.2	31.9	42.5	135.2	-4,5
24	L-167 x L-5440	30.7	-0.70	-	19.4	-2.80	-	-36.8
25	L-5440 x L-167	33.4	-0.37	-	19.2	-2.90	86.9	-42,5
26	L-167 x L-10	34.8	-0.11	-	28.6	3.06	113.0	-17,8
27	L-10 x L-167	22.9	-1.45	85.1	19.1	-2.88	86.4	-16,6
28	Surkh.-14 x L-2006	31.2	4.20	-	21.5	-0.19	-	-31,1
29	L-2006 x Surkh.-14	24.8	-8.60	86.7	17.5	-1.23	-	-29,4
30	L-5445 x L-10	29.2	3.09	-	19.9	-4.14	85.8	-31,8
31	L-10 x L-5445	27.3	1.36	-	18.4	-5.57	79.3	-32,6
32	L-450 xL-1	28.8	1.63	-	27.2	9.92	127.1	-5,6
33	L-167 x L-1	32.3	-0.02	-	17.5	-3.6	-	-45,8
	LSD_{05}	2,5			2.7			

Thus, the trait of leaf WRC in F_1 hybrids was mainly inherited according to the type of overdominance (of which 13 were positive, 4 were negative). In 7 combinations, positive heterosis was noted, the degree of which ranged from 110.2% (L-5445 x L-663) to 15.6% (L-5445 x L-450). Negative heterosis was found in combinations L-10 x L-167 (85.1%), L-2006 x Surkhan-14 (86.7%) and L-663 x L-167 - 88.4%.

In case of water deficit in the group of parental forms, the highest rates of leaf WRC were noted in the variety Surkhan-14 and line L-1 (respectively, 18.4% and 20.1%), and the lowest rates - in the lines L-2006, L- 5440 and L-10 (respectively 26.1%, 25.3% and 25.1%). In F1 hybrids L-2006 x Surkhan-14 and L-167 x L-1, L-10 x L-5445 and L-5440 x L-450, the leaf WRC was high (respectively, 17.5%, 17.5% , 18.4% and 18.7%), and combinations L-5445 x L-663 and L-663 x L-5445 had the lowest leaf WRC (Table 3).

On this stressful background, the trait of leaf WRC from 24 combinations of F_1 was inherited as positive overdominance in 10 combinations, negative overdominance in 11 combinations, positive incomplete dominance in 2, and negative incomplete dominance in 1 combination. Thus, the trait of leaf WRC under water deficiency was inherited mainly by the type of negative and positive overdominance. Under stress conditions, the number of F_1 combinations with negative overdominance sharply increased (from 4 on the optimal background to 11 with water deficit). In 7 F_1 combinations, positive heterosis was noted, the degree of which ranged from 113.0% (L-167 x L-10) to 135.2% (L-5445 x L-663). Negative heterosis was found in 4 F_1 combinations, i.e. in L-10 x L-5445 (79.3%), L-5445 x L-10 (85.8%), L-10 x L-167 (86.4%) and L-5440 x L-167 (86.9%). According to the indicators of the adaptability coefficient (Cad.), the WRC of leaves under conditions of water deficit increased in the group of parental forms from 2.0% to 50.6%, and in the F_1 group - from 4.5% to 45.8%. In the group of parental forms on the basis of leaf WRC to water deficiency, a strong reaction was observed in line L-167 and the variety Surkhan-14, and a weak reaction in lines L-1, L-10 and L-5445; L-167 x L-1, L-5440 x L-450, L-5440 x L-167, Surkhan-14 x L-450, while combinations of L-5445 x L-663 and L-450 x L-1 showed weak sensitivity.

4 Conclusions

As a result of the research, the presence of genotypic polymorphism in groups of varieties and lines of the species *G.barbadense* L., their F_1 hybrids was established in terms of physiological indicators of water metabolism of plants under conditions of optimal water supply and water deficiency. The studied parental forms (cultivars, lines) and their F_1 hybrids reacted to soil moisture deficiency by reducing the water content and intensity of leaf transpiration and increasing the water-retaining capacity of leaves, which serve as physiological mechanisms for adapting fine-staple cotton to this stress. In F_1 hybrids under conditions of water deficiency, the trait of water content of leaf was inherited mainly by the type of positive overdominance and incomplete dominance of the paternal or maternal form with a high index, the trait of the intensity of leaf transpiration was inherited mainly by the type of positive overdominance, the trait of the water-retaining capacity of leaves - mainly by according to the type of negative and positive overdominance.

References

1. Abdalla A.M., Reddy O.U.,El-Zik K.M. and Pepper A.E.. Genetic diversity and relationships of diploid and tetraploid cottons revealed using AFLP//Theor. app. genet. 2001.102:-P.222-229.
2. Abdel – Hafez A.G., El-Keredy M.S., El-Okkia A.F. and Gooda B.M. Estimates of heterosis and combining ability for yield, yield components and fiber properties in Egyptian cotton (*G.barbadense* L.)// Egyptian J. Plant Breed.Agronomy, Department, Giza, Egypt. 2007.11(1).-P. 423-435.
3. Gamal I.A., Abd-El-Halen S.H., Ibrahim E.M.. A genetic analysis of yield and its components of Egyptian cotton (*G.barbadense* L.) under divergent environments. //American-Eurasian J.Agric.&Environ.Sci. 2009.5(1),-P.5-13.
4. Eberhart S.A., Russel W.A. Stability parameters for comparing parameters.// Crop.Sci.-1966.- V.6.- p. 36 - 40.
5. Steven D.W., Robert B.H., Gerardo B., Sonia I.R., Kelly A.H., Daniel S.M., Katherine A.W., Jonathan F.W. and Mark P.K. Impact of Pima defoliation timings on lint yield and quality// The Journal of Cotton Science, 18:48-58 (2014),http://journal.cotton.org.
6. Stahel, J. 2012. Extra Long Stable (ELS) Cotton. http://www.reinhardt.com/our-business/long-stable-cotton/.
7. Percy R.G.Comparison of bulk F_2 performance testing and pedigree selection in thirty Pima cotton populations// J. Cotton Science, 2003, N 7.-P. 170-178.
8. Alimova R. A. et al. Plant growth stimulants and the effect of foliar feeding on the yield of some vegetable crops //Eurasian Research Bulletin. – 2022. – T. 9. – C. 30-38.
9. Alimova R., Mirkhamidova P., Tuychieva D. Effect of pesticides "Butylcaptax (Russia)" and "Droppa (Russia)" on respiration and oxidative phosphorylation of liver mitochondria of pregnant rats and their embryos //E3S Web of Conferences. – EDP Sciences, 2021. – T. 244. – C. 02026.
10. Allayarov, Abdurakhman & Zuparov, Mirakbar & Khakimov, Albert & Omonlikov,

Alisher. (2021). Application of the biopreparation 'Orgamika F' against fusarium disease of cabbage and other cole vegetables. E3S Web of Conferences. 284. 03011. 10.1051/e3sconf/202128403011.

11. Khakimov, Albert & Salakhutdinov, I & Omonlikov, Alisher & Utaganov, Samad. (2022). Traditional and current-prospective methods of agricultural plant diseases detection: A review. IOP Conference Series: Earth and Environmental Science. 951. 012002. 10.1088/1755-1315/951/1/012002.
12. Khujanazarov U. et al. Current state of Cenopopulations Iris Magnifica Vved and Tulipa Fosteriana W. Irving in Uzbekistan //E3S Web of Conferences. – EDP Sciences, 2021. – T. 244. – C. 02027.
13. Sagdiev M. T., Amanova M. M., Omonlikov A. U. The influence of growth regulators on tomato productivity in the conditions of the Tashkent region //ISJ Theoretical & Applied Science, 11 (79). – 2019. – C. 241-244.
14. Sagdiev M. T., Amanova M. M., Omonlikov A. U. The influence of plant growth stimulators on tomato productivity in the conditions of Tashkent region. World Journal of Pharmaceutical and Life Sciences,(WJPLS //India. – 2020. – T. 6. – №. 8. – C. 04.
15. Sodikov B. et al. Soil-borne plant pathogenic fungi biodiversity of sunflower //IOP Conference Series: Earth and Environmental Science. – IOP Publishing, 2022. – T. 1068. – №. 1. – C. 012018.
16. Usmanov R. M. et al. Influence of physicochemical factors on water exchange in cotton leaves under the moisture deficit conditions //IOP Conference Series: Earth and Environmental Science. – IOP Publishing, 2021. – T. 939. – №. 1. – C. 012051.
17. Usmanov R. et al. Assessment of drought stress impact on winter wheat using landsat 8 images: comparison of field data and vegetation indices //IOP Conference Series: Earth and Environmental Science. – IOP Publishing, 2023. – T. 1142. – №. 1. – C. 012082.
18. FAS/USDA. 2007. Cotton: world markets and trade. USDA, Foreing Agric.Serv.Circular Series FOP 05-07 May 2007. http:www.fas.usda.gov/cotton/circular/2007/May/cottonfull0507.pdf (accessed 4Apr.2008).
19. Tayjanov, K., Khojimatov, O., Gafforov, Y., Makhkamov, T., Normakhamatov, N., & Bussmann, R. W. (2021). Plants and fungi in the ethnomedicine of the medieval East-a review. Ethnobotany Research and Applications, 22, 1-20. DOI:0.32859/ERA.22.46.1-20
20. Jabeen, S., Zafar, M., Ahmad, M., Althobaiti, A. T., Ozdemir, F. A., Kutlu, M. A., ... & Majeed, S. (2023). Ultra-sculpturing of seed morphotypes in selected species of genus Salvia L. and their taxonomic significance. Plant Biology, 25(1), 96-106. DOI:10.1111/plb.13473
21. Majeed, S., Ahmad, M., Ozdemir, F. A., Demirpolat, A., Şahan, Z., Makhkamov, T., ... & Nabila. (2023). Micromorphological characterization of seeds of dicot angiosperms from the Thal desert (Pakistan). Plant Biosystems-An International Journal Dealing with all Aspects of Plant Biology, 157(2), 392-418. DOI:10.1080/11263504.2023.2165553
22. Majeed, S., Ahmad, M., Ali, A., Althobaiti, A. T., Ramadan, M. F., Kilic, O., ... & Sultana, S. (2023). Pollen micromorphology among amaranthaceous species from desert rangeland: Exine stratification and their taxonomic significance. BioMed Research International, 2023, 4967771. DOI:10.1155/2023/4967771
23. Makhkamov, T., Sotiboldiyeva, D., Mamarakhimov, O., Yuldashov, Y., & Botirova, L. (2022, May). Morphogenesis and Seasonal Developmental Rhythm Under the Conditions of Introduction of Curcuma Longa L. In International Scientific Conference

on Agricultural Machinery Industry “Interagromash"” (pp. 1460-1469). Cham: Springer International Publishing. DOI:10.1007/978-3-031-21432-5_155

Investigation of the water of Aydarkul Lake into components and the scope of their application

Dilmurod Azimov[1,*], *Sadritdin* Turabdjanov[1], *Elmurod* Egamberdiyev[1], *Shodiyaxon* Azimova[2], *Rano* Nazirova[2], *Baxrom* Dautov[3], *Sharafutdin* Arslanov[4] and *Olimjon* Muratkulov[1]

[1]Tashkent State Technical University, Tashkent, 100095, Uzbekistan
[2]Tashkent Chemical Technical Institute, Tashkent, 100011, Uzbekistan
[3]Central Customs Laboratory of the Customs Committee of the Republic of Uzbekistan, Tashkent, 100003, Uzbekistan
[4]National University of Uzbekistan, Tashkent, 100174, Uzbekistan

Abstract. To assess the suitability of Aydarkul Lake's water for various purposes such as industrial, agricultural, domestic, or drinking usage, a comprehensive analysis was conducted encompassing all pertinent physico-chemical parameters of the water. The study aimed to ascertain whether the water met the requisite standards for safe utilization. The results of the analysis demonstrated that the water in Aydarkul Lake exhibited basic physico-chemical characteristics that exceeded the permissible concentrations for drinking water, as defined by the state standards. These findings underscore a significant deviation from the criteria established for safe drinking water consumption. These conclusions illuminate the need for thorough consideration when contemplating the utilization of Aydarkul Lake's water for various purposes. The data generated from this analysis serves as an important resource for decision-makers and stakeholders, guiding informed choices regarding the potential applications of the lake's water while also highlighting the importance of maintaining water quality standards for public health and environmental sustainability.
Keywords. Aydarkul, Kizilkum, Kili, Akbulak Tuzkan, drinking water, salty, hard, sulfate, nitrate, carbonate, bicarbonate, pH.

1 Introduction

Aydarkul Lake is a kind of ecological antipode of the Aral Sea, a huge drainless reservoir [1]. A large artificial reservoir in the Kyzylkum desert, located in the northeast of Uzbekistan [2]. The reservoir is located at an altitude of 247 meters above sea level of the Baltic system and is part of the Arnasai system of lakes, which occupies a huge salt basin [3]. Aydarkul contains 44.3 km^3 of slightly salted water [4]. The mineralization of the reservoir in the east is only 1.5-2%, and in the western part – 8%. The lake stretches for 160 km and is about 34.8 km wide. Its average depth reaches 12.5 m, and the maximum reaches 33.6 m [5].

* Corresponding author: prof-azimov@mail.ru

More than 100 species of flora and fauna have found shelter here, some of which are listed in the Red Book [6-9]. On the reservoir you can meet rare birds – pink pelicans, swans, dives, white herons, curly pelican, white-tailed and red barracks. Of the fish, carp, asp, carp, walleye, bream, catfish, snakehead, chehon, etc. live here. On the southern shore of the reservoir there is the Nuratau-Qyzylkum Biosphere Reserve, created to preserve endangered and endemic species of flora and fauna [10-15].

Lake Aydarkul is characterized by the absence of undercurrents, resulting in relatively tranquil water conditions [16]. A distinct pattern emerges as one moves westward across the lake: the water's salinity progressively increases. In the western regions, mineralization levels exceed 8 grams per liter, signifying a higher concentration of dissolved salts [17]. In contrast, the estuaries situated in the eastern parts of the lake exhibit mineralization levels ranging from 3 to 5 grams per liter [18]. This variance in mineral content contributes to the creation of favorable environmental conditions that support biological life [19].

The decreasing salinity from east to west within Lake Aydarkul's expanse influences the suitability of the water for various forms of aquatic life [20]. The relatively lower mineralization levels in the eastern estuaries provide a hospitable environment that fosters diverse biological activity. This ecological distinction underscores the intricate interplay between geographical factors, mineral composition, and the nurturing of life forms within the lake's ecosystem.

2 Research methods

Calcium (Ca^{2+}) and magnesium (Mg^{2+}) were determined by titric method with trilon "B". Chlorides (Cl^-) were determined by the argentometric method. Sulfate ions (SO_4^{2-}) were determined by the weight method [21], precipitating by passing through a filter and burning 1:50 hours at a temperature of 800 ^{0}C, the remaining ash was weighed on a scale and counted [22]. The pH was determined on a pH meter device, and nitrate (NO_3^-) on a nitratomer device. HCO_3^-, that is, bicarbonate was determined by titration, titrimetric method with a solution of HCl 0.05 N. Determination of iron, (Fe^{3+}, Fe^{2+}), ammonium ions (NH_4^+) and nitrites (NO_2^-) was carried out in a refractometer. Sodium (Na^+) and potassium (K^+) were determined in a flame apparatus. CO_2 free mg/l by titration. The oxidability was determined by the permanganate titration method [23]. The general severity and aggressiveness can be found by the formula by calculations. Silicon (Si) was determined by the comparison method.

3 Results and discussion

The results of the study. By comparing the results of samples from three different points of the Aidar-Arnasoy lake system with the established state standard standards of Uzbekistan for drinking water, the data obtained are given in detail in Tables 1 and 2.

According to Tables 1 and 2, the following conclusions can be drawn:

- from the point of view of the total content of trace elements, the water at point 1 (≈10,000 mg/l) can be classified as saltier water, and the water at points 2 and 3 (≈5,000 mg/l) can be classified as brackish;
- at all points of the content of nitrate anions (NO_3^-), the indicator is within the normal range, the level is below 45 mg/l;
- at all points of the content of nitrite anions (NO_2^-), the indicator is within the normal range, the level is below 3 mg/l;
- the chlorine anion content (Cl^-) (250 mg/l) is 2-8 times higher than normal at all points;
- the content of sulfate (SO_4^{2-}) is 10-15 times higher than the norm (400 mg/l) by 10-15 times at each point;

- amonium (NH_4^+) cations are within the normal range, the level is below 1 mg/l;
- the indicator for iron cations (Fe^{2+} and Fe^{3+}) at all points, the indicator is within the normal range, the level is below 0.3 mg/l;

The water hardness level is high at all points and, according to the standard, belongs to the class of very hard water.

Table 1. Analysis results for all physico – chemical parameters of water (Qizilkum 4-Collector).

#	Name	Normative indicators in a liter of drinking water	Qizilkum 4-Collector		
			Content in liter		
			mg/l	mg-eq/l	%- eq/l
Anions					
1	Na^+		2182	94.86	49
2	NH_4^+	1 mg/l	0.4	0.02	-
3	Ca^{2+}		661	33.00	17
4	Mg^{2+}		790	65.00	34
5	Fe^{2+}	0,3 mg/l	< 0.3	-	-
6	Fe^{3+}	0,3 mg/l	< 0.3	-	-
Total				192.88	100
Cations					
7	Cl^-	250 mg/l	2127	60.00	31
8	SO_4^{2-}	400 mg/l	6189	128.94	67
9	NO_2^-	3 mg/l	< 0.01	-	-
10	NO_3^-	45 mg/l	15	0.24	-
11	CO_3^-		no	-	-
12	HCO_3^-		226	3.70	2
Total				192.88	100
Formula of the salt composition of water			$12{,}6\frac{SO_4^{67}\ Cl^{31}}{(Na+K)^{49}Mg^{34}Ca^{17}}$		

Table 2. Analysis results for all physico – chemical parameters of water (Qili Collector and Akbulak Collector).

#	Name	Normative indicators in a liter of drinking water	Qili Collector			Akbulak Collector		
			Content in liter			Content in liter		
			mg/l	mg-eq/l	%- eq/l	mg/l	mg-eq/l	%- eq/l
Anions								
1	Na^+		988	42.97	50	594	24.54	30

2	NH_4^+	1 mg/l	0.4	0.02	-	2.1	0.12	-
3	Ca^{2+}		341	17.00	20	301	15.00	19
4	Mg^{2+}		316	26.00	30	492	40.50	51
5	Fe^{2+}	0.3 mg/l	< 0.3	-	-	< 0.3	-	-
6	Fe^{3+}	0.3 mg/l	< 0.3	-	-	0.3	0.01	-
Total				85,99	100		80.17	100
Cations								
7	Cl^-	250 mg/l	576	16.25	19	401	20,00	25
8	SO_4^{2-}	400 mg/l	3094	64.47	75	2806	58,47	73
9	NO_2^-	3 mg/l	0.5	0.01	-	< 0.01	-	-
10	NO_3^-	45 mg/l	10	0.16	-	6	0,10	-
11	CO_3^-		no	-	-	no	-	-
12	HCO_3^-		311	5.10	6	98	1,60	2
Total				85,99	100		80.17	100
Formula of the salt composition of water			$5{,}7\frac{SO_4^{75}\,Cl^{19}}{(Na+K)^{50}Mg^{30}Ca^{20}}$			$4{,}7\frac{SO_4^{73}\,Cl^{25}}{(Na+K)^{30}Mg^{51}Ca^{19}}$		

We have proposed ways to solve the problem: for each case of water use, it is necessary to approach individually, that is why it is advisable to apply a set of measures to purify the waters of the Aydarkul lakes.

For example, to use the waters of Aydarkul for drinking purposes, deep purification is necessary. After the initial mechanical cleaning with coarse dispersed filters, it is necessary to continue cleaning the water with a sand filter, an aerator, a degreasing filter and a carbon filter (adsorption). Each cleaning process is an equipment (cylinder) with a special filler, since there are electronic and mechanical devices on each filter.

In water treatment plants, quartz sand for water filtration is poured into a durable housing that is resistant to corrosion and the action of chemically aggressive media. The water is supplied by a pump and, at excessive pressure, passes through a sand backfill with a grain size of 0.4-0.8 mm (Table 3).

Modern aerators easily remove gases dissolved in water, thereby improving its properties. In addition, they contribute to the oxidation of iron present in the water, prevent the reproduction of bacteria.

The aerator is capable of purifying water from any dissolved metals — whether it is manganese or iron. Organic compounds are no exception. To do this, the impurities are first oxidized. The most affordable and environmentally friendly oxidizer is air oxygen. Therefore, the use of aerators is very effective.

Aerators for water purification pump air into the pressure pipeline. This is done by means of a compressor, the interfering ions are oxidized and remain in the form of a precipitate on the filter material. An important advantage of aerators used in water treatment systems is minimal operating costs. At the same time, the load on water treatment equipment is significantly reduced. As a result, water purification is dramatically accelerated, and without the use of complex, expensive equipment.

The oxygen-enriched water enters a filter based on a catalytically active granular filter material catalytic material. The use of a catalytically active granular filter material is an effective and inexpensive way to remove dissolved iron and manganese from the source water. The catalytically active granular filter material acts as a catalyst that accelerates the reaction between oxygen dissolved in water and divalent iron contained in water. Divalent iron is present in our test water in a dissolved state and therefore does not lend itself to mechanical filtration. The catalytically active granular filter media, acting as a catalyst, promotes the oxidation of divalent iron to trivalent. As a result, insoluble particles of iron hydroxide (rust) are formed, which are quite easily retained in the layer of the filter medium.

The activated carbon filter cleans the water from harmful impurities, primarily chlorine, as well as from unpleasant odor, taste, bacteria and organic substances in small quantities by adsorption into the water.

Table 3. Analysis results for other water indicators.

#	Other definitions	Qizilkum 4-Collector	Qili Collector	Akbulak Collector
Hardness mg-eq/l				
1	Total water hardness	98.00	43.00	55.50
2	Carbonate	3.70	5.10	1.60
3	Non - carbonate	94.30	37.90	53.90
4	pH	7.70	7.00	7.00
5	CO_2 is free. mg/l	35	44	15
6	CO_3 unit mg/l	no	no	9
7	Oxidizability mg O_2/l	3.1	2.9	3.04
8	SiO_2 mg	6	20	8
9	H_2S mg/l	no	no	no
10	Dry residue experiment mg/l	12600	5700	4760
11	Dry residue calculated mg/l	12083	5501	4621
Physical properties				
12	Transparency	Transparent	Transparent	Transparent
13	Taste	Bitterly salty	Very salty	Brackish
14	Colour	Without color	Without color	Without color
15	Smell	Without smell	Without smell	Without smell
16	Sediment	Without sediment	Slight sediment	Without sediment

But in addition to these purification processes, in order to achieve the goal, it is necessary to soften and remove salts from the water. To do this, you must first soften the water by the ion exchange method. The ion-exchange method of water purification refers to the complex of water treatment. It is used to remove hardness salts and is called softening. In the process, calcium, magnesium, manganese salts, as well as heavy metals, radionuclides, organic pollutants are removed.

The new polycondensation anionite obtained by us is called: ANDM, which is effective for sorption and ions of copper, nickel, molybdenum and other elements, the degree of dissociation of ionogenic functional groups from the chemical composition, sorption and physico-chemical properties of the initial monomers suitable for use.

And last of all, you can put reverse osmosis for desalination of water. A process in which, at a certain pressure, water passes through a semipermeable membrane from a more concentrated to a less concentrated solution, that is, in the reverse direction for osmosis. The waters purified after reverse osmosis are very clean and their use for drinking purposes has a beneficial effect on the human body.

It is possible to continue water purification by adding distiller and water deionizer equipment. Distilled water is considered sterile and is used mainly for pharmaceutical, laboratory and other needs for the preparation of injection solutions. Deionized waters of types 1 and 2 are especially pure liquids obtained by filtration. The raw material of the process is distilled water, the output is deionized water (demineralized, desalinated). Very sensitive laboratory equipment, such as gas and liquid chromatographs and other devices requiring especially pure waters use deionized waters of types 1 and 2. The purity of the water affects the results of the tests.

The sequence of application of these devices for purification will allow obtaining clean and high-quality drinking water, distilled water and water of the first and second types, while the phasing of water purification will allow the equipment to be used for a long time in a continuous cycle. The right approach to research will allow you to solve the tasks without extra effort and without wasting time and money.

4 Conclusions

The conclusion was made on the physico-chemical composition and properties of the samples obtained, based on the results of the obtained analyses of the water systems of the Aidar-Arnasay lakes. It was found that the water is very hard, has a very high level of salinity, the level of minerals is higher than the permissible concentrations for drinking water according to GOST.

These conclusions do not apply to the biological-bacteriological and sanitary-epidemiological conditions for the conclusion on fisheries, resort recreation areas, tourist recreation areas and therapeutic properties of the waters of the Aidar-Arnasay lake system.

The right approach to research will allow you to solve the tasks without extra effort and without wasting time and money. The waters of the Aidar-Arnasay lake systems can be used in the above directions, but only first they need to be cleaned before use.

References

1. Ramsar Nature Reserve. Convention on the Conservation of Wetlands. Registration by the Secretariat for No. 1841 dated October 20, 2008
2. Mustafayeva Z.A., Mirzaev U.T., Kamilov B.G. Methods of hydrobiological monitoring of water bodies of Uzbekistan. Tashkent: Navruz, 2017. 112 p.
3. Talskikh V.N. Methods of hydrobiological monitoring of water bodies in the Central Asian region (Recommendations – RUz 52.25.32-97). Tashkent: Glavgidromet RUz, 2017. 67 p.
4. Fedorov V.D., Kapkov V.I. Practical hydrobiology of freshwater ecosystems. Moscow: MSU, 2006. 365 p.
5. Zinoviev E.A., Mandritsa S.A. Methods of freshwater fish research: textbook. stipend. Perm, 2003. 113 p.
6. Lakin G.F. Biometrics: studies. manual for biol. spec. universities. 4th ed. Moscow: Higher School, 2010. 352 p.
7. Alekseev V.R., Dobrynina T.I., Glagoleva S.M. et al. Determinant of freshwater invertebrates of Russia and adjacent territories / under the general editorship of S.Ya. Tsalolikhin. Vol. 2. Crustaceans. Saint Petersburg: The science, 2005. 632 p.
8. Khalilov S.A., Shoyakubov R.S., Mustafayeva Z.A., Ergasheva H.E., Karimov B.K., Tozhibaev T.Zh., Alimzhanova H.A. Determinant of volvox algae of Uzbekistan. Namangan, 2014. 215 p.
9. Mirabdullaev I.M., Mirzaev U.T., Hegai V.N. Determinant of fishes of Uzbekistan. 2nd

ed. Tashkent: Khorezm, 2002. 102 p.19. Holynska M., Mirabdullaev I., Ueda H, Reid W.J. Genera Mesocyclops and Thermocyclops. Copepoda: Cyclopoida // Backhuys Publ.; Leiden. 2013. 318 p.

10. Mustafayeva Z.A., Kuvatov A.K., Azizov N.Ya., Mirzaev U.T., 2021 Lake Aidarkul – the current state of aquatic biocenoses // Scientific works of Dalrybvtuz. 2021. Vol. 56, No. 2. pp. 5-14.
11. Yearbook of surface water quality and efficiency of water protection measures carried out on the territory of Glavgidromet activity for 2007. Part 1.–Tashkent, 2008.
12. A.A. Yuldashev, S.A. Mutalov, R.A. Nazirov, T.T. Tursunov. D.M. Azimov "Weakly basic anion exchange polymers" Chemistry and chemical technology Tashkent-2018 (2) pp. 31-33.
13. Mutalov Sh.A., Azimov D.M., Nazirova R.A., Tursunov T.T., Igitov F.B., A method for obtaining heat-resistant anionite selective to molybdenum ions for wastewater treatment of oil refineries Journal "O'zbekiston neft va gaz jurnali". No. 1. Tashkent, 2020, pp. 48-51.
14. Azimova Sh.A., Arslanov Sh.S., Aripdzhanov O.Yu., Azimov D.M., "Modification of lithium lubricants" Journal "Chemistry and Chemical Technology". No. 1. Tashkent, 2020, pp. 27-30.
15. Azimova Sh.A., Arslanov Sh.S., Turakhuzhaev S.A., Azimov D.M. Regeneration of used motor oils by bentonite of the Navbakhor deposit. The journal "Chemistry and Chemical Technology". No.3. Tashkent, 2020, pp. 50-55.
16. Turabzhanov S.M., Azimov D.M., Saidkobilova F.G., Nazirova R.A., Tursunov T.T., Sharipova U.I. Investigation of the structure of weakly basic anionite of polycondensation type Uzbek Scientific, Technical and Production Journal Composite Materials Tashkent -2020 No. 4 pp. 29-30
17. Azimov Dilmurod., Nazirova, Rano; Tursunov, Tulkin, Igitov, Farrukh; Sorption of cations of some metals on weak basic ionites Euro asia 8th. international congress on applied sciences March 15-16, 2021 Tashkent. Uzbekistan.
18. Egamberdiev E.; Turabdjanov S.; Akmalova G.; Mukhtarova N.; Ayubova I.; Mirzakhmedova M.; Rakhmonberdiev G.: Obtaining paper from composition of different fibers and its analysis. E3S Web of Conferences, 2023, 371, 01004, DOI 10.1051/e3sconf/202337101004
19. Ergashev Y.; Egamberdiev E.; Turabdzhanov S.; Akmalova G.; Isanova R.; Rashidov R.; Sobitov O.: Obtaining filter material from natural fiber composition and areas of its use. E3S Web of Conferences, 2023, 371, 01047, DOI 10.1051/e3sconf/202337101047
20. Egamberdiev E.; Akmalova G.; Rahmonberdiev G. Obtaining paper products from cellulose-containing plants and researching its field of application. 3rd International Conference on Energetics, Civil and Agricultural Engineering, ICECAE, 187394, DOI 10.1088/1755-1315/1142/1/012054
21. Azimov; M. Berdiyeva; R. Nazirova; A. Yuldashev; U. Sharipova; "Study of sorption of some metal ions by the obtained phosphoric acid cation exchanger of polycondensation type" Chemistry and Chemical engineering: vol. 2022: no. 1, article 6. Doi: 10.34920/cce202216 Available at: https://uzjournals.edu.uz/cce/vol2022/iss1/6 14-18 pepers
22. Azimova Sh., Arslanov Sh., Turakhujayev S., Azimov D. Waste engine oil regeneration with bentonite from the navbahor deposit. Chemistry and Chemical Engineering, 2020, no. 3, pp. 50-55. DOI: 10.51348/HQJT6823
23. Arslanov S., Turabdjanov S., Azimova S., Azimov D., Sultankhojaeva N., Egamberdiev E. Physico-chemical properties and research of acids contained in oils of Uzbekistan (2023) E3S Web of Conferences, 371, 01021.

AUTHOR INDEX